W9-CGS-145

COMPLETE SOLUTIONS GUIDE

TO ACCOMPANY

Calculus   LARSON / HOSTETLER

THIRD EDITION

Volume 1
(Chapters 1-7)

Dianna L. Zook
The Pennsylvania State University
The Behrend College

D.C. Heath and Company
Lexington, Massachusetts / Toronto

International Standard Book Number: 0-669-10099-4

Preface

This solutions guide is a supplement to *Calculus with Analytic Geometry, Third Edition,* by Roland E. Larson and Robert P. Hostetler. Solutions to every exercise in the text are given with all essential algebraic steps included. There are three volumes in the complete set of solutions guides. Volume I contains Chapters 1-7, Volume II contains Chapters 8-13, and Volume III contains Chapters 14-17. Also available is a one-volume *Study and Solutions Guide to Accompany Calculus, Third Edition,* written by David E. Heyd, which contains worked-out solutions to *selected* representative exercises from the text.

I have made every effort to see that the solutions are correct. However, I would appreciate very much hearing about any errors or other suggestions for improvement.

I would like to thank several people who helped in the production of this guide: David E. Heyd, who assisted the authors of the text and double checked my solutions, Linda L. Matta, who was in charge of the production of the guide, and Timothy R. Larson, who produced the art for the guide. The typing for the guide was done by Nancy K. Stout, of Stout's Secretarial Services. I would also like to thank the students in my mathematics classes. Finally, I would like to thank my husband Ed Schlindwein for his support during the many months I have worked on this project.

Dianna L. Zook
The Pennsylvania State University
The Behrend College
Erie, Pennsylvania 16563

Contents

Integration

Inverse Functions

Applications of integration

1

The Cartesian plane and functions

1.1

The real line

1. 0.7 Rational

2. −3678 Rational

3. $3\pi/2$ Irrational
   (since π is irrational)

4. $3\sqrt{2} - 1$ Irrational

5. $4.345\overline{1451}$ Rational

6. 22/7 Rational

7. $\sqrt[3]{64} = 4$ Rational

8. $0.8177\overline{8177}$ Rational

9. $4\frac{5}{8}$ Rational

10. $(\sqrt{2})^3$ Irrational

11. Let $x = 0.36\overline{36}$

   $100x = 36.36\overline{36}$

   $\underline{-x = -0.36\overline{36}}$
   $99x = 36$

   $x = \dfrac{36}{99} = \dfrac{4}{11}$

12. Let $x = 0.318\overline{18}$

   $1000x = 318.18\overline{18}$

   $\underline{-10x = -3.18\overline{18}}$
   $990x = 315$

   $x = \dfrac{315}{990} = \dfrac{7}{22}$

13. Let $x = 0.297\overline{297}$

   $1000x = 297.297\overline{297}$

   $\underline{-x = -0.297\overline{297}}$

   $999x = 297$

   $x = \dfrac{297}{999} = \dfrac{11}{37}$

14. Let $x = 0.9900\overline{9900}$

   $10,000x = 9900.9900\overline{9900}$

   $\underline{-x = -0.9900\overline{9900}}$

   $9999x = 9900$

   $x = \dfrac{9900}{9999} = \dfrac{100}{101}$

15. Given $a < b$

    (a) $a + 2 < b + 2$ True   (b) $5b < 5a$ False

    (c) $5 - a > 5 - b$ True   (d) $\dfrac{1}{a} < \dfrac{1}{b}$ False

    (e) $(a - b)(b - a) > 0$ False   (f) $a^2 < b^2$ False

16. $A = \{x: 0 < x \}$, $B = \{ x: -2 \leq x \leq 2\}$, and $C = \{ x: x < 1\}$

    (a) $A \cup B = \{x: -2 \leq x \}$ $[-2, \infty)$

    (b) $A \cap B = \{x: 0 < x \leq 2\}$ $(0, 2]$

    (c) $B \cap C = \{x: -2 \leq x < 1\}$ $[-2, 1)$

    (d) $A \cup C = \{x: x \text{ is a real number}\}$ $(-\infty, \infty)$

    (e) $A \cap B \cap C = \{x: 0 < x < 1\}$ $(0, 1)$

17.

Interval Notation	Set Notation	Graph
$[-2, 0)$	$\{x: -2 \leq x < 0\}$	
$(-\infty, -4]$	$\{x: x \leq -4\}$	
$[3, 11/2]$	$\{x: 3 \leq x \leq 11/2\}$	
$(-1, 7)$	$\{x: -1 < x < 7\}$	

18.

Interval Notation	Set Notation	Graph
$[100, \infty)$	$\{x: x \geq 100\}$	
$(10, \infty)$	$\{x: 10 < x\}$	
$(\sqrt{2}, 8]$	$\{x: \sqrt{2} < x \leq 8\}$	
$(1/3, 22/7]$	$\{x: 1/3 < x \leq 22/7\}$	

19. x − 5 ≥ 7
        x ≥ 12

20. 2x > 3
        x > 3/2

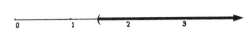

21. 4x + 1 < 2x
        4x < 2x − 1
        2x < −1
         x < −1/2

22. 2x + 7 < 3
        2x < −4
         x < −2

23. 2x − 1 ≥ 0
        2x ≥ 1
         x ≥ 1/2

24. 3x + 1 ≥ 2x + 2
        3x ≥ 2x + 1
         x ≥ 1

25. 4 − 2x < 3x − 1
        −2x < 3x − 5
        −5x < −5
         x > 1

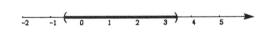

26. x − 4 ≤ 2x + 1
        x ≤ 2x + 5
        −x ≤ 5
         x ≥ −5

27.   −4 < 2x − 3 < 4
        −1 <   2x   < 7
      −1/2 <   x   < 7/2

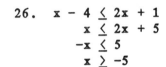

28.   0 ≤ x + 3 < 5
     −3 ≤   x   < 2

29. 3x/4 > x + 1
     −x/4 > 1
        x < −4

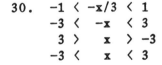

30. −1 < −x/3 < 1
     −3 < −x < 3
      3 >   x > −3
     −3 <   x < 3

31.   x/2 + x/3 > 5

     3x + 2x > 30
         5x > 30
          x > 6

32. x > 1/x
     If x > 0: x² > 1 ⟹ x > 1
     If x < 0: x² < 1 ⟹ −1 < x < 0

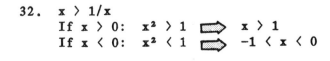

33. $|x| < 1$ ⟹ $-1 < x < 1$

34. $\dfrac{x}{2} - \dfrac{x}{3} > 5$

$$3x - 2x > 30$$
$$x > 30$$

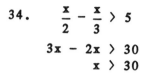

35. $\left|\dfrac{x - 3}{2}\right| \geq 5$

$x - 3 \geq 10$ or $x - 3 \leq -10$
    $x \geq 13$ $x \leq -7$

36. $\left|\dfrac{x}{2}\right| > 3$ ⟹ $x > 6$ or $x < -6$

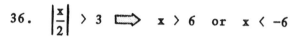

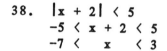

37. $|x - a| < b$
     $-b < x - a < b$
$a - b < \quad x \quad < a + b$

38. $|x + 2| < 5$
    $-5 < x + 2 < 5$
    $-7 < \quad x \quad < 3$

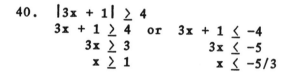

39. $|2x + 1| < 5$
$-5 < 2x + 1 < 5$
$-6 < \quad 2x \quad < 4$
$-3 < \quad x \quad < 2$

40. $|3x + 1| \geq 4$
$3x + 1 \geq 4$ or $3x + 1 \leq -4$
   $3x \geq 3$ $3x \leq -5$
    $x \geq 1$ $x \leq -5/3$

41. $\left|1 - \dfrac{2x}{3}\right| < 1$

$$-1 < 1 - \dfrac{2x}{3} < 1$$

$$-2 < \quad -\dfrac{2x}{3} \quad < 0$$

$$3 > \quad x \quad > 0$$

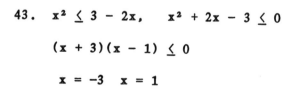

42. $|9 - 2x| < 1$
    $-1 < 9 - 2x < 1$
    $-10 < \quad -2x \quad < -8$
     $5 > \quad\quad x \quad > 4$

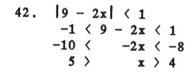

43. $x^2 \leq 3 - 2x$, $x^2 + 2x - 3 \leq 0$

$(x + 3)(x - 1) \leq 0$

$x = -3 \quad x = 1$

Solution: $-3 \leq x \leq 1$

Test Intervals
$(-)(-) > 0$ $(+)(-) < 0$ $(+)(+) > 0$

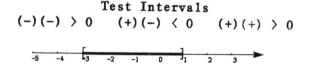

$x < -3,$ $-3 < x < 1,$ $x > 1$

44. $x^4 - x \leq 0$

    $x(x^3 - 1) \leq 0$

    $x = 0 \quad x = 1$

    Solution: $0 \leq x \leq 1$

Test Intervals

$(-)(-) > 0 \quad (+)(-) < 0 \quad (+)(+) > 0$

$x < 0, \qquad 0 < x < 1, \qquad x > 1$

45. $x^2 + x - 1 \leq 5, \quad x^2 + x - 6 \leq 0$

    $(x + 3)(x - 2) \leq 0$

    $x = -3 \quad x = 2$

    Solution: $-3 \leq x \leq 2$

Test Intervals

$(-)(-) > 0 \quad (+)(-) < 0 \quad (+)(+) > 0$

$x < -3 \qquad -3 < x < 2 \qquad x > 2$

46. $2x^2 + 1 < 9x - 3, \quad 2x^2 - 9x + 4 < 0$

    $(2x - 1)(x - 4) < 0$

    $x = 1/2 \quad x = 4$

    Solution: $1/2 < x < 4$

Test Intervals

$(-)(-) > 0 \quad (+)(-) < 0 \quad (+)(+) > 0$

$x < 1/2 \qquad 1/2 < x < 4 \qquad x > 4$

47. (a) $a = -1, \quad b = 3$
    Directed distance from a to b: 4
    Directed distance from b to a: −4
    Distance between a and b: 4

    (b) $a = 1/4, \quad b = 11/4$
    Directed distance from a to b: 5/2
    Directed distance from b to a: −5/2
    Distance between a and b: 5/2

48. (a) $a = -5/2, \quad b = 13/4$
    Directed distance from a to b: 23/4
    Directed distance from b to a: −23/4
    Distance between a and b: 23/4

    (b) $a = -4, \quad b = -3/2$
    Directed distance from a to b: 5/2
    Directed distance from b to a: −5/2
    Distance between a and b: 5/2

49. (a) $a = 126, \quad b = 75$
    Directed distance from a to b: −51
    Directed distance from b to a: 51
    Distance between a and b: 51

(b) a = −126,   b = −75
Directed distance from a to b: 51
Directed distance from b to a: −51
Distance between a and b: 51

50. (a) a = 9.34,   b = −5.65
Directed distance from a to b: −14.99
Directed distance from b to a: 14.99
Distance between a and b: 14.99

(b) a = 16/5,   b = 112/75
Directed distance from a to b: −128/75
Directed distance from b to a: 128/75
Distance between a and b: 128/75

51. (a) a = −1, b = 3,   midpoint: 1
(b) a = −6, b = −1,   midpoint: −7/2

52. (a) a = −5, b = −3/2, midpoint: −13/4
(b) a = 3/4, b = 11/2, midpoint: 25/8

53. (a) [7, 21],   midpoint: 14
(b) [8.6, 11.4],   midpoint: 10

54. (a) [−6.85, 9.35],   midpoint: 1.25
(b) [−4.6, −1.3],   midpoint: −2.95

55. (a) a = −2, b = 2, $|x| \leq 2$ (b) a = −3, b = 3, $|x| < 3$

56. (a) a = −2, b = 2, $|x| > 2$ (b) a = −3, b = 3, $|x| \geq 3$

57. (a) a = 2, b = 6, $|x - 4| \leq 2$ (b) a = −7, b = −1, $|x + 4| < 3$

58. (a) a = 0, b = 4, $|x - 2| > 2$ (b) a = 20, b = 24, $|x - 22| \geq 2$

59. (a) All numbers less than 10 units from 12 $|x - 12| < 10$
(b) All numbers more than 6 units from 3 $|x - 3| > 6$

60. (a) y is at most 2 units from a $|y - a| \leq 2$
(b) y is less than δ units from c $|y - c| < \delta$

61. A = P + Prt
P = 1000, A > 1250, t = 2

1000 + 1000r (2) > 1250
2000r > 250
r > 0.1250
r > 12.5%

62. R = 115.95x, C = 95x + 750, R > C

115.95x > 95x + 750
20.95x > 750
x > 35.7995
$x \geq 36$ units

63. $C = 0.32m + 2300$, $C < 10,000$

$$0.32m + 2300 < 10,000$$
$$0.32m < 7700$$
$$m < 24,062.5 \text{ miles}$$

64. $\left| \dfrac{h - 68.5}{2.7} \right| \leq 1$

$$-1 \leq \dfrac{h - 68.5}{2.7} \leq 1$$
$$-2.7 \leq h - 68.5 \leq 2.7$$
$$65.8'' \leq h \leq 71.2''$$

65. $\left| \dfrac{x - 50}{5} \right| \geq 1.645$

$\dfrac{x - 50}{5} \leq -1.645$ or $\dfrac{x - 50}{5} \geq 1.645$

$$x - 50 \leq -8.225 \qquad\qquad x - 50 \geq 8.225$$
$$x \leq 41.775 \qquad\qquad\qquad x \geq 58.225$$
$$x \leq 41 \qquad\qquad\qquad\quad x \geq 59$$

66. $|p - 2,250,000| < 125,000$

$$-125,000 < p - 2,250,000 < 125,000$$

$$2,125,000 < p < 2,375,000$$

high $= 2,375,000$ barrels
low $= 2,125,000$ barrels

67. (a) $\pi \approx 3.1415926535$

$$\dfrac{355}{113} = 3.141592920$$

$$\dfrac{355}{113} > \pi$$

(b) $\pi \approx 3.1415926535$

$$\dfrac{22}{7} \approx 3.142857143$$

$$\dfrac{22}{7} > \pi$$

68. (a) $\dfrac{224}{151} \approx 1.483443709$

$$\dfrac{144}{97} \approx 1.484536082$$

$$\dfrac{144}{97} > \dfrac{224}{151}$$

(b) $\dfrac{73}{81} \approx 0.901234568$

$$\dfrac{6427}{7132} \approx 0.901149748$$

$$\dfrac{73}{81} > \dfrac{6427}{7132}$$

1.2
The Cartesian plane, the Distance Formula, and circles

1.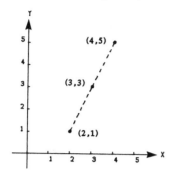

$$d = \sqrt{(4 - 2)^2 + (5 - 1)^2}$$
$$= \sqrt{4 + 16} = \sqrt{20} = 2\sqrt{5}$$
$$x = (4 + 2)/2 = 3$$
$$y = (5 + 1)/2 = 3$$
midpoint: $(3, 3)$

2.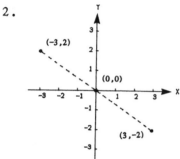

$$d = \sqrt{(3 + 3)^2 + (-2 - 2)^2}$$
$$= \sqrt{36 + 16} = 2\sqrt{13}$$
$$x = (-3 + 3)/2 = 0$$
$$y = (2 + -2)/2 = 0$$
midpoint: $(0, 0)$

3.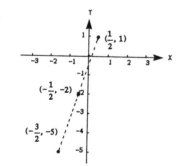

$$d = \sqrt{(\frac{1}{2} + \frac{3}{2})^2 + (1 + 5)^2}$$
$$= \sqrt{4 + 36} = \sqrt{40} = 2\sqrt{10}$$
$$x = (-\frac{3}{2} + \frac{1}{2})/2 = -\frac{1}{2}$$
$$y = (-5 + 1)/2 = -2$$
midpoint: $(-\frac{1}{2}, -2)$

4.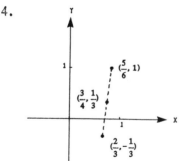

$$d = \sqrt{(\frac{5}{6} - \frac{4}{6})^2 + (\frac{3}{3} + \frac{1}{3})^2}$$
$$= \sqrt{\frac{1}{36} + \frac{64}{36}} = \frac{\sqrt{65}}{6}$$
$$x = (\frac{2}{3} + \frac{5}{6})/2 = \frac{9}{12} = \frac{3}{4}$$
$$y = (-\frac{1}{3} + 1)/2 = \frac{1}{3}$$
midpoint: $(\frac{3}{4}, \frac{1}{3})$

5.

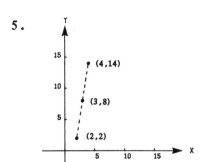

$d = \sqrt{(4 - 2)^2 + (14 - 2)^2}$

$= \sqrt{4 + 144} = 2\sqrt{1 + 36} = 2\sqrt{37}$

$x = (4 + 2)/2 = 3$

$y = (2 + 14)/2 = 8$

midpoint: $(3, 8)$

6.

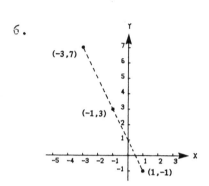

$d = \sqrt{(1 + 3)^2 + (-1 - 7)^2}$

$= \sqrt{16 + 64} = \sqrt{80} = 4\sqrt{5}$

$x = (-3 + 1)/2 = -1$

$y = (7 - 1)/2 = 3$

midpoint: $(-1, 3)$

7.

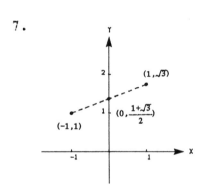

$d = \sqrt{(-1 - 1)^2 + (1 - \sqrt{3})^2}$

$= \sqrt{4 + 1 - 2\sqrt{3} + 3} = \sqrt{8 - 2\sqrt{3}}$

$x = (-1 + 1)/2 = 0$

$y = (1 + \sqrt{3})/2$

midpoint: $(0, (1 + \sqrt{3})/2)$

8.

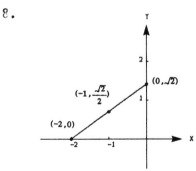

$d = \sqrt{(-2 + 0)^2 + (0 - \sqrt{2})^2} = \sqrt{4 + 2} = \sqrt{6}$

$x = (-2 + 0)/2 = -1$

$y = (0 + \sqrt{2})/2 = \sqrt{2}/2$

midpoint: $(-1, \sqrt{2}/2)$

9.

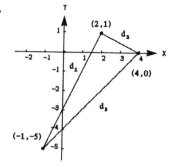

$d_1 = \sqrt{9 + 36} = \sqrt{45}$

$d_2 = \sqrt{4 + 1} = \sqrt{5}$

$d_3 = \sqrt{25 + 25} = \sqrt{50}$

$(d_1)^2 + (d_2)^2 = (d_3)^2$

right triangle

10.

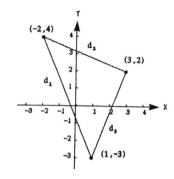

$d_1 = \sqrt{9 + 49} = \sqrt{58}$

$d_2 = \sqrt{25 + 4} = \sqrt{29}$

$d_3 = \sqrt{4 + 25} = \sqrt{29}$

$d_2 = d_3$

isosceles triangle

11.

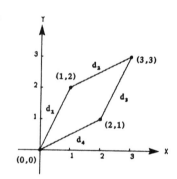

$d_1 = d_2 = d_3 = d_4 = \sqrt{5}$

rhombus

12.

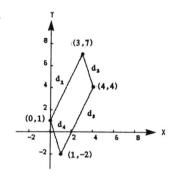

$d_1 = \sqrt{9 + 36} = \sqrt{45} = d_3$

$d_2 = \sqrt{1 + 9} = \sqrt{10} = d_4$

parallelogram

13.

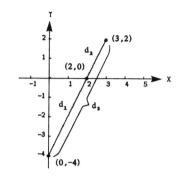

$d_1 = \sqrt{4 + 16} = \sqrt{20} = 2\sqrt{5}$

$d_2 = \sqrt{1 + 4} = \sqrt{5}$

$d_3 = \sqrt{9 + 36} = 3\sqrt{5}$

$d_1 + d_2 = d_3$

collinear

14.

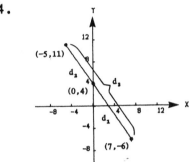

$d_1 = \sqrt{49 + 100} = \sqrt{149} = 12.2066$

$d_2 = \sqrt{25 + 49} = \sqrt{74} = 8.6023$

$d_3 = \sqrt{144 + 289} = \sqrt{433} = 20.8087$

$d_1 + d_2 \neq d_3$

not collinear

15.

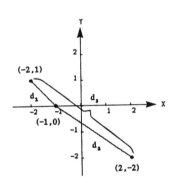

$d_1 = \sqrt{1 + 1} = \sqrt{2}$

$d_2 = \sqrt{9 + 4} = \sqrt{13}$

$d_3 = \sqrt{16 + 9} = 5$

$d_1 + d_2 \neq d_3$

not collinear

16. $\quad 5 = \sqrt{3^2 + y^2}$
$\quad\quad 25 = 9 + y^2$
$\quad\quad 16 = y^2$
$\quad\quad y = \pm 4$

17. $\quad 5 = \sqrt{x^2 + 16}$
$\quad\quad 25 = x^2 + 16$
$\quad\quad\; 9 = x^2$
$\quad\quad\; x = \pm 3$

18. $\quad 5 = \sqrt{(x - 2)^2 + 9}$
$\quad\quad 25 = (x - 2)^2 + 9$
$\quad\quad 16 = (x - 2)^2$
$\quad\quad \pm 4 = x - 2$

$\quad\quad x = 2 \pm 4 = -2, \; 6$

19. $\quad \sqrt{(x + 2)^2 + (y - 3)^2} = \sqrt{(x - 4)^2 + (y + 1)^2}$
$\quad\quad x^2 + 4x + 4 + y^2 - 6y + 9 = x^2 - 8x + 16 + y^2 + 2y + 1$
$\quad\quad 12x - 8y - 4 = 0$
$\quad\quad 3x - 2y - 1 = 0$

20. $\sqrt{(x + 7)^2 + (y + 1)^2} = \sqrt{(x - 3)^2 + (y - 5/2)^2}$

$x^2 + 14x + 49 + y^2 + 2y + 1 = x^2 - 6x + 9 + y^2 - 5y + \dfrac{25}{4}$

$20x + 7y + 50 - \dfrac{61}{4} = 0$

$80x + 28y + 200 - 61 = 0$

$80x + 28y + 139 = 0$

21. The midpoint of the given line segment is $((x_1 + x_2)/2, (y_1 + y_2)/2)$.

The midpoint between (x_1, y_1) and $((x_1 + x_2)/2, (y_1 + y_2)/2)$ is

$(\dfrac{x_1 + (x_1 + x_2)/2}{2}, \dfrac{y_1 + (y_1 + y_2)/2}{2}) = (\dfrac{3x_1 + x_2}{4}, \dfrac{3y_1 + y_2}{4})$.

The midpoint between $((x_1 + x_2)/2, (y_1 + y_2)/2)$ and (x_2, y_2) is

$(\dfrac{(x_1 + x_2)/2 + x_2}{2}, \dfrac{(y_1 + y_2)/2 + y_2}{2}) = (\dfrac{x_1 + 3x_2}{4}, \dfrac{y_1 + 3y_2}{4})$

Thus the three points are $(\dfrac{3x_1 + x_2}{4}, \dfrac{3y_1 + y_2}{4})$, $(\dfrac{x_1 + x_2}{2}, \dfrac{y_1 + y_2}{2})$,

and $(\dfrac{x_1 + 3x_2}{4}, \dfrac{y_1 + 3y_2}{4})$.

22. (a) $(\dfrac{3(1)+4}{4}, \dfrac{3(-2)+(-1)}{4}) = (\dfrac{7}{4}, -\dfrac{7}{4})$ (b) $(\dfrac{3(-2)+0}{4}, \dfrac{3(-3)+0}{4}) = (-\dfrac{3}{2}, -\dfrac{9}{4})$

$(\dfrac{1+4}{2}, \dfrac{-2+(-1)}{2}) = (\dfrac{5}{2}, -\dfrac{3}{2})$ $(\dfrac{-2+0}{2}, \dfrac{-3+0}{2}) = (-1, -\dfrac{3}{2})$

$(\dfrac{1+3(4)}{4}, \dfrac{-2+3(-1)}{4}) = (\dfrac{13}{4}, -\dfrac{5}{4})$ $(\dfrac{-2+3(0)}{4}, \dfrac{-3+3(0)}{4}) = (-\dfrac{1}{2}, -\dfrac{3}{4})$

23. (a) $x^2 + 5x = x^2 + 5x + (\dfrac{5}{2})^2 - (\dfrac{5}{2})^2$

$= x^2 + 5x + \dfrac{25}{4} - \dfrac{25}{4} = (x + \dfrac{5}{2})^2 - \dfrac{25}{4}$

(b) $x^2 + 8x + 7 = x^2 + 8x + (\dfrac{8}{2})^2 - (\dfrac{8}{2})^2 + 7$

$= x^2 + 8x + 16 - 16 + 7$

$= (x + 4)^2 - 9$

24. (a) $4x^2 - 4x - 39 = 4\left[x^2 - x - \dfrac{39}{4}\right]$

$$= 4\left[x^2 - x + \dfrac{1}{4} - \dfrac{1}{4} - \dfrac{39}{4}\right]$$

$$= 4\left[(x - \dfrac{1}{2})^2 - 10\right]$$

$$= 4(x - \dfrac{1}{2})^2 - 40$$

(b) $5x^2 + x = 5\left[x^2 + \dfrac{1}{5}x\right] = 5\left[x^2 + \dfrac{1}{5}x + \dfrac{1}{100} - \dfrac{1}{100}\right]$

$$= 5\left[(x + \dfrac{1}{10})^2 - \dfrac{1}{100}\right]$$

$$= 5(x + \dfrac{1}{10})^2 - \dfrac{1}{20}$$

25. Center $(0, 0)$, Radius 1   Matches graph (d)

26. Center $(1, 3)$, Radius 2   Matches graph (c)

27. Center $(1, 0)$, Radius 0   Matches graph (b)

28. Center $(-\dfrac{1}{2}, \dfrac{3}{4})$, Radius $\dfrac{1}{2}$ Matches graph (e)

29. Center $(-3, 1)$, Radius 4   Matches graph (a)

30. Center $(0, 1)$, Radius 1   Matches graph (f)

31. $(x - 0)^2 + (y - 0)^2 = (3)^2$ 32. $(x - 0)^2 + (y - 0)^2 = (5)^2$
$\qquad\quad x^2 + y^2 - 9 = 0$ $\qquad\quad x^2 + y^2 - 25 = 0$

33. $(x - 2)^2 + (y + 1)^2 = (4)^2$
$\quad x^2 + y^2 - 4x + 2y - 11 = 0$

34. $(x + 4)^2 + (y - 3)^2 = (\dfrac{5}{8})^2$

$\quad 64\,(x + 4)^2 + 64\,(y - 3)^2 = 25$
$\quad 64x^2 + 64y^2 + 512x - 384y + 1575 = 0$

35. radius $= \sqrt{(-1 - 0)^2 + (2 - 0)^2} = \sqrt{5}$
$\quad (x + 1)^2 + (y - 2)^2 = 5, \qquad x^2 + 2x + 1 + y^2 - 4y + 4 = 5,$
$\quad x^2 + y^2 + 2x - 4y = 0$

36. radius = $\sqrt{[3 - (-1)]^2 + (-2 - 1)^2} = 5$
$(x - 3)^2 + (y + 2)^2 = 25$
$x^2 - 6x + 9 + y^2 + 4y + 4 = 25$
$x^2 + y^2 - 6x + 4y - 12 = 0$

37. center = $(3, 2)$
radius = $\sqrt{10}$
$(x - 3)^2 + (y - 2)^2 = 10$
$x^2 - 6x + 9 + y^2 - 4y + 4 = 10$
$x^2 + y^2 - 6x - 4y + 3 = 0$

38. center = $(0, 0)$
radius = $\sqrt{2}$
$(x - 0)^2 + (y - 0)^2 = (\sqrt{2})^2$
$x^2 + y^2 - 2 = 0$

39. $(0, 0), (0, 8), (6, 0)$
$(0 - h)^2 + (0 - k)^2 = r^2$, $h^2 + k^2 = r^2$ Eq. 1
$(0 - h)^2 + (8 - k)^2 = r^2$, $h^2 + (8 - k)^2 = r^2$ Eq. 2
$(6 - h)^2 + (0 - k)^2 = r^2$, $(6 - h)^2 + k^2 = r^2$ Eq. 3

$(8 - k)^2 - k^2 = 0$, $64 - 16k = 0$, $k = 0$ Eq. 2 - Eq. 1
$(6 - h)^2 - h^2 = 0$, $36 - 12h = 0$, $h = 3$ Eq. 3 - Eq. 1
$(3)^2 + (4)^2 = r^2$, $25 = r^2$ Eq. 1
$(x - 3)^2 + (y - 4)^2 = 25$
$x^2 + y^2 - 6x - 8y = 0$

Alternate Solution
Note that the given points form the vertices of a right triangle. Thus
the hypotenuse of the triangle must lie on the diameter of the circle.
Thus the center must be the midpoint of the line segment joining $(0, 8)$
and $(6, 0)$.

$(h, k) = (3, 4)$
$r = \sqrt{(3 - 6)^2 + (4 - 0)^2} = \sqrt{9 + 16} = 5$
$(x - 3)^2 + (y - 4)^2 = 25$, $x^2 - 6x + 9 + y^2 - 8y + 16 = 25$
$x^2 + y^2 - 6x - 8y = 0$

40. $(1, -1), (2, -2), (0, -2)$
$(1 - h)^2 + (-1 - k)^2 = r^2$ Eq. 1
$(2 - h)^2 + (-2 - k)^2 = r^2$ Eq. 2
$h^2 + (-2 - k)^2 = r^2$ Eq. 3
$(2 - h)^2 - h^2 = 0$, $4 - 4h = 0$, $h = 1$ Eq. 2 - Eq. 3
$(-1 - k)^2 = r^2$ Eq. 1
$(2 - 1)^2 + (-2 - k)^2 = r^2$ Eq. 2

$(-1 - k)^2 = 1 + (-2 - k)^2$
$1 + 2k + k^2 = 1 + 4 + 4k + k^2$, $-4 = 2k$, $k = -2$

$(1 - 1)^2 + (-1 + 2)^2 = r^2$, $1 = r^2$
$(x - 1)^2 + (y + 2)^2 = 1$, $x^2 + y^2 - 2x + 4y + 4 = 0$

41. $x^2 + y^2 - 2x + 6y = -6$

    $(x^2 - 2x + 1) + (y^2 + 6y + 9) = -6 + 1 + 9$

    $(x - 1)^2 + (y + 3)^2 = 4$

    center $(1, -3)$

    radius 2

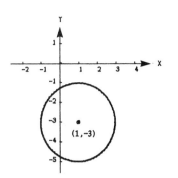

42. $x^2 + y^2 - 2x + 6y = 15$

    $(x^2 - 2x + 1) + (y^2 + 6y + 9) = 15 + 1 + 9$

    $(x - 1)^2 + (y + 3)^2 = 25$

    center $(1, -3)$

    radius 5

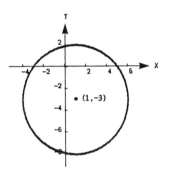

43. $x^2 + y^2 - 2x + 6y = -10$

    $(x^2 - 2x + 1) + (y^2 + 6y + 9) = -10 + 1 + 9$

    $(x - 1)^2 + (y + 3)^2 = 0$

    only a point $(1, -3)$

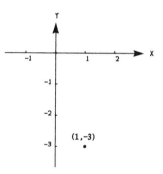

44. $3x^2 + 3y^2 - 6y = 1$

    $3x^2 + 3(y^2 - 2y + 1) = 1 + 3$

    $x^2 + (y - 1)^2 = 4/3$

    center $(0, 1)$

    radius $2\sqrt{3}/3$

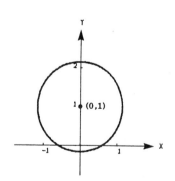

45. $2x^2 + 2y^2 - 2x - 2y - 3 = 0$

$2(x^2 - x + \frac{1}{4}) + 2(y^2 - y + \frac{1}{4}) = 3 + \frac{1}{2} + \frac{1}{2}$

$(x - \frac{1}{2})^2 + (y - \frac{1}{2})^2 = 2$

center $(1/2, 1/2)$

radius $\sqrt{2}$

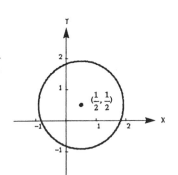

46. $4x^2 + 4y^2 - 4x + 2y - 1 = 0$

$4(x^2 - x + \frac{1}{4}) + 4(y^2 + \frac{y}{2} + \frac{1}{16}) = 1 + 1 + \frac{1}{4}$

$4(x - \frac{1}{2})^2 + 4(y + \frac{1}{4})^2 = \frac{9}{4}$

$(x - \frac{1}{2})^2 + (y + \frac{1}{4})^2 = \frac{9}{16}$

center $(1/2, -1/4)$

radius $3/4$

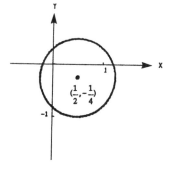
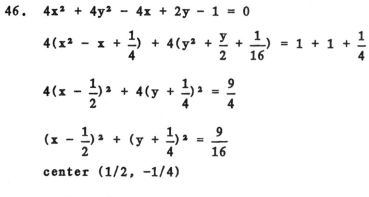

47. $16x^2 + 16y^2 + 16x + 40y - 7 = 0$

$16(x^2 + x + \frac{1}{4}) + 16(y^2 + \frac{5y}{2} + \frac{25}{16})$

$= 7 + 4 + 25$

$16(x + \frac{1}{2})^2 + 16(y + \frac{5}{4})^2 = 36$

$(x + \frac{1}{2})^2 + (y + \frac{5}{4})^2 = \frac{9}{4}$

center $(-1/2, -5/4)$

radius $3/2$

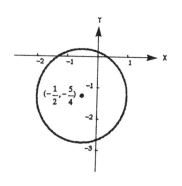

48. $x^2 + y^2 - 4x + 2y + 3 = 0$

$(x^2 - 4x + 4) + (y^2 + 2y + 1) = -3 + 4 + 1$

$(x - 2)^2 + (y + 1)^2 = 2$

center $(2, -1)$

radius $\sqrt{2}$

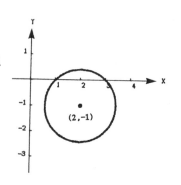

16

49. Assume that the center of the earth is $(0, 0)$

$$x^2 + y^2 = (22{,}000 + 4000)^2, \qquad x^2 + y^2 - 676{,}000{,}000 = 0$$

50. Substitution of the points $(1, 2)$, $(-1, 2)$, and $(2, 1)$ into the equation

$$x^2 + y^2 + Cx + Dy + E = 0$$

yields the following system of equations:

  (let $x = 1$ and $y = 2$) $\qquad C + 2D + E = -5$

  (let $x = -1$ and $y = 2$) $\quad -C + 2D + E = -5$

  (let $x = 2$ and $y = 1$) $\qquad 2C + D + E = -5$

The simultaneous solution is $C = 0$, $D = 0$, and $E = -5$. Thus, the equation of the circle is $x^2 + y^2 = 5$

51. The given points form the vertices of a right triangle. From geometry, the hypotenuse of the triangle must lie on a diameter of the circle. Thus, the center must be the midpoint of the line segment joining $(4, 3)$ and $(-2, -5)$.

$$(h, k) = [(4 - 2)/2, \ (3 - 5)/2] = (1, -1)$$

$$r = \sqrt{(4 - 1)^2 + [3 - (-1)]^2} = 5, \qquad (x - 1)^2 + [y - (-1)]^2 = 5^2$$

$$(x - 1)^2 + (y + 1)^2 = 25$$

52. Substitution of the points $(4, 1)$ and $(6, 3)$ into the equation

$$(x - h)^2 + (y - k)^2 = 10$$

yields the following system of equations:

  (let $x = 4$ and $y = 1$) $\qquad (4 - h)^2 + (1 - k)^2 = 10$

  (let $x = 6$ and $y = 3$) $\qquad (6 - h)^2 + (3 - k)^2 = 10$

Solving these two equations for h and k yields $h = 7$ and $k = 0$ or $h = 3$ and $k = 4$. Therefore, there are two possible solutions:

$$(x - 7)^2 + (y - 0)^2 = 10 \qquad \text{and} \qquad (x - 3)^2 + (y - 4)^2 = 10$$

$$x^2 + y^2 - 14x + 39 = 0 \qquad\qquad x^2 + y^2 - 6x - 8y + 15 = 0$$

53. $x^2 + y^2 - 4x + 2y + 1 \leq 0$

$(x^2 - 4x + 4) +$
$(y^2 + 2y + 1) \leq -1 + 4 + 1$

$(x - 2)^2 + (y + 1)^2 \leq 4$

center $(2, -1)$
radius $= 2$

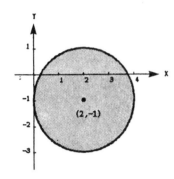

54. $x^2 + y^2 - 4x + 2y + 1 > 0$

$(x^2 - 4x + 4) +$
$(y^2 + 2y + 1) > -1 + 4 + 1$

$(x - 2)^2 + (y + 1)^2 > 4$

center $(2, -1)$
radius $= 2$

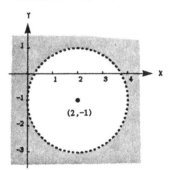

55. $(x + 3)^2 + (y - 1)^2 < 9$

center $(-3, 1)$
radius $= 3$

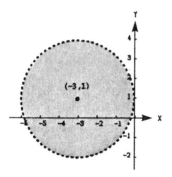

56. $(x - 1)^2 + (y - \frac{1}{2})^2 > 1$

center $(1, 1/2)$
radius $= 1$

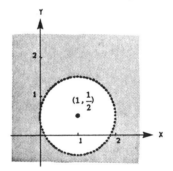

58. (a) $\left(\dfrac{2(1) + 4}{3}, \dfrac{2(-2) + 1}{3}\right) = (2, -1)$

$\left(\dfrac{1 + 2(4)}{3}, \dfrac{-2 + 2(1)}{3}\right) = (3, 0)$

(b) $\left(\dfrac{2(-2) + 0}{3}, \dfrac{2(-3) + 0}{3}\right) = (-4/3, -2)$

$\left(\dfrac{-2 + 2(0)}{3}, \dfrac{-3 + 2(0)}{3}\right) = (-2/3, -1)$

1.3
Graphs of equations

1. $y = x - 2$

    x-intercept: (2, 0)
    y-intercept: (0, −2)
    Matches graph (c)

2. $y = -\frac{1}{2}x + 2$

    x-intercept: (4, 0)
    y-intercept: (0, 2)
    Matches graph (d)

3. $y = x^2 + 2x$
    x-intercepts: (0, 0) and (−2, 0)
    y-intercept: (0, 0)
    Matches graph (b)

4. $y = \sqrt{9 - x^2}$
    x-intercepts: (−3, 0) and (3, 0)
    y-intercept: (0, 3)
    Matches graph (f)

5. $y = |x| - 2$
    x-intercepts: (2, 0) and (−2, 0)

    y-intercept: (0, −2)
    Matches graph (a)

6. $y = x^3 - x$
    x-intercepts: (0, 0), (−1, 0)
    and (1, 0)
    y-intercept: (0, 0)
    Matches graph (e)

7. Symmetric to the y-axis since
    $y = (-x)^2 - 2 = x^2 - 2$

8. Symmetric to the y-axis since
    $y = (-x)^4 - (-x)^2 + 3$
    $= x^4 - x^2 + 3$

9. Symmetric to the y-axis since $(-x)^2y - (-x)^2 + 4y = x^2y - x^2 + 4y = 0$

10. Symmetric to the y-axis since $(-x)^2y - (-x^2) - 4y = x^2y - x^2 - 4y = 0$

11. Symmetric to the x-axis since
    $(-y)^2 = y^2 = x^3 - 4x$

12. Symmetric to the x-axis since
    $x(-y)^2 = xy^2 = -10$

13. Symmetric to the origin since
    $(-y) = (-x)^3 + (-x)$
    $-y = -x^3 - x$
    $y = x^3 + x$

14. Symmetric to the origin since
    $(-x)(-y) = xy = 1$

15. Symmetric to the origin since
    $-y = -x/[(-x)^2 + 1]$ and
    $y = x/(x^2 + 1)$

16. No symmetry with respect to
    either axis or the origin.

17. $y = 2x - 3$
    y-intercept: $y = 2(0) - 3 = -3$
    x-intercept: $0 = 2x - 3$
    $\qquad\qquad x = 3/2$

18. $y = (x - 1)(x - 3)$
    y-intercept: $y = (0 - 1)(0 - 3)$
    $\qquad\qquad y = 3$
    x-intercepts: $0 = (x - 1)(x - 3)$
    $\qquad\qquad x = 1, 3$

19. $y = x^2 + x - 2$
    y-intercept: $y = 0^2 + 0 - 2$
    $\qquad\qquad y = -2$
    x-intercepts: $0 = x^2 + x - 2$
    $\qquad\qquad 0 = (x + 2)(x - 1)$
    $\qquad\qquad x = -2, 1$

20. $y^2 = x^3 - 4x$
    y-intercept: $y^2 = 0^3 - 4(0)$
    $\qquad\qquad y = 0$
    x-intercepts: $0 = x^3 - 4x$
    $\qquad\qquad 0 = x(x - 2)(x + 2)$
    $\qquad\qquad x = 0, \pm 2$

21. $y = x^2\sqrt{9 - x^2}$
    y-intercept: $y = 0^2\sqrt{9 - 0^2}$
    $\qquad\qquad y = 0$
    x-intercepts: $0 = x^2\sqrt{9 - x^2}$
    $\qquad\qquad 0 = x^2\sqrt{(3 - x)(3 + x)}$
    $\qquad\qquad x = 0, \pm 3$

22. No intercepts

23. $y = \dfrac{x - 1}{x - 2}$
    y-intercept: $y = \dfrac{0 - 1}{0 - 2}$
    $\qquad\qquad y = 1/2$
    x-intercept: $0 = \dfrac{x - 1}{x - 2}$
    $\qquad\qquad x = 1$

24. $y = \dfrac{x^2 + 3x}{(3x + 1)^2}$
    y-intercept: $y = \dfrac{0^2 + 3(0)}{[3(0) + 1]^2}$
    $\qquad\qquad y = 0$
    x-intercepts: $0 = \dfrac{x^2 + 3x}{(3x + 1)^2}$
    $\qquad\qquad 0 = \dfrac{x(x + 3)}{(3x + 1)^2}$
    $\qquad\qquad x = 0, -3$

25. $x^2y - x^2 + 4y = 0$
    y-intercept: $0^2(y) - 0^2 + 4y = 0$
    $\qquad\qquad y = 0$
    x-intercept: $x^2(0) - x^2 + 4(0) = 0$
    $\qquad\qquad x = 0$

26. $y = 2x - \sqrt{x^2 + 1}$
    y-intercept: $y = 2(0) - \sqrt{0^2 + 1}$
    $\qquad\qquad y = -1$
    x-intercepts: $0 = 2x - \sqrt{x^2 + 1}$
    $\qquad\qquad 2x = \sqrt{x^2 + 1}$
    $\qquad\qquad 4x^2 = x^2 + 1$
    $\qquad\qquad 3x^2 = 1$
    $\qquad\qquad x^2 = 1/3$
    $\qquad\qquad x = \pm\sqrt{3}/3$
    Since $x > 0$, $x = \sqrt{3}/3$

27. **y = x**
    Intercept: (0, 0)
    Symmetry: origin

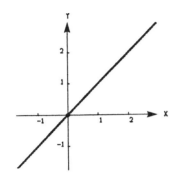

28. **y = x - 2**
    Intercepts: (0, -2), (2, 0)
    Symmetry: none

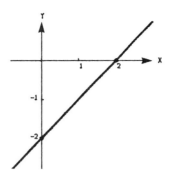

29. **y = x + 3**
    Intercepts: (-3, 0), (0, 3)
    Symmetry: none

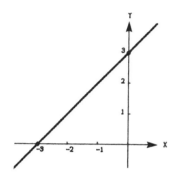

30. **y = 2x - 3**
    Intercepts: (3/2, 0), (0, -3)
    Symmetry: none

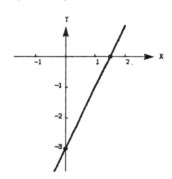

31. **y = -3x + 2**

    Intercepts: (2/3, 0), (0, 2)
    Symmetry: none

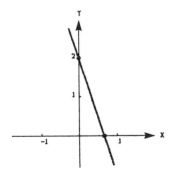

32. $y = -\dfrac{x}{2} + 2$

    Intercepts: (4, 0), (0, 2)
    Symmetry: none

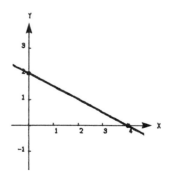

33. $y = \frac{x}{2} - 4$

Intercepts: $(8, 0)$, $(0, -4)$
Symmetry: none

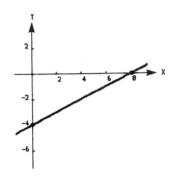

34. $y = x^2 + 3$

Intercept: $(0, 3)$
Symmetry: y-axis

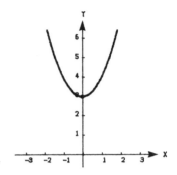

35. $y = 1 - x^2$
Intercepts: $(1, 0)$, $(-1, 0)$,
   $(0, 1)$
Symmetry: y-axis

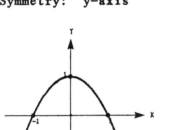

36. $y = 2x^2 + x = x(2x + 1)$
Intercepts: $(0, 0)$, $(-1/2, 0)$
Symmetry: none

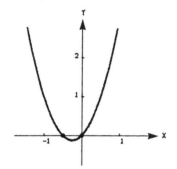

37. $y = -2x^2 + x + 1$
   $= (2x + 1)(-x + 1)$
Intercepts: $(-1/2, 0)$, $(1, 0)$,
   $(0, 1)$
Symmetry: none

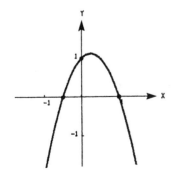

38. $y = x^3 - 1$
Intercepts: $(1, 0)$, $(0, -1)$
Symmetry: none

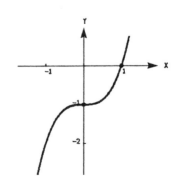

39. $y = x^3 + 2$
Intercepts: $(-\sqrt[3]{2}, 0)$, $(0, 2)$
Symmetry: none

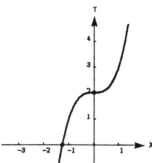

40. $y = \sqrt{9 - x^2}$
Intercepts: $(-3, 0)$, $(3, 0)$,
$\qquad\qquad\quad (0, 3)$
Symmetry: y-axis

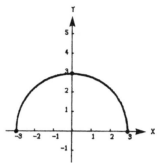

41. $x^2 + 4y^2 = 4$
Intercepts: $(-2, 0)$, $(2, 0)$,
$\qquad\qquad\quad (0, -1)$, $(0, 1)$
Symmetry: origin and both axes

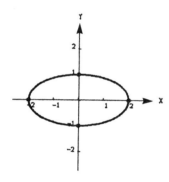

42. $x = y^2 - 4$
Intercepts: $(0, 2)$, $(0, -2)$,
$\qquad\qquad\quad (-4, 0)$
Symmetry: x-axis

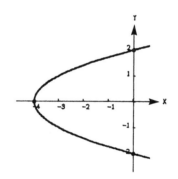

43. $y = (x + 2)^2$
Intercepts: $(-2, 0)$, $(0, 4)$
Symmetry: none

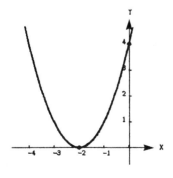

44. $y = \dfrac{1}{x^2 + 1}$
Intercept: $(0, 1)$
Symmetry: y-axis

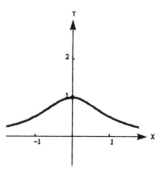

45. $y = 1/x$
     Intercepts: none
     Symmetry: origin

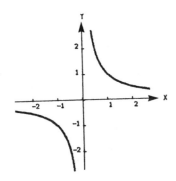

46. $y = 2x^4$
     Intercept: $(0, 0)$
     Symmetry: y-axis

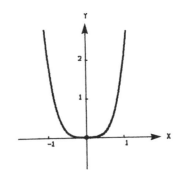

47. $x + y = 2$
     (+) $\underline{2x - y = 1}$
           $3x \quad\;\; = 3$
                  $x = 1$

     Substituting $x = 1$ into the
     first equation:

            $1 + y = 2$
                 $y = 1 \quad (1, 1)$

48. $2x - 3y = 13$
     (+) $\underline{5x + 3y = 1}$
           $7x \quad\quad = 14$
                  $x = 2$

     Substituting $x = 2$ into the
     first equation:

            $4 - 3y = 13$
                 $y = -3 \quad (2, -3)$

49. $x + y = 7 \Rightarrow 2x + 2y = 14$
     $3x - 2y = 11 \Rightarrow \underline{3x - 2y = 11}$
                        $5x \quad\quad = 25$
                               $x = 5$

     Substituting $x = 5$ into the
     first equation:

            $5 + y = 7$
                 $y = 2 \quad (5, 2)$

50. $x^2 + y^2 = 25$
     $2x + y = 10$

     From equation two, $y = 10 - 2x$
     therefore the first equation
     yields

            $5x^2 - 40x + 75 = 0$
            $5(x - 3)(x - 5) = 0$
            $x = 3, 5$
            $(3, 4), (5, 0)$

51. $x^2 + y^2 = 5$
     $x - y = 1$
     From equation two, $x = y + 1$
     therefore equation one yields

            $(y + 1)^2 + y^2 = 5$
            $2y^2 + 2y - 4 = 0$
            $2(y + 2)(y - 1) = 0$
            $y = -2, 1$
            $(-1, -2), (2, 1)$

52. $x^2 + y = 4$
     (+) $\underline{2x - y = 1}$
            $x^2 + 2x = 5$
            $x^2 + 2x - 5 = 0$

            $x = \dfrac{-2 \pm \sqrt{4 + 20}}{2} = -1 \pm \sqrt{6}$

            $(-1 + \sqrt{6}, -3 + 2\sqrt{6})$,
            $(-1 - \sqrt{6}, -3 - 2\sqrt{6})$

53. $y = x^3$, $y = x$
    $x = x^3$
    $0 = x^3 - x$
    $0 = x(x + 1)(x - 1)$
    $x = 0, \pm 1$
    $(-1, -1), (0, 0), (1, 1)$

54. $y = x^4 - 2x^2 + 1$, $y = 1 - x^2$
    $1 - x^2 = x^4 - 2x^2 + 1$
    $0 = x^4 - x^2$
    $0 = x^2(x + 1)(x - 1)$
    $x = -1, 0, 1$
    $(-1, 0), (0, 1), (1, 0)$

55. $y = x^3 - 2x^2 + x - 1$
    $y = -x^2 + 3x - 1$
    $x^3 - 2x^2 + x - 1 = -x^2 + 3x - 1$
    $x^3 - x^2 - 2x = 0$
    $x(x - 2)(x + 1) = 0$
    $x = -1, 0, 2$
    $(-1, -5), (0, -1), (2, 1)$

56. $x = 3 - y^2$, $y = x - 1$
    $y = (3 - y^2) - 1$
    $y^2 + y - 2 = 0$
    $(y + 2)(y - 1) = 0$
    $y = -2, 1$
    $(-1, -2), (2, 1)$

57. $8650x + 250{,}000 = 9{,}950x$
         $250{,}000 = 1300x$
              $x \approx 192$ units

58. $5.5\sqrt{x} + 10{,}000 = 3.29x$
    $(5.5\sqrt{x})^2 = (3.29x - 10{,}000)^2$
    $30.25x = 10.8241x^2 - 65800x + 100{,}000{,}000$
    $0 = 10.8241x^2 - 65830.25x + 100{,}000{,}000$ (Use the Quadratic Formula)
    $x \approx 3133$ units

59. $y = 2x - 3$

    $(1, 2)$: $2 \neq 2(1) - 3$ not on the graph
    $(1, -1)$: $-1 = 2(1) - 3$ on the graph
    $(4, 5)$: $5 = 2(4) - 3$ on the graph

60. $x^2 + y^2 = 4$

    $(1, -\sqrt{3})$: $1^2 + (-\sqrt{3})^2 = 4$ on the graph
    $(1/2, -1)$: $(1/2)^2 + (-1)^2 \neq 4$ not on the graph
    $(3/2, 7/2)$: $(3/2)^2 + (7/2)^2 \neq 4$ not on the graph

61. $x^2y - x^2 + 4y = 0$ $(1, 1/5)$: $1/5 = 1/(1^2 + 4)$ on the graph
                              $(2, 1/2)$: $1/2 = 4/(2^2 + 4)$ on the graph
    $y = \dfrac{x^2}{x^2 + 4}$ $(-1, -2)$: $-2 \neq 1/[(-1)^2 + 4]$ not on the graph

62. $x^2 - xy + 4y = 3$ $(0, 2)$: $2 \neq (0 - 3)/(0 - 4)$ not on the graph

    $y = \dfrac{x^2 - 3}{x - 4}$ $(-2, -1/6)$: $-1/6 = \dfrac{(-2)^2 - 3}{(-2) - 4}$ on the graph

                              $(3, -6)$: $-6 = (3^2 - 3)/(3 - 4)$ on the graph

63. (a) $4 = k(1)^3$ $k = 4$
   (b) $1 = k(-2)^3$ $k = -1/8$
   (c) $0 = k(0)^3$ $k =$ any real number
   (d) $-1 = k(-1)^3$ $k = 1$

64. (a) $1 = 4k(1)$ $k = 1/4$
   (b) $16 = 4k(2)$ $k = 2$
   (c) $0 = 4k(0)$ $k =$ any real number
   (d) $9 = 4k(3)$ $k = 3/4$

65. (a)
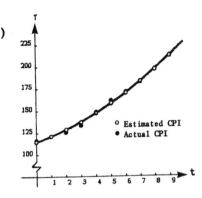

(b) $t = 15$ corresponds to 1985

$y = 0.55(15)^2 + 5.85(15) + 114.41$

$y = 325.91$

66.

t	0	2	4	6	8	10	12
v	0.8740	0.7794	0.6821	0.5906	0.5091	0.4388	0.3791

$$v = \frac{100}{0.55t^2 + 5.85t + 114.41}$$

67. (a)
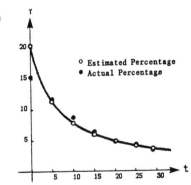

(b) $t = 40$ corresponds to 1990

$$y = \frac{100}{4.90 + 0.79(40)} \approx 2.74$$

68. (a)

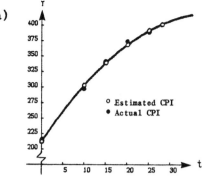

(b) $y = -0.13(35)^2 + 10.43(35) + 211.3$

$= 417.1$

1.4
Lines in the plane

1. $m = 1$

2. $m = 2$

3. $m = 0$

4. $m = -1$

5. $m = -3$

6. $m = 3/2$

7. $m = \dfrac{2 - (-4)}{5 - 3} = \dfrac{6}{2} = 3$

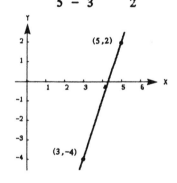

8. $m = \dfrac{1 - (-3)}{-2 - 4} = \dfrac{4}{-6} = -\dfrac{2}{3}$

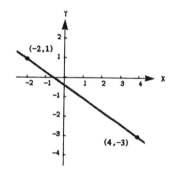

9. $m = \dfrac{2 - 2}{6 - (1/2)} = 0$

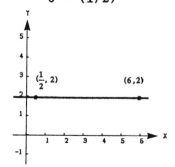

10. $m = \dfrac{4 - (-5)}{(5/6) - (-3/2)} = \dfrac{9}{14/6} = \dfrac{27}{7}$

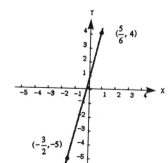

11. $\quad m = \dfrac{-4 - (-1)}{-6 - (-6)} = \dfrac{-3}{0} \quad$ undefined

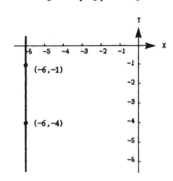

12. $\quad m = \dfrac{5 - 1}{2 - 2} = \dfrac{4}{0} \quad$ undefined

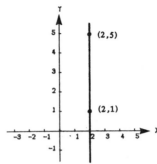

13. $\quad m = \dfrac{2 - 2}{-2 - 1} = 0$

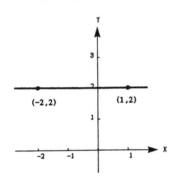

14. $\quad m = \dfrac{(3/4) - (-1/4)}{(7/8) - (5/4)} = \dfrac{1}{-3/8} = -\dfrac{8}{3}$

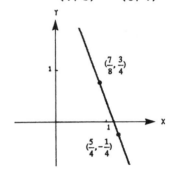

15. $\quad m = \dfrac{1 - (-3)}{2 - 0} = 2$

$y - 1 = 2(x - 2)$
$y - 1 = 2x - 4$
$0 = 2x - y - 3$

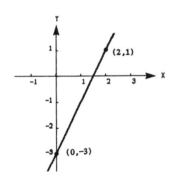

16. $\quad m = \dfrac{4 - (-4)}{1 - (-3)} = \dfrac{8}{4} = 2$

$y - 4 = 2(x - 1)$
$y - 4 = 2x - 2$
$0 = 2x - y + 2$

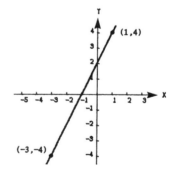

17. $m = \dfrac{3 - 0}{-1 - 0} = -3$

$y - 0 = -3(x - 0)$

$y = -3x$

$3x + y = 0$

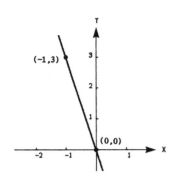

18. $m = \dfrac{6 - 2}{-3 - 1} = \dfrac{4}{-4} = -1$

$y - 2 = -1(x - 1)$

$y - 2 = -x + 1$

$x + y - 3 = 0$

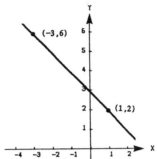

19. **The slope is undefined because x = 2.**

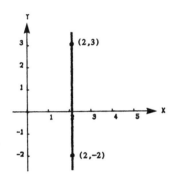

20. $m = 0$

$y = 1$

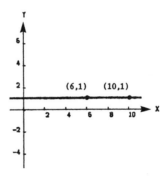

21. $m = 0$

$y = -2$

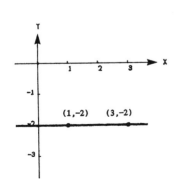

22. $m = \dfrac{(3/4) - (-1/4)}{(7/8) - (5/4)} = \dfrac{1}{-3/8} = -\dfrac{8}{3}$

$y + \dfrac{1}{4} = \dfrac{-8}{3}\left(x - \dfrac{5}{4}\right)$

$12y + 3 = -32x + 40$

$32x + 12y - 37 = 0$

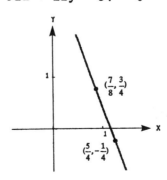

23. $y = \dfrac{3}{4} x + 3$

$4y = 3x + 12$
$0 = 3x - 4y + 12$

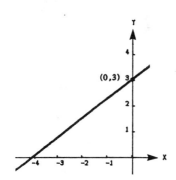

24. $x = -1$

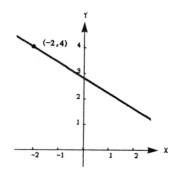

25. $y = \dfrac{2}{3} x$

$3y = 2x$
$2x - 3y = 0$

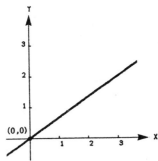

26. $y + 4 = \dfrac{1}{4}(x + 1)$

$4y + 16 = x + 1$
$x - 4y - 15 = 0$

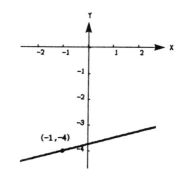

27. $y = -2x + 5$
$2x + y - 5 = 0$

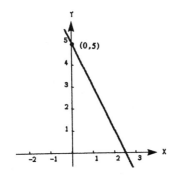

28. $y - 4 = -\dfrac{3}{5}(x + 2)$

$5y - 20 = -3x - 6$
$3x + 5y - 14 = 0$

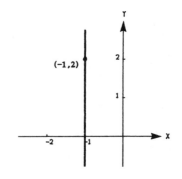

29. $y = 4x + 2$
     $0 = 4x - y + 2$

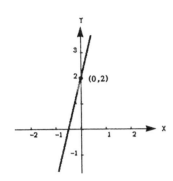

30. $y = \dfrac{1}{6}x - \dfrac{2}{3}$

     $6y = x - 4$
     $x - 6y - 4 = 0$

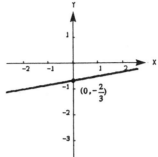

31. $y = \dfrac{3}{4}x + \dfrac{2}{3}$

     $12y = 9x + 8$
     $9x - 12y + 8 = 0$

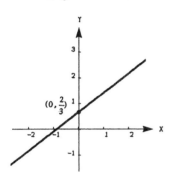

32. $y = 4$

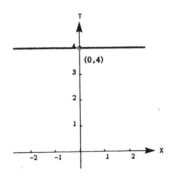

33. $x = 3$

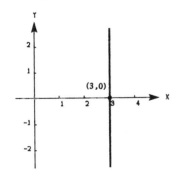

34. $m = -b/a$

$y = \dfrac{-b}{a} x + b$

$\dfrac{b}{a} x + y = b$

$\dfrac{x}{a} + \dfrac{y}{b} = 1$

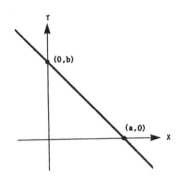

35. $\dfrac{x}{2} + \dfrac{y}{3} = 1$

$3x + 2y = 6$

36. $\dfrac{x}{-3} + \dfrac{y}{4} = 1$

$4x - 3y = -12$

37. $\dfrac{x}{-1/6} + \dfrac{y}{-2/3} = 1$

$-6x - \dfrac{3}{2} y = 1$

$12x + 3y = -2$

38. $\dfrac{x}{-2/3} + \dfrac{y}{-2} = 1$

$\dfrac{-3x}{2} - \dfrac{y}{2} = 1$

$3x + y = -2$

39. $\dfrac{x}{a} + \dfrac{y}{a} = 1$

$\dfrac{1}{a} + \dfrac{2}{a} = 1$

$\dfrac{3}{a} = 1$

$a = 3 \implies x + y = 3$

40. $\dfrac{x}{a} + \dfrac{y}{a} = 1$

$\dfrac{-3}{a} + \dfrac{4}{a} = 1$

$\dfrac{1}{a} = 1$

$a = 1 \implies x + y = 1$

41. $\dfrac{x}{2a} + \dfrac{y}{a} = 1$

$\dfrac{3/2}{2a} + \dfrac{1/2}{a} = 1$

$\dfrac{5}{4a} = 1$

$a = \dfrac{5}{4} \implies 2x + 4y = 5$

42. $\dfrac{x}{a} + \dfrac{y}{-a} = 1$

$\dfrac{-3}{a} + \dfrac{1}{-a} = 1$

$a = -4 \implies \dfrac{x}{-4} + \dfrac{y}{4} = 1$

$x - y = -4$

43. (a) $y - 1 = 2(x - 2)$
       $y - 1 = 2x - 4$
       $2x - y - 3 = 0$

    (b) $y - 1 = -\dfrac{1}{2}(x - 2)$
       $2y - 2 = -x + 2$
       $x + 2y - 4 = 0$

44. (a) $y - 2 = -1(x + 3)$
       $y - 2 = -x - 3$
       $x + y + 1 = 0$

    (b) $y - 2 = 1(x + 3)$
       $y - 2 = x + 3$
       $x - y + 5 = 0$

45. (a) $y - \dfrac{3}{4} = -\dfrac{5}{3}(x - \dfrac{7}{8})$
       $24y - 18 = -40x + 35$
       $40x + 24y - 53 = 0$

    (b) $y - \dfrac{3}{4} = \dfrac{3}{5}(x - \dfrac{7}{8})$
       $40y - 30 = 24x - 21$
       $24x - 40y + 9 = 0$

46. (a) $y - 4 = -\dfrac{3}{4}(x + 6)$
       $4y - 16 = -3x - 18$
       $3x + 4y + 2 = 0$

    (b) $y - 4 = \dfrac{4}{3}(x + 6)$
       $3y - 12 = 4x + 24$
       $4x - 3y + 36 = 0$

47. (a) $x = 2$
    (b) $y = 5$

48. (a) $y = 0$
    (b) $x = -1$

49. (a) $y = -3$ (b) $2x - y - 3 = 0$ (c) $x + 2y + 6 = 0$
                    $y = 2x - 3$ $y = -\dfrac{1}{2}x - 3$

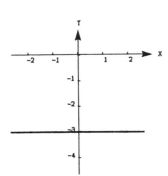
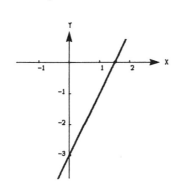
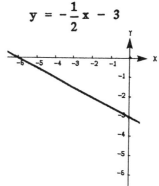

50. $x^2 - 4x + 3 = -x^2 + 2x + 3$
    $2x^2 - 6x = 0$
    $2x(x - 3) = 0$
    $x = 0 \qquad x = 3$
    $y = 3 \qquad y = 0$
    Points of intersection: $(0, 3)$ and $(3, 0)$
    $m = -1$
    $y - 3 = -(x - 0)$
    $y = -x + 3$

51. $m_1 = \dfrac{-4 - 0}{0 - 2} = 2, \quad m_2 = \dfrac{2 - 0}{3 - 2} = 2$

   $m_1 = m_2$, therefore the lines are collinear.

52. $m_1 = \dfrac{-6 - 4}{7 - 0} = -\dfrac{10}{7}, \quad m_2 = \dfrac{11 - 4}{-5 - 0} = -\dfrac{7}{5}$

   $m_1 \neq m_2$, therefore the lines are not collinear.

53. $m_1 = \dfrac{1 - 0}{-2 - (-1)} = -1, \quad m_2 = \dfrac{-2 - 0}{2 - (-1)} = -\dfrac{2}{3}$

   $m_1 \neq m_2$, therefore the lines are not collinear.

54. Equations of the medians:

   $y = \dfrac{c}{b}x$

   $y = \dfrac{c}{3a + b}(x + a)$

   $y = \dfrac{c}{-3a + b}(x - a)$

   Solving simultaneously, the point of intersection is $(b/3, c/3)$.

   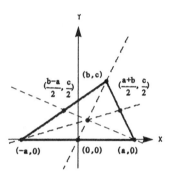

55. Equations of perpendicular bisectors:

   $x = 0$

   $y - \dfrac{c}{2} = \dfrac{a - b}{c}\left(x - \dfrac{a + b}{2}\right)$

   $y - \dfrac{c}{2} = \dfrac{a + b}{-c}\left(x - \dfrac{b - a}{2}\right)$

   Solving simultaneously, the point of intersection is

   $\left(0, \dfrac{-a^2 + b^2 + c^2}{2c}\right).$

   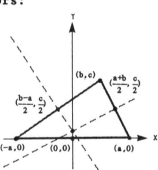

56. Equations of altitudes:

$$y = \frac{a - b}{c}(x + a)$$

$$x = b$$

$$y = -\frac{a + b}{c}(x - a)$$

Solving simultaneously, the point of intersection is

$$(b, \frac{a^2 - b^2}{c}).$$

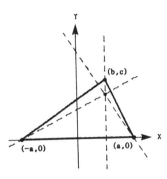

57. The slope of the line segment from $(\frac{b}{3}, \frac{c}{3})$ to $(b, \frac{a^2 - b^2}{c})$ is:

$$m_1 = \frac{[(a^2 - b^2)/c] - (c/3)}{b - (b/3)} = \frac{(3a^2 - 3b^2 - c^2)/(3c)}{(2b)/3} = \frac{3a^2 - 3b^2 - c^2}{2bc}$$

The slope of the line segment from $(\frac{b}{3}, \frac{c}{3})$ to $(0, \frac{-a^2 + b^2 + c^2}{2c})$ is:

$$m_2 = \frac{[(-a^2 + b^2 + c^2)/(2c)] - (c/3)}{0 - (b/3)} = \frac{(-3a^2 + 3b^2 + 3c^2 - 2c^2)/(6c)}{-b/3}$$

$$= \frac{3a^2 - 3b^2 - c^2}{2bc}$$

$m_1 = m_2$, therefore the points are collinear.

58. Find the equation of the line through the points (0, 32) and (100,212):

m = 180/100 = 9/5
F = (9/5)C + 32
5F - 9C - 160 = 0

59.

C	-17.8	-10	10	20	32.2	177
F	0	14	50	68	90	350.6

60. w = 0.75x + 4.50

61. Depreciation per year:

    $\dfrac{875}{5} = \$175$

    $V = 875 - 175t$
    where $0 \leq t \leq 5$

62. Depreciation per year:

    $\dfrac{825{,}000 - 75{,}000}{25} = \$30{,}000$

    $V = 825{,}000 - 30{,}000t$
    where $0 \leq t \leq 25$

63. (a) The equation of the line through $(280, 50)$ and $(325, 47)$ is

    $x - 50 = \dfrac{3}{-45}(p - 280)$ ⟹ $x - 50 = -\dfrac{1}{15}(p - 280)$

    $x = -\dfrac{1}{15}p + \dfrac{280}{15} + 50$ ⟹ $x = \dfrac{1}{15}(1030 - p)$

    (b) $x = \dfrac{1}{15}(1030 - 355)$ ⟹ $x = 45$ units

    (c) $x = \dfrac{1}{15}(1030 - 295)$ ⟹ $x = 49$ units

64. (a) $m = \dfrac{172 - 96}{3 - 0} = \dfrac{76}{3}$ ⟹ $y - 96 = \dfrac{76}{3}t$ ⟹ $y = \dfrac{76}{3}t + 96$

    (b) In 1976 $t = 1$ ⟹ $y = \dfrac{76}{3} + 96$ ⟹ $y = \$121.3\overline{3}$

    In 1977 $t = 2$ ⟹ $y = \dfrac{76}{3}(2) + 96$ ⟹ $y = \$146.6\overline{6}$

    (c) In 1980 $t = 5$ ⟹ $y = \dfrac{76}{3}(5) + 96$ ⟹ $y = \$222.6\overline{6}$

    (d) Every three years the amount spent by the U.S. on energy imports increases by 76 billion dollars.

65. (a)

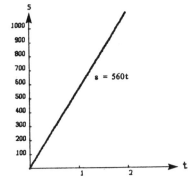

    (b) The speed of the plane is 560 miles per hour.

66. $4x + 3y - 10 = 0$ $\Rightarrow$ $d = \dfrac{|4(0) + 3(0) - 10|}{\sqrt{4^2 + 3^2}} = \dfrac{10}{5} = 2$

67. $4x + 3y - 10 = 0$ $\Rightarrow$ $d = \dfrac{|4(2) + 3(3) - 10|}{\sqrt{4^2 + 3^2}} = \dfrac{7}{5}$

68. $x - y - 2 = 0$ $\Rightarrow$ $d = \dfrac{|1(-2) + (-1)(1) - 2|}{\sqrt{1^2 + 1^2}} = \dfrac{5}{\sqrt{2}} = \dfrac{5\sqrt{2}}{2}$

69. $x + 1 = 0$ $\Rightarrow$ $d = \dfrac{|1(6) + (0)(2) + 1|}{\sqrt{1^2 + 0^2}} = 7$

70. A point on the line $x + y = 1$ is $(0, 1)$. The distance from the point $(0, 1)$ to $x + y = 5$ is:

    $d = \dfrac{|1 - 5|}{\sqrt{2}} = \dfrac{4}{\sqrt{2}} = 2\sqrt{2}$

71. A point on the line $3x - 4y = 1$ is $(-1, -1)$. The distance from the point $(-1, -1)$ to $3x - 4y - 10 = 0$ is:

    $d = \dfrac{|-3 + 4 - 10|}{5} = \dfrac{9}{5}$

72. Find y such that $d_1 = d_2$:

    $2 - y = 4y/5$
    $10 - 5y = 4y$
    $y = 10/9$

    The equation of the line through $(0, 10/9)$ and $(8/3, 2)$ is
    $3x - 9y + 10 = 0$.

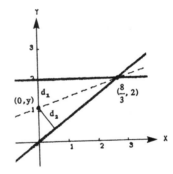

73. The slope of the given line is $-4/3$ so the slope of the perpendicular line is $3/4$.
    The equation of the perpendicular line through the origin is $y = (3/4)x$.
    The point of intersection between $4x + 3y = 10$ and $y = (3/4)x$ is $(8/5, 6/5)$.
    The distance between this point and the origin is 2.

1.5
Functions

1. (a) $f(0) = 2(0) - 3 = -3$ (b) $f(-3) = 2(-3) - 3 = -9$
   (c) $f(b) = 2b - 3$ (d) $f(x - 1) = 2(x - 1) - 3 = 2x - 5$

2. (a) $f(\frac{1}{2}) = (\frac{1}{2})^2 - 2(\frac{1}{2}) + 2 = \dfrac{5}{4}$ (b) $f(-1) = (-1)^2 - 2(-1) + 2 = 5$

   (c) $f(c) = c^2 - 2c + 2$
   (d) $f(x + \Delta x) = (x + \Delta x)^2 - 2(x + \Delta x) + 2$
   $\qquad\qquad\quad = x^2 + 2x\Delta x + (\Delta x)^2 - 2x - 2\Delta x + 2$

3. (a) $f(-2) = \sqrt{-2 + 3} = \sqrt{1} = 1$ (b) $f(6) = \sqrt{6 + 3} = \sqrt{9} = 3$

    (c) $f(c) = \sqrt{c + 3}$ (d) $f(x + \Delta x) = \sqrt{x + \Delta x + 3}$

4. (a) $f(2) = \dfrac{1}{\sqrt{2}} = \dfrac{\sqrt{2}}{2}$ (b) $f(1/4) = \dfrac{1}{\sqrt{(1/4)}} = 2$

    (c) $f(x + \Delta x) = \dfrac{1}{\sqrt{x + \Delta x}}$

    (d) $f(x + \Delta x) - f(x) = \dfrac{1}{\sqrt{x + \Delta x}} - \dfrac{1}{\sqrt{x}} = \dfrac{\sqrt{x} - \sqrt{x + \Delta x}}{\sqrt{x}\sqrt{x + \Delta x}} = \dfrac{\sqrt{x} - \sqrt{x + \Delta x}}{\sqrt{x^2 + x(\Delta x)}}$

5. (a) $f(2) = \dfrac{|2|}{2} = 1$ (b) $f(-2) = \dfrac{|-2|}{-2} = -1$

    (c) $f(x^2) = \dfrac{|x^2|}{x^2} = 1$ (d) $f(x - 1) = \dfrac{|x - 1|}{x - 1}$

6. (a) $f(2) = 6$ (b) $f(-2) = 6$

    (c) $f(x^2) = x^2 + 4$

    (d) $f(x + \Delta x) - f(x) = |x + \Delta x| + 4 - (|x| + 4) = |x + \Delta x| - |x|$

7. $\dfrac{f(2 + \Delta x) - f(2)}{\Delta x} = \dfrac{(2 + \Delta x)^2 - (2 + \Delta x) + 1 - [(2)^2 - 2 + 1]}{\Delta x}$

   $= \dfrac{4 + 4\Delta x + (\Delta x)^2 - 2 - \Delta x + 1 - 4 + 2 - 1}{\Delta x} = 4 - 1 + \Delta x = 3 + \Delta x$

8. $\dfrac{f(1 + \Delta x) - f(1)}{\Delta x} = \dfrac{1/(1 + \Delta x) - 1}{\Delta x} = \dfrac{1 - (1 + \Delta x)}{\Delta x(1 + \Delta x)} = \dfrac{-1}{1 + \Delta x}$

9. $\dfrac{f(x + \Delta x) - f(x)}{\Delta x} = \dfrac{(x + \Delta x)^3 - x^3}{\Delta x}$

   $= \dfrac{x^3 + 3x^2\Delta x + 3x(\Delta x)^2 + (\Delta x)^3 - x^3}{\Delta x} = 3x^2 + 3x\Delta x + (\Delta x)^2$

10. $\dfrac{f(x) - f(1)}{x - 1} = \dfrac{3x - 1 - (3 - 1)}{x - 1} = \dfrac{3(x - 1)}{x - 1} = 3$

11. $\dfrac{f(x) - f(2)}{x - 2} = \dfrac{(1/\sqrt{x - 1}) - 1}{x - 2} = \dfrac{1 - \sqrt{x - 1}}{(x - 2)\sqrt{x - 1}} \cdot \dfrac{1 + \sqrt{x - 1}}{1 + \sqrt{x - 1}}$

   $= \dfrac{2 - x}{(x - 2)\sqrt{x - 1}(1 + \sqrt{x - 1})} = \dfrac{-1}{\sqrt{x - 1}(1 + \sqrt{x - 1})}$

12. $\dfrac{f(x) - f(1)}{x - 1} = \dfrac{x^3 - x - 0}{x - 1} = \dfrac{x(x + 1)(x - 1)}{x - 1} = x(x + 1)$

13. (a) $f(g(1)) = f(0) = 0$ (b) $g(f(1)) = g(1) = 0$
    (c) $g(f(0)) = g(0) = -1$ (d) $f(g(-4)) = f(15) = \sqrt{15}$
    (e) $f(g(x)) = f(x^2 - 1)$ (f) $g(f(x)) = g(\sqrt{x})$
    $\qquad = \sqrt{x^2 - 1}$ $\qquad = (\sqrt{x})^2 - 1 = x - 1$

14. (a) $f(g(2)) = f(3) = 1/3$ (b) $g(f(2)) = g(1/2) = -3/4$

    (c) $f(g(\frac{1}{\sqrt{2}})) = f(-\frac{1}{2}) = -2$ (d) $g(f(\frac{1}{\sqrt{2}})) = g(\sqrt{2}) = 1$

    (e) $g(f(x)) = g(\frac{1}{x}) = (\frac{1}{x})^2 - 1 = \dfrac{1 - x^2}{x^2}$

    (f) $f(g(x)) = f(x^2 - 1) = \dfrac{1}{x^2 - 1}$

15. $f(x) = x^2 - 9 = (x + 3)(x - 3)$ 16. $f(x) = x^3 - x = x(x + 1)(x - 1)$
    Zeros: $-3, 3$ Zeros: $0, -1, 1$

17. $f(x) = \dfrac{3}{x - 1} + \dfrac{4}{x - 2}$ 18. $f(x) = a + \dfrac{b}{x}$

    $3(x - 2) + 4(x - 1) = 0$ $ax + b = 0$
    $7x - 10 = 0$
    $x = 10/7$ $x = -\dfrac{b}{a}$
    Zero: $10/7$
    $\qquad\qquad$ Zero: $-\dfrac{b}{a}$

19. $f(x) = \sqrt{x - 1}$ 20. $f(x) = \sqrt{1 - x}$
    Domain: $[1, \infty)$ Domain: $(-\infty, 1]$
    Range: $[0, \infty)$ Range: $[0, \infty)$

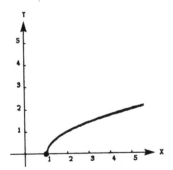   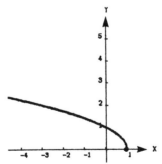

21. $f(x) = x^2$
     Domain: $(-\infty, \infty)$
     Range: $[0, \infty)$

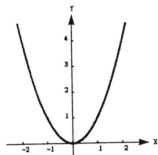

22. $f(x) = 4 - x^2$
     Domain: $(-\infty, \infty)$
     Range: $(-\infty, 4]$

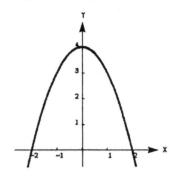

23. $f(x) = \sqrt{9 - x^2}$
     Domain: $[-3, 3]$
     Range: $[0, 3]$

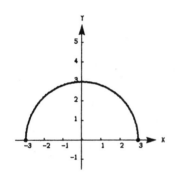

24. $f(x) = \sqrt{25 - x^2}$
     Domain: $[-5, 5]$
     Range: $[0, 5]$

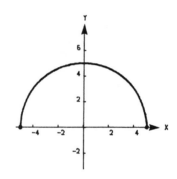

25. $f(x) = 1/|x|$
     Domain: $(-\infty, 0), (0, \infty)$
     Range: $(0, \infty)$

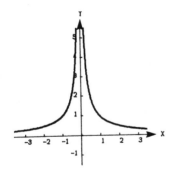

26. $f(x) = |x - 2|$
     Domain: $(-\infty, \infty)$
     Range: $[0, \infty)$

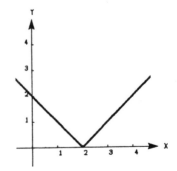

27. $f(x) = |x|/x$
    Domain: $(-\infty, 0), (0, \infty)$
    Range: -1 and 1

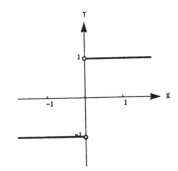

28. $f(x) = \sqrt{x^2 - 4}$
    Domain: $(-\infty, -2], [2, \infty)$
    Range: $[0, \infty)$

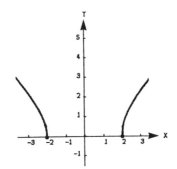

29. $y = x^2$
    y is a function of x

30. $y = x^3 - 1$
    y is a function of x

31. $x - y^2 = 0$
    y is not a function of x

32. $x^2 + y^2 = 9$
    y is not a function of x

33. $\sqrt{x^2 - 4} - y = 0$
    y is a function of x

34. $x - xy + y + 1 = 0$
    y is a function of x

35. $x^2 = xy - 1$
    y is a function of x

36. $x = |y|$
    y is not a function of x

37. $x^2 - 4y^2 + 4 = 0$
    y is not a function of x

38. $y = |x - 3| + |x - 1|$
    y is a function of x

39. $x^2 + y^2 = 4 \implies y = \pm\sqrt{4 - x^2}$
    y is not a function of x since there are two values of y for some x.

40. $x = y^2 \implies y = \pm\sqrt{x}$
    y is not a function of x since there are two values of y for some x.

41. $x^2 + y = 4 \implies y = 4 - x^2$
    y is a function of x since there is one value of y for each x.

42. $x + y^2 = 4 \implies y = \pm\sqrt{4 - x}$
    y is not a function of x since there are two values for y for some x.

43. $2x + 3y = 4 \implies y = (4 - 2x)/3$
    y is a function of x since there is one value of y for each x.

44. $x^2 + y^2 - 2x - 4y + 1 = 0 \implies y = 2 \pm\sqrt{3 + 2x - x^2}$
    y is not a function of x since there are two values of y for some x.

45. $y^2 = x^2 - 1 \implies y = \pm\sqrt{x^2 - 1}$
    y is not a function of x since there are two values of y for some x.

46. $x^2y - x^2 + 4y = 0$ ⟹ $y = x^2/(x^2 + 4)$
y is a function of x since there is one value of y for each x.

47. (a) $y = \sqrt{x} + 2$

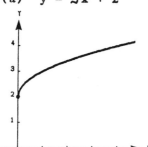

(b) $y = -\sqrt{x}$

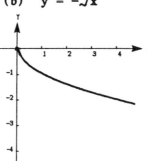

(c) $y = \sqrt{x - 2}$

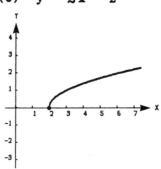

(d) $y = \sqrt{x + 3}$

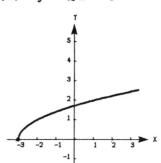

(e) $y = \sqrt{x - 4}$

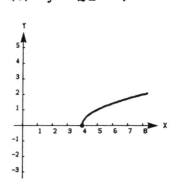

(f) $y = 2\sqrt{x}$

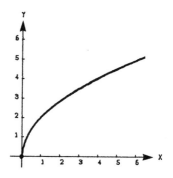

48. (a) $y = \sqrt[3]{x} - 1$

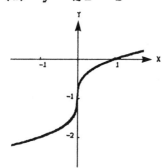

(b) $y = \sqrt[3]{x + 1}$

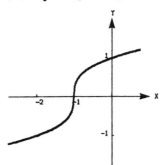

(c) $y = \sqrt[3]{x - 1}$

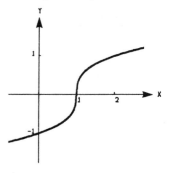

(d) $y = -\sqrt[3]{x - 2}$

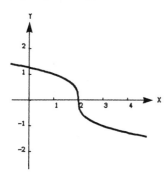

(e) $y = \dfrac{1}{2} \sqrt[3]{x}$

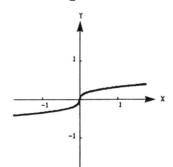

(f) $y = \sqrt[3]{x + 1} - 1$

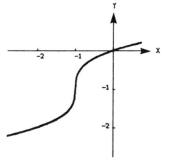

49. (a) $y = (x + 1)\sqrt{x + 4}$ (b) $y = x\sqrt{x + 3} + 2$
    (c) $y = -x\sqrt{x + 3}$ (d) $y = -(x - 1)\sqrt{x + 2}$

50. (a) $f(x) = \dfrac{1}{x^2 + 1} - 1 = \dfrac{-x^2}{x^2 + 1}$

    (b) $f(x) = \dfrac{1}{(x - 2)^2 + 1} = \dfrac{1}{x^2 - 4x + 5}$

    (c) $f(x) = -\left[\dfrac{1}{x^2 + 1} - 1\right] = \dfrac{x^2}{x^2 + 1}$

    (d) $f(x) = \dfrac{1}{(x - 2)^2 + 1} - 1 = \dfrac{-x^2 + 4x - 4}{x^2 - 4x + 5} = \dfrac{-(x - 2)^2}{x^2 - 4x + 5}$

51. $f(-x) = 4 - (-x)^2$
    $\qquad = 4 - x^2 = f(x)$
    Even

52. $f(-x) = \sqrt[3]{-x}$
    $\qquad = -\sqrt[3]{x} = -f(x)$
    Odd

53. $f(-x) = (-x)(4 - (-x)^2)$
    $\qquad = -x(4 - x^2) = -f(x)$
    Odd

54. $f(-x) = \sqrt[3]{-x} + 1$
    $\qquad = -\sqrt[3]{x} + 1$
    Neither odd nor even

55. $f(-x) = 4(-x) - (-x)^2$
    $\qquad = -4x - x^2$
    Neither odd nor even

56. $f(-x) = (-x)^{2/3} = x^{2/3} = f(x)$
    Even

57. y-axis symmetry

58. Origin symmetry

59. $f(-x) = a_{2n+1}(-x)^{2n+1} + \ldots + a_3(-x)^3 + a_1(-x)$

    $\qquad = -[a_{2n+1}x^{2n+1} + \ldots + a_3x^3 + a_1x]$

    $\qquad = -f(x)$
    Odd

60. $f(-x) = a_{2n}(-x)^{2n} + a_{2n-2}(-x)^{2n-2} + \ldots + a_2(-x)^2 + a_0$

    $\qquad = a_{2n}x^{2n} + a_{2n-2}x^{2n-2} + \ldots + a_2x^2 + a_0$

    $\qquad = f(x)$
    Even

61. Let $F(x) = f(x)g(x)$ where f and g are even, then $F(-x) = f(-x)g(-x)$
    $= f(x)g(x) = F(x)$, thus $F(x)$ is even.

    Let $F(x) = f(x)g(x)$ where f and g are odd, then $F(-x) = f(-x)g(-x)$
    $= [-f(x)][-g(x)] = f(x)g(x) = F(x)$, thus $F(x)$ is even.

62. Let $F(x) = f(x)g(x)$ where f is even and g is odd, then $F(-x)$
    $= f(-x)g(-x) = f(x)[-g(x)] = -f(x)g(x) = -F(x)$, thus $F(x)$ is odd.

63. $2x + 2y = 100$ ➥ $A = xy = \dfrac{x(100 - 2x)}{2} = x(50 - x)$

64. $4x + 3y = 200$ ➥ $A = 2xy = 2x\left[\dfrac{200 - 4x}{3}\right] = \dfrac{8}{3}(50x - x^2)$

65. $V = lwh = x(12 - 2x)^2$

66. $y = \sqrt{25 - x^2}$
    $A = 2xy = 2x\sqrt{25 - x^2}$

67. $4x + y = 108$
    $V = x^2y = x^2(108 - 4x)$
    $\quad = 108x^2 - 4x^3$

68. $2x^2 + 4xy = 100$

    $V = x^2y = x^2(\dfrac{100 - 2x^2}{4x})$

    $\quad = 25x - \dfrac{x^3}{2}$

69. $T = \dfrac{D}{R}$

    $T = T_{row} + T_{walk}$

    $\quad = \dfrac{\sqrt{x^2 + 4}}{2} + \dfrac{\sqrt{1 + (3 - x)^2}}{4}$

    $\quad = \dfrac{\sqrt{x^2 + 4}}{2} + \dfrac{\sqrt{x^2 - 6x + 10}}{4}$

70. $A = \pi r^2 = \pi y^2 = \pi(\sqrt{x})^2 = \pi x$

1.6
Review of trigonometric functions

1. (a) $396°$, $-324°$ (b) $315°$, $-405°$
   (c) $240°$, $-480°$ (d) $30°$, $-330°$

2. (a) $660°$, $-60°$ (b) $20°$, $-340°$
   (c) $300°$, $-60°$ (d) $590°$, $-130°$

3. (a) $\dfrac{19\pi}{9}, \quad -\dfrac{17\pi}{9}$ (b) $\dfrac{10\pi}{3}, \quad -\dfrac{2\pi}{3}$

   (c) $\dfrac{23\pi}{6}, \quad -\dfrac{\pi}{6}$ (d) $\dfrac{5\pi}{6}, \quad -\dfrac{19\pi}{6}$

4. (a) $\dfrac{7\pi}{4}, \quad -\dfrac{\pi}{4}$ (b) $\dfrac{28\pi}{15}, \quad -\dfrac{32\pi}{15}$

   (c) $\dfrac{26\pi}{9}, \quad -\dfrac{10\pi}{9}$ (d) $\dfrac{98\pi}{45}, \quad -\dfrac{82\pi}{45}$

5. (a) $30\left(\dfrac{\pi}{180}\right) = \dfrac{\pi}{6}$ (b) $150\left(\dfrac{\pi}{180}\right) = \dfrac{5\pi}{6}$

   (c) $315\left(\dfrac{\pi}{180}\right) = \dfrac{7\pi}{4}$ (d) $120\left(\dfrac{\pi}{180}\right) = \dfrac{2\pi}{3}$

6. (a) $-20\left(\dfrac{\pi}{180}\right) = -\dfrac{\pi}{9}$ (b) $-240\left(\dfrac{\pi}{180}\right) = -\dfrac{4\pi}{3}$

   (c) $-270\left(\dfrac{\pi}{180}\right) = -\dfrac{3\pi}{2}$ (d) $144\left(\dfrac{\pi}{180}\right) = \dfrac{4\pi}{5}$

7. (a) $\dfrac{3\pi}{2}\left(\dfrac{180}{\pi}\right) = 270^\circ$ (b) $\dfrac{7\pi}{6}\left(\dfrac{180}{\pi}\right) = 210^\circ$

   (c) $-\dfrac{7\pi}{12}\left(\dfrac{180}{\pi}\right) = -105^\circ$ (d) $\dfrac{\pi}{9}\left(\dfrac{180}{\pi}\right) = 20^\circ$

8. (a) $\dfrac{7\pi}{3}\left(\dfrac{180}{\pi}\right) = 420^\circ$ (b) $-\dfrac{11\pi}{30}\left(\dfrac{180}{\pi}\right) = -66^\circ$

   (c) $\dfrac{11\pi}{6}\left(\dfrac{180}{\pi}\right) = 330^\circ$ (d) $\dfrac{34\pi}{15}\left(\dfrac{180}{\pi}\right) = 408^\circ$

9.

r	8 ft	15 in	85 cm	24 in	$\dfrac{12963}{\pi}$ mi
s	12 ft	24 in	63.75π cm	96 in	8642 mi
θ	1.5	1.6	$3\pi/4$	4	$2\pi/3$

10. $S = r\theta = (3\frac{1}{2})(\frac{5\pi}{6}) = \frac{35\pi}{12} \approx 2.9167\pi$ inches

11. (a) $75^\circ = \frac{5\pi}{12}$ radians   (b) $S = (18.75)(\frac{5\pi}{12}) = 7.8125\pi$ inches

12. (a) $80^\circ = \frac{4\pi}{9}$ (b) $S = (2.5)(\frac{4\pi}{9}) \approx 3.491$ feet

13. (a) $x = 3, \quad y = 4, \quad r = 5$ (b) $x = 8, \quad y = -15, \quad r = 17$

$\sin\theta = 4/5 \quad \csc\theta = 5/4$	$\sin\theta = -15/17 \quad \csc\theta = -17/15$
$\cos\theta = 3/5 \quad \sec\theta = 5/3$	$\cos\theta = 8/17 \quad \sec\theta = 17/8$
$\tan\theta = 4/3 \quad \cot\theta = 3/4$	$\tan\theta = -15/8 \quad \cot\theta = -8/15$

14. (a) $x = -12, \quad y = -5, \quad r = 13$ (b) $x = 1, \quad y = -1, \quad r = \sqrt{2}$

$\sin\theta = -5/13 \quad \csc\theta = -13/5$	$\sin\theta = -\sqrt{2}/2 \quad \csc\theta = -\sqrt{2}$
$\cos\theta = -12/13 \quad \sec\theta = -13/12$	$\cos\theta = \sqrt{2}/2 \quad \sec\theta = \sqrt{2}$
$\tan\theta = 5/12 \quad \cot\theta = 12/5$	$\tan\theta = -1 \quad \cot\theta = -1$

15. Quadrant III   16. Quadrant II   17. $\csc\theta = \frac{1}{\sin\theta} = 2$

18. $y = 1, \quad r = 3, \quad x = \sqrt{8} = 2\sqrt{2}, \quad \tan\theta = \frac{1}{2\sqrt{2}} = \frac{\sqrt{2}}{4}$

19. $x = 4, \quad r = 5, \quad y = 3, \quad \cot\theta = 4/3$

20. $r = 13, \quad x = 5, \quad y = 12, \quad \cot\theta = 5/12$

21. $x = 15, \quad y = 8, \quad r = 17, \quad \sec\theta = 17/15$

22. $y = 1, \quad x = 2, \quad r = \sqrt{5}, \quad \sin\theta = \frac{1}{\sqrt{5}} = \frac{\sqrt{5}}{5}$

23. (a) $\sin 60^\circ = \frac{\sqrt{3}}{2}$ (b) $\sin\frac{2\pi}{3} = \frac{\sqrt{3}}{2}$

        $\cos 60^\circ = \frac{1}{2}$ $\cos\frac{2\pi}{3} = -\frac{1}{2}$

        $\tan 60^\circ = \sqrt{3}$ $\tan\frac{2\pi}{3} = -\sqrt{3}$

   (c) $\sin\frac{\pi}{4} = \frac{\sqrt{2}}{2}$ (d) $\sin\frac{5\pi}{4} = -\frac{\sqrt{2}}{2}$

        $\cos\frac{\pi}{4} = \frac{\sqrt{2}}{2}$ $\cos\frac{5\pi}{4} = -\frac{\sqrt{2}}{2}$

        $\tan\frac{\pi}{4} = 1$ $\tan\frac{5\pi}{4} = 1$

24. (a) $\sin(-\frac{\pi}{6}) = -\frac{1}{2}$

$\cos(-\frac{\pi}{6}) = \frac{\sqrt{3}}{2}$

$\tan(-\frac{\pi}{6}) = -\frac{\sqrt{3}}{3}$

(b) $\sin 150^\circ = \frac{1}{2}$

$\cos 150^\circ = -\frac{\sqrt{3}}{2}$

$\tan 150^\circ = -\frac{\sqrt{3}}{3}$

(c) $\sin(-\frac{\pi}{2}) = -1$

$\cos(-\frac{\pi}{2}) = 0$

$\tan(-\frac{\pi}{2})$ is undefined

(d) $\sin\frac{\pi}{2} = 1$

$\cos\frac{\pi}{2} = 0$

$\tan\frac{\pi}{2}$ is undefined

25. (a) $\sin 225^\circ = -\frac{\sqrt{2}}{2}$

$\cos 225^\circ = -\frac{\sqrt{2}}{2}$

$\tan 225^\circ = 1$

(b) $\sin(-225^\circ) = \frac{\sqrt{2}}{2}$

$\cos(-225^\circ) = -\frac{\sqrt{2}}{2}$

$\tan(-225^\circ) = -1$

(c) $\sin 300^\circ = -\frac{\sqrt{3}}{2}$

$\cos 300^\circ = \frac{1}{2}$

$\tan 300^\circ = -\sqrt{3}$

(d) $\sin 330^\circ = -\frac{1}{2}$

$\cos 330^\circ = \frac{\sqrt{3}}{2}$

$\tan 330^\circ = -\frac{\sqrt{3}}{3}$

26. (a) $\sin 750^\circ = \frac{1}{2}$

$\cos 750^\circ = \frac{\sqrt{3}}{2}$

$\tan 750^\circ = \frac{\sqrt{3}}{3}$

(b) $\sin 510^\circ = \frac{1}{2}$

$\cos 510^\circ = -\frac{\sqrt{3}}{2}$

$\tan 510^\circ = -\frac{\sqrt{3}}{3}$

(c) $\sin\frac{10\pi}{3} = -\frac{\sqrt{3}}{2}$

$\cos\frac{10\pi}{3} = -\frac{1}{2}$

$\tan\frac{10\pi}{3} = \sqrt{3}$

(d) $\sin\frac{17\pi}{3} = -\frac{\sqrt{3}}{2}$

$\cos\frac{17\pi}{3} = \frac{1}{2}$

$\tan\frac{17\pi}{3} = -\sqrt{3}$

27. (a) $\sin 10^\circ = 0.1736$ (b) $\csc 10^\circ = 5.7588$

28. (a) $\sec 225^\circ = -1.4142$ (b) $\sec 135^\circ = -1.4142$

29. (a) $\tan \dfrac{\pi}{9} = 0.3640$ (b) $\tan \dfrac{10\pi}{9} = 0.3640$

30. (a) $\cot 1.35 = 0.2245$ (b) $\tan 1.35 = 4.4552$

31. (a) $\cos \theta = \dfrac{\sqrt{2}}{2}$ (b) $\cos \theta = -\dfrac{\sqrt{2}}{2}$

$\theta = \dfrac{\pi}{4}, \dfrac{7\pi}{4}$ $\theta = \dfrac{3\pi}{4}, \dfrac{5\pi}{4}$

32. (a) $\sec \theta = 2$ (b) $\sec \theta = -2$

$\theta = \dfrac{\pi}{3}, \dfrac{5\pi}{3}$ $\theta = \dfrac{2\pi}{3}, \dfrac{4\pi}{3}$

33. (a) $\tan \theta = 1$ (b) $\cot \theta = -\sqrt{3}$

$\theta = \dfrac{\pi}{4}, \dfrac{5\pi}{4}$ $\theta = \dfrac{5\pi}{6}, \dfrac{11\pi}{6}$

34. (a) $\sin \theta = \dfrac{\sqrt{3}}{2}$ (b) $\sin \theta = -\dfrac{\sqrt{3}}{2}$

$\theta = \dfrac{\pi}{3}, \dfrac{2\pi}{3}$ $\theta = \dfrac{4\pi}{3}, \dfrac{5\pi}{3}$

35. $2 \sin^2 \theta = 1$

$\sin \theta = \pm \dfrac{\sqrt{2}}{2}$

$\theta = \dfrac{\pi}{4}, \dfrac{3\pi}{4}, \dfrac{5\pi}{4}, \dfrac{7\pi}{4}$

36. $\tan^2 \theta = 3$
$\tan \theta = \pm \sqrt{3}$

$\theta = \dfrac{\pi}{3}, \dfrac{2\pi}{3}, \dfrac{4\pi}{3}, \dfrac{5\pi}{3}$

37. $\tan^2 \theta = \tan \theta$
$\tan \theta (\tan \theta - 1) = 0$
$\tan \theta = 0 \quad \tan \theta = 1$

$\theta = 0, \pi \quad \theta = \dfrac{\pi}{4}, \dfrac{5\pi}{4}$

38. $2 \cos^2 \theta - \cos \theta - 1 = 0$
$(2 \cos \theta + 1)(\cos \theta - 1) = 0$

$\cos \theta = -\dfrac{1}{2} \quad\quad \cos \theta = 1$

$\theta = \dfrac{2\pi}{3}, \dfrac{4\pi}{3} \quad\quad \theta = 0$

39. $\sec \theta \csc \theta - 2 \csc \theta = 0$
$\csc \theta (\sec \theta - 2) = 0$
$(\csc \theta \neq 0$ for any value of $\theta)$
$\sec \theta = 2$

$\theta = \dfrac{\pi}{3}, \dfrac{5\pi}{3}$

40. $\sin \theta = \cos \theta$
$\tan \theta = 1$

$\theta = \dfrac{\pi}{4}, \dfrac{5\pi}{4}$

41. $\cos^2\theta + \sin\theta = 1$
$1 - \sin^2\theta + \sin\theta = 1$
$\sin^2\theta - \sin\theta = 0$
$\sin\theta(\sin\theta - 1) = 0$
$\sin\theta = 0 \qquad \sin\theta = 1$
$\theta = 0, \pi \qquad \theta = \pi/2$

42. $\cos(\theta/2) - \cos\theta = 1$
$\cos(\theta/2) = \cos\theta + 1$
$\sqrt{(1/2)(1 + \cos\theta)} = \cos\theta + 1$
$(1/2)(1 + \cos\theta) = \cos^2\theta + 2\cos\theta + 1$
$0 = \cos^2\theta + 3/2\cos\theta + 1/2$
$0 = (1/2)(2\cos^2\theta + 3\cos\theta + 1)$
$0 = (1/2)(2\cos\theta + 1)(\cos\theta + 1)$
$\cos\theta = -1/2 \qquad \cos\theta = -1$
$\theta = 2\pi/3 \qquad \theta = \pi$
$(\theta = 4\pi/3$ is extraneous$)$

43. $\tan 30^o = \dfrac{y}{100}$

$y = 100(\dfrac{1}{\sqrt{3}}) = \dfrac{100\sqrt{3}}{3}$

44. $\cos 60^o = \dfrac{x}{10}$

$x = 10\ (\dfrac{1}{2}) = 5$

45. $\tan 60^o = \dfrac{25}{x}$

$x = \dfrac{25}{\sqrt{3}} = \dfrac{25\sqrt{3}}{3}$

46. $\sin 45^o = \dfrac{30}{r}$

$r = \dfrac{30}{1/\sqrt{2}} = 30\sqrt{2}$

47. $\sin 75^o = \dfrac{x}{20}$

$x = 20\sin 75^o$
≈ 19.32 feet

48. $\tan 50^o = \dfrac{w}{100}$

$w = 100\tan 50^o$
≈ 119.175 feet

49. $\tan 4^o = \dfrac{150}{x}$

$x = \dfrac{150}{\tan 4^o}$

≈ 2145.1 feet

50. $\sin\theta = \dfrac{3\frac{1}{2}}{17\frac{1}{2}} = \dfrac{1}{5}$

$\theta \approx 11.5^o$

51. (a) Period: π
    Amplitude: 2

  (b) Period: 2
    Amplitude: 1/2

52. (a) Period: $2\pi/3$
    Amplitude: 3

  (b) Period: 4
    Amplitude: 5/2

53. (a) Period: 4π
    Amplitude: 3/2

  (b) Period: 2π
    Amplitude: 2

54. (a) Period: 6π
    Amplitude: 2

  (b) Period: 3π
    Amplitude: 1

55. Period: $\pi/5$
    Amplitude: 2

56. Period: 3π
    Amplitude: 1/2

57. Period: 1/2
    Amplitude: 3

58. Period: 20
    Amplitude: 2/3

59. Period: $\pi/2$

60. Period: 1/2

61. Period: $2\pi/5$

62. Period: $\pi/2$

63. $y = \sin(x/2)$
    Period: 4π
    Amplitude: 1

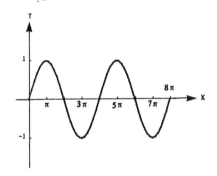

64. $y = 2\cos 2x$
    Period: π
    Amplitude: 2

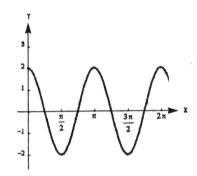

65. $y = -2\sin 6x$
    Period: $\pi/3$
    Amplitude: 2

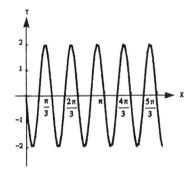

66. $y = \cos 2\pi x$
    Period: 1
    Amplitude: 1

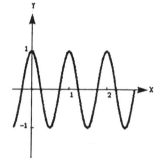

67. $y = -\sin(2\pi x/3)$
    Period: 3
    Amplitude: 1

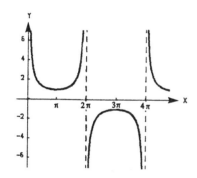

68. $y = 2\tan x$
    Period: π

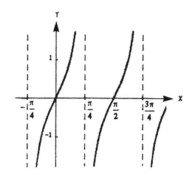

69. $y = \csc(x/2)$
    Period: 4π

70. $y = \tan 2x$
    Period: $\pi/2$

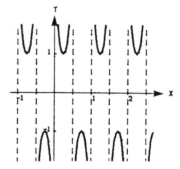

71. $y = 2\sec 2x$
    Period: π

72. $y = \csc 2\pi x$
    Period: 1

73. $y = \sin(x + \pi)$
Period: 2π
Amplitude: 1

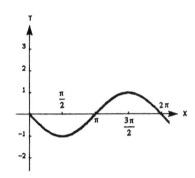

74. $y = \cos\left(x - \dfrac{\pi}{3}\right)$
Period: 2π
Amplitude: 1

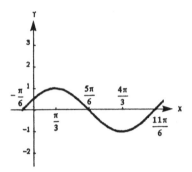

75. $y = 1 + \cos\left(x - \dfrac{\pi}{2}\right)$

Period: 2π
Amplitude: 1

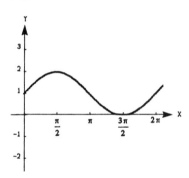

76. $y = 1 + \sin\left(x + \dfrac{\pi}{2}\right)$

Period: 2π
Amplitude: 1

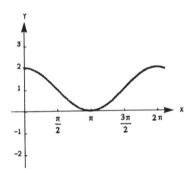

77. $v = 0.85 \sin\dfrac{\pi t}{3}$

(a) Period = 6 sec

(b) $\dfrac{60}{6} = 10$ cycles per min

(c)

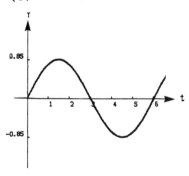

78. $v = 1.75 \sin\dfrac{\pi t}{2}$

(a) Period = 4

(b) $\dfrac{60}{4} = 15$ cycles per min

(c)

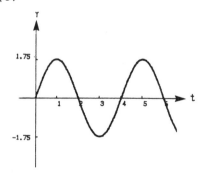

79. $y = 0.001 \sin 880\pi t$

(a) $p = \dfrac{2\pi}{880\pi} = \dfrac{1}{440}$

(c)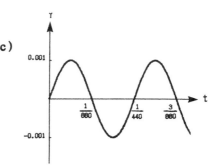

(b) $f = \dfrac{1}{p} = 440$

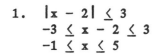 ## Review Exercises for Chapter 1

1. $|x - 2| \leq 3$
    $-3 \leq x - 2 \leq 3$
    $-1 \leq x \leq 5$

2. $|3x - 2| \leq 0,\ x = 2/3$

3. $4 < (x + 3)^2$
    $4 < x^2 + 6x + 9$
    $0 < x^2 + 6x + 5$
    $0 < (x + 1)(x + 5)$
    $x < -5 \text{ or } x > -1$

4. $\dfrac{1}{|x|} < 1,\ |x| > 1,$
    $x > 1 \text{ or } x < -1$

5. $\dfrac{(7/8) + (10/4)}{2} = \dfrac{27/8}{2} = \dfrac{27}{16}$

6. $\dfrac{-1 + (3/2)}{2} = \dfrac{1}{4}$

7. The length of the interval is: $|6 - (-2)| = 8.$ Therefore the points of trisection are: $-2 + \dfrac{8}{3} = \dfrac{2}{3},$ and $\dfrac{2}{3} + \dfrac{8}{3} = \dfrac{10}{3}.$

8. The length of the interval is: $|1 - 5| = 4.$ Therefore the points of trisection are: $1 + \dfrac{4}{3} = \dfrac{7}{3}$ and $\dfrac{7}{3} + \dfrac{4}{3} = \dfrac{11}{3}.$

9. $(\dfrac{1 - 3}{2}, \dfrac{4 + 2}{2}) = (-1,\ 3)$

    $(\dfrac{1 + 5}{2}, \dfrac{4 + 0}{2}) = (3,\ 2)$

    $(\dfrac{5 - 3}{2}, \dfrac{0 + 2}{2}) = (1,\ 1)$

10. $(\dfrac{0 + 1}{2}, \dfrac{2 - 1}{2}) = (\dfrac{1}{2},\ \dfrac{1}{2})$

    $(\dfrac{0 + 2}{2}, \dfrac{2 + 1}{2}) = (1,\ \dfrac{3}{2})$

    $(\dfrac{2 + 1}{2}, \dfrac{1 - 1}{2}) = (\dfrac{3}{2},\ 0)$

11. $(x^2 + 6x + 9) + (y^2 - 2y + 1) = -1 + 9 + 1$

    $(x + 3)^2 + (y - 1)^2 = 9$

    center: $(-3, 1)$ radius: 3

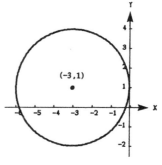

12. $4(x^2 - x + \frac{1}{4}) + 4(y^2 + 2y + 1) = 11 + 1 + 4$

    $(x - \frac{1}{2})^2 + (y + 1)^2 = 4$

    center: $(1/2, -1)$ radius $= 2$

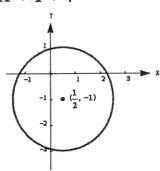

13. $(x^2 + 6x + 9) + (y^2 - 2y + 1) = -10 + 9 + 1$

    $(x + 3)^2 + (y - 1)^2 = 0$

    point $(-3, 1)$

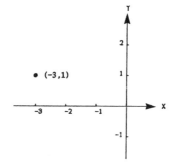

14. $(x^2 - 6x + 9) + (y^2 + 8y + 16) = 9 + 16$

    $(x - 3)^2 + (y + 4)^2 = 25$

    center: $(3, -4)$ radius: 5

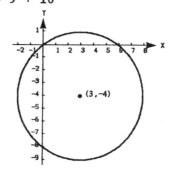

15. $(x^2 - 6x + 9) + (y^2 + 8y + 16) = c + 9 + 16$

    $(x - 3)^2 + (y + 4)^2 = c + 25$

    If the radius is 2 then

    $c + 25 = 4$

    $c = -21$

16. $\sqrt{(x + 2)^2 + (y - 0)^2} = 2\sqrt{(x - 3)^2 + (y - 1)^2}$

    $(x + 2)^2 + y^2 = 4 [(x - 3)^2 + (y - 1)^2]$

    $x^2 + 4x + 4 + y^2 = 4x^2 - 24x + 4y^2 - 8y + 40$

    $0 = 3x^2 + 3y^2 - 28x - 8y + 36$

    $(x - \frac{14}{3})^2 + (y - \frac{4}{3})^2 = \frac{104}{9}$

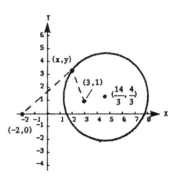

17. $(x - 1)^2 + (y - 2)^2 = 9$
    $x^2 - 2x + 1 + y^2 - 4y + 4 = 9$
    $x^2 + y^2 - 2x - 4y = 4$

    (a) (1, 5)   on the circle since $1^2 + 5^2 - 2(1) - 4(5) = 4$
    (b) (0, 0)   inside the circle since $0^2 + 0^2 - 2(0) - 4(0) < 4$
    (c) (-2, 1) outside the circle since $(-2)^2 + 1^2 - 2(-2) - 4(1) > 4$
    (d) (0, 4)   inside the circle since $0^2 + 4^2 - 2(0) - 4(4) < 4$

18. $(x - 2)^2 + (y - 1)^2 = 4$
    $x^2 + y^2 - 4x - 2y = -1$

    (a) (1, 1)   inside the circle since $(1)^2 + (1)^2 - 4(1) - 2(1) < -1$
    (b) (4, 2)   outside the circle since $(4)^2 + (2)^2 - 4(4) - 2(2) > -1$
    (c) (0, 1)   on the circle since $(0)^2 + (1)^2 - 4(0) - 2(1) = -1$
    (d) (3, 1)   inside the circle since $(3)^2 + (1)^2 - 4(3) - 2(1) < -1$

19. $y = -\frac{1}{2}x + \frac{3}{2}$

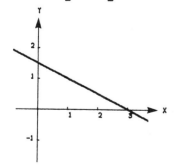

20. $y = 1 + \frac{1}{x}$

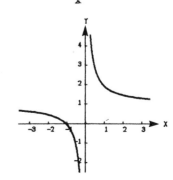

21. $y = 7 - 6x - x^2$

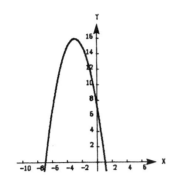

22. $y = x(6 - x)$

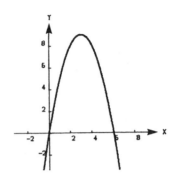

23. The slope of the line segment from $(-1, 3)$ to $(2, 9)$ is $\dfrac{9 - 3}{2 + 1} = 2$.

The slope of the line segment from $(2, 9)$ to $(3, 1)$ is $\dfrac{1 - 9}{3 - 2} = -8$.

The points do not lie on the same line.

24. The slope of the line segment from $(2, 5)$ to $(4, 10)$ is $\dfrac{10 - 5}{4 - 2} = \dfrac{5}{2}$.

The slope of the line segment from $(4, 10)$ to $(6, 20)$ is $\dfrac{20 - 10}{6 - 4} = 5$.

The points do not lie on the same line.

25. $4x - 2y = 6$
   $y = 2x - 3$

   slope: 2
   y-intercept: −3

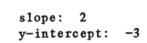

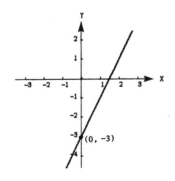

26. $0.02x + 0.15y = 0.25$
   $2x + 15y = 25$

   $y = -\dfrac{2}{15}x + \dfrac{5}{3}$

   slope: −2/15
   y-intercept: 5/3

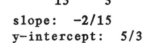

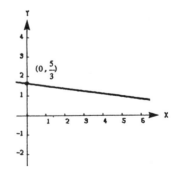

27. $-\dfrac{1}{3}x + \dfrac{5}{6}y = 1$

$-\dfrac{2}{5}x + y = \dfrac{6}{5}$

$y = \dfrac{2}{5}x + \dfrac{6}{5}$

slope: 2/5
y-intercept: 6/5

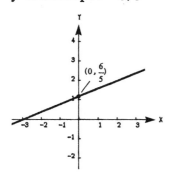

28. $51x + 17y = 102$

$y = -3x + 6$

slope: -3
y-intercept: 6

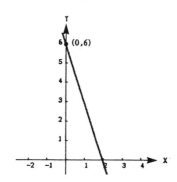

29. (a) $y - 4 = \dfrac{7}{16}(x + 2)$

$y = \dfrac{7}{16}x + \dfrac{39}{8}$

(b) $y - 4 = \dfrac{5}{3}(x + 2)$

$y = \dfrac{5}{3}x + \dfrac{22}{3}$

(c) $m = \dfrac{4 - 0}{-2 - 0} = -2$

$y = -2x$

(d) $x = -2$

30. (a) $y - 3 = -\dfrac{2}{3}(x - 1)$

$y = -\dfrac{2}{3}x + \dfrac{11}{3}$

(b) $y - 3 = 1\,(x - 1)$

$y = x + 2$

(c) $m = \dfrac{4 - 3}{2 - 1} = 1$

$y - 3 = 1(x - 1)$
$y = x + 2$

(d) $y = 3$

31. $\left(\dfrac{x + 2}{2}, \dfrac{y + 3}{2}\right) = (-1, 4)$

$x = -4, \qquad y = 5$
Other endpoint: $(-4, 5)$

32. $\sqrt{(x-0)^2 + (y-0)^2} = \sqrt{(x-2)^2 + (y-3)^2} = \sqrt{(x-3)^2 + (y+2)^2}$

$x^2 + y^2 = x^2 + y^2 - 4x - 6y + 13 = x^2 + y^2 - 6x + 4y + 13$

$\quad\quad\quad 4x + 6y = 13 \quad\quad\quad\quad\quad\quad 2x - 10y = 0$

$\quad\quad 4x + 6y = 13 \quad\quad\quad\quad\quad\quad 2x - 10(1/2) = 0$
$\quad\quad \underline{-4x + 20y = 0} \quad\quad\quad\quad\quad\quad\quad\quad x = 5/2$
$\quad\quad\quad\quad\quad 26y = 13$
$\quad\quad\quad\quad\quad\quad y = 1/2 \quad\quad\quad$ Point: (5/2, 1/2)

33. $3x - 4y = 8$
$\quad \underline{4x + 4y = 20}$
$\quad 7x \quad\quad\; = 28$
$\quad\quad\quad x = 4$
$\quad\quad\quad y = 1$
Point: (4, 1)

34. $y = x + 1$
$\quad (x + 1) - x^2 = 7$
$\quad 0 = x^2 - x + 6$
No real solution
No points of intersection

35. $V = 850a + 300,000$
Domain: $\{a \mid a \geq 0\}$

36. $V = 3.25b$
Domain: $\{b \mid b \geq 0\}$

37. $S = 6x^2$
Domain: $\{x \mid x \geq 0\}$

38. $S = 4\pi r^2$
Domain: $\{r \mid r \geq 0\}$

39. $d = 45t$
Domain: $\{t \mid t \geq 0\}$

40. $A = \dfrac{\sqrt{3}x^2}{4}$
Domain: $\{x \mid x \geq 0\}$

41. $R = 4 - \dfrac{x^2}{2}$
$r = 2$

42. $R = x^2$
$r = x^3$

43. $h = x^2$
$p = x$

44. $h = \sqrt{4 - x^2}$
$p = x$

45. $x + y = 500$
$p = xy$
$\quad = x(500 - x)$
$\quad = 500x - x^2$

46. $xy = 120$
$S = x + y$
$\quad = x + \dfrac{120}{x}$
$\quad = \dfrac{x^2 + 120}{x}$

47. $x^2 - y = 0$

    Function since there is one
    value of y for each x.

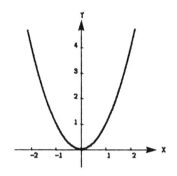

48. $x^2 + 4y^2 = 16$

    $y = \pm \dfrac{\sqrt{16 - x^2}}{2}$

    Not a function since there are
    two values of y for some x.

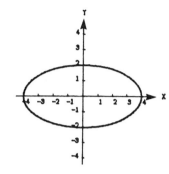

49. $x - y^2 = 0$
    $y = \pm\sqrt{x}$

    Not a function since there are
    two values of y for some x.

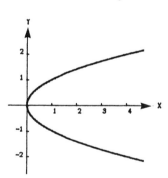

50. $x^3 - y^2 + 1 = 0$
    $y = \pm\sqrt{x^3 + 1}$

    Not a function since there are
    two values of y for some x.

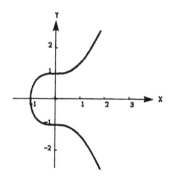

51. $y = x^2 - 2x$

    Function since there is one
    value of y for each x.

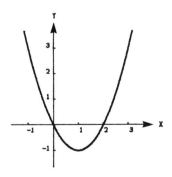

52. $y = 36 - x^2$

    Function since there is one
    value of y for each x.

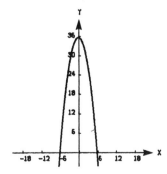

53. $f(x) = 1 - x^2$ $g(x) = 2x + 1$

(a) $f(x) + g(x) = -x^2 + 2x + 2$
(b) $f(x) - g(x) = -x^2 - 2x$
(c) $f(x)g(x) = (1 - x^2)(2x + 1) = -2x^3 - x^2 + 2x + 1$

(d) $\dfrac{f(x)}{g(x)} = \dfrac{1 - x^2}{2x + 1}$

(e) $f[g(x)] = 1 - (2x + 1)^2 = 1 - (4x^2 + 4x + 1) = -4x^2 - 4x$
(f) $g[f(x)] = 2(1 - x^2) + 1 = 3 - 2x^2$

54. $f(x) = 2x - 3$ $g(x) = \sqrt{x + 1}$

(a) $f(x) + g(x) = 2x - 3 + \sqrt{x + 1}$
(b) $f(x) - g(x) = 2x - 3 - \sqrt{x + 1}$
(c) $f(x)g(x) = (2x - 3)\sqrt{x + 1}$

(d) $\dfrac{f(x)}{g(x)} = \dfrac{2x - 3}{\sqrt{x + 1}}$

(e) $f(g(x)) = 2\sqrt{x + 1} - 3$
(f) $g(f(x)) = \sqrt{(2x - 3) + 1} = \sqrt{2x - 2} = \sqrt{2(x - 1)}$

55. $C = 0.25x + 100$

56.   (a) $C = (5.25 + 9.50)t + 26{,}500 = 14.75t + 26{,}500$
(b) $R = 25t$
(c) $25t = 14.75t + 26{,}500$
$10.25t = 26{,}500$
$t \approx 2585.4$ hours

57.   (a) $\tan 240^\circ \approx 1.7321$
(b) $\cot 210^\circ \approx 1.7321$
(c) $\sin(-0.65) \approx -0.6052$

58.   (a) $\sin(5.63) \approx -0.6077$
(b) $\csc(2.62) \approx 2.0070$
(c) $\csc 150^\circ = 2$

59. $\sin\theta = -1/2$

$\theta = 210^\circ = \dfrac{7\pi}{6}$ or $330^\circ = \dfrac{11\pi}{6}$

60. $\csc\theta = \sqrt{2}$

$\theta = 45^\circ = \dfrac{\pi}{4}$ or $135^\circ = \dfrac{3\pi}{4}$

61. $\cos\theta = -\dfrac{\sqrt{3}}{2}$

$\theta = 150^\circ = \dfrac{5\pi}{6}$ or $\theta = 210^\circ = \dfrac{7\pi}{6}$

62. $\tan\theta = -\dfrac{1}{\sqrt{3}}$

$\theta = 150^\circ = \dfrac{5\pi}{6}$ or $\theta = 330^\circ = \dfrac{11\pi}{6}$

63. $\theta = 60^{\circ}$

$\cos 30^{\circ} = \dfrac{5\sqrt{3}}{C}$ ➡ $C = 10$

64. $\theta = 45^{\circ}$

$\sin 45^{\circ} = \dfrac{a}{288}$ ➡ $a = \dfrac{288}{\sqrt{2}} = 144\sqrt{2}$

65. $\theta = 30^{\circ}$

$\sin 60^{\circ} = \dfrac{a}{8}$ ➡ $a = 4\sqrt{3}$

66. $\theta = 60^{\circ}$

$\sin 60^{\circ} = \dfrac{4}{S}$ ➡ $S = \dfrac{8}{\sqrt{3}} = \dfrac{8\sqrt{3}}{3}$

67. $\theta = 40^{\circ}$

68. $\dfrac{h}{1} = \dfrac{2}{3}$ ➡ $h = \dfrac{2}{3}$

69. $\dfrac{h}{20} = \dfrac{6}{8}$

$h = 15$ feet

70. $c = \sqrt{125^2 + 200^2}$

≈ 235.85 feet

71. $f(x) = 2 \sin \left(\dfrac{2x}{3}\right)$

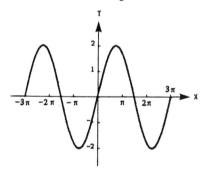

72. $f(x) = \dfrac{1}{2} \cos \left(\dfrac{x}{3}\right)$

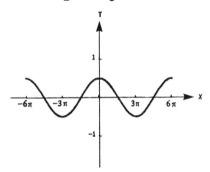

73. $f(x) = \cos \left(2x - \dfrac{\pi}{3}\right)$

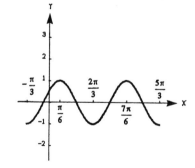

74. $f(x) = -\sin \left(2x + \dfrac{\pi}{2}\right)$

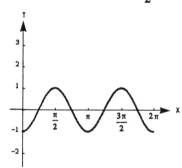

75. $f(x) = \tan\left(\dfrac{x}{2}\right)$

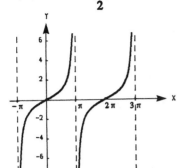

76. $f(x) = \csc 2x$

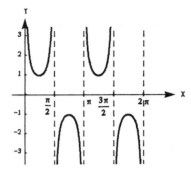

77. $f(x) = \sec\left(x - \dfrac{\pi}{4}\right)$

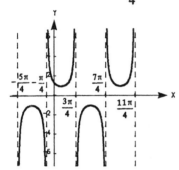

78. $f(x) = \cot 3x$

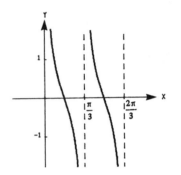

79. $S = 74.50 + 43.75 \sin\dfrac{\pi t}{6}$

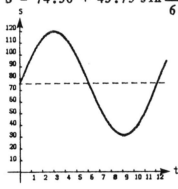

80. $p = 100 - 20 \cos \dfrac{5\pi t}{3}$

   (a) $p = \dfrac{2\pi}{(5\pi/3)} = \dfrac{6}{5}$

   (b) $\dfrac{60}{(6/5)} = 50$ heartbeats per minute

   (c)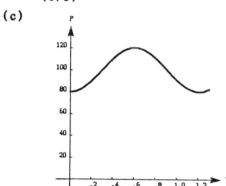

81. (a) $T = 45 - 23 \cos \left[\dfrac{2\pi}{365} (1 - 32) \right] \approx 25.2°F$

   (b) $T = 45 - 23 \cos \left[\dfrac{2\pi}{365} (185 - 32) \right] \approx 65.1°F$

   (c) $T = 45 - 23 \cos \left[\dfrac{2\pi}{365} (291 - 32) \right] \approx 50.8° F$

82. (a) $S = 23.1 + 0.442(2) + 4.3 \sin (\dfrac{2\pi}{6}) \approx 27.7$

   (b) $S = 23.1 + 0.442(14) + 4.3 \sin (\dfrac{14\pi}{6}) \approx 33$

   (c) $S = 23.1 + 0.442(9) + 4.3 \sin (\dfrac{9\pi}{6}) \approx 22.8$

   (d) $S = 23.1 + 0.442(21) + 4.3 \sin (\dfrac{21\pi}{6}) \approx 28.1$

2 Limits and their properties

2.1
An introduction to limits

1.

x	1.9	1.99	1.999	2.001	2.01	2.1
f(x)	0.3448	0.3344	0.3334	0.3332	0.3322	0.3226

$$\lim_{x \to 2} \frac{x - 2}{x^2 - x - 2} \approx 0.3333 \quad \text{(Actual limit is 1/3.)}$$

2.

x	1.9	1.99	1.999	2.001	2.01	2.1
f(x)	0.2564	0.2506	0.2501	0.2499	0.2494	0.2439

$$\lim_{x \to 2} \frac{x - 2}{x^2 - 4} \approx 0.25 \quad \text{(Actual limit is 1/4.)}$$

3.

x	−0.1	−0.01	−0.001	0.001	0.01	0.1
f(x)	0.2911	0.2889	0.2887	0.2887	0.2884	0.2863

$$\lim_{x \to 0} \frac{\sqrt{x + 3} - \sqrt{3}}{x} \approx 0.2887 \quad \text{(Actual limit is } 1/2\sqrt{3}.\text{)}$$

4.

x	−3.1	−3.01	−3.001	−2.999	−2.99	−2.9
f(x)	−0.2485	−0.2498	−0.2500	−0.2500	−0.2502	−0.2516

$$\lim_{x \to -3} \frac{\sqrt{1 - x} - 2}{x + 3} \approx -0.25 \quad \text{(Actual limit is } -1/4.\text{)}$$

5.

x	2.9	2.99	2.999	3.001	3.01	3.1
f(x)	−0.0641	−0.0627	−0.0625	−0.0625	−0.0623	−0.0610

$$\lim_{x \to 3} \frac{[1/(x + 1)] - (1/4)}{x - 3} \approx -0.0625 \quad \text{(Actual limit is } -1/16.\text{)}$$

6.

x	3.9	3.99	3.999	4.001	4.01	4.1
f(x)	0.0408	0.0401	0.0400	0.0400	0.0399	0.0392

$$\lim_{x \to 4} \frac{[x/(x + 1)] - (4/5)}{x - 4} \approx 0.04 \quad \text{(Actual limit is } 1/25.\text{)}$$

7.

x	0.70	0.75	0.784	0.786	0.80	0.85
f(x)	1.8468	1.9324	1.9972	2.0012	2.0298	2.1413

$$\lim_{x \to \pi/4} \frac{\tan x - 1}{x - (\pi/4)} \approx 2 \quad \text{(Actual limit is } 2.\text{)}$$

8.

x	−0.1	−0.01	−0.001	0.001	0.01	0.1
f(x)	−0.0502	−0.0050	−0.0005	0.0005	0.0050	0.0502

$$\lim_{x \to 0} \frac{\sec x - 1}{x} \approx 0.0000 \quad \text{(Actual limit is } 0.\text{)}$$

9.

x	−0.1	−0.01	−0.001	0.001	0.01	0.1
f(x)	0.9983	0.99998	1.0000	1.0000	0.99998	0.9983

$$\lim_{x \to 0} \frac{\sin x}{x} \approx 1.0000 \quad \text{(Actual limit is } 1.\text{)}$$

10.

x	−0.1	−0.01	−0.001	0.001	0.01	0.1
f(x)	0.0500	0.0050	0.0005	−0.0005	−0.0050	−0.0500

$$\lim_{x \to 0} \frac{\cos x - 1}{x} \approx 0.0000 \quad \text{(Actual limit is 0.)}$$

11. 1   12. 3   13. 2   14. 3

15. Limit does not exist   16. Limit does not exist

17. Limit does not exist   18. 1

19. Limit does not exist   20. 0

21. $\lim_{x \to 2} (3x + 2) = 8$

   $|(3x + 2) - 8| < 0.01$
   $|3x - 6| < 0.01$
   $3|x - 2| < 0.01$
   $0 < |x - 2| < (0.01)/3 = 0.0033$
   $\qquad\qquad\qquad\qquad = \delta$

22. $\lim_{x \to 4} [4 - (x/2)] = 2$

   $|[4 - (x/2)] - 2| < 0.01$
   $|2 - (x/2)| < 0.01$
   $\left|-\frac{1}{2}(x - 4)\right| < 0.01$
   $0 < |x - 4| < 0.02 = \delta$

23. $\lim_{x \to 2} (x^2 - 3) = 1$

   $|(x^2 - 3) - 1| < 0.01$
   $|x^2 - 4| < 0.01$
   $|(x + 2)(x - 2)| < 0.01$
   $|x + 2|\,|x - 2| < 0.01$

   $|x - 2| < \dfrac{0.01}{|x + 2|}$

   If we assume $1 < x < 3$, then

   $\delta = \dfrac{0.01}{5} = 0.002$

24. $\lim_{x \to 5} \sqrt{x - 4} = 1$

   $|\sqrt{x - 4} - 1| < 0.01$
   $1 < \sqrt{x - 4} < 1.01$
   $1 < x - 4 < 1.0201$
   $0 < x - 5 < 0.0201 = \delta$

25. $\lim_{x \to 2} (x + 3) = 5$

   Given $\varepsilon > 0$:
   $|(x + 3) - 5| < \varepsilon$
   $|x - 2| < \varepsilon = \delta$

26. $\lim_{x \to -3} (2x + 5) = -1$

   Given $\varepsilon > 0$:
   $|(2x + 5) - (-1)| < \varepsilon$
   $|2x + 6| < \varepsilon$
   $2|x + 3| < \varepsilon$
   $|x + 3| < \varepsilon/2 = \delta$

27. $\lim_{x \to 6} 3 = 3$

Given $\varepsilon > 0$:
$|3 - 3| < \varepsilon$
$0 < \varepsilon$
any δ will work

28. $\lim_{x \to 2} -1 = -1$

Given $\varepsilon > 0$:
$|-1 - (-1)| < \varepsilon$
$0 < \varepsilon$
any δ will work

29. $\lim_{x \to 0} \sqrt[3]{x} = 0$

Given $\varepsilon > 0$:
$|\sqrt[3]{x} - 0| < \varepsilon$
$|\sqrt[3]{x}| < \varepsilon$
$|x| < \varepsilon^3 = \delta$

30. $\lim_{x \to 3} |x - 3| = 0$

Given $\varepsilon > 0$:
$|(x - 3) - 0| < \varepsilon$
$|x - 3| < \varepsilon = \delta$

2.2
Properties of limits

1. $\lim_{x \to 2} x^2 = 2^2 = 4$

2. $\lim_{x \to -3} (3x + 2) = 3(-3) + 2 = -7$

3. $\lim_{x \to 0} (2x - 1) = 2(0) - 1 = -1$

4. $\lim_{x \to 1} (-x^2 + 1) = -(1)^2 + 1 = 0$

5. $\lim_{x \to 2} (-x^2 + x - 2) = -(2)^2 + (2) - 2 = -4$

6. $\lim_{x \to 1} (3x^3 - 2x^2 + 4) = 3(1)^3 - 2(1)^2 + 4 = 5$

7. $\lim_{x \to 3} \sqrt{x + 1} = \sqrt{3 + 1} = 2$

8. $\lim_{x \to 4} \sqrt[3]{x + 4} = \sqrt[3]{4 + 4} = 2$

9. $\lim_{x \to -4} (x + 3)^2 = (-4 + 3)^2 = 1$

10. $\lim_{x \to 0} (2x - 1)^3 = [2(0) - 1]^3 = -1$

11. $\lim_{x \to 2} \frac{1}{x} = \frac{1}{2}$

12. $\lim_{x \to -3} \frac{2}{x + 2} = \frac{2}{-3 + 2} = -2$

13. $\lim\limits_{x \to -1} \dfrac{x^2 + 1}{x} = \dfrac{(-1)^2 + 1}{-1} = -2$

14. $\lim\limits_{x \to 3} \dfrac{\sqrt{x + 1}}{x - 4} = \dfrac{\sqrt{3 + 1}}{3 - 4} = -2$

15. $\lim\limits_{x \to \pi/2} \sin x = \sin\dfrac{\pi}{2} = 1$

16. $\lim\limits_{x \to \pi} \tan x = \tan \pi = 0$

17. $\lim\limits_{x \to 1} \cos \pi x = \cos \pi = -1$

18. $\lim\limits_{x \to 1} \sin\dfrac{\pi x}{2} = \sin\dfrac{\pi}{2} = 1$

19. $\lim\limits_{x \to 0} \sec 2x = \sec 0 = 1$

20. $\lim\limits_{x \to \pi} \cos 3x = \cos 3\pi = -1$

21. $\lim\limits_{x \to 5\pi/6} \sin x = \sin\dfrac{5\pi}{6} = \dfrac{1}{2}$

22. $\lim\limits_{x \to 5\pi/3} \cos x = \cos\dfrac{5\pi}{3} = \dfrac{1}{2}$

23. $\lim\limits_{x \to 3} \tan\dfrac{\pi x}{4} = \tan\dfrac{3\pi}{4} = -1$

24. $\lim\limits_{x \to 7} \sec\dfrac{\pi x}{6} = \sec\dfrac{7\pi}{6} = \dfrac{-2\sqrt{3}}{3}$

25. (a) $\lim\limits_{x \to c} [f(x) + g(x)] = \lim\limits_{x \to c} f(x) + \lim\limits_{x \to c} g(x) = 2 + 3 = 5$

(b) $\lim\limits_{x \to c} [f(x)g(x)] = \left[\lim\limits_{x \to c} f(x)\right]\left[\lim\limits_{x \to c} g(x)\right] = (2)(3) = 6$

(c) $\lim\limits_{x \to c} \dfrac{f(x)}{g(x)} = \dfrac{\lim\limits_{x \to c} f(x)}{\lim\limits_{x \to c} g(x)} = \dfrac{2}{3}$

26. (a) $\lim\limits_{x \to c} [f(x) + g(x)] = \lim\limits_{x \to c} f(x) + \lim\limits_{x \to c} g(x) = \dfrac{3}{2} + \dfrac{1}{2} = 2$

(b) $\lim\limits_{x \to c} [f(x)g(x)] = \left[\lim\limits_{x \to c} f(x)\right]\left[\lim\limits_{x \to c} g(x)\right] = \left(\dfrac{3}{2}\right)\left(\dfrac{1}{2}\right) = \dfrac{3}{4}$

(c) $\lim\limits_{x \to c} \dfrac{f(x)}{g(x)} = \dfrac{\lim\limits_{x \to c} f(x)}{\lim\limits_{x \to c} g(x)} = \dfrac{3/2}{1/2} = 3$

2.3
Techniques for evaluating limits

1. (a) $\lim\limits_{x \to 0} f(x) = 1$ (b) $\lim\limits_{x \to -1} f(x) = 3$

2. (a) $\lim\limits_{x \to 1} f(x) = -2$ (b) $\lim\limits_{x \to 3} f(x) = 0$

3. (a) $\lim\limits_{x \to 0} g(x) = 1$ (b) $\lim\limits_{x \to -1} g(x) = 3$

4. (a) $\lim\limits_{x \to -2} h(x) = -5$ (b) $\lim\limits_{x \to 0} h(x) = -3$

5. (a) $\lim\limits_{x \to 1} g(x) = 2$ (b) $\lim\limits_{x \to -1} g(x) = 0$

6. (a) $\lim\limits_{x \to \pi/3} F(x) = \dfrac{\sqrt{3}}{2}$ (b) $\lim\limits_{x \to \pi/2} F(x) = 1$

7. (a) $\lim\limits_{x \to 1} f(x)$ does not exist.   (b) $\lim\limits_{x \to 2} f(x) = 1$

8. (a) $\lim\limits_{x \to 0} f(x) = 0$ (b) $\lim\limits_{x \to \pi/2} f(x)$ does not exist.

9. $\lim\limits_{x \to -1} \dfrac{x^2 - 1}{x + 1} = \lim\limits_{x \to -1} \dfrac{(x + 1)(x - 1)}{(x + 1)} = \lim\limits_{x \to -1} (x - 1) = -2$

10. $\lim\limits_{x \to -1} \dfrac{2x^2 - x - 3}{x + 1} = \lim\limits_{x \to -1} \dfrac{(2x - 3)(x + 1)}{x + 1} = \lim\limits_{x \to -1} (2x - 3) = -5$

11. $\lim\limits_{x \to 3} \dfrac{x - 3}{x^2 - 9} = \lim\limits_{x \to 3} \dfrac{x - 3}{(x + 3)(x - 3)} = \lim\limits_{x \to 3} \dfrac{1}{x + 3} = \dfrac{1}{6}$

12. $\lim\limits_{x \to -1} \dfrac{x^3 + 1}{x + 1} = \lim\limits_{x \to -1} \dfrac{(x + 1)(x^2 - x + 1)}{x + 1} = \lim\limits_{x \to -1} (x^2 - x + 1) = 3$

13. $\lim\limits_{x \to -2} \dfrac{x^3 + 8}{x + 2} = \lim\limits_{x \to -2} \dfrac{(x + 2)(x^2 - 2x + 4)}{x + 2} = \lim\limits_{x \to -2} (x^2 - 2x + 4) = 12$

Complete Solutions Guide to Accompany Calculus

14. $\lim\limits_{\Delta x \to 0} \dfrac{(x + \Delta x)^2 - x^2}{\Delta x} = \lim\limits_{\Delta x \to 0} \dfrac{x^2 + 2x\Delta x + (\Delta x)^2 - x^2}{\Delta x}$

$= \lim\limits_{\Delta x \to 0} \dfrac{\Delta x(2x + \Delta x)}{\Delta x} = 2x$

15. $\lim\limits_{\Delta x \to 0} \dfrac{2(x + \Delta x) - 2x}{\Delta x} = \lim\limits_{\Delta x \to 0} \dfrac{2x + 2\Delta x - 2x}{\Delta x} = \lim\limits_{\Delta x \to 0} 2 = 2$

16. $\lim\limits_{\Delta x \to 0} \dfrac{(x + \Delta x)^3 - x^3}{\Delta x} = \lim\limits_{\Delta x \to 0} \dfrac{x^3 + 3x^2\Delta x + 3x(\Delta x)^2 + (\Delta x)^3 - x^3}{\Delta x}$

$= \lim\limits_{\Delta x \to 0} [3x^2 + 3x\Delta x + (\Delta x)^2] = 3x^2$

17. $\lim\limits_{\Delta x \to 0} \dfrac{(x + \Delta x)^2 - 2(x + \Delta x) + 1 - (x^2 - 2x + 1)}{\Delta x}$

$= \lim\limits_{\Delta x \to 0} \dfrac{x^2 + 2x\Delta x + (\Delta x)^2 - 2x - 2\Delta x + 1 - x^2 + 2x - 1}{\Delta x}$

$= \lim\limits_{\Delta x \to 0} (2x + \Delta x - 2) = 2x - 2$

18. $\lim\limits_{\Delta x \to 0} \dfrac{(1 + \Delta x)^3 - 1}{\Delta x} = \lim\limits_{\Delta x \to 0} \dfrac{1 + 3\Delta x + 3(\Delta x)^2 + (\Delta x)^3 - 1}{\Delta x}$

$= \lim\limits_{\Delta x \to 0} [3 + 3\Delta x + (\Delta x)^2] = 3$

19. $\lim\limits_{x \to 5} \dfrac{x - 5}{x^2 - 25} = \lim\limits_{x \to 5} \dfrac{x - 5}{(x + 5)(x - 5)} = \lim\limits_{x \to 5} \dfrac{1}{x + 5} = \dfrac{1}{10}$

20. $\lim\limits_{x \to 2} \dfrac{-(x - 2)}{x^2 - 4} = \lim\limits_{x \to 2} \dfrac{-1}{x + 2} = -\dfrac{1}{4}$

21. $\lim\limits_{x \to 1} \dfrac{x^2 + x - 2}{x^2 - 1} = \lim\limits_{x \to 1} \dfrac{(x + 2)(x - 1)}{(x + 1)(x - 1)} = \lim\limits_{x \to 1} \dfrac{x + 2}{x + 1} = \dfrac{3}{2}$

22. $\lim\limits_{x \to 0} \dfrac{\sqrt{2 + x} - \sqrt{2}}{x} = \lim\limits_{x \to 0} \dfrac{\sqrt{2 + x} - \sqrt{2}}{x} \cdot \dfrac{\sqrt{2 + x} + \sqrt{2}}{\sqrt{2 + x} + \sqrt{2}}$

$= \lim\limits_{x \to 0} \dfrac{2 + x - 2}{(\sqrt{2 + x} + \sqrt{2})x} = \lim\limits_{x \to 0} \dfrac{1}{\sqrt{2 + x} + \sqrt{2}} = \dfrac{1}{2\sqrt{2}} = \dfrac{\sqrt{2}}{4}$

23. $\lim\limits_{x \to 0} \dfrac{\sqrt{3+x} - \sqrt{3}}{x} = \lim\limits_{x \to 0} \dfrac{\sqrt{3+x} - \sqrt{3}}{x} \cdot \dfrac{\sqrt{3+x} + \sqrt{3}}{\sqrt{3+x} + \sqrt{3}}$

$= \lim\limits_{x \to 0} \dfrac{3 + x - 3}{x(\sqrt{3+x} + \sqrt{3})} = \lim\limits_{x \to 0} \dfrac{1}{\sqrt{3+x} + \sqrt{3}} = \dfrac{1}{2\sqrt{3}} = \dfrac{\sqrt{3}}{6}$

24. $\lim\limits_{x \to 0} \dfrac{[1/(x+4)] - (1/4)}{x} = \lim\limits_{x \to 0} \dfrac{\frac{4-(x+4)}{4(x+4)}}{x} = \lim\limits_{x \to 0} \dfrac{-1}{4(x+4)} = -\dfrac{1}{16}$

25. $\lim\limits_{x \to 0} \dfrac{\frac{1}{2+x} - \frac{1}{2}}{x} = \lim\limits_{x \to 0} \dfrac{\frac{2-(2+x)}{2(2+x)}}{x} = \lim\limits_{x \to 0} \dfrac{-1}{2(2+x)} = -\dfrac{1}{4}$

26. $\lim\limits_{x \to 3} \dfrac{\sqrt{x+1} - 2}{x-3} = \lim\limits_{x \to 3} \dfrac{\sqrt{x+1} - 2}{x-3} \cdot \dfrac{\sqrt{x+1} + 2}{\sqrt{x+1} + 2}$

$= \lim\limits_{x \to 3} \dfrac{x-3}{(x-3)[\sqrt{x+1} + 2]}$

$= \lim\limits_{x \to 3} \dfrac{1}{\sqrt{x+1} + 2} = \dfrac{1}{4}$

27.

x	−0.1	−0.01	−0.001	0	0.001	0.01	0.1
f(x)	0.358	0.354	0.354	?	0.354	0.353	0.349

$\lim\limits_{x \to 0} \dfrac{\sqrt{x+2} - \sqrt{2}}{x} \approx 0.354$ (Actual limit is $\dfrac{1}{2\sqrt{2}}$)

28.

x	0.9	0.99	0.999	1	1.001	1.01	1.1
f(x)	2.130	2.012	2.001	?	1.999	1.988	1.879

$\lim\limits_{x \to 1} \dfrac{1-x}{\sqrt{5 - x^2} - 2} = 2$

29.

x	−0.1	−0.01	−0.001	0	0.001	0.01	0.1
f(x)	−0.263	−0.251	−0.250	?	−0.250	−0.249	−0.238

$\lim\limits_{x \to 0} \dfrac{\frac{1}{2+x} - \frac{1}{2}}{x} = -\dfrac{1}{4}$

30.

x	1.9	1.99	1.999	1.9999	2.0	2.0001	2.001	2.01	2.1
f(x)	72.39	79.20	79.92	79.99	?	80.01	80.08	80.80	88.41

$$\lim_{x \to 2} \frac{x^5 - 32}{x - 2} = 80$$

31. $\lim\limits_{x \to 0} \dfrac{\sin x}{5x} = \lim\limits_{x \to 0} \left[\left(\dfrac{\sin x}{x}\right)\left(\dfrac{1}{5}\right) \right] = (1)\left(\dfrac{1}{5}\right) = \dfrac{1}{5}$

32. $\lim\limits_{x \to 0} \dfrac{3(1 - \cos x)}{x} = \lim\limits_{x \to 0} \left[3\left(\dfrac{1 - \cos x}{x}\right) \right] = (3)(0) = 0$

33. $\lim\limits_{\theta \to 0} \dfrac{\sec \theta - 1}{\theta \sec \theta} = \lim\limits_{\theta \to 0} \dfrac{1 - \cos \theta}{\theta} = 0$

34. $\lim\limits_{\theta \to 0} \dfrac{\cos \theta \tan \theta}{\theta} = \lim\limits_{\theta \to 0} \dfrac{\sin \theta}{\theta} = 1$

35. $\lim\limits_{\theta \to 0} \dfrac{\sin^2 x}{x} = \lim\limits_{\theta \to 0} \left[\dfrac{\sin x}{x} \sin x \right] = (1)\sin 0 = 0$

36. $\lim\limits_{\phi \to \pi} \phi \sec \phi = -\pi$

37. $\lim\limits_{x \to \pi/2} \dfrac{\cos x}{\cot x} = \lim\limits_{x \to \pi/2} \sin x = 1$

38. $\lim\limits_{x \to \pi/4} \dfrac{1 - \tan x}{\sin x - \cos x} = \lim\limits_{x \to \pi/4} \dfrac{\cos x - \sin x}{\sin x \cos x - \cos^2 x}$

$$= \lim\limits_{x \to \pi/4} \dfrac{-(\sin x - \cos x)}{\cos x(\sin x - \cos x)}$$

$$= \lim\limits_{x \to \pi/4} \dfrac{-1}{\cos x} = \lim\limits_{x \to \pi/4} (-\sec x) = -\sqrt{2}$$

39. $\lim\limits_{t \to 0} \dfrac{\sin^2 t}{t^2} = \lim\limits_{t \to 0} \left(\dfrac{\sin t}{t}\right)^2 = (1)^2 = 1$

40. $\lim\limits_{t \to 0} \dfrac{\sin 3t}{t} = \lim\limits_{t \to 0} \left[3\left(\dfrac{\sin 3t}{3t}\right)\right] = 3(1) = 3$

41. $\lim\limits_{x \to 0} \dfrac{\sin 2x}{\sin 3x} = \lim\limits_{x \to 0} \left[2\left(\dfrac{\sin 2x}{2x}\right)\left(\dfrac{1}{3}\right)\left(\dfrac{3x}{\sin 3x}\right)\right] = 2(1)\left(\dfrac{1}{3}\right)(1) = \dfrac{2}{3}$

42. $\lim\limits_{x \to 0} \dfrac{\tan^2 x}{x} = \lim\limits_{x \to 0} \dfrac{\sin^2 x}{x \cos^2 x} = \lim\limits_{x \to 0} \left[\dfrac{\sin x}{x} \cdot \dfrac{\sin x}{\cos^2 x}\right] = (1)(0) = 0$

43. $\lim\limits_{h \to 0} \dfrac{(1 - \cos h)^2}{h} = \lim\limits_{h \to 0} \left[\dfrac{1 - \cos h}{h}(1 - \cos h)\right] = (0)(0) = 0$

44. $\lim\limits_{h \to 0} (1 + \cos 2h) = 1 + 1 = 2$

45. $\lim\limits_{x \to 0} f(x) = 4$
by the Squeeze Theorem.

46. $\lim\limits_{x \to a} f(x) = b$
by the Squeeze Theorem.

2.4 Continuity and one-sided limits

1. (a) $\lim\limits_{x \to 3^+} f(x) = 1$

   (b) $\lim\limits_{x \to 3^-} f(x) = 1$

   (c) $\lim\limits_{x \to 3} f(x) = 1$

2. (a) $\lim\limits_{x \to -2^+} f(x) = -2$

   (b) $\lim\limits_{x \to -2^-} f(x) = -2$

   (c) $\lim\limits_{x \to -2} f(x) = -2$

3. (a) $\lim\limits_{x \to 3^+} f(x) = 0$

   (b) $\lim\limits_{x \to 3^-} f(x) = 0$

   (c) $\lim\limits_{x \to 3} f(x) = 0$

4. (a) $\lim\limits_{x \to -2^+} f(x) = 2$

   (b) $\lim\limits_{x \to -2^-} f(x) = 2$

   (c) $\lim\limits_{x \to -2} f(x) = 2$

5. (a) $\lim\limits_{x \to 3^+} f(x) = 3$

    (b) $\lim\limits_{x \to 3^-} f(x) = -3$

    (c) $\lim\limits_{x \to 3} f(x)$ does not exist

6. (a) $\lim\limits_{x \to -1^+} f(x) = 0$

    (b) $\lim\limits_{x \to -1^-} f(x) = 2$

    (c) $\lim\limits_{x \to -1} f(x)$ does not exist

7. $\lim\limits_{x \to 5^+} \dfrac{x - 5}{x^2 - 25} = \lim\limits_{x \to 5^+} \dfrac{1}{x + 5} = \dfrac{1}{10}$

8. $\lim\limits_{x \to 2^+} \dfrac{2 - x}{x^2 - 4} = \lim\limits_{x \to 2^+} -\dfrac{1}{x + 2} = -\dfrac{1}{4}$

9. $\lim\limits_{x \to 2^+} \dfrac{x}{\sqrt{x^2 - 4}}$ does not exist

10. $\lim\limits_{x \to 4^-} \dfrac{\sqrt{x} - 2}{x - 4} = \lim\limits_{x \to 4^-} \dfrac{\sqrt{x} - 2}{x - 4} \cdot \dfrac{\sqrt{x} + 2}{\sqrt{x} + 2}$

$$= \lim\limits_{x \to 4^-} \dfrac{x - 4}{(x - 4)(\sqrt{x} + 2)} = \lim\limits_{x \to 4^-} \dfrac{1}{\sqrt{x} + 2} = \dfrac{1}{4}$$

11. $\lim\limits_{\Delta x \to 0^+} \dfrac{2(x + \Delta x) - 2x}{\Delta x} = \lim\limits_{\Delta x \to 0^+} \dfrac{2\Delta x}{\Delta x} = \lim\limits_{\Delta x \to 0} 2 = 2$

12. $\lim\limits_{x \to 1^-} \dfrac{x^2 - 2x + 1}{x - 1} = \lim\limits_{x \to 1^-} (x - 1) = 0$

13. $\lim\limits_{x \to 0} \dfrac{|x|}{x}$ does not exist since $\lim\limits_{x \to 0^+} \dfrac{|x|}{x} = 1$ and $\lim\limits_{x \to 0^-} \dfrac{|x|}{x} = -1$

14. $\lim\limits_{x \to 2^+} \dfrac{|x - 2|}{x - 2} = \lim\limits_{x \to 2^+} \dfrac{x - 2}{x - 2} = 1$

    $\lim\limits_{x \to 2^-} \dfrac{|x - 2|}{x - 2} = \lim\limits_{x \to 2^-} \dfrac{-(x - 2)}{x - 2} = -1$

    $\lim\limits_{x \to 2^+} \dfrac{|x - 2|}{x - 2} \neq \lim\limits_{x \to 2^-} \dfrac{|x - 2|}{x - 2}$
Thus the limit does not exist.

15. $\lim\limits_{x\rightarrow 3^+} f(x) = \lim\limits_{x\rightarrow 3^+} \dfrac{12 - 2x}{3} = 2$

$\lim\limits_{x\rightarrow 3^-} f(x) = \lim\limits_{x\rightarrow 3^-} \dfrac{x + 2}{2} = \dfrac{5}{2}$

Thus $\lim\limits_{x\rightarrow 3} f(x)$ does not exist.

16. $\lim\limits_{x\rightarrow 2^+} f(x) = \lim\limits_{x\rightarrow 2^+} (-x^2 + 4x - 2) = 2$

$\lim\limits_{x\rightarrow 2^-} f(x) = \lim\limits_{x\rightarrow 2^-} (x^2 - 4x + 6) = 2$

Thus $\lim\limits_{x\rightarrow 2} f(x) = 2$

17. $\lim\limits_{x\rightarrow 1^+} f(x) = \lim\limits_{x\rightarrow 1^+} (x + 1) = 2$

$\lim\limits_{x\rightarrow 1^-} f(x) = \lim\limits_{x\rightarrow 1^-} (x^3 + 1) = 2$

Thus $\lim\limits_{x\rightarrow 1} f(x) = 2$

18. $\lim\limits_{x\rightarrow 1^+} f(x) = \lim\limits_{x\rightarrow 1^+} (1 - x) = 0$

$\lim\limits_{x\rightarrow 1^-} f(x) = \lim\limits_{x\rightarrow 1^-} (x) = 1$

Thus $\lim\limits_{x\rightarrow 1} f(x)$ does not exist.

19. $\lim\limits_{x\rightarrow \pi} \cot x$ does not exist.

20. $\lim\limits_{x\rightarrow \pi/2} \sec x$ does not exist.

21.   Continuous for all real x

22.   Discontinuous at x = 0

23.   Discontinuous at x = -1

24.   Discontinuous at x = -2 and x = 2

25.   Discontinuous at x = 1

26.   Discontinuous at each integer k

27.   Continuous for all real x

28.   Continuous for all real x

29.   Continuous for all real x

30.   Continuous for all real x

31.   Nonremovable discontinuity at x = 1

32.   Nonremovable discontinuities at x = 1 and x = -1

33. Continuous for all real x

34. Removable discontinuity at x = 3
    Nonremovable discontinuity at
    x = −3

35. Removable discontinuity at
    x = −2
    Nonremovable discontinuity
    at x = 5

36. Removable discontinuity at x = 1
    Nonremovable discontinuity at
    x = −2

37. Continuous for all real x

38. Continuous for all real x

39. Nonremovable discontinuity
    at x = 2

40. Nonremovable discontinuity
    at x = 2

41. Nonremovable discontinuity
    at x = −2

42. Nonremovable discontinuity
    at x = 3

43. Continuous for all real x

44. Continuous for all real x

45. Continuous for all real x

46. Continuous for all real x

47. Nonremovable discontinuities at integer multiples of $\pi/2$

48. Nonremovable discontinuities at each integer k

49. Nonremovable discontinuities at each integer k

50. Nonremovable discontinuities at each integer k

51. $f(g(x)) = (x - 1)^2$ Continuous for all real x

52. $f(g(x)) = \dfrac{1}{\sqrt{x - 1}}$ Nonremovable discontinuity at x = 1

53. $f(g(x)) = \dfrac{1}{(x^2 + 5) - 1} = \dfrac{1}{x^2 + 4}$ Continuous for all real x

54. $f(g(x)) = \sin x^2$ Continuous for all real x

55. $y = \dfrac{x^2 - 16}{x - 4}$

Removable discontinuity at
$x = 4$

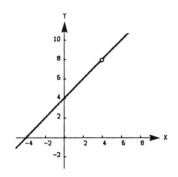

56. $y = \dfrac{x^3 - 8}{x - 2}$

Removable discontinuity at
$x = 2$

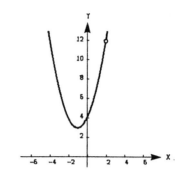

57. $y = [\![x]\!] - x$

A nonremovable discontinuity
at each integer.

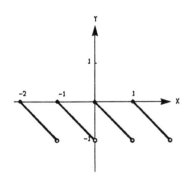

58. $f(x) = \dfrac{\sin x}{x}$

Removable discontinuity at
$x = 0$

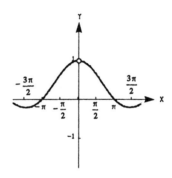

59. $f(x) = \dfrac{x^2}{x^2 - 36}$ Continuous on $(-\infty, -6)$, $(-6, 6)$, $(6, \infty)$

60. $f(x) = x\sqrt{x + 3}$ Continuous on $[-3, \infty)$

61. $f(x) = \dfrac{x}{x^2 + 1}$ Continuous on $(-\infty, \infty)$

62. $f(x) = \dfrac{x + 1}{\sqrt{x}}$ Continuous on $(0, \infty)$

63. $f(x) = \csc \dfrac{x}{2}$ Continuous on $\dots$, $(-2\pi, 0)$, $(0, 2\pi)$, $(2\pi, 4\pi)$, $\dots$

64. $f(x) = \cot x$ Continuous on $\dots$, $(-\pi, 0)$, $(0, \pi)$, $(\pi, 2\pi)$, $\dots$

65. f(x) is continuous on [2, 4]
     f(2) = −1 and f(4) = 3
     By the Intermediate Value
     Theorem f(x) = 0 for at least
     one value c between 2 and 4

66. f(x) is continuous on [0, 1]
     f(0) = −2 and f(1) = 2
     By the Intermediate Value
     Theorem f(x) = 0 for at least
     one value c between 0 and 1

67. $f(x) = x^3 + x - 1$

(a)

x	0	0.1	0.2	0.3	0.4	0.5	0.6	0.7	0.8	0.9	1
f(x)	−	−	−	−	−	−	−	+	+	+	+

Zero is in the subinterval (0.6, 0.7)

(b)

x	0.6	0.61	0.62	0.63	0.64	0.65	0.66	0.67	0.68	0.69	0.7
f(x)	−	−	−	−	−	−	−	−	−	+	+

Zero is in the subinterval (0.68, 0.69)

68. $f(x) = x^3 + 3x - 2$

(a)

x	0	0.1	0.2	0.3	0.4	0.5	0.6	0.7	0.8	0.9	1
f(x)	−	−	−	−	−	−	+	+	+	+	+

Zero is in the subinterval (0.5, 0.6)

(b)

x	0.5	0.51	0.52	0.53	0.54	0.55	0.56	0.57	0.58	0.59	0.6
f(x)	−	−	−	−	−	−	−	−	−	−	+

Zero is in the subinterval (0.59, 0.6)

69. f is a polynomial and therefore continuous; f(3) = 11

70. f is a polynomial and therefore continuous; f(2) = 0

71. f is a polynomial and therefore continuous; f(2) = 4

72. f has a non−removable discontinuity at x = 1, outside the interval;
     f(3) = 6

73. $f(2) = 8$; find a so that $\lim_{x \to 2^+} ax^2 = 8$ ⟹ $a = \dfrac{8}{2^2} = 2$

74. Find a and b such that $\lim\limits_{x \to -1^+} (ax + b) = -a + b = 2$ and

   $\lim\limits_{x \to 3^-} (ax + b) = 3a + b = -2.$

$$\begin{array}{r} a - b = -2 \\ (+)\ \underline{3a + b = -2} \\ 4a\qquad = -4 \\ a = -1 \end{array}$$

$$f(x) = \begin{cases} 2, & x \le -1 \\ -x + 1, & -1 < x < 3 \\ -2, & x \ge 3 \end{cases}$$

   $b = 2 + (-1) = 1$

75. Yes, $f(x) = \sqrt{1 - x^2}$ is continuous on $[-1, 1]$.
   Note that $\lim\limits_{x \to 1^-} f(x) = 0 = f(1)$

76. $S(t) = 28,500(1.09)^{[t]}$
   Discontinuous at every positive integer

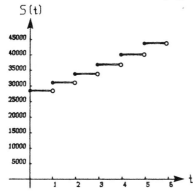

77. $C = \begin{cases} 1.04, & 0 < t \le 2 \\ 1.04 + 0.36[t - 1], & t > 2,\ t \text{ not an integer} \\ 1.04 + 0.36(t - 2), & t > 2,\ t \text{ is an integer} \end{cases}$

   Nonremovable discontinuity at each integer greater than 2.

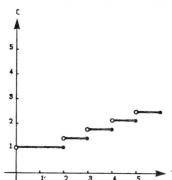

78. $N(t) = 25(2\left[\dfrac{t + 2}{2}\right] - t)$

t	0	1	1.8	2	3	3.8
N(t)	50	25	5	50	25	5

Discontinuous at every positive even integer. The company replenishes its inventory every two months.

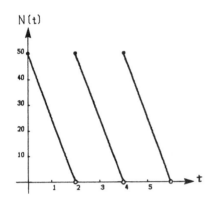

2.5
Infinite limits

1. $\lim\limits_{x \to -3^+} \dfrac{1}{x^2 - 9} = -\infty$ $\lim\limits_{x \to -3^-} \dfrac{1}{x^2 - 9} = \infty$

2. $\lim\limits_{x \to -3^+} \dfrac{x}{x^2 - 9} = \infty$ $\lim\limits_{x \to -3^-} \dfrac{x}{x^2 - 9} = -\infty$

3. $\lim\limits_{x \to -3^+} \dfrac{x^2}{x^2 - 9} = -\infty$ $\lim\limits_{x \to -3^-} \dfrac{x^2}{x^2 - 9} = \infty$

4. $\lim\limits_{x \to -3^+} \sec \dfrac{\pi x}{6} = \infty$ $\lim\limits_{x \to -3^-} \sec \dfrac{\pi x}{6} = -\infty$

5. $\lim\limits_{x \to -2^+} \dfrac{1}{(x + 2)^2} = \infty$ $\lim\limits_{x \to -2^-} \dfrac{1}{(x + 2)^2} = \infty$

6. $\lim\limits_{x \to -2^+} \dfrac{1}{x + 2} = \infty$ $\lim\limits_{x \to -2^-} \dfrac{1}{x + 2} = -\infty$

7. $\lim\limits_{x \to -2^+} \tan \dfrac{\pi x}{4} = -\infty$ $\lim\limits_{x \to -2^-} \tan \dfrac{\pi x}{4} = \infty$

8. $\lim\limits_{x \to -2^+} \sec \dfrac{\pi x}{4} = \infty$ $\lim\limits_{x \to -2^-} \sec \dfrac{\pi x}{4} = -\infty$

9. $\lim\limits_{x \to 0^+} \dfrac{1}{x^2} = \infty = \lim\limits_{x \to 0^-} \dfrac{1}{x^2}$

Therefore x = 0 is a vertical asymptote.

10. $\lim_{x \to 2^+} \frac{4}{(x - 2)^3} = \infty, \quad \lim_{x \to 2^-} \frac{4}{(x - 2)^3} = -\infty$
Therefore x = 2 is a vertical asymptote.

11. $\lim_{x \to 2^+} \frac{x^2 - 2}{(x - 2)(x + 1)} = \infty, \quad \lim_{x \to 2^-} \frac{x^2 - 2}{(x - 2)(x + 1)} = -\infty$
Therefore x = 2 is a vertical asymptote.

$\lim_{x \to -1^+} \frac{x^2 - 2}{(x - 2)(x + 1)} = \infty, \quad \lim_{x \to -1^-} \frac{x^2 - 2}{(x - 2)(x + 1)} = -\infty$
Therefore x = -1 is a vertical asymptote.

12. $\lim_{x \to 1^+} \frac{2 + x}{1 - x} = -\infty, \quad \lim_{x \to 1^-} \frac{2 + x}{1 - x} = \infty$
Therefore x = 1 is a vertical asymptote.

13. $\lim_{x \to -1^+} \frac{x^3}{x^2 - 1} = \infty, \quad \lim_{x \to -1^-} \frac{x^3}{x^2 - 1} = -\infty$
Therefore x = -1 is a vertical asymptote.

$\lim_{x \to 1^+} \frac{x^3}{x^2 - 1} = \infty, \quad \lim_{x \to 1^-} \frac{x^3}{x^2 - 1} = -\infty$
Therefore x = 1 is a vertical asymptote.

14. No vertical asymptote since the denominator is never zero.

15. $\lim_{x \to 0^+} (1 - \frac{4}{x^2}) = -\infty = \lim_{x \to 0^-} (1 - \frac{4}{x^2})$
Therefore x = 0 is a vertical asymptote.

16. $\lim_{x \to 2^+} \frac{-2}{(x - 2)^2} = -\infty = \lim_{x \to 2^-} \frac{-2}{(x - 2)^2}$
Therefore x = 2 is a vertical asymptote.

17. $\lim_{x \to -2^+} \frac{x}{(x + 2)(x - 1)} = \infty, \quad \lim_{x \to -2^-} \frac{x}{(x + 2)(x - 1)} = -\infty$
Therefore x = -2 is a vertical asymptote.

$\lim_{x \to 1^+} \frac{x}{(x + 2)(x - 1)} = \infty, \quad \lim_{x \to 1^-} \frac{x}{(x + 2)(x - 1)} = -\infty$
Therefore x = 1 is a vertical asymptote.

18. $\lim_{x \to -3^+} \frac{1}{(x + 3)^4} = \infty = \lim_{x \to -3^-} \frac{1}{(x + 3)^4}$
Therefore x = -3 is a vertical asymptote.

19. $\lim\limits_{x \to -1} \dfrac{x^2 - 1}{x + 1} = \lim\limits_{x \to -1} (x - 1) = -2$

Removable discontinuity at $x = -1$.

20. $\lim\limits_{x \to -1} \dfrac{x^2 - 6x - 7}{x + 1} = \lim\limits_{x \to -1} (x - 7) = -8$

Removable discontinuity at $x = -1$.

21. $\lim\limits_{x \to -1^+} \dfrac{x^2 + 1}{x + 1} = \infty, \qquad \lim\limits_{x \to -1^-} \dfrac{x^2 + 1}{x + 1} = -\infty$

Vertical asymptote at $x = -1$.

22. $\lim\limits_{x \to -1} \dfrac{\sin(x + 1)}{x + 1} = 1$

Removable discontinuity at $x = -1$.

23. $\lim\limits_{x \to 2^+} \dfrac{x - 3}{x - 2} = -\infty$

24. $\lim\limits_{x \to 1^+} \dfrac{2 + x}{1 - x} = -\infty$

25. $\lim\limits_{x \to 4^+} \dfrac{x^2}{x^2 - 16} = \infty$

26. $\lim\limits_{x \to 4} \dfrac{x^2}{x^2 + 16} = \dfrac{1}{2}$

27. $\lim\limits_{x \to 0^-} \left(1 + \dfrac{1}{x}\right) = -\infty$

28. $\lim\limits_{x \to 0^-} \left(x^2 - \dfrac{1}{x}\right) = \infty$

29. $\lim\limits_{x \to 0^+} \dfrac{2}{\sin x} = \infty$

30. $\lim\limits_{x \to (\pi/2)^+} \dfrac{-2}{\cos x} = \infty$

31. $\lim\limits_{x \to 1} \dfrac{x^2 - x}{(x^2 + 1)(x - 1)} = \lim\limits_{x \to 1} \dfrac{x}{x^2 + 1} = \dfrac{1}{2}$

32. $\lim\limits_{x \to 1} \dfrac{x^3 - 1}{x^2 + x + 1} = \lim\limits_{x \to 1} (x - 1) = 0$

33. $\lim\limits_{x \to 1^+} \dfrac{x^2 + x + 1}{x^3 - 1} = \lim\limits_{x \to 1^+} \dfrac{1}{x - 1} = \infty$

34. $\lim\limits_{x \to 0^-} \dfrac{x^2 - 2x}{x^3} = \lim\limits_{x \to 0^-} \dfrac{x - 2}{x^2} = -\infty$

35. (a) $r = \dfrac{2(7)}{\sqrt{625 - 49}} = \dfrac{7}{12}$ ft/sec

36. (a) $r = 50\pi \sec^2 \dfrac{\pi}{6} = \dfrac{200\pi}{3}$ ft/sec

(b) $r = \dfrac{2(15)}{\sqrt{625 - 225}} = \dfrac{3}{2}$ ft/sec

(b) $r = 50\pi \sec^2 \dfrac{\pi}{3} = 200\pi$ ft/sec

(c) $\lim\limits_{x \to 25^-} \dfrac{2x}{\sqrt{625 - x^2}} = \infty$

(c) $\lim\limits_{\theta \to (\pi/2)^-} [50\pi \sec^2 \theta] = \infty$

37. $C = \dfrac{528x}{100 - x}$, $0 \le x < 100$

     (a) $C(25) = \$176$ million

     (b) $C(50) = \$528$ million

     (c) $C(75) = \$1584$ million

     (d) $\displaystyle \lim_{x \to 100^-} \frac{528x}{100 - x} = \infty$

        Thus it is not possible.

38. $C = \dfrac{80{,}000p}{100 - p}$, $0 \le p < 100$

     (a) $C(15) = \$14{,}117.65$

     (b) $C(50) = \$80{,}000$

     (c) $C(90) = \$720{,}000$

     (d) $\displaystyle \lim_{p \to 100^-} \frac{80{,}000p}{100 - p} = \infty$

Review Exercises for Chapter 2

1. $\displaystyle \lim_{x \to 2} (5x - 3) = 5(2) - 3 = 7$

2. $\displaystyle \lim_{x \to 2} (3x + 5) = 3(2) + 5 = 11$

3. $\displaystyle \lim_{x \to 2} (5x - 3)(3x + 5) = [5(2) - 3][3(2) + 5] = 7 \cdot 11 = 77$

4. $\displaystyle \lim_{x \to 2} \left(\frac{3x + 5}{5x - 3} \right) = \frac{3(2) + 5}{5(2) - 3} = \frac{11}{7}$

5. $\displaystyle \lim_{t \to 3} \frac{t^2 + 1}{t} = \frac{3^2 + 1}{3} = \frac{10}{3}$

6. $\displaystyle \lim_{t \to 3} \frac{t^2 - 9}{t - 3} = \lim_{t \to 3} (t + 3) = 6$

7. $\displaystyle \lim_{t \to -2} \frac{t + 2}{t^2 - 4} = \lim_{t \to -2} \frac{1}{t - 2} = -\frac{1}{4}$

8. $\displaystyle \lim_{x \to 0} \frac{\sqrt{4 + x} - 2}{x} = \lim_{x \to 0} \frac{\sqrt{4 + x} - 2}{x} \cdot \frac{\sqrt{4 + x} + 2}{\sqrt{4 + x} + 2}$

           $\displaystyle = \lim_{x \to 0} \frac{1}{\sqrt{4 + x} + 2} = \frac{1}{4}$

9. $\displaystyle \lim_{x \to 0} \frac{[1/(x + 1)] - 1}{x} = \lim_{x \to 0} \frac{-1}{x + 1} = -1$

10. $\displaystyle \lim_{s \to 0} \frac{(1/\sqrt{1 + s}) - 1}{s} = \lim_{s \to 0} \left[\frac{(1/\sqrt{1 + s}) - 1}{s} \cdot \frac{(1/\sqrt{1 + s}) + 1}{(1/\sqrt{1 + s}) + 1} \right]$

       $\displaystyle = \lim_{s \to 0} \frac{[1/(1 + s)] - 1}{s(1/\sqrt{1 + s} + 1)} = \lim_{s \to 0} \frac{-1}{(1 + s)(1/\sqrt{1 + s} + 1)} = -\frac{1}{2}$

11. $\displaystyle \lim_{x \to -1} \frac{x^3 + 1}{x + 1} = \lim_{x \to -1} \frac{(x + 1)(x^2 - x + 1)}{x + 1} = \lim_{x \to -1} (x^2 - x + 1) = 3$

12. $\displaystyle\lim_{x \to -2} \frac{x^2-4}{x^3+8} = \lim_{x \to -2}\frac{(x+2)(x-2)}{(x+2)(x^2-2x+4)}$

$\displaystyle = \lim_{x \to -2}\frac{x-2}{x^2-2x+4} = -\frac{4}{12} = -\frac{1}{3}$

13. $\displaystyle\lim_{x \to 0^+}(x - \frac{1}{x^3}) = -\infty$

14. $\displaystyle\lim_{x \to 2^+}\frac{1}{\sqrt[3]{x^2-4}} = \infty, \quad \lim_{x \to 2^-}\frac{1}{\sqrt[3]{x^2-4}} = -\infty$

Thus $\displaystyle\lim_{x \to 2}\frac{1}{\sqrt[3]{x^2-4}}$ does not exist

15. $\displaystyle\lim_{x \to -2^-}\frac{2x^2+x+1}{x+2} = -\infty$

16. $\displaystyle\lim_{x \to 1/2}\frac{2x-1}{6x-3} = \lim_{x \to 1/2}\frac{1}{3} = \frac{1}{3}$

17. $\displaystyle\lim_{x \to -1}\frac{x+1}{x^3+1} = \lim_{x \to -1}\frac{1}{x^2-x+1} = \frac{1}{3}$

18. $\displaystyle\lim_{x \to -1}\frac{x+1}{x^4-1} = \lim_{x \to -1}\frac{1}{(x^2+1)(x-1)} = -\frac{1}{4}$

19. $\displaystyle\lim_{x \to 1}\frac{x^2-2x+1}{x+1} = 0$

20. $\displaystyle\lim_{x \to -1^+}\frac{x^2-2x+1}{x+1} = \infty$

21. $\displaystyle\lim_{x \to 0}\frac{\sin 4x}{5x} = \frac{4}{5}$

22. $\displaystyle\lim_{x \to 0^+}\frac{\sec x}{x} = \infty$

23. $\displaystyle\lim_{x \to 0^+}\frac{\csc 2x}{x} = \infty$

24. $\displaystyle\lim_{x \to 0^-}\frac{\cos^2 x}{x} = -\infty$

25. $\displaystyle\lim_{\Delta x \to 0}\frac{\sin(\pi/6+\Delta x)-1/2}{\Delta x} = \lim_{\Delta x \to 0}\frac{\sin\pi/6\cos\Delta x + \cos\pi/6\sin\Delta x - 1/2}{\Delta x}$

$\displaystyle = \lim_{\Delta x \to 0}\frac{1}{2}\cdot\frac{(\cos\Delta x-1)}{\Delta x} + \lim_{\Delta x \to 0}\frac{\sqrt{3}}{2}\cdot\frac{\sin\Delta x}{\Delta x} = 0 + \frac{\sqrt{3}}{2}(1) = \frac{\sqrt{3}}{2}$

26. $\displaystyle\lim_{\Delta x \to 0}\frac{\cos(\pi+\Delta x)+1}{\Delta x} = \lim_{\Delta x \to 0}\frac{\cos\pi\cos\Delta x - \sin\pi\sin\Delta x + 1}{\Delta x}$

$\displaystyle = \lim_{\Delta x \to 0}\left[-\frac{(\cos\Delta x-1)}{\Delta x}\right] - \lim_{\Delta x \to 0}\left[\sin\pi\frac{\sin\Delta x}{\Delta x}\right] = -0 - (0)(1) = 0$

27. $f(x) = \dfrac{\sqrt{2x + 1} - \sqrt{3}}{x - 1}$

x	1.1	1.01	1.001	1.0001
f(x)	0.5680	0.5764	0.5772	0.5773

$\displaystyle\lim_{x \to 1^+} \dfrac{\sqrt{2x + 1} - \sqrt{3}}{x - 1} \approx 0.577$

28. $f(x) = \dfrac{1 - \sqrt[3]{x}}{x - 1}$

x	1.1	1.01	1.001	1.0001
f(x)	−0.3228	−0.3322	−0.3332	−0.3333

$\displaystyle\lim_{x \to 1^+} \dfrac{1 - \sqrt[3]{x}}{x - 1} \approx -0.333$

29. $\displaystyle\lim_{x \to 1^+} \dfrac{\sqrt{2x + 1} - \sqrt{3}}{x - 1} = \lim_{x \to 1^+} \dfrac{\sqrt{2x + 1} - \sqrt{3}}{x - 1} \cdot \dfrac{\sqrt{2x + 1} + \sqrt{3}}{\sqrt{2x + 1} + \sqrt{3}}$

$\qquad\qquad = \displaystyle\lim_{x \to 1^+} \dfrac{(2x + 1) - 3}{(x - 1)[\sqrt{2x + 1} + \sqrt{3}]}$

$\qquad\qquad = \displaystyle\lim_{x \to 1^+} \dfrac{2}{\sqrt{2x + 1} + \sqrt{3}} = \dfrac{2}{2\sqrt{3}} = \dfrac{1}{\sqrt{3}}$

30. $\displaystyle\lim_{x \to 1^+} \dfrac{1 - \sqrt[3]{x}}{x - 1} = \lim_{x \to 1^+} \dfrac{1 - \sqrt[3]{x}}{x - 1} \cdot \dfrac{1 + \sqrt[3]{x} + (\sqrt[3]{x})^2}{1 + \sqrt[3]{x} + (\sqrt[3]{x})^2}$

$\qquad\qquad = \displaystyle\lim_{x \to 1^+} \dfrac{1 - x}{(x - 1)[1 + \sqrt[3]{x} + (\sqrt[3]{x})^2]}$

$\qquad\qquad = \displaystyle\lim_{x \to 1^+} \dfrac{-1}{1 + \sqrt[3]{x} + (\sqrt[3]{x})^2} = -\dfrac{1}{3}$

31. $\displaystyle\lim_{x \to 0} \dfrac{|x|}{x} = 1$ is false since $\displaystyle\lim_{x \to 0} \dfrac{|x|}{x}$ does not exist

32. $\displaystyle\lim_{x \to 0} x^3 = 0$ is true

33. $\displaystyle\lim_{x \to 2} f(x) = 3$ is false since $\displaystyle\lim_{x \to 2^-} f(x) = 3$ and $\displaystyle\lim_{x \to 2^+} f(x) = 0$

34. $\lim_{x \to 3} f(x) = 1$ is true

35. $\lim_{x \to 0^+} \sqrt{x} = 0$ is true   36. $\lim_{x \to 0} \sqrt[3]{x} = 0$ is true

37. $\lim_{x \to k^+} [x + 3] = k + 3$ where k is an integer

     $\lim_{x \to k^-} [x + 3] = k + 2$ where k is an integer

     A nonremovable discontinuity at each integer k.

38. $f(x) = \dfrac{3x^2 - x - 2}{x - 1} = \dfrac{(3x + 2)(x - 1)}{x - 1} = 3x + 2$

     A removable discontinuity at x = 1

39. $\lim_{x \to 1} \dfrac{(3x + 2)(x - 1)}{x - 1} = 5$

     A removable discontinuity at x = 1

40. $\lim_{x \to 2^-} (5 - x) = 3, \quad \lim_{x \to 2^+} (2x - 3) = 1$

     A nonremovable discontinuity at x = 2

41. $f(x) = \dfrac{1}{(x - 2)^2}$ 42. $f(x) = \sqrt{\dfrac{x + 2}{x}}$

     A nonremovable discontinuity   A nonremovable discontinuity
     at x = 2   at x = 0

43. $f(x) = \dfrac{3}{x + 1}$

     A nonremovable discontinuity
     at x = −1

44. $f(x) = \dfrac{x + 1}{2x + 2} = \dfrac{x + 1}{2(x + 1)}, \quad \lim_{x \to -1} \dfrac{x + 1}{2(x + 1)} = \dfrac{1}{2}$

     A removable discontinuity at x = −1

45. $f(x) = \csc \dfrac{\pi x}{2}$ 46. $f(x) = \tan 2x$

     Nonremovable discontinuities   Nonremovable discontinuities at
     at each even integer   each odd multiple of $\pi/4$

47. $f(2) = 5$; Therefore find c so that $\lim\limits_{x \to 2^+} (cx + 6) = 5$

 $c(2) + 6 = 5$, $2c = -1$, $c = -\dfrac{1}{2}$

48. $\lim\limits_{x \to 1^+} (x + 1) = 2$, $\lim\limits_{x \to 3^-} (x + 1) = 4$

 Find b and c so that $\lim\limits_{x \to 1^-} (x^2 + bx + c) = 2$ and $\lim\limits_{x \to 3^+} (x^2 + bx + c) = 4$

 Consequently we get $1 + b + c = 2$ and $9 + 3b + c = 4$
 Solving simultaneously $b = -3$ and $c = 4$

49. $A = 5000 \, (1.06)^{[2t]}$

 Nonremovable discontinuity
 every 6 months

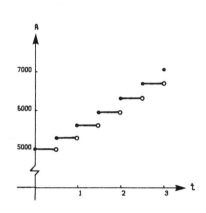

50. $A = 1000 \, (1.14)^{[t]}$

 Nonremovable discontinuity
 every year

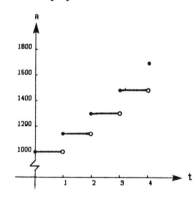

3
Differentiation

1.(a)

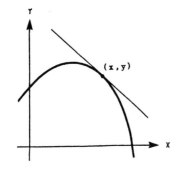

(b)

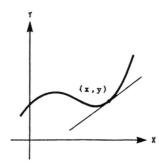

2.(a)

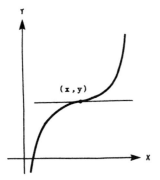

(b)

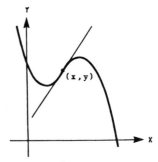

3.(a) $m = 0$

(b) $m = -2$

4.(a) $m = \dfrac{1}{4}$

(b) $m = 1$

5. $f(x) = 3$

(i) $f(x + \Delta x) = 3$

(ii) $f(x + \Delta x) - f(x) = 0$

(iii) $\dfrac{f(x + \Delta x) - f(x)}{\Delta x} = 0$

(iv) $\lim\limits_{\Delta x \to 0} \dfrac{f(x + \Delta x) - f(x)}{\Delta x} = 0$

6. $f(x) = 3x + 2$

(i) $f(x + \Delta x) = 3x + 3\Delta x + 2$

(ii) $f(x + \Delta x) - f(x) = 3\Delta x$

(iii) $\dfrac{f(x + \Delta x) - f(x)}{\Delta x} = 3$

(iv) $\lim\limits_{\Delta x \to 0} \dfrac{f(x + \Delta x) - f(x)}{\Delta x} = 3$

7. $f(x) = -5x$

    (i) $f(x + \Delta x) = -5(x + \Delta x) = -5x - 5\Delta x$

    (ii) $f(x + \Delta x) - f(x) = -5\Delta x$

    (iii) $\dfrac{f(x + \Delta x) - f(x)}{\Delta x} = -5$

    (iv) $\lim\limits_{\Delta x \to 0} \dfrac{f(x + \Delta x) - f(x)}{\Delta x} = -5$

8. $f(x) = 1 - x^2$

    (i) $f(x + \Delta x) = 1 - x^2 - 2x\Delta x - (\Delta x)^2$

    (ii) $f(x + \Delta x) - f(x) = -2x\Delta x - (\Delta x)^2$

    (iii) $\dfrac{f(x + \Delta x) - f(x)}{\Delta x} = -2x - \Delta x$

    (iv) $\lim\limits_{\Delta x \to 0} \dfrac{f(x + \Delta x) - f(x)}{\Delta x} = -2x$

9. $f(x) = 2x^2 + x - 1$

    (i) $f(x + \Delta x) = 2(x + \Delta x)^2 + (x + \Delta x) - 1$
                  $= 2x^2 + 4x\Delta x + 2(\Delta x)^2 + x + \Delta x - 1$

    (ii) $f(x + \Delta x) - f(x) = 4x\Delta x + 2(\Delta x)^2 + \Delta x = \Delta x(4x + 2\Delta x + 1)$

    (iii) $\dfrac{f(x + \Delta x) - f(x)}{\Delta x} = 4x + 2\Delta x + 1$

    (iv) $\lim\limits_{\Delta x \to 0} \dfrac{f(x + \Delta x) - f(x)}{\Delta x} = 4x + 1$

10. $f(x) = \sqrt{x - 4}$

    (i) $f(x + \Delta x) = \sqrt{x + \Delta x - 4}$

    (ii) $f(x + \Delta x) - f(x) = \dfrac{\Delta x}{\sqrt{x + \Delta x - 4} + \sqrt{x - 4}}$ (Rationalize)

    (iii) $\dfrac{f(x + \Delta x) - f(x)}{\Delta x} = \dfrac{1}{\sqrt{x + \Delta x - 4} + \sqrt{x - 4}}$

    (iv) $\lim\limits_{\Delta x \to 0} \dfrac{f(x + \Delta x) - f(x)}{\Delta x} = \dfrac{1}{2\sqrt{x - 4}}$

11. $f(x) = \dfrac{1}{x - 1}$

    (i) $f(x + \Delta x) = \dfrac{1}{x + \Delta x - 1}$

    (ii) $f(x + \Delta x) - f(x) = \dfrac{1}{x + \Delta x - 1} - \dfrac{1}{x - 1} = \dfrac{(x - 1) - (x + \Delta x - 1)}{(x + \Delta x - 1)(x - 1)}$

$$= \dfrac{-\Delta x}{(x + \Delta x - 1)(x - 1)}$$

    (iii) $\dfrac{f(x + \Delta x) - f(x)}{\Delta x} = \dfrac{-1}{(x + \Delta x - 1)(x - 1)}$

    (iv) $\lim\limits_{\Delta x \to 0} \dfrac{f(x + \Delta x) - f(x)}{\Delta x} = \dfrac{-1}{(x - 1)^2}$

12. $f(x) = 1/x^2$

    (i) $f(x + \Delta x) = \dfrac{1}{(x + \Delta x)^2}$

    (ii) $f(x + \Delta x) - f(x) = \dfrac{-2x\Delta x - (\Delta x)^2}{x^2(x + \Delta x)^2}$

    (iii) $\dfrac{f(x + \Delta x) - f(x)}{\Delta x} = \dfrac{-2x - \Delta x}{x^2(x + \Delta x)^2}$

    (iv) $\lim\limits_{\Delta x \to 0} \dfrac{f(x + \Delta x) - f(x)}{\Delta x} = -\dfrac{2}{x^3}$

13. $f(t) = t^3 - 12t$

    (i) $f(t + \Delta t) = (t + \Delta t)^3 - 12(t + \Delta t)$
$$= t^3 + 3t^2\Delta t + 3t(\Delta t)^2 + (\Delta t)^3 - 12t - 12\Delta t$$

    (ii) $f(t + \Delta t) - f(t) = 3t^2\Delta t + 3t(\Delta t)^2 + (\Delta t)^3 - 12\Delta t$

    (iii) $\dfrac{f(t + \Delta t) - f(t)}{\Delta t} = 3t^2 + 3t\Delta t + (\Delta t)^2 - 12$

    (iv) $\lim\limits_{\Delta t \to 0} \dfrac{f(t + \Delta t) - f(t)}{\Delta t} = 3t^2 - 12$

14. $f(t) = t^3 + t^2$

    (i) $f(t + \Delta t) = (t + \Delta t)^3 + (t + \Delta t)^2$

    (ii) $f(t + \Delta t) - f(t) = (3t^2 + 2t)(\Delta t) + (3t + 1)(\Delta t)^2 + (\Delta t)^3$

    (iii) $\dfrac{f(t + \Delta t) - f(t)}{\Delta t} = 3t^2 + 2t + (3t + 1)(\Delta t) + (\Delta t)^2$

    (iv) $\displaystyle\lim_{\Delta t \to 0} \dfrac{f(t + \Delta t) - f(t)}{\Delta t} = 3t^2 + 2t$

15. $f(x) = x^2 + 1,$ (2, 5)

    (i) $f(x + \Delta x) = (x + \Delta x)^2 + 1 = x^2 + 2x\Delta x + (\Delta x)^2 + 1$

    (ii) $f(x + \Delta x) - f(x) = 2x\Delta x + (\Delta x)^2$ (iii) $\dfrac{f(x + \Delta x) - f(x)}{\Delta x} = 2x + \Delta x$

    (iv) $\displaystyle\lim_{\Delta x \to 0} \dfrac{f(x + \Delta x) - f(x)}{\Delta x} = 2x$

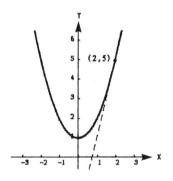

At (2, 5) the slope of the tangent
line is m = 2(2) = 4. The equation
of the tangent line is

$y - 5 = 4(x - 2)$
$y - 5 = 4x - 8$
   $y = 4x - 3$

16. $f(x) = x^2 + 2x + 1,$ (−3, 4)

    (i) $f(x + \Delta x) = (x + \Delta x)^2 + 2(x + \Delta x) + 1$
                   $= x^2 + 2x\Delta x + (\Delta x)^2 + 2x + 2\Delta x + 1$

    (ii) $f(x + \Delta x) - f(x) = 2x\Delta x + (\Delta x)^2 + 2(\Delta x)$

    (iii) $\dfrac{f(x + \Delta x) - f(x)}{\Delta x} = 2x + \Delta x + 2$

    (iv) $\displaystyle\lim_{\Delta x \to 0} \dfrac{f(x + \Delta x) - f(x)}{\Delta x} = 2x + 2$

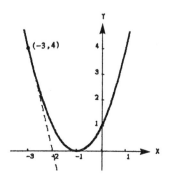

At (−3, 4) the slope of the tangent line
is m = 2(−3) + 2 = −4. The equation of
the tangent line is

$y - 4 = -4(x + 3)$
$y = -4x - 8$

17. $f(x) = x^3$,   (2, 8)

    (i) $f(x + \Delta x) = x^3 + 3x^2\Delta x + 3x(\Delta x)^2 + (\Delta x)^3$

    (ii) $f(x + \Delta x) - f(x) = 3x^2\Delta x + 3x(\Delta x)^2 + (\Delta x)^3$

    (iii) $\dfrac{f(x + \Delta x) - f(x)}{\Delta x} = 3x^2 + 3x\Delta x + (\Delta x)^2$

    (iv) $\displaystyle\lim_{\Delta x \to 0} \dfrac{f(x + \Delta x) - f(x)}{\Delta x} = 3x^2$

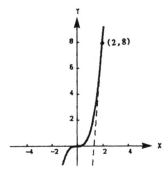

At (2, 8) the slope of the tangent line is $m = 3(2)^2 = 12$. The equation of the tangent line is

$y - 8 = 12(x - 2)$
$\quad y = 12x - 16$

18. $f(x) = x^3$,   (−2, −8)

$f'(x) = 3x^2$ (See Exercise 17)

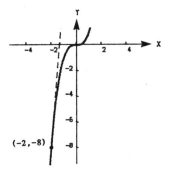

At (−2, −8) the slope of the tangent line is $m = 3(-2)^2 = 12$. The equation of the tangent line is

$y + 8 = 12(x + 2)$
$\quad y = 12x + 16$

19. $f(x) = \sqrt{x + 1}$,   (3, 2)

    (i) $f(x + \Delta x) = \sqrt{x + \Delta x + 1}$

    (ii) $f(x + \Delta x) - f(x) = \sqrt{x + \Delta x + 1} - \sqrt{x + 1}$

$$= \frac{\sqrt{x + \Delta x + 1} - \sqrt{x + 1}}{1} \cdot \frac{\sqrt{x + \Delta x + 1} + \sqrt{x + 1}}{\sqrt{x + \Delta x + 1} + \sqrt{x + 1}}$$

$$= \frac{\Delta x}{\sqrt{x + \Delta x + 1} + \sqrt{x + 1}}$$

    (iii) $\dfrac{f(x + \Delta x) - f(x)}{\Delta x} = \dfrac{1}{\sqrt{x + \Delta x + 1} + \sqrt{x + 1}}$

    (iv) $\displaystyle\lim_{\Delta x \to 0} \dfrac{f(x + \Delta x) - f(x)}{\Delta x} = \dfrac{1}{2\sqrt{x + 1}}$

At (3, 2) the slope of the tangent
line is

$$m = \frac{1}{2\sqrt{3 + 1}} = \frac{1}{4}$$

The equation of the tangent line is

$$y - 2 = \frac{1}{4}(x - 3)$$

$$y = \frac{1}{4}x + \frac{5}{4}$$

$$4y = x + 5$$

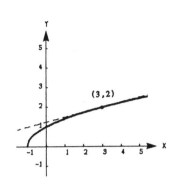

20. $f(x) = \dfrac{1}{x + 1},$ (0, 1)

  (i) $f(x + \Delta x) = \dfrac{1}{x + \Delta x + 1}$

  (ii) $f(x + \Delta x) - f(x) = \dfrac{1}{x + \Delta x + 1} - \dfrac{1}{x + 1}$

$$= \frac{-\Delta x}{(x + \Delta x + 1)(x + 1)}$$

  (iii) $\dfrac{f(x + \Delta x) - f(x)}{\Delta x} = -\dfrac{1}{(x + \Delta x + 1)(x + 1)}$

  (iv) $\displaystyle\lim_{\Delta x \to 0} \dfrac{f(x + \Delta x) - f(x)}{\Delta x} = -\dfrac{1}{(x + 1)^2}$

At (0, 1) the slope of the tangent line is

$$m = \frac{-1}{(0 + 1)^2} = -1$$

The equation of the tangent line is

$$y = -x + 1$$

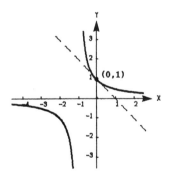

21. $f(x) = x^2 - 1,$ $c = 2$

    $f'(2) = \displaystyle\lim_{x \to 2} \dfrac{f(x) - f(2)}{x - 2}$

           $= \displaystyle\lim_{x \to 2} \dfrac{(x^2 - 1) - 3}{x - 2}$

           $= \displaystyle\lim_{x \to 2} (x + 2)$

           $= 4$

22. $f(x) = x^3 + 2x,$ $c = 1$

    $f'(1) = \displaystyle\lim_{x \to 1} \dfrac{f(x) - f(1)}{x - 1}$

           $= \displaystyle\lim_{x \to 1} \dfrac{x^3 + 2x - 3}{x - 1}$

           $= \displaystyle\lim_{x \to 1} (x^2 + x + 3)$

           $= 5$

23. $f(x) = x^3 + 2x^2 + 1$, $c = -2$

$f'(-2) = \lim_{x \to -2} \dfrac{f(x) - f(-2)}{x + 2}$

$= \lim_{x \to -2} \dfrac{(x^3 + 2x^2 + 1) - 1}{x + 2}$

$= \lim_{x \to -2} x^2$

$= 4$

24. $f(x) = \dfrac{1}{x}$, $c = 3$

$f'(3) = \lim_{x \to 3} \dfrac{f(x) - f(3)}{x - 3}$

$= \lim_{x \to 3} \dfrac{1/x - 1/3}{x - 3}$

$= \lim_{x \to 3} \dfrac{3 - x}{3x} \cdot \dfrac{1}{x - 3}$

$= \lim_{x \to 3} -\dfrac{1}{3x}$

$= -\dfrac{1}{9}$

25. $f(x) = (x - 1)^{2/3}$, $c = 1$

$f'(1) = \lim_{x \to 1} \dfrac{f(x) - f(1)}{x - 1}$

$= \lim_{x \to 1} \dfrac{(x - 1)^{2/3} - 0}{x - 1}$

$= \lim_{x \to 1} \dfrac{1}{(x - 1)^{1/3}}$

The limit does not exist, thus
f is not differentiable at x = 1

26. $f(x) = |x - 2|$, $c = 2$

$f'(2) = \lim_{x \to 2} \dfrac{f(x) - f(2)}{x - 2}$

$= \lim_{x \to 2} \dfrac{|x - 2|}{x - 2}$

The limit does not exist, thus
f is not differentiable at x = 2

27. x = -3,
sharp turn in the graph

28. x = ± 3,
sharp turn in the graph

29. x = -1, discontinuity

30. x = 1, discontinuity

31. x = 3, vertical tangent

32. x = 0, vertical tangent

33. x = 1, vertical tangent

34. x = ± 2, discontinuities

35. x = 0, discontinuity

36. x = 1, discontinuity

37. $f(x) = \sqrt{1 - x^2}$

The derivative from the left is

$$\lim_{x \to 1^-} \frac{f(x) - f(1)}{x - 1} = \lim_{x \to 1^-} \frac{\sqrt{1 - x^2} - 0}{x - 1}$$

$$= \lim_{x \to 1^-} \frac{\sqrt{1 - x^2}}{x - 1} \cdot \frac{\sqrt{1 - x^2}}{\sqrt{1 - x^2}}$$

$$= \lim_{x \to 1^-} -\frac{1 + x}{\sqrt{1 - x^2}} = -\infty$$

The limit from the right does not exist since f is undefined for $x > 1$. Therefore f is not differentiable at $x = 1$.

38. $f(x) = \begin{cases} x - 1, & x \leq 1 \\ (x - 1)^2, & x > 1 \end{cases}$

The derivative from the left is

$$\lim_{x \to 1^-} \frac{f(x) - f(1)}{x - 1} = \lim_{x \to 1^-} \frac{(x - 1) - 0}{x - 1} = \lim_{x \to 1^-} 1 = 1$$

The derivative from the right is

$$\lim_{x \to 1^+} \frac{f(x) - f(1)}{x - 1} = \lim_{x \to 1^+} \frac{(x - 1)^2 - 0}{x - 1} = \lim_{x \to 1^+} (x - 1) = 0$$

These one-sided limits are not equal, therefore f is not differentiable at $x = 1$.

39. $f(x) = \begin{cases} (x - 1)^3, & x \leq 1 \\ (x - 1)^2, & x > 1 \end{cases}$

The derivative from the left is

$$\lim_{x \to 1^-} \frac{f(x) - f(1)}{x - 1} = \lim_{x \to 1^-} \frac{(x - 1)^3 - 0}{x - 1} = \lim_{x \to 1^-} (x - 1)^2 = 0$$

The derivative from the right is

$$\lim_{x \to 1^+} \frac{f(x) - f(1)}{x - 1} = \lim_{x \to 1^+} \frac{(x - 1)^2 - 0}{x - 1} = \lim_{x \to 1^+} (x - 1) = 0$$

These one-sided limits are equal, therefore f is differentiable at $x = 1$ ($f'(1) = 0$).

40. $f(x) = \begin{cases} x, & x \leq 1 \\ x^2, & x > 1 \end{cases}$

The derivative from the left is

$$\lim_{x \to 1^-} \frac{f(x) - f(1)}{x - 1} = \lim_{x \to 1^-} \frac{x - 1}{x - 1} = \lim_{x \to 1^-} 1 = 1$$

The derivative from the right is

$$\lim_{x \to 1^+} \frac{f(x) - f(1)}{x - 1} = \lim_{x \to 1^+} \frac{x^2 - 1}{x - 1} = \lim_{x \to 1^+} (x + 1) = 2$$

These one-sided limits are not equal, therefore f is not differentiable at x = 1.

41. From Exercise 17 we know that $f'(x) = 3x^2$. Since the slope of the given line is 3, we have

$$3x^2 = 3$$
$$x = \pm 1$$

Therefore, at the points (1, 1) and (−1, −1) the tangent lines are parallel to 3x − y + 1 = 0. These lines have equations

$$y - 1 = 3(x - 1)$$
$$y = 3x - 2$$
and
$$y + 1 = 3(x + 1)$$
$$y = 3x + 2$$

42. By the four-step process we have

$$y = \frac{-1}{2x\sqrt{x}}$$

Since the slope of the given line is −1/2, we have

$$-\frac{1}{2x\sqrt{x}} = -\frac{1}{2}$$
$$x = 1$$

Therefore at the point (1, 1) the tangent line is parallel to x + 2y − 6 = 0. The equation of this line is

$$y - 1 = -\frac{1}{2}(x - 1)$$

$$2y - 2 = -x + 1$$
$$x + 2y - 3 = 0$$

43. Given the equation y = 4x − x² and the point (2, 5), by the four-step process we have

$$\frac{dy}{dx} = 4 - 2x$$

The equation of the tangent line is y − 5 = (4 − 2x)(x − 2). Substituting the given equation for y, we have

$$(4x - x^2) - 5 = (4 - 2x)(x - 2)$$
$$4x - x^2 - 5 = -2x^2 + 8x - 8$$
$$x^2 - 4x + 3 = 0$$
$$(x - 1)(x - 3) = 0$$

Therefore, the tangent lines intersect
the parabola at (1, 3) and (3, 3) and
their equations are

$$y - 5 = 2(x - 2)$$
$$y = 2x + 1$$
and
$$y - 5 = -2(x - 2)$$
$$y = -2x + 9$$

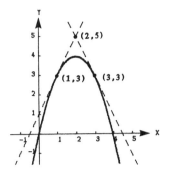

44. Given the equation $y = x^2$ and the point (1, -3), by the four-step
process we have

$$\frac{dy}{dx} = 2x$$

The equation of the tangent line is

$$y + 3 = 2x(x - 1)$$

Substituting the given equation for y we have

$$x^2 + 3 = 2x(x - 1)$$
$$x^2 - 2x - 3 = 0$$
$$(x + 1)(x - 3) = 0$$

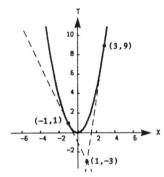

Therefore, the tangent lines intersect
the curve at (-1, 1) and (3, 9) and
their equations are

$$2x + y + 1 = 0 \quad \text{and} \quad 6x - y - 9 = 0$$

3.2
Velocity, acceleration, and other rates of change

1. $f(t) = 2t + 7$

 (i) $f(t + \Delta t) = 2(t + \Delta t) + 7 = 2t + 2\Delta t + 7$

 (ii) $f(t + \Delta t) - f(t) = 2\Delta t$

 (iii) $\dfrac{f(t + \Delta t) - f(t)}{\Delta t} = 2$

 (iv) $\lim_{\Delta t \to 0} \dfrac{f(t + \Delta t) - f(t)}{\Delta t} = 2$

Instantaneous rate of change is the constant 2. Average rate of change
is

$$\frac{f(2) - f(1)}{2 - 1} = \frac{[2(2) + 7] - [2(1) + 7]}{1} = 2$$

2. $f(t) = 3t - 1$

    (i) $f(t + \Delta t) = 3(t + \Delta t) - 1 = 3t + 3\Delta t - 1$

    (ii) $f(t + \Delta t) - f(t) = 3\Delta t$

    (iii) $\dfrac{f(t + \Delta t) - f(t)}{\Delta t} = 3$

    (iv) $\displaystyle\lim_{\Delta t \to 0} \dfrac{f(t + \Delta t) - f(t)}{\Delta t} = 3$

Instantaneous rate of change is the constant 3. Average rate of change is

$$\frac{f(1/3) - f(0)}{(1/3) - 0} = \frac{[3(1/3) - 1] - [3(0)-1]}{1/3} = 3$$

3. $f(x) = \dfrac{1}{(x + 1)}$

    (i) $f(x + \Delta x) = \dfrac{1}{x + \Delta x + 1}$

    (ii) $f(x + \Delta x) - f(x) = \dfrac{1}{x + \Delta x + 1} - \dfrac{1}{x + 1} = -\dfrac{\Delta x}{(x + 1)(x + \Delta x + 1)}$

    (iii) $\dfrac{f(x + \Delta x) - f(x)}{\Delta x} = -\dfrac{1}{(x + 1)(x + \Delta x + 1)}$

    (iv) $\displaystyle\lim_{\Delta x \to 0} \dfrac{f(x + \Delta x) - f(x)}{\Delta x} = -\dfrac{1}{(x + 1)^2}$

Instantaneous rate of change: $(0, 1) \Longrightarrow f'(0) = \dfrac{-1}{(0 + 1)^2} = -1$

$(3, 1/4) \Longrightarrow f'(3) = \dfrac{-1}{(3 + 1)^2} = -\dfrac{1}{16}$

Average rate of change: $\dfrac{f(3) - f(0)}{3 - 0} = \dfrac{(1/4) - 1}{3 - 0} = \dfrac{-3/4}{3} = -\dfrac{1}{4}$

4. $f(x) = -\dfrac{1}{x}$

    (i) $f(x + \Delta x) = -\dfrac{1}{x + \Delta x}$

    (ii) $f(x + \Delta x) - f(x) = -\dfrac{1}{x + \Delta x} + \dfrac{1}{x} = \dfrac{\Delta x}{(x + \Delta x)x}$

    (iii) $\dfrac{f(x + \Delta x) - f(x)}{\Delta x} = \dfrac{1}{(x + \Delta x)x}$

    (iv) $\displaystyle\lim_{\Delta x \to 0} \dfrac{f(x + \Delta x) - f(x)}{\Delta x} = \dfrac{1}{x^2}$

Instantaneous rate of change: $(1, -1) \implies f'(1) = 1$

$$(2, -\tfrac{1}{2}) \implies f'(2) = \tfrac{1}{4}$$

Average rate of change: $\dfrac{f(2) - f(1)}{2 - 1} = \dfrac{(-1/2) - (-1)}{2 - 1} = \dfrac{1}{2}$

5. $f(t) = t^2 - 3$

(i) $f(t + \Delta t) = (t + \Delta t)^2 - 3 = t^2 + 2t\Delta t + (\Delta t)^2 - 3$

(ii) $f(t + \Delta t) - f(t) = 2t\Delta t + (\Delta t)^2$

(iii) $\dfrac{f(t + \Delta t) - f(t)}{\Delta t} = 2t + \Delta t$

(iv) $\lim\limits_{\Delta t \longrightarrow 0} \dfrac{f(t + \Delta t) - f(t)}{\Delta t} = 2t$

Instantaneous rates of change: $(2, 1) \implies f'(2) = 2(2) = 4$

$$(2.1, 1.41) \implies f'(2.1) = 4.2$$

Average rate of change: $\dfrac{f(2.1) - f(2)}{2.1 - 2} = \dfrac{1.41 - 1}{0.1} = 4.1$

6. $f(x) = x^2 - 6x - 1$

(i) $f(x + \Delta x) = (x + \Delta x)^2 - 6(x + \Delta x) - 1$

$$= x^2 + 2x\Delta x + (\Delta x)^2 - 6x - 6\Delta x - 1$$

(ii) $f(x + \Delta x) - f(x) = 2x(\Delta x) + (\Delta x)^2 - 6\Delta x$

(iii) $\dfrac{f(x + \Delta x) - f(x)}{\Delta x} = 2x + \Delta x - 6$

(iv) $\lim\limits_{\Delta x \longrightarrow 0} \dfrac{f(x + \Delta x) - f(x)}{\Delta x} = 2x - 6$

Instantaneous rates of change: $(-1, 6) \implies f'(-1) = 2(-1) - 6 = -8$

$$(3, -10) \implies f'(3) = 2(3) - 6 = 0$$

Average rate of change: $\dfrac{f(3) - f(-1)}{3 - (-1)} = \dfrac{-10 - 6}{3 - (-1)} = \dfrac{-16}{4} = -4$

7. $s(t) = -16t^2 + 1350$

(a) $\dfrac{s(2) - s(1)}{2 - 1} = 1286 - 1334 = -48$ ft/sec

(b) $s'(t) = -32t$

When $t = 1$, $s'(1) = -32$ ft/sec   When $t = 2$, $s'(2) = -64$ ft/sec

(c) $-16t^2 + 1350 = 0$

$t^2 = \dfrac{1350}{16}$ ➡ $t = \dfrac{15\sqrt{6}}{4} \approx 9.2$ sec

(d) When $t = \dfrac{15\sqrt{6}}{4}$, $s'(\dfrac{15\sqrt{6}}{4}) = -120\sqrt{6} \approx -293.9$ ft/sec

8. $v(t) = \dfrac{100t}{2t + 15}$

(i) $v(t + \Delta t) = \dfrac{100(t + \Delta t)}{2(t + \Delta t) + 15}$

(ii) $v(t + \Delta t) - v(t) = \dfrac{100(t + \Delta t)}{2(t + \Delta t) + 15} - \dfrac{100t}{2t + 15}$

$= \dfrac{100(t + \Delta t)(2t + 15) - 100t[2(t + \Delta t) + 15]}{[2(t + \Delta t) + 15](2t + 15)}$

$= \dfrac{1500\Delta t}{[2(t + \Delta t) + 15](2t + 15)}$

(iii) $\dfrac{v(t + \Delta t) - v(t)}{\Delta t} = \dfrac{1500}{[2(t + \Delta t) + 15](2t + 15)}$

(iv) $\lim\limits_{\Delta t \longrightarrow 0} \dfrac{v(t + \Delta t) - v(t)}{\Delta t} = \dfrac{1500}{(2t + 15)^2}$

(a) $a(5) = \dfrac{1500}{(2(5) + 15)^2} = 2.4$ ft/sec²

(b) $a(10) = \dfrac{1500}{(2(10) + 15)^2} = 1.2$ ft/sec²

(c) $a(20) = \dfrac{1500}{(2(20) + 15)^2} = 0.5$ ft/sec²

9. $s(t) = -16t^2 + 384t$
   $v(t) = s'(t) = -32t + 384$
   $v(5) = 224$ ft/sec
   $v(10) = 64$ ft/sec

10. $s(t) = -16t^2 + 256t$
    $v(t) = s'(t) = -32t + 256$
    $v(5) = 96$ ft/sec
    $v(10) = -64$ ft/sec

11. $s(t) = -16t^2 + 600$

    $-16t^2 + 600 = 0$

           $t^2 = 600/16$

           $t = 10\sqrt{6}/4$

           $t = 5\sqrt{6}/2$ sec

(the instant the pebble hits the ground)

$v(t) = s'(t) = -32t$

$v\left(\dfrac{5\sqrt{6}}{2}\right) = -80\sqrt{6}$ ft/sec

12. $s(t) = -16t^2 - 22t + 220$

    $v = -32t - 22$

    $v(3) = -118$ ft/sec

    $s(t) = -16t^2 - 22t + 220 = 112$

(height after falling 108 ft)

    $s(t) = -16t^2 - 22t + 108 = 0$

          $= -2(t - 2)(8t + 27) = 0$

               $t = 2$

    $v(2) = -32(2) - 22$

           $= -86$ ft/sec

13. $s(t) = -16t^2 + s_0 = 0$ when $t = 6.8$

    $s_0 = 16(6.8)^2 \approx 740$ ft

14. "100 ft" ball: $s(t) = -16t^2 + v_0 t + s_0 = -16t^2 + 100 = 0$, $t = 2.5$

    "75 ft" ball: $s(t) = -16t^2 + 75 = 0$, $t \approx 2.165$

The ball dropped from 100 feet is the first to hit the ground.

15.

t (sec)	0	1	2	3	4
s(t) (ft)	0	57.75	99	123.75	132
v(t) = s'(t) (ft/sec)	66	49.5	33	16.5	0
a(t) = v'(t) (ft/sec²)	-16.5	-16.5	-16.5	-16.5	-16.5

16.

t (sec)	0	2	4	6	8	10
s(t) (ft)	0	40	160	360	640	1000
v(t) = s'(t) (ft/sec)	0	40	80	120	160	200
a(t) = v'(t) (ft/sec²)	20	20	20	20	20	20

17. Average velocity on:

    [0, 1] is $\dfrac{57.75 - 0}{1 - 0} = 57.75$ [1, 2] is $\dfrac{99 - 57.75}{2 - 1} = 41.25$

    [2, 3] is $\dfrac{123.75 - 99}{3 - 2} = 24.75$ [3, 4] is $\dfrac{132 - 123.75}{4 - 3} = 8.25$

18. Average velocity on:

    $[0, 2]$ is $\dfrac{40 - 0}{2 - 0} = 20.0$ $\qquad$ $[2, 4]$ is $\dfrac{160 - 40}{4 - 2} = 60.0$

    $[4, 6]$ is $\dfrac{360 - 160}{6 - 4} = 100.0$ $\qquad$ $[6, 8]$ is $\dfrac{640 - 360}{8 - 6} = 140.0$

    $[8, 10]$ is $\dfrac{1000 - 640}{10 - 8} = 180.0$

19. $f'(x) = 2x^2$
    Using the limit definition of the derivative

    $$f''(x) = \lim_{\Delta x \to 0} \frac{f'(x + \Delta x) - f'(x)}{\Delta x} = \lim_{\Delta x \to 0} \frac{2(x + \Delta x)^2 - 2x^2}{\Delta x}$$

    $$= \lim_{\Delta x \to 0} 4x + \Delta x = 4x$$

20. $f''(x) = 20x^3 - 36x^2$
    Using the limit definition of the derivative

    $$f'''(x) = \lim_{\Delta x \to 0} \frac{f''(x + \Delta x) - f''(x)}{\Delta x}$$

    $$= \lim_{\Delta x \to 0} \frac{20(x + \Delta x)^3 - 36(x + \Delta x)^2 - 20x^3 + 36x^2}{\Delta x}$$

    $$= \lim_{\Delta x \to 0} 60x^2 + 60x\Delta x + 20\Delta x^2 - 72x - 36\Delta x = 60x^2 - 72x$$

21. $f''(x) = \dfrac{2x - 2}{x}$
    Using the limit definition of the derivative

    $$f'''(x) = \lim_{\Delta x \to 0} \frac{f''(x + \Delta x) - f''(x)}{\Delta x} = \lim_{\Delta x \to 0} \frac{1}{\Delta x}\left[\frac{[2(x + \Delta x) - 2]}{x + \Delta x} - \frac{2x - 2}{x}\right]$$

    $$= \lim_{\Delta x \to 0} \frac{2}{x(x + \Delta x)} = \frac{2}{x^2}$$

22. $f'''(x) = 2\sqrt{x - 1}$
    Using the limit definition of the derivative

    $$f^{(4)}(x) = \lim_{\Delta x \to 0} \frac{f'''(x + \Delta x) - f'''(x)}{\Delta x} = \lim_{\Delta x \to 0} \frac{2\sqrt{(x + \Delta x) - 1} - 2\sqrt{x - 1}}{\Delta x}$$

    $$= \lim_{\Delta x \to 0} \frac{2}{\sqrt{x + \Delta x - 1} + \sqrt{x - 1}} = \frac{2}{\sqrt{x - 1} + \sqrt{x - 1}} = \frac{1}{\sqrt{x - 1}}$$

23. $f^{(4)}(x) = 2x + 1$
     Using the limit definition of the derivative, differentiate
     $f^{(4)}(x)$ twice

$$f^{(5)}(x) = \lim_{\Delta x \to 0} \frac{f^{(4)}(x + \Delta x) - f^{(4)}(x)}{\Delta x}$$

$$= \lim_{\Delta x \to 0} \frac{2(x + \Delta x) + 1 - 2x - 1}{\Delta x} = \lim_{\Delta x \to 0} 2 = 2$$

$$f^{(6)}(x) = \lim_{\Delta x \to 0} \frac{f^{(5)}(x + \Delta x) - f^{(5)}(x)}{\Delta x} = \lim_{\Delta x \to 0} \frac{2 - 2}{\Delta x} = \lim_{\Delta x \to 0} 0 = 0$$

24. $f(x) = x^3 - 2x$
     Using the limit definition of the derivative, differentiate
     $f(x)$ twice

$$f'(x) = \lim_{\Delta x \to 0} \frac{f(x + \Delta x) - f(x)}{\Delta x}$$

$$= \lim_{\Delta x \to 0} \frac{(x + \Delta x)^3 - 2(x + \Delta x) - x^3 + 2x}{\Delta x}$$

$$= \lim_{\Delta x \to 0} (3x^2 + 3x\Delta x + \Delta x^2 - 2) = 3x^2 - 2$$

$$f''(x) = \lim_{\Delta x \to 0} \frac{f'(x + \Delta x) - f'(x)}{\Delta x}$$

$$= \lim_{\Delta x \to 0} \frac{3(x + \Delta x)^2 - 2 - 3x^2 + 2}{\Delta x}$$

$$= \lim_{\Delta x \to 0} (6x + 3\Delta x) = 6x$$

25. $C = \dfrac{1,008,000}{Q} + 6.3Q, \qquad \dfrac{dC}{dQ} = -\dfrac{1,008,000}{Q^2} + 6.3$

$C(351) - C(350) = 5083.095 - 5085 \approx -\$1.91$

When $Q = 350$, $\dfrac{dC}{dQ} = -\$1.93$

26. C = (gallons of fuel used)(cost per gallon) = $(\frac{15,000}{x})(1.50)$

Thus C = $\frac{22,500}{x}$

By the limit definition of the derivative, we have $\frac{dC}{dx} = -\frac{22,500}{x^2}$

x	10	15	20	25	30	35	40
C	$2250	$1500	$1125	$900	$750	$642.86	$562.50
$\frac{dC}{dx}$	−225	−100	−56.25	−36	−25	−18.37	−14.06

27. (a) When W = 4, $\frac{dT}{dW} = -1.43(\frac{5}{\sqrt{4}} - 1) = -2.145$

(b) When W = 9, $\frac{dT}{dW} = -1.43(\frac{5}{\sqrt{9}} - 1) = -0.953$

28. F = 200 $\sqrt{T}$

By the limit definition of the derivative, we have $\frac{dF}{dT} = \frac{100}{\sqrt{T}}$

(a) When T = 4, $\frac{dF}{dT} = 50$

(b) When T = 9, $\frac{dF}{dT} = 33.33$

3.3
Differentiation rules for constant multiples, sums, powers, sines, and cosines

1.(a) y = $x^{1/2}$

$y' = \frac{1}{2}x^{-1/2}$

$y' = (1) = \frac{1}{2}$

(b) y = $x^{3/2}$

$y' = \frac{3}{2}x^{1/2}$

$y'(1) = \frac{3}{2}$

(c) y = x^2
y' = 2x
y'(1) = 2

(d) y = x^3
y' = $3x^2$
y'(1) = 3

2.(a) $y = x^{-1/2}$

  $y' = -\dfrac{1}{2}x^{-3/2}$

  $y'(1) = -\dfrac{1}{2}$

(b) $y = x^{-1}$

  $y' = -x^{-2}$

  $y'(1) = -1$

(c) $y = x^{-3/2}$

  $y' = -\dfrac{3}{2}x^{-5/2}$

  $y'(1) = -\dfrac{3}{2}$

(d) $y = x^{-2}$

  $y' = -2x^{-3}$

  $y'(1) = -2$

3. $y = 3$
  $y' = 0$

4. $f(x) = -2$
  $f'(x) = 0$

5. $f(x) = x + 1$
  $f'(x) = 1$

6. $g(x) = 3x - 1$
  $g'(x) = 3$

7. $g(x) = x^2 + 4$
  $g'(x) = 2x$

8. $y = t^2 + 2t - 3$
  $y' = 2t + 2$

9. $f(t) = -2t^2 + 3t - 6$
  $f'(t) = -4t + 3$

10. $y = x^3 - 9$
  $y' = 3x^2$

11. $s(t) = t^3 - 2t + 4$
  $s'(t) = 3t^2 - 2$

12. $f(x) = 2x^3 - x^2 + 3x - 1$
  $f'(x) = 6x^2 - 2x + 3$

13. $y = x^2 - \dfrac{1}{2}\cos x$

  $y' = 2x + \dfrac{1}{2}\sin x$

14. $y = 5 + \sin x$
  $y' = \cos x$

15. $y = \dfrac{1}{x} - 3\sin x$

  $y' = -\dfrac{1}{x^2} - 3\cos x$

16. $g(t) = \pi \cos t$
  $g'(t) = -\pi \sin t$

17. $f(x) = \dfrac{1}{x}$, $(1, 1)$

  $f'(x) = -\dfrac{1}{x^2}$

  $f'(1) = -1$

18. $f(x) = -\dfrac{1}{2} + \dfrac{7}{5}x^3$, $\left(0, -\dfrac{1}{2}\right)$

  $f'(x) = \dfrac{21}{5}x^2$

  $f'(0) = 0$

19. $f(t) = 3 - \dfrac{3}{5t}$, $(\dfrac{3}{5}, 2)$

$f'(t) = \dfrac{3}{5t^2}$

$f'(\dfrac{3}{5}) = \dfrac{5}{3}$

20. $y = 3x(x^2 - \dfrac{2}{x})$, $(2, 18)$

$= 3x^3 - 6$

$y' = 9x^2$

$y'(2) = 36$

21. $y = (2x + 1)^2$, $(0, 1)$

$= 4x^2 + 4x + 1$

$y' = 8x + 4$

$y'(0) = 4$

22. $f(x) = 3(5 - x)^2$, $(5, 0)$

$= 3x^2 - 30x + 75$

$f'(x) = 6x - 30$

$f'(5) = 0$

23. $f(\theta) = 4(\sin \theta) - \theta$, $(0, 0)$

$f'(\theta) = 4(\cos \theta) - 1$

$f'(0) = 4(1) - 1 = 3$

24. $g(t) = 2 + 3 \cos t$, $(\pi, -1)$

$g'(t) = -3 \sin t$

$g'(\pi) = 0$

25. $f(x) = x^2 - \dfrac{4}{x}$

$f'(x) = 2x + \dfrac{4}{x^2}$

26. $f(x) = x^2 - 3x - 3x^{-2} + 5x^{-3}$

$f'(x) = 2x - 3 + 6x^{-3} - 15x^{-4}$

$= 2x - 3 + \dfrac{6}{x^3} - \dfrac{15}{x^4}$

27. $f(x) = x^3 - 3x - \dfrac{2}{x^4}$

$f'(x) = 3x^2 - 3 + \dfrac{8}{x^5}$

28. $f(x) = \dfrac{2x^2 - 3x + 1}{x} = 2x - 3 + \dfrac{1}{x}$

$f'(x) = 2 - \dfrac{1}{x^2} = \dfrac{2x^2 - 1}{x^2}$

29. $f(x) = \dfrac{x^3 - 3x^2 + 4}{x^2}$

$= x - 3 + \dfrac{4}{x^2}$

$f'(x) = 1 - \dfrac{8}{x^3} = \dfrac{x^3 - 8}{x^3}$

30. $f(x) = (x^2 + 2x)(x + 1)$

$= x^3 + 3x^2 + 2x$

$f'(x) = 3x^2 + 6x + 2$

31. $f(x) = x(x^2 + 1)$

$= x^3 + x$

$f'(x) = 3x^2 + 1$

32. $f(x) = x + \dfrac{1}{x^2}$

$f'(x) = 1 - \dfrac{2}{x^3}$

33. $f(x) = x^{4/5}$

$f'(x) = \dfrac{4}{5} x^{-1/5}$

$= \dfrac{4}{5x^{1/5}}$

34. $f(x) = x^{1/3} - 1$

$f'(x) = \dfrac{1}{3} x^{-2/3}$

$= \dfrac{1}{3x^{2/3}}$

35. $f(x) = \sqrt[3]{x} + \sqrt[5]{x}$

$= x^{1/3} + x^{1/5}$

$f'(x) = \dfrac{1}{3} x^{-2/3} + \dfrac{1}{5} x^{-4/5}$

$= \dfrac{1}{3x^{2/3}} + \dfrac{1}{5x^{4/5}}$

36. $f(x) = \dfrac{1}{\sqrt[3]{x^2}}$

$= x^{-2/3}$

$f'(x) = -\dfrac{2}{3} x^{-5/3}$

$= -\dfrac{2}{3x^{5/3}}$

37. $f(x) = 4\sqrt{x} + 3\cos x$

$f'(x) = 2x^{-1/2} - 3\sin x$

$= \dfrac{2}{\sqrt{x}} - 3\sin x$

38. $f(x) = 2\sin x + 3\cos x$

$f'(x) = 2\cos x - 3\sin x$

	Function	Rewrite	Derivative	Simplify
39.	$y = \dfrac{1}{3x^3}$	$y = \dfrac{1}{3} x^{-3}$	$y' = -x^{-4}$	$y' = -\dfrac{1}{x^4}$
40.	$y = \dfrac{2}{3x^2}$	$y = \dfrac{2}{3} x^{-2}$	$y' = -\dfrac{4}{3} x^{-3}$	$y' = -\dfrac{4}{3x^3}$
41.	$y = \dfrac{1}{(3x)^3}$	$y = \dfrac{1}{27} x^{-3}$	$y' = -\dfrac{1}{9} x^{-4}$	$y' = -\dfrac{1}{9x^4}$
42.	$y = \dfrac{\pi}{(3x)^2}$	$y = \dfrac{\pi}{9} x^{-2}$	$y' = -\dfrac{2\pi}{9} x^{-3}$	$y' = -\dfrac{2\pi}{9x^3}$
43.	$y = \dfrac{\sqrt{x}}{x}$	$y = x^{-1/2}$	$y' = -\dfrac{1}{2} x^{-3/2}$	$y' = -\dfrac{1}{2x^{3/2}}$
44.	$y = \dfrac{4}{x^{-3}}$	$y = 4x^3$	$y' = 12x^2$	$y' = 12x^2$

45. $y = x^4 - 3x^2 + 2$
$y' = 4x^3 - 6x$
At $(1, 0)$: $y' = 4 - 6 = -2$
Tangent line: $y - 0 = -2(x - 1)$
$2x + y - 2 = 0$

46. $y = x^3 + x$
$y' = 3x^2 + 1$
At $(-1, -2)$: $y' = 4$
Tangent line: $y + 2 = 4(x + 1)$
$4x - y + 2 = 0$

47. $y = x^4 - 3x^2 + 2$,
    $y' = 4x^3 - 6x = 2x(2x^2 - 3) = 0 \implies x = 0$ or $x = \pm\sqrt{3/2}$.
    At $x = 0$, $y = 2$, and at $x = \pm\sqrt{3/2}$, $y = -1/4$.
    Horizontal tangents at $(0, 2)$ and $(\pm\sqrt{3/2}, -1/4)$

48. $y = x^3 + x$, $y' = 3x^2 + 1 > 0$ for all x. Therefore there are no horizontal tangents.

49. $y = 1/x^2$, $y' = -2/x^3$ cannot equal zero. Therefore there are no horizontal tangents.

50. $y = x^2 + 1$, $y' = 2x = 0 \implies x = 0$.
    At $x = 0$, $y = 1$. Horizontal tangent at $(0, 1)$

51. $y = x + \sin x$, $0 \le x < 2\pi$, $y' = 1 + \cos x = 0$
    $\cos x = -1 \implies x = \pi$
    At $x = \pi$, $y = \pi$. Horizontal tangent at (π, π)

52. $y = \sqrt{3}x + 2\cos x$, $0 \le x < 2\pi$, $y' = \sqrt{3} - 2\sin x = 0$

    $\sin x = \dfrac{\sqrt{3}}{2} \implies x = \dfrac{\pi}{3}$ or $x = \dfrac{2\pi}{3}$

    At $x = \dfrac{\pi}{3}$, $y = \dfrac{\sqrt{3}\pi + 3}{3}$ and at $x = \dfrac{2\pi}{3}$, $y = \dfrac{2\sqrt{3}\pi - 3}{3}$.

    Horizontal tangents at $(\dfrac{\pi}{3}, \dfrac{\sqrt{3}\pi + 3}{3})$ and $(\dfrac{2\pi}{3}, \dfrac{2\sqrt{3}\pi - 3}{3})$

53. Let (x_1, y_1) and (x_2, y_2) be the points of tangency.

    The derivatives of these functions are
    $y = 2x \implies m = 2x_1$ and $y = -2x + 6 \implies m = -2x_2 + 6$
    $m = 2x_1 = -2x_2 + 6$
    $\quad x_1 = -x_2 + 3$

    Since $y_1 = x_2$ and $y_2 = -x_2^2 + 6x_2 - 5$,

    $m = \dfrac{y_2 - y_1}{x_2 - x_1} = \dfrac{(-x_2^2 + 6x_2 - 5) - (x_1^2)}{x_2 - x_1} = -2x_2 + 6$

    $\dfrac{(-x_2^2 + 6x_2 - 5) - (-x_2 + 3)^2}{x_2 - (-x_2 + 3)} = -2x_2 + 6$

    $\qquad -2x_2^2 + 12x_2 - 14 = -4x_2^2 + 18x_2 - 18$

    $\qquad\qquad 2x_2^2 - 6x_2 + 4 = 0$

    $\qquad\qquad (x_2 - 2)(x_2 - 1) = 0$

    $\qquad\qquad\qquad x_2 = 1$ or 2

$x_2 = 1$ ⟹ $y_2 = 0$, $x_1 = 2$ and $y_1 = 4$

Thus the tangent line through $(1, 0)$ and $(2, 4)$ is

$y - 0 = (\dfrac{4 - 0}{2 - 1})(x - 1)$ ⟹ $y = 4x - 4$

$x_2 = 2$ ⟹ $y_2 = 3$, $x_1 = 1$ and $y_1 = 1$

Thus the tangent line through $(2, 3)$ and $(1, 1)$ is

$y - 1 = (\dfrac{3 - 1}{2 - 1})(x - 1)$ ⟹ $y = 2x - 1$

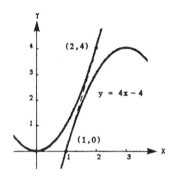
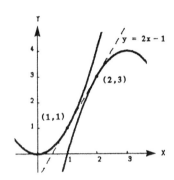

54. m_1 is the slope of the line tangent to $y = x$.
    m_2 is the slope of the line tangent to $y = 1/x$.

Since
$y = x$ ⟹ $y' = 1$ ⟹ $m_1 = 1$
and
$y = 1/x$ ⟹ $y' = -1/x^2$ ⟹ $m_2 = -1/x^2$

The points of intersection of $y = x$ and $y = 1/x$ are
$x = 1/x$ ⟹ $x^2 = 1$ ⟹ $x = \pm 1$

At $x = \pm 1$, $m_2 = -1$. Since $m_2 = -1/m_1$, these tangent lines are perpendicular at the points of intersection.

55. $A = s^2$, $\dfrac{dA}{ds} = 2s$ 56. $V = s^3$, $\dfrac{dV}{ds} = 3s^2$

when $s = 4$, $\dfrac{dA}{ds} = 8$ when $s = 4$, $\dfrac{dV}{ds} = 48$

57. $R = 12{,}000p - 1{,}000p^2$, $\dfrac{dR}{dp} = 12{,}000 - 2{,}000p$

  (a) when $p = 1$, $\dfrac{dR}{dp} = 10{,}000$ (b) when $p = 4$, $\dfrac{dR}{dp} = 4{,}000$

  (c) when $p = 6$, $\dfrac{dR}{dp} = 0$ (d) when $p = 10$, $\dfrac{dR}{dp} = -8{,}000$

58. $P = 50\sqrt{x} - 0.5x - 500$, $\dfrac{dP}{dx} = \dfrac{25}{\sqrt{x}} - 0.5$

   (a) when $x = 900$, $\dfrac{dP}{dx} = \dfrac{1}{3}$ (b) when $x = 1600$, $\dfrac{dP}{dx} = \dfrac{1}{8}$

   (c) when $x = 2500$, $\dfrac{dP}{dx} = 0$ (d) when $x = 3600$, $\dfrac{dP}{dx} = -\dfrac{1}{12}$

59. $E = \dfrac{1}{27}(9t + 3t^2 - t^3)$, $\dfrac{dE}{dt} = \dfrac{1}{9}(3 + 2t - t^2)$

   (a) when $t = 1$, $\dfrac{dE}{dt} = \dfrac{4}{9}$ (b) when $t = 2$, $\dfrac{dE}{dt} = \dfrac{1}{3}$

   (c) when $t = 3$, $\dfrac{dE}{dt} = 0$ (d) when $t = 4$, $\dfrac{dE}{dt} = -\dfrac{5}{9}$

60. $Q(t) = 16t - 4t^2$, $Q'(t) = \dfrac{dQ}{dt} = 16 - 8t$

   (a) $Q'(\dfrac{1}{2}) = 16 - 8(\dfrac{1}{2}) = 12$ (b) $Q'(1) = 16 - 8(1) = 8$

   (c) $Q'(2) = 16 - 8(2) = 0$

3.4
Differentiation rules for products, quotients, secants, and tangents

1. $f(x) = \dfrac{1}{3}(2x^3 - 4)$, $f'(x) = \dfrac{1}{3}(6x^2) = 2x^2$, $f'(0) = 0$

2. $f(x) = \dfrac{5 - 6x^2}{7}$, $f'(x) = \dfrac{1}{7}(-12x) = \dfrac{-12x}{7}$, $f'(1) = -\dfrac{12}{7}$

3. $f(x) = 5x^{-2}(x + 3)$

   $f'(x) = 5[x^{-2}(1) + (x + 3)(-2x^{-3})] = 5\left[\dfrac{1}{x^2} - \dfrac{2(x + 3)}{x^3}\right] = \dfrac{-5}{x^3}(x + 6)$

   $f'(1) = -35$

4. $f(x) = (x^2 - 2x + 1)(x^3 - 1)$
   $f'(x) = (x^2 - 2x + 1)(3x^2) + (x^3 - 1)(2x - 2)$
        $= 3x^2(x - 1)^2 + 2(x - 1)^2(x^2 + x + 1)$
        $= (x - 1)^2(5x^2 + 2x + 2)$
   $f'(1) = 0$

5. $f(x) = (x^3 - 3x)(2x^2 + 3x + 5)$
   $f'(x) = (x^3 - 3x)(4x + 3) + (2x^2 + 3x + 5)(3x^2 - 3)$
        $= 10x^4 + 12x^3 - 3x^2 - 18x - 15$
   $f'(0) = -15$

6. $f(x) = (x - 1)(x^2 - 3x + 2)$
   $f'(x) = (x - 1)(2x - 3) + (x^2 - 3x + 2)(1) = 3x^2 - 8x + 5$
   $\qquad = (3x - 5)(x - 1)$
   $f'(0) = 5$

7. $f(x) = (x^5 - 3x)(\dfrac{1}{x^2})$
   $f'(x) = (x^5 - 3x)(-\dfrac{2}{x^3}) + (\dfrac{1}{x^2})(5x^4 - 3) = 3x^2 + \dfrac{3}{x^2}$
   $f'(-1) = 6$

8. $f(x) = \dfrac{x + 1}{x - 1}$
   $f'(x) = \dfrac{(x - 1)(1) - (x + 1)(1)}{(x - 1)^2} = \dfrac{-2}{(x - 1)^2}$
   $f'(2) = -2$

9. $f(x) = \dfrac{3x - 2}{2x - 3}$
   $f'(x) = \dfrac{(2x - 3)(3) - (3x - 2)(2)}{(2x - 3)^2} = \dfrac{-5}{(2x - 3)^2}$

10. $f(x) = \dfrac{x^3 + 3x + 2}{x^2 - 1}$
    $f'(x) = \dfrac{(x^2 - 1)(3x^2 + 3) - (x^3 + 3x + 2)(2x)}{(x^2 - 1)^2} = \dfrac{x^4 - 6x^2 - 4x - 3}{(x^2 - 1)^2}$

11. $f(x) = \dfrac{3 - 2x - x^2}{x^2 - 1}$
    $f'(x) = \dfrac{(x^2 - 1)(-2 - 2x) - (3 - 2x - x^2)(2x)}{(x^2 - 1)^2}$
    $\qquad = \dfrac{2x^2 - 4x + 2}{(x^2 - 1)^2} = \dfrac{2(x - 1)^2}{(x^2 - 1)^2} = \dfrac{2}{(x + 1)^2}$

12. $f(x) = x^4 \left[1 - \dfrac{2}{x + 1} \right] = x^4 \left[\dfrac{x - 1}{x + 1} \right]$
    $f'(x) = x^4 \left[\dfrac{(x + 1) - (x - 1)}{(x + 1)^2} \right] + \left[\dfrac{x - 1}{x + 1} \right](4x^3) = 2x^3 \left[\dfrac{2x^2 + x - 2}{(x + 1)^2} \right]$

13. $f(x) = \dfrac{x + 1}{\sqrt{x}}$

$f'(x) = \dfrac{\sqrt{x}(1) - (x + 1)[1/(2\sqrt{x})]}{x} = \dfrac{x - 1}{2x^{3/2}}$

14. $f(x) = \sqrt[3]{x}(\sqrt{x} + 3) = x^{1/3}(x^{1/2} + 3)$

$f'(x) = x^{1/3}(\dfrac{1}{2}x^{-1/2}) + (x^{1/2} + 3)(\dfrac{1}{3}x^{-2/3})$

$= \dfrac{5}{6}x^{-1/6} + x^{-2/3} = \dfrac{5}{6x^{1/6}} + \dfrac{1}{x^{2/3}}$

15. $h(t) = \dfrac{t + 1}{t^2 + 2t + 2}$

$h'(t) = \dfrac{(t^2 + 2t + 2)(1) - (t + 1)(2t + 2)}{(t^2 + 2t + 2)^2} = \dfrac{-t^2 - 2t}{(t^2 + 2t + 2)^2}$

16. $h(x) = (x^2 - 1)^2 = x^4 - 2x^2 + 1$

$h'(x) = 4x^3 - 4x = 4x(x^2 - 1)$

17. $h(s) = (s^3 - 2)^2 = s^6 - 4s^3 + 4$

$h'(s) = 6s^5 - 12s^2 = 6s^2(s^3 - 2)$

18. $f(x) = (\dfrac{x^2 - x - 3}{x^2 + 1})(x^2 + x + 1)$

$f'(x) = (\dfrac{x^2 - x - 3}{x^2 + 1})(2x + 1)$

$\quad + (x^2 + x + 1)\left[\dfrac{(x^2 + 1)(2x - 1) - (x^2 - x - 3)(2x)}{(x^2 + 1)^2}\right]$

$= \dfrac{2x^3 - x^2 - 7x - 3}{x^2 + 1} + \dfrac{x^4 + 9x^3 + 8x^2 + 7x - 1}{(x^2 + 1)^2}$

$= \dfrac{2x^5 + 4x^3 + 4x^2 - 4}{(x^2 + 1)^2}$

19. $g(x) = \left[\dfrac{x + 1}{x + 2}\right](2x - 5)$

$g'(x) = \left[\dfrac{x + 1}{x + 2}\right](2) + (2x - 5)\left[\dfrac{(x + 2)(1) - (x + 1)(1)}{(x + 2)^2}\right]$

$\qquad = \dfrac{2x^2 + 6x + 4 + 2x - 5}{(x + 2)^2} = \dfrac{2x^2 + 8x - 1}{(x + 2)^2}$

20. $f(x) = (x^2 - x)(x^2 + 1)(x^2 + x + 1)$

$f'(x) = (x^2 - x)[(x^2 + 1)(2x + 1) + (x^2 + x + 1)(2x)]$

$\qquad + [(x^2 + 1)(x^2 + x + 1)](2x - 1)$

$\qquad = (x^2 - x)(4x^3 + 3x^2 + 4x + 1) + (x^4 + x^3 + 2x^2 + x + 1)(2x - 1)$

$\qquad = 6x^5 + 4x^3 - 3x^2 - 1$

21. $f(x) = (3x^3 + 4x)(x - 5)(x + 1)$

$f'(x) = [(3x^3 + 4x)(x - 5)](1) + (x + 1)[(3x^3 + 4x)(1) + (x - 5)(9x^2 + 4)]$

$\qquad = 3x^4 - 15x^3 + 4x^2 - 20x + 12x^4 - 33x^3 - 37x^2 - 12x - 20$

$\qquad = 15x^4 - 48x^3 - 33x^2 - 32x - 20$

22. $f(x) = \dfrac{x^2 + c^2}{x^2 - c^2}$

$f'(x) = \dfrac{(x^2 - c^2)(2x) - (x^2 + c^2)(2x)}{(x^2 - c^2)^2} = \dfrac{-4xc^2}{(x^2 - c^2)^2}$

23. $f(x) = \dfrac{c^2 - x^2}{c^2 + x^2}$

$f'(x) = \dfrac{(c^2 + x^2)(-2x) - (c^2 - x^2)(2x)}{(c^2 + x^2)^2} = \dfrac{-4xc^2}{(c^2 + x^2)^2}$

24. $f(x) = \dfrac{x(x^2 - 1)}{x + 3} = \dfrac{x^3 - x}{x + 3}$

$f'(x) = \dfrac{(x + 3)(3x^2 - 1) - (x^3 - x)}{(x + 3)^2} = \dfrac{2x^3 + 9x^2 - 3}{(x + 3)^2}$

	Function	Rewrite	Derivative	Simplify
25.	$y = \dfrac{x^2 + 2x}{x}$	$y = x + 2$	$y' = 1$	$y' = 1$
26.	$y = \dfrac{4x^{3/2}}{x}$	$y = 4\sqrt{x}$	$y' = 2x^{-1/2}$	$y' = \dfrac{2}{\sqrt{x}}$
27.	$y = \dfrac{7}{3x^3}$	$y = \dfrac{7}{3}x^{-3}$	$y' = -7x^{-4}$	$y' = -\dfrac{7}{x^4}$
28.	$y = \dfrac{4}{5x^2}$	$y = \dfrac{4}{5}x^{-2}$	$y' = -\dfrac{8}{5}x^{-3}$	$y' = -\dfrac{8}{5x^3}$
29.	$y = \dfrac{3x^2 - 5}{7}$	$y = \dfrac{1}{7}(3x^2 - 5)$	$y' = \dfrac{1}{7}(6x)$	$y' = \dfrac{6x}{7}$
30.	$y = \dfrac{x^2 - 4}{x + 2}$	$y = x - 2$	$y' = 1$	$y' = 1$

31. $f(t) = t^2 \sin t$

   $f'(t) = t^2 \cos t + 2t \sin t$

   $\quad = t(t \cos t + 2 \sin t)$

32. $f(x) = \dfrac{\sin x}{x}$

   $f'(x) = \dfrac{x \cos x - \sin x}{x^2}$

33. $f(t) = \dfrac{\cos t}{t}$

   $f'(t) = \dfrac{-t \sin t - \cos t}{t^2}$

   $\quad = -\dfrac{t \sin t + \cos t}{t^2}$

34. $f(\theta) = (\theta + 1)\cos\theta$

   $f'(\theta) = (\theta + 1)(-\sin\theta) + (\cos\theta)(1)$

   $\quad = \cos\theta - (\theta + 1)\sin\theta$

35. $f(x) = -x + \tan x$
   $f'(x) = -1 + \sec^2 x$
   $\quad = \tan^2 x$

36. $y = x + \cot x$
   $y' = 1 - \csc^2 x$
   $\quad = -\cot^2 x$

37. $y = 5x \csc x$

   $y' = -5x \csc x \cot x + 5 \csc x$

   $\quad = 5 \csc x(-x \cot x + 1)$

   $\quad = 5 \csc x(1 - x \cot x)$

38. $y = \dfrac{\sec x}{x}$

   $y' = \dfrac{x \sec x \tan x - \sec x}{x^2}$

   $\quad = \dfrac{\sec x(x \tan x - 1)}{x^2}$

39. $y = -\csc x - \sin x$
$y' = \csc x \cot x - \cos x$

$\quad = \dfrac{\cos x}{\sin^2 x} - \cos x$

$\quad = \cos x \, (\csc^2 x - 1)$

$\quad = \cos x \cot^2 x$

40. $y = x \sin x + \cos x$
$\quad y' = x \cos x + \sin x - \sin x$
$\quad\quad = x \cos x$

41. $y = x^2 \sin x + 2x \cos x - 2 \sin x$
$y' = x^2 \cos x + 2x \sin x - 2x \sin x + 2 \cos x - 2 \cos x$
$\quad = x^2 \cos x$

42. $y = \dfrac{1 + \csc x}{1 - \csc x}$

$y' = \dfrac{(1 - \csc x)(-\csc x \cot x) - (1 + \csc x)(\csc x \cot x)}{(1 - \csc x)^2}$

$\quad = \dfrac{-2 \csc x \cot x}{(1 - \csc x)^2}$

43. $f(x) = \sin x \cos x$
$f'(x) = \sin x \, (-\sin x) + \cos x \, (\cos x)$
$\quad\quad = \cos 2x$

44. $f(x) = \tan x \cot x = 1$
$f'(x) = 0$

45. $f(x) = x^2 \tan x$
$f'(x) = x^2 \sec^2 x + 2x \tan x$
$\quad\quad = x(x \sec^2 x + 2 \tan x)$

46. $f(\theta) = 5 \sec \theta + \tan \theta$
$\quad f'(\theta) = 5 \sec \theta \tan \theta + \sec^2 \theta$
$\quad\quad\quad = \sec \theta \, (5 \tan \theta + \sec \theta)$

47. $g(\theta) = \dfrac{\theta}{1 - \sin \theta}$

$g'(\theta) = \dfrac{(1 - \sin \theta)(1) - \theta(-\cos \theta)}{(1 - \sin \theta)^2} = \dfrac{1 - \sin \theta + \theta \cos \theta}{(1 - \sin \theta)^2}$

48. $f(\theta) = \dfrac{\sin \theta}{1 - \cos \theta}$

$f'(\theta) = \dfrac{(1 - \cos \theta) \cos \theta - \sin \theta \, (\sin \theta)}{(1 - \cos \theta)^2} = \dfrac{\cos \theta - 1}{(1 - \cos \theta)^2} = \dfrac{1}{\cos \theta - 1}$

49. $h(t) = \dfrac{\sec t}{t}$

$h'(t) = \dfrac{t \sec t \tan t - \sec t\,(1)}{t^2} = \dfrac{\sec t\,(t \tan t - 1)}{t^2}$

50. $f(x) = \sin x\,(\sin x + \cos x)$
$f'(x) = \sin x\,(\cos x - \sin x) + (\sin x + \cos x)\cos x$
$ = \sin x \cos x - \sin^2 x + \sin x \cos x + \cos^2 x$
$ = \sin 2x + \cos 2x$

51. $f(x) = 4x^{3/2}$

$f'(x) = 6x^{1/2}$

$f''(x) = 3x^{-1/2} = \dfrac{3}{\sqrt{x}}$

52. $f(x) = \dfrac{x^2 + 2x - 1}{x} = x + 2 - \dfrac{1}{x}$

$f'(x) = 1 + \dfrac{1}{x^2}$

$f''(x) = -\dfrac{2}{x^3}$

53. $f(x) = \dfrac{x}{x - 1}$

$f'(x) = \dfrac{(x - 1)(1) - x(1)}{(x - 1)^2}$

$ = \dfrac{-1}{(x - 1)^2}$

$f''(x) = \dfrac{2}{(x - 1)^3}$

54. $f(x) = x + \dfrac{32}{x^2}$

$f'(x) = 1 - \dfrac{64}{x^3}$

$f''(x) = \dfrac{192}{x^4}$

55. $f(x) = \sin x$
$f'(x) = \cos x$
$f''(x) = -\sin x$

56. $f(x) = \sec x$
$f'(x) = \sec x \tan x$
$f''(x) = \sec x\,(\sec^2 x) + \tan x\,(\sec x \tan x)$
$ = \sec x\,(\sec^2 x + \tan^2 x)$

57. $f(x) = \dfrac{x}{x - 1}$, $(2, 2)$

$f'(x) = \dfrac{(x - 1)(1) - x(1)}{(x - 1)^2} = \dfrac{-1}{(x - 1)^2}$

$f'(2) = \dfrac{-1}{(2 - 1)^2} = -1 = $ slope at $(2, 2)$

$y - 2 = -1(x - 2) \implies y = -x + 4$

58. $f(x) = (x - 1)(x^2 - 2)$, $(0, 2)$

$f'(x) = (x - 1)(2x) + (x^2 - 2)(1) = 3x^2 - 2x - 2$

$f'(0) = -2 = $ slope at $(0, 2)$

$y - 2 = -2x \implies y + 2x = 2$

59. $f(x) = (x^3 - 3x + 1)(x + 2)$, $(1, -3)$

$f'(x) = (x^3 - 3x + 1)(1) + (x + 2)(3x^2 - 3) = 4x^3 - 6x + 6x^2 - 5$

$f'(1) = -1 = $ slope at $(1, -3)$

$y + 3 = -1(x - 1) \implies y = -x - 2$

60. $f(x) = \dfrac{x - 1}{x + 1}$, $(2, \dfrac{1}{3})$

$f'(x) = \dfrac{(x + 1)(1) - (x - 1)(1)}{(x + 1)^2} = \dfrac{2}{(x + 1)^2}$

$f'(2) = \dfrac{2}{9} = $ slope at $(2, \dfrac{1}{3})$

$y - \dfrac{1}{3} = \dfrac{2}{9}(x - 2) \implies 2x - 9y - 1 = 0$

61. $f(x) = \tan x$, $(\dfrac{\pi}{4}, 1)$

$f'(x) = \sec^2 x$

$f'(\dfrac{\pi}{4}) = 2 = $ slope at $(\dfrac{\pi}{4}, 1)$

$y - 1 = 2(x - \dfrac{\pi}{4})$

$y - 1 = 2x - \dfrac{\pi}{2}$

$4x - 2y - \pi + 2 = 0$

62. $f(x) = \sec x$, $(\dfrac{\pi}{3}, 2)$

$f'(x) = \sec x \tan x$

$f'(\dfrac{\pi}{3}) = 2\sqrt{3} = $ slope at $(\dfrac{\pi}{3}, 2)$

$y - 2 = 2\sqrt{3}(x - \dfrac{\pi}{3})$

$6\sqrt{3}\, x - 3y + 6 - 2\sqrt{3}\, \pi = 0$

63. $f(x) = \dfrac{x^2}{x - 1}$

$f'(x) = \dfrac{(x - 1)(2x) - x^2(1)}{(x - 1)^2} = \dfrac{x^2 - 2x}{(x - 1)^2} = \dfrac{x(x - 2)}{(x - 1)^2}$

$f'(x) = 0$ when $x = 0$ or $x = 2$
Horizontal tangents are at $(0, 0)$ and $(2, 4)$.

64. $f(x) = \dfrac{x^2}{x^2 + 1}$

$f'(x) = \dfrac{(x^2 + 1)(2x) - (x^2)(2x)}{(x^2 + 1)^2} = \dfrac{2x}{(x^2 + 1)^2}$

$f'(x) = 0$ when $x = 0$
Horizontal tangent is at $(0, 0)$.

65. $f(t) = \dfrac{t^2 - t + 1}{t^2 + 1}$ $\Longrightarrow$ $f'(t) = \dfrac{t^2 - 1}{(t^2 + 1)^2}$

(a) when $t = 0.5$, $f'(t) = -0.48$ (b) when $t = 2$, $f'(t) = 0.12$
(c) when $t = 8$, $f'(t) = 0.0149$

66. $T = 10\left[\dfrac{4t^2 + 16t + 75}{t^2 + 4t + 10}\right]$, $T_0 = 75$, $\lim\limits_{t \to \infty} T = 40$

$T' = 10\left[\dfrac{(t^2 + 4t + 10)(8t + 16) - (4t^2 + 16t + 75)(2t + 4)}{(t^2 + 4t + 10)^2}\right]$

$= 10\left[\dfrac{-70t - 140}{(t^2 + 4t + 10)^2}\right] = -700\left[\dfrac{t + 2}{(t^2 + 4t + 10)^2}\right]$

(a) when $t = 1$, $T' = -9.33$ (b) when $t = 3$, $T' = -3.64$
(c) when $t = 5$, $T' = -1.62$ (d) when $t = 10$, $T' = -0.37$

67. $P(t) = 500\left[1 + \dfrac{4t}{50 + t^2}\right]$

$P'(t) = 500\left[\dfrac{(50 + t^2)(4) - (4t)(2t)}{(50 + t^2)^2}\right]$

$= 500\left[\dfrac{200 - 4t^2}{(50 + t^2)^2}\right] = 2000\left[\dfrac{50 - t^2}{(50 + t^2)^2}\right]$

$P'(2) \approx 31.55$

68. $f(x) = \sec x$, $g(x) = \csc x$, $[0, 2\pi]$, $f'(x) = g'(x)$

$\sec x \tan x = -\csc x \cot x$ $\Longrightarrow$ $\dfrac{\sec x \tan x}{\csc x \cot x} = -1$ $\Longrightarrow$

$\dfrac{\frac{1}{\cos x}\frac{\sin x}{\cos x}}{\frac{1}{\sin x}\frac{\cos x}{\sin x}} = -1$ $\Longrightarrow$ $\dfrac{\sin^3 x}{\cos^3 x} = -1$ $\Longrightarrow$ $\tan^3 x = -1$ $\Longrightarrow$ $\tan x = -1$

$x = \dfrac{3\pi}{4}, \dfrac{7\pi}{4}$

118

69.(a) $y = \sec x = \dfrac{1}{\cos x}$

$y' = \dfrac{(\cos x)(0) - (1)(-\sin x)}{(\cos x)^2} = \dfrac{\sin x}{\cos x \cos x} = \dfrac{1}{\cos x}\dfrac{\sin x}{\cos x}$

$\quad = \sec x \tan x$

(b) $y = \csc x = \dfrac{1}{\sin x}$

$y' = \dfrac{(\sin x)(0) - (1)(\cos x)}{(\sin x)^2} = -\dfrac{\cos x}{\sin x \sin x} = -\dfrac{1}{\sin x}\dfrac{\cos x}{\sin x}$

$\quad = -\csc x \cot x$

(c) $y = \cot x = \dfrac{\cos x}{\sin x}$

$y' = \dfrac{\sin x(-\sin x) - (\cos x)(\cos x)}{(\sin x)^2} = -\dfrac{\sin^2 x + \cos^2 x}{\sin^2 x}$

$\quad = -\dfrac{1}{\sin^2 x} = -\csc^2 x$

3.5
The Chain Rule

$y = f(g(x))$	$u = g(x)$	$y = f(u)$
1. $y = \sqrt{x^2 - 1}$	$u = x^2 - 1$	$y = \sqrt{u}$
2. $y = \cos \dfrac{3x}{2}$	$u = \dfrac{3x}{2}$	$y = \cos u$
3. $y = \csc^3 x$	$u = \csc x$	$y = u^3$
4. $y = (6x - 5)^4$	$u = 6x - 5$	$y = u^4$
5. $y = \tan(\pi x + 1)$	$u = \pi x + 1$	$y = \tan u$
6. $y = \dfrac{1}{\sqrt{x + 1}}$	$u = x + 1$	$y = u^{-1/2}$

7. $y = (2x - 7)^3$
$y' = 3(2x - 7)^2(2)$
$\quad = 6(2x - 7)^2$

8. $y = (3x^2 + 1)^4$
$y' = 4(3x^2 + 1)^3(6x)$
$\quad = 24x(3x^2 + 1)^3$

9. $f(x) = 2(x^2 - 1)^3$
$f'(x) = 6(x^2 - 1)^2(2x)$
$\quad = 12x(x^2 - 1)^2$

10. $g(x) = 3(9x - 4)^4$
$g'(x) = 12(9x - 4)^3(9)$
$\quad = 108(9x - 4)^3$

11. $y = (x - 2)^{-1}$

    $y' = -1(x - 2)^{-2}(1)$

    $\quad = \dfrac{-1}{(x - 2)^2}$

12. $s(t) = (t^2 + 3t - 1)^{-1}$

    $s'(t) = -1(t^2 + 3t - 1)^{-2}(2t + 3)$

    $\quad = \dfrac{-(2t + 3)}{(t^2 + 3t - 1)^2}$

13. $f(t) = (t - 3)^{-2}$

    $f'(t) = -2(t - 3)^{-3} = \dfrac{-2}{(t - 3)^3}$

14. $y = -4(t + 2)^{-2}$

    $y' = 8(t + 2)^{-3} = \dfrac{8}{(t + 2)^3}$

15. $f(x) = 3(x^3 - 4)^{-1}$

    $f'(x) = -3(x^3 - 4)^{-2}(3x^2)$

    $\quad = \dfrac{-9x^2}{(x^3 - 4)^2}$

16. $f(x) = (x^2 - 3x)^{-2}$

    $f'(x) = -2(x^2 - 3x)^{-3}(2x - 3)$

    $\quad = \dfrac{6 - 4x}{(x^2 - 3x)^3}$

17. $f(x) = x^2(x - 2)^4$

    $f'(x) = x^2[4(x - 2)^3(1)] + (x - 2)^4(2x) = 2x(x - 2)^3[2x + (x - 2)]$

    $\quad = 2x(x - 2)^3(3x - 2)$

18. $f(x) = x(3x - 9)^3$

    $f'(x) = x[3(3x - 9)^2(3)] + (3x - 9)^3(1) = (3x - 9)^2[9x + 3x - 9]$

    $\quad = 27(x - 3)^2(4x - 3)$

19. $f(t) = (1 - t)^{1/2}$

    $f'(t) = \dfrac{1}{2}(1 - t)^{-1/2}(-1)$

    $\quad = -\dfrac{1}{2\sqrt{1 - t}}$

20. $g(x) = (3 - 2x)^{1/2}$

    $g'(x) = \dfrac{1}{2}(3 - 2x)^{-1/2}(-2)$

    $\quad = -\dfrac{1}{\sqrt{3 - 2x}}$

21. $s(t) = (t^2 + 2t - 1)^{1/2}$

    $s'(t) = \dfrac{1}{2}(t^2 + 2t - 1)^{-1/2}(2t + 2)$

    $\quad = \dfrac{t + 1}{\sqrt{t^2 + 2t - 1}}$

22. $y = (3x^3 + 4x)^{1/3}$

    $y' = \dfrac{1}{3}(3x^3 + 4x)^{-2/3}(9x^2 + 4)$

    $\quad = \dfrac{9x^2 + 4}{3(3x^3 + 4x)^{2/3}}$

23. $g(x) = (9x^2 + 4)^{1/3}$

$g'(x) = \frac{1}{3}(9x^2 + 4)^{-2/3}(18x)$

$= \frac{6x}{(9x^2 + 4)^{2/3}}$

24. $g(x) = \sqrt{x^2 - 2x + 1}$

$= \sqrt{(x - 1)^2} = |x - 1|$

$g'(x) = \begin{cases} 1, & x > 1 \\ -1, & x < 1 \end{cases}$

25. $y = 2(4 - x^2)^{1/2}$

$y' = (4 - x^2)^{-1/2}(-2x)$

$= -\frac{2x}{\sqrt{4 - x^2}}$

26. $f(x) = -3(2 - 9x)^{1/4}$

$f'(x) = -\frac{3}{4}(2 - 9x)^{-3/4}(-9)$

$= \frac{27}{4(2 - 9x)^{3/4}}$

27. $f(x) = (9 - x^2)^{2/3}$

$f'(x) = \frac{2}{3}(9 - x^2)^{-1/3}(-2x)$

$= -\frac{4x}{3(9 - x^2)^{1/3}}$

28. $f(t) = (9t + 2)^{2/3}$

$f'(t) = \frac{2}{3}(9t + 2)^{-1/3}(9)$

$= \frac{6}{\sqrt[3]{9t + 2}}$

29. $y = (x + 2)^{-1/2}$

$\frac{dy}{dx} = -\frac{1}{2}(x + 2)^{-3/2}$

$= -\frac{1}{2(x + 2)^{3/2}}$

30. $g(t) = (t^2 - 2)^{-1/2}$

$g'(t) = -\frac{1}{2}(t^2 - 2)^{-3/2}(2t)$

$= -\frac{t}{(t^2 - 2)^{3/2}}$

31. $g(x) = 3(x^3 - 1)^{-1/3}$

$g'(x) = -(x^3 - 1)^{-4/3}(3x^2)$

$= -\frac{3x^2}{(x^3 - 1)^{4/3}}$

32. $s(x) = (x^2 - 3x + 4)^{-1/2}$

$s'(x) = -\frac{1}{2}(x^2 - 3x + 4)^{-3/2}(2x - 3)$

$= \frac{3 - 2x}{2(x^2 - 3x + 4)^{3/2}}$

33. $y = -1(\sqrt{x} + 1)^{-1}$

$\frac{dy}{dx} = (\sqrt{x} + 1)^{-2}(\frac{1}{2}x^{-1/2})$

$= \frac{1}{2\sqrt{x}(\sqrt{x} + 1)^2}$

34. $y = \frac{1}{2}(t - 3)^{-1/2}$

$\frac{dy}{dx} = -\frac{1}{4}(t - 3)^{-3/2}(1)$

$= -\frac{1}{4(t - 3)^{3/2}}$

35. $y = \dfrac{\sqrt{x} + 1}{x^2 + 1}$

$$\dfrac{dy}{dx} = \dfrac{(x^2 + 1)(1/2)x^{-1/2} - (\sqrt{x} + 1)(2x)}{(x^2 + 1)^2}$$

$$= \dfrac{(x^2 + 1)/(2\sqrt{x}) - 2x(\sqrt{x} + 1)}{(x^2 + 1)^2} = \dfrac{(1 - 3x^2 - 4x^{3/2})}{2\sqrt{x}(x^2 + 1)^2}$$

36. $f(x) = \dfrac{x + 1}{2x - 3}, \qquad f'(x) = \dfrac{(2x - 3) - (x + 1)(2)}{(2x - 3)^2} = -\dfrac{5}{(2x - 3)^2}$

37. $f(t) = \dfrac{3t + 2}{t - 1}, \qquad f'(t) = \dfrac{(t - 1)(3) - (3t + 2)}{(t - 1)^2} = -\dfrac{5}{(t - 1)^2}$

38. $y = (\dfrac{2x}{x + 1})^{1/2}, \qquad \dfrac{dy}{dx} = \dfrac{1}{2}(\dfrac{2x}{x + 1})^{-1/2}\left[\dfrac{2(x + 1) - 2x}{(x + 1)^2}\right] = \dfrac{1}{\sqrt{2x}(x + 1)^{3/2}}$

39. $g(t) = 3t^2(t^2 + 2t - 1)^{-1/2}$

$$g'(t) = 3t^2\left[-\dfrac{1}{2}(t^2 + 2t - 1)^{-3/2}(2t + 2)\right] + (t^2 + 2t - 1)^{-1/2}(6t)$$

$$= 3t\left[\dfrac{-t(t + 1)}{(t^2 + 2t - 1)^{3/2}} + \dfrac{2}{(t^2 + 2t - 1)^{1/2}}\right] = \dfrac{3t(t^2 + 3t - 2)}{(t^2 + 2t - 1)^{3/2}}$$

40. $f(x) = \sqrt{x}(2 - x)^2$

$$f'(x) = \sqrt{x}\,[2(2 - x)(-1)] + (2 - x)^2(\dfrac{1}{2}x^{-1/2}) = (2 - x)\left[-2\sqrt{x} + \dfrac{2 - x}{2\sqrt{x}}\right]$$

$$= \dfrac{2 - x}{2\sqrt{x}}\,[-4x + 2 - x] = \dfrac{(x - 2)(5x - 2)}{2\sqrt{x}}$$

41. $y = \sqrt{\dfrac{x + 1}{x}}$

$$y' = \dfrac{1}{2}(\dfrac{x + 1}{x})^{-1/2}\left[\dfrac{x - (x + 1)}{x^2}\right] = \dfrac{1}{2}(\dfrac{x + 1}{x})^{-1/2}(-\dfrac{1}{x^2}) = -\dfrac{1}{2x^{3/2}\sqrt{x + 1}}$$

42. $y = (t^2 - 9)(t + 2)^{1/2}$

$$y' = (t^2 - 9)\left[\dfrac{1}{2}(t + 2)^{-1/2}\right] + (t + 2)^{1/2}(2t) = \dfrac{t^2 - 9}{2\sqrt{t + 2}} + 2t\sqrt{t + 2}$$

$$= \dfrac{1}{2\sqrt{t + 2}}\,[t^2 - 9 + 4t^2 + 8t] = \dfrac{5t^2 + 8t - 9}{2\sqrt{t + 2}}$$

43. $s(t) = -\dfrac{2}{3}(2 - t)(1 + t)^{1/2}$

$s'(t) = -\dfrac{2}{3}\left[(2 - t)\,\dfrac{1}{2\sqrt{1 + t}} + (-1)\sqrt{1 + t} \right]$

$ = -\dfrac{1}{3\sqrt{1 + t}}(2 - t - 2 - 2t) = \dfrac{t}{\sqrt{1 + t}}$

44. $g(x) = (x - 1)^{1/2} + (x + 1)^{1/2}$

$g'(x) = \dfrac{1}{2}(x - 1)^{-1/2} + \dfrac{1}{2}(x + 1)^{-1/2} = \dfrac{1}{2\sqrt{x - 1}} + \dfrac{1}{2\sqrt{x + 1}}$

45. $f(x) = \sqrt{3x^2 - 2}, \quad (3, 5)$

$f'(x) = \dfrac{1}{2}(3x^2 - 2)^{-1/2}(6x) = \dfrac{3x}{\sqrt{3x^2 - 2}}$

$f'(3) = \dfrac{9}{5}, \quad$ Tangent line: $\; y - 5 = \dfrac{9}{5}(x - 3) \implies 9x - 5y - 2 = 0$

46. $f(x) = x\sqrt{x^2 + 5}, \quad (2, 6)$

$f'(x) = x\left[\dfrac{1}{2}(x^2 + 5)^{-1/2}(2x) \right] + (x^2 + 5)^{1/2}$

$ = \dfrac{x^2}{\sqrt{x^2 + 5}} + \sqrt{x^2 + 5} = \dfrac{2x^2 + 5}{\sqrt{x^2 + 5}}$

$f'(2) = \dfrac{13}{3}, \quad$ Tangent line: $\; y - 6 = \dfrac{13}{3}(x - 2) \implies 13x - 3y - 8 = 0$

47. $f(x) = \sin 2x, \quad (\pi, 0)$
$f'(x) = 2\cos 2x, \quad f'(\pi) = 2$
Tangent line: $\; y = 2(x - \pi) \implies 2x - y - 2\pi = 0$

48. $f(x) = \tan^2 x, \quad (\dfrac{\pi}{4}, 1)$

$f'(x) = 2\tan x \sec^2 x, \quad f'(\dfrac{\pi}{4}) = 2(1)(2) = 4$

Tangent line: $\; y - 1 = 4(x - \dfrac{\pi}{4}) \implies 4x - y + (1 - \pi) = 0$

49. $f(x) = 2(x^2 - 1)^3$

$f'(x) = 6(x^2 - 1)^2(2x) = 12x(x^4 - 2x^2 + 1) = 12x^5 - 24x^3 + 12x$

$f''(x) = 60x^4 - 72x^2 + 12 = 12(5x^2 - 1)(x^2 - 1)$

50. $f(x) = (x - 2)^{-1}$

$f'(x) = -(x - 2)^{-2} = \dfrac{-1}{(x - 2)^2}$

$f''(x) = 2(x - 2)^{-3} = \dfrac{2}{(x - 2)^3}$

51. $f(x) = \sin x^2$
$f'(x) = 2x \cos x^2$
$f''(x) = 2x[2x(-\sin x^2)] + 2\cos x^2 = 2[\cos x^2 - 2x^2 \sin x^2]$

52. $f(t) = \dfrac{\sqrt{t^2 + 1}}{t}$

$f'(t) = \dfrac{(t(1/2)(t^2 + 1)^{-1/2} 2t) - \sqrt{t^2 + 1}}{t^2} = \dfrac{-1}{t^2\sqrt{t^2 + 1}} = -(t^6 + t^4)^{-1/2}$

$f''(t) = \dfrac{1}{2}(t^6 + t^4)^{-3/2}(6t^5 + 4t^3) = \dfrac{3t^2 + 2}{t^3(t^2 + 1)^{3/2}}$

53.(a) $y = \sin x$
$y' = \cos x$
$y'(0) = 1$
1 cycle in $[0, 2\pi]$

(b) $y = \sin 2x$
$y' = 2\cos 2x$
$y'(0) = 2$
2 cycles in $[0, 2\pi]$

54.(a) $y = \sin 3x$
$y' = 3\cos 3x$
$y'(0) = 3$
3 cycles in $[0, 2\pi]$

(b) $y = \sin (1/2) x$
$y' = (1/2) \cos (1/2) x$
$y'(0) = 1/2$
1/2 cycle in $[0, 2\pi]$

55. $y = \cos 3x$

$\dfrac{dy}{dx} = -3\sin 3x$

56. $y = \sin 2x$

$\dfrac{dy}{dx} = 2\cos 2x$

57. $y = 3\tan 4x$

$\dfrac{dy}{dx} = 12\sec^2 4x$

58. $y = 2\cos(\dfrac{x}{2})$

$\dfrac{dy}{dx} = -\sin(\dfrac{x}{2})$

59. $y = \sin \pi x$

$\dfrac{dy}{dx} = \pi \cos \pi x$

60. $y = \sec x^2$

$\dfrac{dy}{dx} = 2x \sec x^2 \tan x^2$

61. $y = \frac{1}{4} \sin^2 2x = \frac{1}{4}(\sin 2x)^2$

$\frac{dy}{dx} = 2(\frac{1}{4})(\sin 2x)(2)\cos 2x$

$= \sin 2x \cos 2x = \frac{1}{2} \sin 4x$

62. $y = 5 \cos^2 \pi x$

$\frac{dy}{dx} = 10 \cos \pi x (-\sin \pi x)(\pi)$

$= -10\pi(\sin \pi x)(\cos \pi x)$

$= -5\pi \sin 2\pi x$

63. $y = \frac{1}{4} \sin (2x)^2 = \frac{1}{4} \sin 4x^2$

$\frac{dy}{dx} = \frac{1}{4}(\cos 4x^2)(8x) = 2x \cos (2x)^2$

64. $y = 5 \cos (\pi x)^2 = 5 \cos \pi^2 x^2$

$\frac{dy}{dx} = -5(\sin \pi^2 x^2)(2\pi^2 x)$

$= -10\pi^2 x \sin (\pi x)^2$

65. $y = \sqrt{\sin x}$

$\frac{dy}{dx} = \frac{1}{2} \sin^{-1/2} x \cos x$

$= \frac{\cos x}{2\sqrt{\sin x}} = \frac{1}{2} \cot x \sqrt{\sin x}$

66. $y = \csc^2 x$

$\frac{dy}{dx} = 2 \csc x(-\csc x \cot x)$

$= -2 \csc^2 x \cot x$

67. $y = \tan (\pi x - \frac{\pi}{2})$

$\frac{dy}{dx} = \pi \sec^2 (\pi x - \frac{\pi}{2}) = \pi \csc^2 (\pi x)$

68. $y = \cot (\frac{x}{2} + \frac{\pi}{4})$

$\frac{dy}{dx} = -\frac{1}{2} \csc^2 (\frac{x}{2} + \frac{\pi}{4})$

69. $y = x \sin (\frac{1}{x})$

$\frac{dy}{dx} = x \cos (\frac{1}{x})(\frac{-1}{x^2}) + \sin (\frac{1}{x})$

$= \sin (\frac{1}{x}) - \frac{1}{x} \cos (\frac{1}{x})$

70. $y = x^2 \sin (\frac{1}{x})$

$\frac{dy}{dx} = x^2 \cos (\frac{1}{x})(\frac{-1}{x^2}) + 2x \sin (\frac{1}{x})$

$= 2x \sin (\frac{1}{x}) - \cos (\frac{1}{x})$

71. $y = \sec^3 2x$

$\frac{dy}{dx} = 3 \sec^2 2x \sec 2x \tan 2x(2)$

$= 6 \sec^3 2x \tan 2x$

72. $y = \frac{\cos x + 1}{x}$

$\frac{dy}{dx} = \frac{x(-\sin x) - (\cos x + 1)}{x^2}$

$= \frac{-x \sin x - \cos x - 1}{x^2}$

73.(a) $F = 132400(331 - v)^{-1}$

$F' = (-1)(132400)(331 - v)^{-2}(-1) = \dfrac{132400}{(331 - v)^2}$

When $v = 30$, $F' \approx 1.461$

(b) $F = 132400(331 + v)^{-1}$

$F' = (-1)(132400)(331 + v)^{-2}(1) = \dfrac{-132400}{(331 + v)^2}$

When $v = 30$, $F' \approx -1.016$

74. $V = \dfrac{7500}{1 + 0.4t + 0.1t^2}$, $V' = \dfrac{0 - 7500(0.4 + 0.2t)}{(1 + 0.4t + 0.1t^2)^2} = - \dfrac{3000 + 1500t}{(1 + 0.4t + 0.1t^2)^2}$

(a) when $t = 1$, $V' = -2000$ (b) when $t = 2$, $V' = -1239.67$

75. $S = C(R^2 - r^2)$, $\dfrac{dS}{dt} = C\left(2R\dfrac{dR}{dt} - 2r\dfrac{dr}{dt}\right)$

Since r is constant, we have $\dfrac{dr}{dt} = 0$ and

$\dfrac{dS}{dt} = (1.76 \times 10^5)(2)(1.2 \times 10^{-2})(10^{-5}) = 4.224 \times 10^{-2} = 0.04224$

76. $f(x) = \sec^2 x$
$f'(x) = 2 \sec x \sec x \tan x = 2 \sec^2 x \tan x$
$g(x) = \tan^2 x$
$g'(x) = 2 \tan x \sec^2 x$

Thus $f'(x) = g'(x)$

3.6
Implicit differentiation

1. $x^2 + y^2 = 16$
$2x + 2yy' = 0$

$y' = \dfrac{-x}{y}$

At $(3, \sqrt{7})$: $y' = \dfrac{-3}{\sqrt{7}} = \dfrac{-3\sqrt{7}}{7}$

2. $x^2 - y^2 = 16$
$2x - 2yy' = 0$

$y' = \dfrac{x}{y}$

At $(4, 0)$: $y' = \dfrac{4}{0}$

undefined (vertical tangent)

3. $xy = 4$
$xy' + y(1) = 0$
$xy' = -y$

$y' = \dfrac{-y}{x}$

At $(-4, -1)$: $y' = -\dfrac{1}{4}$

4. $x^2 - y^3 = 0$
$2x - 3y^2y' = 0$

$y' = \dfrac{2x}{3y^2}$

At $(-1, 1)$: $y' = -\dfrac{2}{3}$

5. $x^{1/2} + y^{1/2} = 9$

$\dfrac{1}{2}x^{-1/2} + \dfrac{1}{2}y^{-1/2}y' = 0$

$y' = -\dfrac{x^{-1/2}}{y^{-1/2}} = -\sqrt{\dfrac{y}{x}}$

At $(16, 25)$: $y' = -\dfrac{5}{4}$

6. $x^3 + y^3 = 8$
$3x^2 + 3y^2y' = 0$

$y' = -\dfrac{x^2}{y^2}$

At $(0, 2)$: $y' = 0$

7. $x^3 - xy + y^2 = 4$
$3x^2 - xy' - y + 2yy' = 0$
$(2y - x)y' = y - 3x^2$

$y' = \dfrac{y - 3x^2}{2y - x}$

At $(0, -2)$: $y' = \dfrac{1}{2}$

8. $x^2y + y^2x = -2$
$x^2y' + 2xy + y^2 + 2yxy' = 0$
$(x^2 + 2xy)y' = -(y^2 + 2xy)$

$y' = \dfrac{-y(y + 2x)}{x(x + 2y)}$

At $(2, -1)$: $y' = \dfrac{3}{0}$

undefined (vertical tangent)

9. $y^2 = \dfrac{x^2 - 9}{x^2 + 9}$, $\quad 2yy' = \dfrac{(x^2 + 9)(2x) - (x^2 - 9)2x}{(x^2 + 9)^2}$

$y' = \dfrac{18x}{(x^2 + 9)^2 y}$ $\qquad$ At $(3, 0)$: $y' = \dfrac{54}{0}$ undefined (vertical tangent)

10. $(x + y)^3 = x^3 + y^3$, $\quad x^3 + 3x^2y + 3xy^2 + y^3 = x^3 + y^3$,

$3x^2y + 3xy^2 = 0$, $\quad x^2y + xy^2 = 0$, $\quad x^2y' + 2xy + 2xyy' + y^2 = 0$,

$(x^2 + 2xy)y' = -(y^2 + 2xy)$, $\quad y' = -\dfrac{y(y + 2x)}{x(x + 2y)}$ $\quad$ At $(-1, 1)$: $y' = -1$

11. $x^3y^3 - y - x = 0$, $\quad 3x^3y^2y' + 3x^2y^3 - y' - 1 = 0$,

$(3x^3y^2 - 1)y' = 1 - 3x^2y^3$, $\quad y' = \dfrac{1 - 3x^2y^3}{3x^3y^2 - 1}$ $\qquad$ At $(0, 0)$: $y' = -1$

12. $(xy)^{1/2} - x + 2y = 0$, $\quad \frac{1}{2}(xy)^{-1/2}(xy' + y) - 1 + 2y' = 0$,

$\frac{x}{2\sqrt{xy}}y' + \frac{y}{2\sqrt{xy}} - 1 + 2y' = 0$, $\quad xy' + y - 2\sqrt{xy} + 4\sqrt{xy}y' = 0$,

$y' = \frac{2\sqrt{xy} - y}{4\sqrt{xy} + x}$ $\quad$ At $(4, 1)$: $y' = \frac{1}{4}$

13. $x^{2/3} + y^{2/3} = 5$, $\quad \frac{2}{3}x^{-1/3} + \frac{2}{3}y^{-1/3}y' = 0$, $\quad y' = \frac{-x^{-1/3}}{y^{-1/3}} = -\sqrt[3]{\frac{y}{x}}$

At $(8, 1)$: $y' = -\frac{1}{2}$

14. $x^3 + y^3 - 2xy = 0$, $\quad 3x^2 + 3y^2y' - 2xy' - 2y = 0$,

$(3y^2 - 2x)y' = 2y - 3x^2$, $\quad y' = \frac{2y - 3x^2}{3y^2 - 2x}$ $\quad$ At $(1, 1)$: $y' = -1$

15. $x^3 - 2x^2y + 3xy^2 = 38$, $\quad 3x^2 - 2x^2y' - 4xy + 6xyy' + 3y^2 = 0$,

$2x(3y - x)y' = 4xy - 3x^2 - 3y^2$, $\quad y' = \frac{4xy - 3x^2 - 3y^2}{2x(3y - x)}$

At $(2, 3)$: $y' = \frac{-15}{28}$

16. $2\sin x \cos y = 1$, $\quad 2[\sin x(-\sin y)y' + \cos y(\cos x)] = 0$

$y' = \frac{\cos x \cos y}{\sin x \sin y} = \tan x \cot y$ $\quad$ At $(\frac{\pi}{4}, \frac{\pi}{4})$: $y' = 1$

17. $\sin x + \cos 2y = 2$, $\quad \cos x - 2\sin 2y \, y' = 0$

$y' = \frac{\cos x}{2\sin 2y}$ $\quad$ At $(\frac{\pi}{2}, 0)$: y' is undefined

18. $(\sin \pi x + \cos \pi y)^2 = 2$
$2(\sin \pi x + \cos \pi y)[\pi \cos \pi x - \pi(\sin \pi y)y'] = 0$
$(\sin \pi x + \cos \pi y)[\pi \cos \pi x - \pi(\sin \pi y)y'] = 0$

$y' = \frac{\cos \pi x}{\sin \pi y}$ $\quad$ At $(\frac{1}{4}, \frac{1}{4})$: $y' = 1$

19. $\sin x = x(1 + \tan y)$, $\quad \cos x = x(\sec^2 y)y' + (1 + \tan y)(1)$

$y' = \frac{\cos x - \tan y - 1}{x \sec^2 y}$ $\quad$ At $(\pi, \frac{3\pi}{4})$: $y' = \frac{-1 + 1 - 1}{\pi(2)} = -\frac{1}{2\pi}$

20. $\cot y = x - y$, $(-\csc^2 y)y' = 1 - y'$

$y' = \dfrac{1}{1 - \csc^2 y} = \dfrac{1}{-\cot^2 y} = -\tan^2 y$ At $(\dfrac{\pi}{2}, \dfrac{\pi}{2})$: y' is undefined

21. $\tan (x + y) = x$, $(1 + y') \sec^2 (x + y) = 1$

$y' = \dfrac{1 - \sec^2 (x + y)}{\sec^2 (x + y)} = \dfrac{-\tan^2 (x + y)}{\tan^2 (x + y) + 1} = -\dfrac{x^2}{x^2 + 1}$

At $(0, 0)$: $y' = 0$

22. $x \cos y = 1$, $x[-y' \sin y] + \cos y = 0$

$y' = \dfrac{\cos y}{x \sin y} = \dfrac{1}{x} \cot y = \dfrac{\cot y}{x}$ At $(2, \dfrac{\pi}{3})$: $y' = \dfrac{1}{2\sqrt{3}}$

23. $y = \sin (xy)$, $y' = [xy' + y] \cos (xy)$, $y' = \dfrac{y \cos (xy)}{1 - x \cos (xy)}$

At $(\dfrac{\pi}{2}, 1)$: $y' = 0$

24. $x = \sec \dfrac{1}{y}$, $1 = -\dfrac{y'}{y^2} \sec \dfrac{1}{y} \tan \dfrac{1}{y}$, $y' = \dfrac{-y^2}{\sec (1/y) \tan (1/y)}$

At $(\sqrt{2}, \dfrac{4}{\pi})$: $y' = \dfrac{-16/\pi^2}{(\sqrt{2})(1)} = -\dfrac{8\sqrt{2}}{\pi^2}$

25. $x^2 + y^2 = 16$
$y^2 = 16 - x^2$
$y = \pm \sqrt{16 - x^2}$

Explicitly: $\dfrac{dy}{dx} = \pm\dfrac{1}{2}(16 - x^2)^{-1/2}(-2x)$

$= \dfrac{\mp x}{\sqrt{16 - x^2}} = -\dfrac{x}{y}$

Implicitly: $2x + 2yy' = 0$

$y' = -\dfrac{x}{y}$

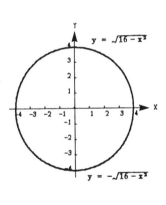

26. $(x^2 - 4x + 4) + (y^2 + 6y + 9) = -9 + 4 + 9$
    $(x - 2)^2 + (y + 3)^2 = 4$
    $(y + 3)^2 = 4 - (x - 2)^2$
    $y = -3 \pm \sqrt{4 - (x - 2)^2}$

    Explicitly:

    $$\frac{dy}{dx} = \mp \frac{1}{2} [4 - (x - 2)^2]^{-1/2} (-2)(x - 2)$$

    $$= \frac{\mp (x - 2)}{(\sqrt{4 - (x - 2)^2} - 3) + 3} = \frac{-(x - 2)}{y + 3}$$

    Implicitly: $2x + 2yy' - 4 + 6y' = 0$,
    $\quad\quad\quad (2y + 6)y' = -2(x - 2)$

    $$y' = \frac{-(x - 2)}{y + 3}$$

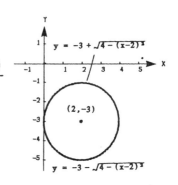

27. $16y^2 = 144 - 9x^2$, $\quad y^2 = \frac{1}{16}(144 - 9x^2)$,

    $y = \pm \frac{1}{4} \sqrt{144 - 9x^2}$

    Explicitly:

    $$\frac{dy}{dx} = \pm \frac{1}{8}(144 - 9x^2)^{-1/2}(-18x) = \frac{\mp 9x}{4\sqrt{144 - 9x^2}}$$

    $$= \frac{\mp 9x}{16(1/4)\sqrt{144 - 9x^2}} = \frac{-9x}{16y}$$

    Implicitly: $18x + 32yy' = 0$, $\quad y' = \frac{-9x}{16y}$

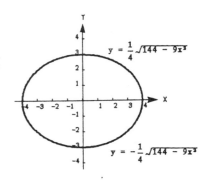

28. $y^2 = 1 + \frac{x^2}{4} = \frac{x^2 + 4}{4}$, $\quad y = \pm \frac{1}{2}\sqrt{x^2 + 4}$

    Explicitly:

    $$\frac{dy}{dx} = \pm \frac{1}{4}(x^2 + 4)^{-1/2}(2x) = \frac{\pm x}{2\sqrt{x^2 + 4}}$$

    $$= \frac{\pm x}{4(1/2)\sqrt{x^2 + 4}} = \frac{x}{4y}$$

    Implicitly:

    $$2yy' - \frac{1}{2}x = 0, \quad y' = \frac{x}{4y}$$

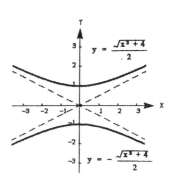

29. $x^2 + xy = 5$, $\quad 2x + xy' + y = 0$ $\quad$ or $\quad y' = \dfrac{-(2x + y)}{x}$

$2 + xy'' + y' + y' = 0$, $\quad xy'' = -2(1 + y')$,

$y'' = \dfrac{-2[1 - (2x + y)/x]}{x} = \dfrac{2(x + y)}{x^2}$, $\qquad y'' = \dfrac{2(x + y)}{x^2} \cdot \dfrac{x}{x} = \dfrac{10}{x^3}$

30. $x^2y^2 - 2x = 3$, $\quad 2x^2yy' + 2xy^2 - 2 = 0$, $\quad x^2yy' + xy^2 - 1 = 0$

or $y' = \dfrac{1 - xy^2}{x^2y}$, $\quad 2xyy' + x^2y'^2 + x^2yy'' + 2xyy' + y^2 = 0$,

$4xyy' + x^2y'^2 + x^2yy'' + y^2 = 0$,

$\dfrac{4 - 4xy^2}{x} + \dfrac{(1 - xy^2)^2}{x^2y^2} + x^2yy'' + y^2 = 0$,

$4xy^2 - 4x^2y^4 + 1 - 2xy^2 + x^2y^4 + x^4y^3y'' + x^2y^4 = 0$

$x^4y^3y'' = 2x^2y^4 - 2xy^2 - 1$, $\qquad y'' = \dfrac{2x^2y^4 - 2xy^2 - 1}{x^4y^3}$

31. $x^2 - y^2 = 16$, $\quad 2x - 2yy' = 0$ $\quad$ or $\quad y' = \dfrac{x}{y}$

$x - yy' = 0$, $\quad 1 - yy'' - y'^2 = 0$, $\quad y'' = \dfrac{y^2 - x^2}{y^3} = \dfrac{-16}{y^3}$

32. $1 - xy = x - y$, $\quad y - xy = x - 1$, $\quad y = \dfrac{x - 1}{1 - x} = -1$

$y' = 0$, $\quad y'' = 0$

33. $y^2 = x^3$, $\quad 2yy' = 3x^2$, $\quad y' = \dfrac{3x^2}{2y} = \dfrac{3x^2}{2y} \cdot \dfrac{xy}{xy} = \dfrac{3y}{2x} \cdot \dfrac{x^3}{y^2} = \dfrac{3y}{2x}$

$y'' = \dfrac{2x(3y') - 3y(2)}{4x^2} = \dfrac{3y}{4x^2} = \dfrac{3x}{4y}$

34. $y^2 = 4x$, $\quad 2yy' = 4$, $\quad y' = \dfrac{2}{y}$, $\quad y'' = -2y^{-2}y' = \left[\dfrac{-2}{y^2}\right] \cdot \dfrac{2}{y} = \dfrac{-4}{y^3}$

35. $x^2 + y^2 = 25$, $\quad y' = \dfrac{-x}{y}$

(a) At $(4, 3)$ Tangent line: $y - 3 = \dfrac{-4}{3}(x - 4)$ ⟹ $4x + 3y - 25 = 0$

$\qquad\qquad$ Normal line: $y - 3 = \dfrac{3}{4}(x - 4)$ ⟹ $3x - 4y = 0$

(b) At $(-3, 4)$ Tangent line: $y - 4 = \dfrac{3}{4}(x + 3)$ ⟹ $3x - 4y + 25 = 0$

$\qquad\qquad$ Normal line: $y - 4 = \dfrac{-4}{3}(x + 3)$ ⟹ $4x + 3y = 0$

36. $x^2 + y^2 = 9$, $y' = \dfrac{-x}{y}$

    (a) At (0, 3) Tangent line: $y = 3$
                           Normal line: $x = 0$

    (b) At $(2, \sqrt{5})$

          Tangent line: $y - \sqrt{5} = \dfrac{-2}{\sqrt{5}}(x - 2)$ ➡ $2x + \sqrt{5}y - 9 = 0$

          Normal line: $y - \sqrt{5} = \dfrac{\sqrt{5}}{2}(x - 2)$ ➡ $\sqrt{5}x - 2y = 0$

37. $25x^2 + 16y^2 + 200x - 160y + 400 = 0$

    $50x + 32yy' + 200 - 160y' = 0$, $y' = \dfrac{200 + 50x}{160 - 32y}$

    Horizontal tangents occur when $x = -4$:
    $25(16) + 16y^2 + 200(-4) - 160y + 400 = 0$
    $y(y - 10) = 0$ ➡ $y = 0, 10$

    Horizontal tangents: $(-4, 0)$, $(-4, 10)$

    Vertical tangents occur when $y = 5$:
    $25x^2 + 400 + 200x - 800 + 400 = 0$
    $25x(x + 8) = 0$ ➡ $x = 0, -8$

    Vertical tangents: $(0, 5)$, $(-8, 5)$

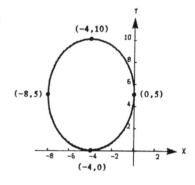

38. $4x^2 + y^2 - 8x + 4y + 4 = 0$

    $8x + 2yy' - 8 + 4y' = 0$, $y' = \dfrac{8 - 8x}{2y + 4} = \dfrac{4 - 4x}{y + 2}$

    Horizontal tangents occur when $x = 1$
    $4(1)^2 + y^2 - 8(1) + 4y + 4 = 0$
    $y^2 + 4y = y(y + 4) = 0$ ➡ $y = 0, -4$

    Horizontal tangents: $(1, 0)$, $(1, -4)$

    Vertical tangents occur when $y = -2$
    $4x^2 + (-2)^2 - 8x + 4(-2) + 4 = 0$
    $4x^2 - 8x = 4x(x - 2) = 0$ ➡ $x = 0, 2$

    Vertical tangents: $(0, -2)$, $(2, -2)$

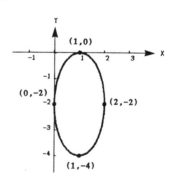

39. Find the points of intersection by letting $y^2 = 4x$ in the equation

    $2x^2 + y^2 = 6$: $2x^2 + 4x = 6$, $(x + 3)(x - 1) = 0$

    The curves intersect at $(1, \pm 2)$

    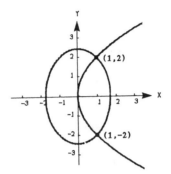

Ellipse:	Parabola:
$4x + 2yy' = 0$	$2yy' = 4$
$y' = -2x/y$	$y' = 2/y$

    At $(1, 2)$ the slopes are:
      $y' = -1$ $y' = 1$

    At $(1, -2)$ the slopes are:
      $y' = 1$ $y' = -1$

    Tangents are perpendicular

40. Find the points of intersection by letting $y^2 = x^3$ in the equation

    $2x^2 + 3y^2 = 5$: $2x^2 + 3x^3 = 5$, $3x^3 + 2x^2 - 5 = 0$

    Intersect when $x = 1$ Points of Intersection: $(1, \pm 1)$

    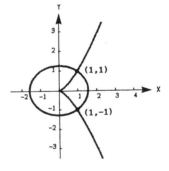

$y^2 = x^3$:	$2x^2 + 3y^2 = 5$:
$2yy' = 3x^2$	$4x + 6yy' = 0$
$y' = 3x^2/(2y)$	$y' = -2x/(3y)$

    At $(1, 1)$ the slopes are:
      $y' = 3/2$ $y' = -2/3$

    At $(1, -1)$ the slopes are:
      $y' = -3/2$ $y' = 2/3$

    Tangents are perpendicular

·41. $y = -x$ and $x = \sin y$

    Point of intersection: $(0, 0)$

    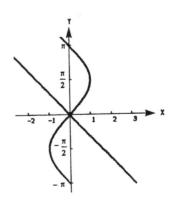

$y = -x$:	$x = \sin y$:
$y' = -1$	$1 = y' \cos y$
	$y' = \sec y$

    At $(0, 0)$ the slopes are:
      $y' = -1$ $y' = 1$

    Tangents are perpendicular

42. Rewriting each equation and differentiating,

$$x^3 = 3(y - 1) \qquad x(3y - 29) = 3$$

$$y = \frac{x^3}{3} + 1 \qquad y = \frac{1}{3}(\frac{3}{x} + 29)$$

$$y' = x^2 \qquad y' = -\frac{1}{x^2}$$

For each value of x, the derivatives are negative reciprocals of each other. Thus the tangent lines are orthogonal at all points of intersection.

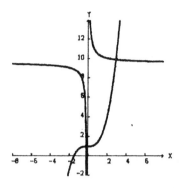

43. $y^2 = 4x$, $2yy' = 4$, $y' = 2/y = 1$ at $(1, 2)$

Equation of normal at $(1, 2)$ is $y - 2 = -1(x - 1)$, $y = 3 - x$
The centers of the circles must be on the normal and at a distance of 4 units from $(1, 2)$. Therefore:

$$(x - 1)^2 + [(3 - x) - 2]^2 = 16, \quad 2(x - 1)^2 = 16, \quad x = 1 \pm 2\sqrt{2}$$

Centers of the circles: $(1 + 2\sqrt{2}, 2 - 2\sqrt{2})$ and $(1 - 2\sqrt{2}, 2 + 2\sqrt{2})$

Equations: $(x - 1 - 2\sqrt{2})^2 + (y - 2 + 2\sqrt{2})^2 = 16$
$(x - 1 + 2\sqrt{2})^2 + (y - 2 - 2\sqrt{2})^2 = 16$

44. $2x + 2yy' = 0$, $y' = \frac{-x}{y}$ = slope of tangent line

$\frac{y}{x}$ = slope of normal line. Any point on the circumference of the circle

will have the form $(x_0, \pm\sqrt{r^2 - x_0^2})$

Equation of normal line: $y \mp \sqrt{r^2 - x_0^2} = \frac{\pm\sqrt{r^2 - x_0^2}}{x_0}(x - x_0)$

$$y = \frac{\pm\sqrt{r^2 - x_0}}{x_0}x \mp \frac{\sqrt{r^2 - x_0^2}}{1} \pm \sqrt{r^2 - x_0^2} = \frac{\pm\sqrt{r^2 - x_0^2}}{x_0}x$$

which passes through $(0, 0)$.

45. $\tan y = x$

$y' \sec^2 y = 1$

$y' = \frac{1}{\sec^2 y}, \quad -\frac{\pi}{2} < y < \frac{\pi}{2}$

$\sec^2 y = 1 + \tan^2 y = 1 + x^2$

$y' = \frac{1}{1 + x^2}$

46. $(x - h)^2 + (y - k)^2 = r^2$

$2(x - h) + 2(y - k)y' = 0$

$y' = \frac{-2(x - h)}{2(y - k)}$

$= -\frac{x - h}{y - k}$

3.7
Related rates

1. $y = \sqrt{x}$, $\dfrac{dy}{dt} = (\dfrac{1}{2\sqrt{x}})\dfrac{dx}{dt}$, $\dfrac{dx}{dt} = 2\sqrt{x}\dfrac{dy}{dt}$

    (a) When $x = 4$ and $\dfrac{dx}{dt} = 3$,   (b) When $x = 25$ and $\dfrac{dy}{dt} = 2$,

        $\dfrac{dy}{dt} = \dfrac{1}{2\sqrt{4}}(3) = \dfrac{3}{4}$ $\dfrac{dx}{dt} = 2\sqrt{25}(2) = 20$

2. $y = x^2 - 3x$, $\dfrac{dy}{dt} = (2x - 3)\dfrac{dx}{dt}$, $\dfrac{dx}{dt} = (\dfrac{1}{2x - 3})\dfrac{dy}{dt}$

    (a) When $x = 3$ and $\dfrac{dx}{dt} = 2$,   (b) When $x = 1$ and $\dfrac{dy}{dt} = 5$,

        $\dfrac{dy}{dt} = [2(3) - 3](2) = 6$ $\dfrac{dx}{dt} = \dfrac{1}{2(1) - 3}(5) = -5$

3. $xy = 4$, $x\dfrac{dy}{dt} + y\dfrac{dx}{dt} = 0$, $\dfrac{dy}{dt} = (-\dfrac{y}{x})\dfrac{dx}{dt}$, $\dfrac{dx}{dt} = (-\dfrac{x}{y})\dfrac{dy}{dt}$

    (a) When $x = 8$, $y = \dfrac{1}{2}$, and $\dfrac{dx}{dt} = 10$, $\dfrac{dy}{dt} = -\dfrac{1/2}{8}(10) = -\dfrac{5}{8}$

    (b) When $x = 1$, $y = 4$, and $\dfrac{dy}{dt} = -6$, $\dfrac{dx}{dt} = -\dfrac{1}{4}(-6) = \dfrac{3}{2}$

4. $x^2 + y^2 = 25$, $2x\dfrac{dx}{dt} + 2y\dfrac{dy}{dt} = 0$, $\dfrac{dy}{dt} = (-\dfrac{x}{y})\dfrac{dx}{dt}$, $\dfrac{dx}{dt} = (-\dfrac{y}{x})\dfrac{dy}{dt}$

    (a) When $x = 3$, $y = 4$, and $\dfrac{dx}{dt} = 8$, $\dfrac{dy}{dt} = -\dfrac{3}{4}(8) = -6$

    (b) When $x = 4$, $y = 3$, and $\dfrac{dy}{dt} = -2$, $\dfrac{dx}{dt} = -\dfrac{3}{4}(-2) = \dfrac{3}{2}$

5. $A = \pi r^2$, $\dfrac{dr}{dt} = 2$, $\dfrac{dA}{dt} = 2\pi r\dfrac{dr}{dt}$

    (a) When $r = 6$, $\dfrac{dA}{dt} = 2\pi(6)(2) = 24\pi$ in²/min

    (b) When $r = 24$, $\dfrac{dA}{dt} = 2\pi(24)(2) = 96\pi$ in²/min

6. $V = \frac{4}{3}\pi r^3$, $\frac{dr}{dt} = 2$, $\frac{dV}{dt} = 4\pi r^2 \frac{dr}{dt}$

   (a) When $r = 6$, $\frac{dV}{dt} = 4\pi(6)^2(2) = 288\pi$ in³/min

   (b) When $r = 24$, $\frac{dV}{dt} = 4\pi(24)^2(2) = 4608\pi$ in³/min

7. $A = \pi r^2$, $\frac{dA}{dt} = 2\pi r \frac{dr}{dt}$ If $\frac{dr}{dt}$ is constant, $\frac{dA}{dt}$ is proportional to r

8. $V = \frac{4}{3}\pi r^3$, $\frac{dV}{dt} = 4\pi r^2 \frac{dr}{dt}$ If $\frac{dr}{dt}$ is constant, $\frac{dV}{dt}$ is proportional to r²

9. $V = \frac{4}{3}\pi r^3$, $\frac{dV}{dt} = 20$

   $\frac{dV}{dt} = 4\pi r^2 \frac{dr}{dt}$, $\frac{dr}{dt} = \left(\frac{1}{4\pi r^2}\right)\frac{dV}{dt}$

   (a) When $r = 1$, $\frac{dr}{dt} = \frac{1}{4\pi(1)^2}(20) = \frac{5}{\pi}$ ft/min

   (b) When $r = 2$, $\frac{dr}{dt} = \frac{1}{4\pi(2)^2}(20) = \frac{5}{4\pi}$ ft/min

10. $V = \frac{1}{3}\pi r^2 h = \frac{1}{3}\pi r^2(3r) = \pi r^3$, $\frac{dr}{dt} = 2$

    $\frac{dV}{dt} = 3\pi r^2 \frac{dr}{dt}$

    (a) When $r = 6$, $\frac{dV}{dt} = 3\pi(6)^2(2) = 216\pi$ in³/min

    (b) When $r = 24$, $\frac{dV}{dt} = 3\pi(24)^2(2) = 3456\pi$ in³/min

11. $V = \frac{1}{3}\pi r^2 h = \frac{1}{3}\pi(\frac{9}{4}h^2)h$ [since $2r = 3h$]

    $= \frac{3\pi}{4}h^3$, $\frac{dV}{dt} = 10$

    $\frac{dV}{dt} = \frac{9\pi}{4}h^2 \frac{dh}{dt}$ ⟹ $\frac{dh}{dt} = \frac{4(dV/dt)}{9\pi h^2}$

    When $h = 15$, $\frac{dh}{dt} = \frac{4(10)}{9\pi(15)^2} = \frac{8}{405\pi}$ ft/min

12. $V = \frac{1}{3}\pi r^2 h = \frac{1}{3}\pi \frac{25}{144} h^3 = \frac{25\pi}{3(144)} h^3$

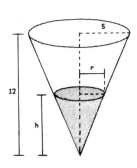

$r = \frac{5}{12}h, \quad \frac{dV}{dt} = 10$

$\frac{dV}{dt} = \frac{25\pi}{144} h^2 \frac{dh}{dt} \implies \frac{dh}{dt} = (\frac{144}{25\pi h^2})\frac{dV}{dt}$

When $h = 8$, $\frac{dh}{dt} = \frac{144}{25\pi(64)}(10) = \frac{9}{10\pi}$ ft/min

13. $V = x^3, \quad \frac{dx}{dt} = 3, \quad \frac{dV}{dt} = 3x^2 \frac{dx}{dt}$

 (a) When $x = 1$, $\frac{dV}{dt} = 3(1)^2(3) = 9$ cm³/sec

 (b) When $x = 10$, $\frac{dV}{dt} = 3(10)^2(3) = 900$ cm³/sec

14. $s = 6x^2, \quad \frac{dx}{dt} = 3, \quad \frac{ds}{dt} = 12x \frac{dx}{dt}$

 (a) When $x = 1$, $\frac{ds}{dt} = 12(1)(3) = 36$ cm²/sec

 (b) When $x = 10$, $\frac{ds}{dt} = 12(10)(3) = 360$ cm²/sec

15. $y = x^2, \quad \frac{dx}{dt} = 2, \quad \frac{dy}{dt} = 2x \frac{dx}{dt}$

 (a) When $x = 0$, $\frac{dy}{dt} = 2(0)(2) = 0$ cm/min

 (b) When $x = 3$, $\frac{dy}{dt} = 2(3)(2) = 12$ cm/min

16. $y = x^2$, $\quad \dfrac{dx}{dt} = 2$, $\quad L^2 = x^2 + y^2 = x^2 + x^4$, $\qquad 2L\dfrac{dL}{dt} = (2x + 4x^3)\dfrac{dx}{dt}$

$\dfrac{dL}{dt} = \dfrac{(x + 2x^3)\,dx/dt}{L}$

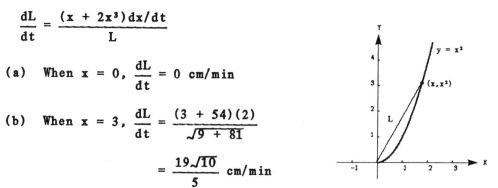

(a) When $x = 0$, $\dfrac{dL}{dt} = 0$ cm/min

(b) When $x = 3$, $\dfrac{dL}{dt} = \dfrac{(3 + 54)(2)}{\sqrt{9 + 81}}$

$\qquad\qquad = \dfrac{19\sqrt{10}}{5}$ cm/min

17. $y = \dfrac{1}{1 + x^2}$, $\quad \dfrac{dx}{dt} = 2$

$\dfrac{dy}{dt} = \left[\dfrac{-2x}{(1 + x^2)^2}\right]\dfrac{dx}{dt}$

(a) When $x = -2$, $\dfrac{dy}{dt} = \dfrac{-2(-2)(2)}{25} = \dfrac{8}{25}$ cm/min

(b) When $x = 0$, $\dfrac{dy}{dt} = 0$ cm/min

(c) When $x = 2$, $\dfrac{dy}{dt} = \dfrac{-2(2)(2)}{25} = \dfrac{-8}{25}$ cm/min

(d) When $x = 10$, $\dfrac{dy}{dt} = \dfrac{-2(10)(2)}{(101)^2} = \dfrac{-40}{10201}$

$\qquad\qquad = -0.0039$ cm/min

18. $y = \sin x$, $\quad \dfrac{dx}{dt} = 2$

$\dfrac{dy}{dt} = \cos x \dfrac{dx}{dt}$

(a) When $x = \dfrac{\pi}{6}$, $\dfrac{dy}{dt} = \left(\cos\dfrac{\pi}{6}\right)(2) = \sqrt{3}$ cm/min

(b) When $x = \dfrac{\pi}{3}$, $\dfrac{dy}{dt} = \left(\cos\dfrac{\pi}{3}\right)(2) = 1$ cm/min

(c) When $x = \dfrac{\pi}{4}$, $\dfrac{dy}{dt} = \left(\cos\dfrac{\pi}{4}\right)(2) = \sqrt{2}$ cm/min

(d) When $x = \dfrac{\pi}{2}$, $\dfrac{dy}{dt} = \left(\cos\dfrac{\pi}{2}\right)(2) = 0$ cm/min

19.(a) Total volume = $(1/2)(40)(5)(20) + (40)(4)(20) = 5200$ ft^3

Volume of 4 ft. of water = $(1/2)(32)(4)(20) = 1280$ ft^3

% of pool filled = $\dfrac{1280}{5200}(100\%) = 24.6\%$

(b) $(b = 8h)$, $0 \le h \le 5$

$V = (1/2)bh(20) = 10bh = 80h^2$

$\dfrac{dV}{dt} = 160h\dfrac{dh}{dt}$ ⇨ $\dfrac{dh}{dt} = (\dfrac{1}{160h})\dfrac{dV}{dt}$

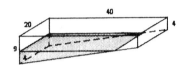

When $h = 4$ and $\dfrac{dV}{dt} = 10$,

$\dfrac{dh}{dt} = \dfrac{1}{160(4)}(10) = \dfrac{1}{64}$ ft/min

20. $V = (1/2)bh(12) = 6bh = 6h^2$ (since $b = h$)

$\dfrac{dV}{dt} = 12h\dfrac{dh}{dt}$ ⇨ $\dfrac{dh}{dt} = (\dfrac{1}{12h})\dfrac{dV}{dt}$

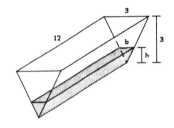

When $h = 1$ and $\dfrac{dV}{dt} = 2$,

$\dfrac{dh}{dt} = \dfrac{1}{12}(2) = \dfrac{1}{6}$ ft/min = 2 in/min

21. $x^2 + y^2 = 25^2$, $2x\dfrac{dx}{dt} + 2y\dfrac{dy}{dt} = 0$,

$\dfrac{dy}{dt} = \dfrac{-x}{y}\dfrac{dx}{dt} = \dfrac{-2x}{y}$ since $\dfrac{dx}{dt} = 2$

(a) When $x = 7$, $y = \sqrt{576} = 24$ ft

$\dfrac{dy}{dt} = \dfrac{-2(7)}{24} = \dfrac{-7}{12}$ ft/sec

(b) When $x = 15$, $y = \sqrt{400} = 20$

$\dfrac{dy}{dt} = \dfrac{-2(15)}{20} = \dfrac{-3}{2}$ ft/sec

(c) When $x = 24$, $y = 7$

$\dfrac{dy}{dt} = \dfrac{-2(24)}{7} = \dfrac{-48}{7}$ ft/sec

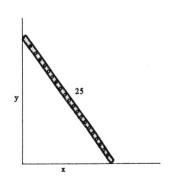

22. $L^2 = 144 + x^2$, $\quad 2L\dfrac{dL}{dt} = 2x\dfrac{dx}{dt}$,

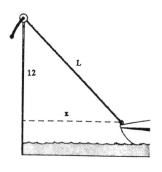

$\dfrac{dx}{dt} = \dfrac{L}{x}\dfrac{dL}{dt} = -\dfrac{4L}{x}$ $\quad$ since $\dfrac{dL}{dt} = -4$ ft/sec

When $L = 13$, $\quad x = \sqrt{L^2 - 144}$
$\qquad\qquad\qquad\qquad = \sqrt{169 - 144} = 5$

$\dfrac{dx}{dt} = -\dfrac{4(13)}{5} = -\dfrac{52}{5} = -10.4$ ft/sec

Speed of the boat increases as it approaches the dock.

23.(a) $L^2 = x^2 + y^2$, $\quad \dfrac{dx}{dt} = -450$, $\quad \dfrac{dy}{dt} = -600$

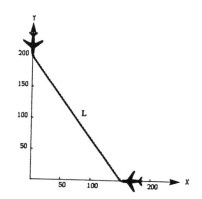

$\dfrac{dL}{dt} = \dfrac{x(dx/dt) + y(dy/dt)}{L}$

When $x = 150$ and $y = 200$,

$\dfrac{dL}{dt} = \dfrac{150(-450) + 200(-600)}{250} = -750$ mph

(b) $\quad t = \dfrac{250}{750} = \dfrac{1}{3}$ hr $= 20$ min

24. $x^2 + y^2 = L^2$, $\quad 2x\dfrac{dx}{dt} + 2y\dfrac{dy}{dt} = 2L\dfrac{dL}{dt}$,

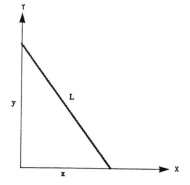

$\dfrac{dL}{dt} = \dfrac{x(dx/dt) + y(dy/dt)}{L} = \dfrac{rx + Ry}{L}$

25. $\quad s^2 = 90^2 + x^2$, $\quad x = 30$, $\quad \dfrac{dx}{dt} = 28$

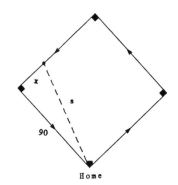

$2s\dfrac{ds}{dt} = 2x\dfrac{dx}{dt}$ $\quad\Longrightarrow\quad$ $\dfrac{ds}{dt} = \dfrac{x}{s}\dfrac{dx}{dt}$

When $x = 30$, $s = \sqrt{90^2 + 30^2} = 30\sqrt{10}$

$\dfrac{ds}{dt} = \dfrac{30}{30\sqrt{10}}(28) = \dfrac{28}{\sqrt{10}} \approx 8.85$ ft/sec

26. $s^2 = 90^2 + x^2$, $\quad x = 60$, $\quad \dfrac{dx}{dt} = 28$

$\dfrac{ds}{dt} = \dfrac{x}{s} \dfrac{dx}{dt}$

When $x = 60$, $s = \sqrt{90^2 + 60^2} = 30\sqrt{13}$

$\dfrac{ds}{dt} = \dfrac{60}{30\sqrt{13}}(28) = \dfrac{56}{\sqrt{13}} \approx 15.5$ ft/sec

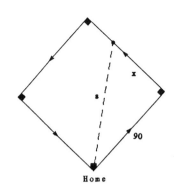

Home

27.(a) $\dfrac{15}{6} = \dfrac{y}{y - x}$, $\quad \dfrac{dx}{dt} = 5$

$y = \dfrac{5}{3} x$

$\dfrac{dy}{dt} = \dfrac{5}{3} \dfrac{dx}{dt} = \dfrac{5}{3}(5) = \dfrac{25}{3}$ ft/sec

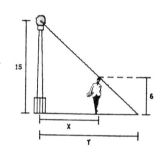

(b) $\dfrac{d(y - x)}{dt} = \dfrac{dy}{dt} - \dfrac{dx}{dt}$

$\qquad\qquad = \dfrac{25}{3} - 5 = \dfrac{10}{3}$ ft/sec

28. $x^2 + y^2 = s^2$, $\quad x = \sqrt{s^2 - y^2}$

$\dfrac{ds}{dt} = -240$ mph

$y = 6$ mi

$x = \sqrt{s^2 - 36}$

$\dfrac{dx}{dt} = (\dfrac{1}{2})(s^2 - 36)^{-1/2}(2s\dfrac{ds}{dt})$

$\qquad = \dfrac{s}{\sqrt{s^2 - 36}} \dfrac{ds}{dt}$

When $s = 10$, $\dfrac{dx}{dt} = \dfrac{10}{\sqrt{10^2 - 36}}(-240)$

$\dfrac{dx}{dt} = \dfrac{10}{8}(-240) = -300$ mph

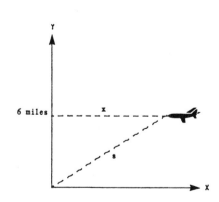

29. Since the evaporation rate is proportional to the surface area

$\dfrac{dV}{dt} = k(4\pi r^2)$. However, since $V = \dfrac{4}{3}\pi r^3$ we have $\dfrac{dV}{dt} = 4\pi r^2 \dfrac{dr}{dt}$.

Therefore $k(4\pi r^2) = 4\pi r^2 \dfrac{dr}{dt} \implies k = \dfrac{dr}{dt}$

30. $\dfrac{1}{R} = \dfrac{1}{R_1} + \dfrac{1}{R_2}$, $\dfrac{dR_1}{dt} = 1$ and $\dfrac{dR_2}{dt} = 1.5$

$\dfrac{1}{R^2}\dfrac{dR}{dt} = \dfrac{1}{R_1{}^2}\dfrac{dR_1}{dt} + \dfrac{1}{R_2{}^2}\dfrac{dR_2}{dt}$

When $R_1 = 50$ and $R_2 = 75$, $R = 30$

$\dfrac{dR}{dt} = (30)^2\left[\dfrac{1}{(50)^2}(1) + \dfrac{1}{(75)^2}(1.5)\right] = 0.6$ ohms/sec

31. $\tan\theta = \dfrac{x}{100}$, $\dfrac{dx}{dt} = 10$ ft/sec

$(\sec^2\theta)\dfrac{d\theta}{dt} = \dfrac{1}{100}\dfrac{dx}{dt}$

$\dfrac{d\theta}{dt} = \dfrac{1}{100}\cos^2\theta\dfrac{dx}{dt}$

When $x = 100$, $\theta = \dfrac{\pi}{4}$

$\dfrac{d\theta}{dt} = \dfrac{1}{100}(\dfrac{1}{2})(10) = \dfrac{1}{20}$ rad/sec

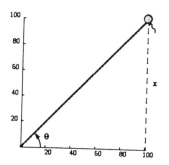

32. $\sin\theta = \dfrac{15}{x}$, $\dfrac{dx}{dt} = -1$ ft/sec

$\cos\theta(\dfrac{d\theta}{dt}) = \dfrac{-15}{x^2}\dfrac{dx}{dt}$

$\dfrac{d\theta}{dt} = \dfrac{-15}{x^2}(\sec\theta)\dfrac{dx}{dt}$

$= \dfrac{-15}{625}(\dfrac{25}{20})(-1)$

$= \dfrac{3}{100}$ rad/sec

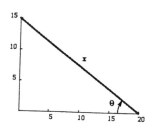

33. $\tan\theta = \dfrac{y}{x}$, $\qquad \dfrac{dx}{dt} = -600$ mi/hr

$(\sec^2\theta)\dfrac{d\theta}{dt} = -\dfrac{y}{x^2}\dfrac{dx}{dt}$

$\dfrac{d\theta}{dt} = \cos^2\theta(-\dfrac{y}{x^2})\dfrac{dx}{dt}$

$\qquad = \dfrac{x^2}{L^2}(-\dfrac{y}{x^2})\dfrac{dx}{dt} = (-\dfrac{y^2}{L^2})(\dfrac{1}{y})\dfrac{dx}{dt}$

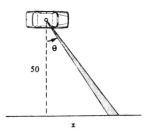

$\qquad = (-\sin^2\theta)(\dfrac{1}{5})(-600) = 120\sin^2\theta$

(a) When $\theta = 30^\circ$, $\quad \dfrac{d\theta}{dt} = \dfrac{120}{4} = 30$ rad/hr $= \dfrac{1}{2}$ rad/min

(b) When $\theta = 60^\circ$, $\quad \dfrac{d\theta}{dt} = 120(\dfrac{3}{4}) = 90$ rad/hr $= \dfrac{3}{2}$ rad/min

(c) When $\theta = 75^\circ$, $\quad \dfrac{d\theta}{dt} = 120\sin^2 75^\circ \approx 111.96$ rad/hr ≈ 1.87 rad/min

34. $\tan\theta = \dfrac{x}{50}$, $\qquad \dfrac{d\theta}{dt} = 30(2\pi) = 60\pi$ rad/min $= \pi$ rad/sec

$\sec^2\theta(\dfrac{d\theta}{dt}) = \dfrac{1}{50}(\dfrac{dx}{dt})$, $\dfrac{dx}{dt} = 50\sec^2\theta(\dfrac{d\theta}{dt})$

(a) When $\theta = 30^\circ$, $\dfrac{dx}{dt} = \dfrac{200\pi}{3}$ ft/sec

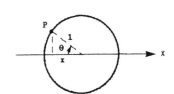

(b) When $\theta = 60^\circ$, $\dfrac{dx}{dt} = 200\pi$ ft/sec

(c) When $\theta = 70^\circ$, $\dfrac{dx}{dt} = 427.43\pi$ ft/sec

35. $x = -\cos\theta$ and $\dfrac{d\theta}{dt} = 10(2\pi) = 20\pi$ rad/sec

$\dfrac{dx}{dt} = \sin\theta\dfrac{d\theta}{dt} = 20\pi\sin\theta$

(a) When $\theta = 0^\circ$, $\dfrac{dx}{dt} = 0$

(b) When $\theta = 30^\circ$, $\dfrac{dx}{dt} = 10\pi$ ft/sec

(c) When $\theta = 60^\circ$, $\dfrac{dx}{dt} = 10\sqrt{3}\,\pi$ ft/sec

Review Exercises for Chapter 3

1. $f(x) = x^3 - 3x^2,$ $f'(x) = 3x^2 - 6x = 3x(x - 2)$

2. $f(x) = 2x - x^{-2},$ $f'(x) = 2 + 2x^{-3} = 2(1 + \dfrac{1}{x^3}) = \dfrac{2(x^3 + 1)}{x^3}$

3. $f(x) = x^{1/2} - x^{-1/2},$ $f'(x) = \dfrac{1}{2}x^{-1/2} + \dfrac{1}{2}x^{-3/2} = \dfrac{x + 1}{2x^{3/2}}$

4. $f(x) = \dfrac{x + 1}{x - 1},$ $f'(x) = \dfrac{(x - 1)(1) - (x + 1)(1)}{(x - 1)^2} = \dfrac{-2}{(x - 1)^2}$

5. $g(t) = \dfrac{2}{3}t^{-2},$ $g'(t) = \dfrac{-4}{3}t^{-3} = \dfrac{-4}{3t^3}$

6. $h(x) = \dfrac{2}{9}x^{-2},$ $h'(x) = \dfrac{-4}{9}x^{-3} = \dfrac{-4}{9x^3}$

7. $f(x) = (x^3 + 1)^{1/2},$ $f'(x) = \dfrac{1}{2}(x^3 + 1)^{-1/2}(3x^2) = \dfrac{3x^2}{2\sqrt{x^3 + 1}}$

8. $f(x) = (x^2 - 1)^{1/3},$ $f'(x) = \dfrac{1}{3}(x^2 - 1)^{-2/3}(2x) = \dfrac{2x}{3(x^2 - 1)^{2/3}}$

9. $f(x) = (3x^2 + 7)(x^2 - 2x + 3)$

   $f'(x) = (3x^2 + 7)(2x - 2) + (x^2 - 2x + 3)(6x) = 2(6x^3 - 9x^2 + 16x - 7)$

10. $f(x) = (x^2 + \dfrac{1}{x})^5,$ $f'(x) = 5(x^2 + \dfrac{1}{x})^4(2x - \dfrac{1}{x^2})$

11. $f(s) = (s^2 - 1)^{5/2}(s^3 + 5)^{5/3},$

   $f'(s) = (s^2 - 1)^{5/2}[\dfrac{5}{3}(s^3 + 5)^{2/3}(3s^2)] + (s^3 + 5)^{5/3}[\dfrac{5}{2}(s^2 - 1)^{3/2}(2s)]$

   $= 5s(s^2 - 1)^{3/2}(s^3 + 5)^{2/3}[s(s^2 - 1) + (s^3 + 5)]$

   $= 5s(s^2 - 1)^{3/2}(s^3 + 5)^{2/3}(2s^3 - s + 5)$

12. $h(\theta) = \dfrac{\theta}{(1 - \theta)^3}$

   $h'(\theta) = \dfrac{(1 - \theta)^3 - \theta[3(1 - \theta)^2(-1)]}{(1 - \theta)^6} = \dfrac{(1 - \theta)^2(1 - \theta + 3\theta)}{(1 - \theta)^6} = \dfrac{2\theta + 1}{(1 - \theta)^4}$

13. $g(x) = \sqrt{x^3 + x} = (x^3 + x)^{1/2}$

   $g'(x) = \frac{1}{2}(x^3 + x)^{-1/2}(3x^2 + 1) = \frac{3x^2 + 1}{2\sqrt{x^3 + x}}$

14. $f(x) = \frac{6x - 5}{x^2 + 1}$

   $f'(x) = \frac{(x^2 + 1)(6) - (6x - 5)(2x)}{(x^2 + 1)^2} = \frac{2(3 + 5x - 3x^2)}{(x^2 + 1)^2}$

15. $f(x) = \frac{x^2 + x - 1}{x^2 - 1}$

   $f'(x) = \frac{(x^2 - 1)(2x + 1) - (x^2 + x - 1)(2x)}{(x^2 - 1)^2} = \frac{-(x^2 + 1)}{(x^2 - 1)^2}$

16. $f(t) = t^2(t - 1)^5(t + 2)^3$

   $f'(t) = 2t(t - 1)^5(t + 2)^3 + t^2(5)(t - 1)^4(t + 2)^3 + t^2(t - 1)^5(3)(t + 2)^2$

   $= t(t - 1)^4(t + 2)^2[2(t - 1)(t + 2) + 5t(t + 2) + 3t(t - 1)]$

   $= t(t - 1)^4(t + 2)^2(10t^2 + 9t - 4)$

17. $f(x) = -2(1 - 4x^2)^2$

   $f'(x) = -4(1 - 4x^2)(-8x) = 32x(1 - 4x^2) = 32x - 128x^3$

18. $f(x) = (x^2 + 2x - 8)^2$

   $f'(x) = 2(x^2 + 2x - 8)(2x + 2) = 4(x^3 + 3x^2 - 6x - 8)$

19. $f(x) = (4 - 3x^2)^{-1}$, $f'(x) = -(4 - 3x^2)^{-2}(-6x) = \frac{6x}{(4 - 3x^2)^2}$

20. $f(x) = 9(3x^2 - 2x)^{-1}$, $f'(x) = -9(3x^2 - 2x)^{-2}(6x - 2) = \frac{18(1 - 3x)}{(3x^2 - 2x)^2}$

21. $g(x) = (2x)(x + 1)^{-1/2}$

   $g'(x) = (2x)[-\frac{1}{2}(x + 1)^{-3/2}(1)] + (x + 1)^{-1/2}(2)$

   $= \frac{-x}{(x + 1)^{3/2}} + \frac{2}{(x + 1)^{1/2}} \frac{(x + 1)}{(x + 1)} = \frac{x + 2}{(x + 1)^{3/2}}$

22. $f(x) = \dfrac{-2x^2}{x - 1}$

$f'(x) = \dfrac{(x - 1)(-4x) - (-2x^2)(1)}{(x - 1)^2} = \dfrac{4x - 2x^2}{(x - 1)^2} = \dfrac{2x(2 - x)}{(x - 1)^2}$

23. $f(t) = (t + 1)^{1/2}(t + 1)^{1/3} = (t + 1)^{5/6}$

$f'(t) = \dfrac{5}{6}(t + 1)^{-1/6}(1) = \dfrac{5}{6(t + 1)^{1/6}}$

24. $y = \sqrt{3x}\,(x + 2)^3$

$y' = \sqrt{3x}\,[3(x + 2)^2(1)] + (x + 2)^3[\frac{1}{2}(3x)^{-1/2}(3)]$

$= \dfrac{3}{2}(3x)^{-1/2}(x + 2)^2[6x + (x + 2)] = \dfrac{3(x + 2)^2(7x + 2)}{2\sqrt{3x}}$

25. $y = \dfrac{1}{2}\csc 2x, \qquad y' = \dfrac{1}{2}(-\csc 2x \cot 2x)(2) = -\csc 2x \cot 2x$

26. $y = \dfrac{1 + \sin x}{1 - \sin x}$

$y' = \dfrac{(1 - \sin x)\cos x - (1 + \sin x)(-\cos x)}{(1 - \sin x)^2} = \dfrac{2\cos x}{(1 - \sin x)^2}$

27. $y = \tan x - x, \qquad y' = \sec^2 x - 1 = \tan^2 x$

28. $y = \dfrac{2}{3}\sin^{3/2} x - \dfrac{2}{7}\sin^{7/2} x, \qquad y' = \sin^{1/2} x \cos x - \sin^{5/2} x \cos x$

$= (\cos x)\sqrt{\sin x}\,(1 - \sin^2 x) = (\cos^3 x)\sqrt{\sin x}$

29. $y = \dfrac{x}{2} - \dfrac{\sin 2x}{4}, \qquad y' = \dfrac{1}{2} - \dfrac{1}{4}\cos 2x(2) = \dfrac{1}{2}(1 - \cos 2x) = \sin^2 x$

30. $y = \dfrac{\sec^7 x}{7} - \dfrac{\sec^5 x}{5}, \qquad y' = \sec^6 x(\sec x \tan x) - \sec^4 x(\sec x \tan x)$

$= \sec^5 x \tan x(\sec^2 x - 1) = \sec^5 x \tan^3 x$

31. $y = 3\cos(3x + 1), \qquad y' = -9\sin(3x + 1)$

32. $y = 2\csc^3 \sqrt{x}$

$y' = -6(\csc^2 \sqrt{x})(\dfrac{1}{2\sqrt{x}}\csc \sqrt{x} \cot \sqrt{x}) = -\dfrac{3}{\sqrt{x}}\csc^3 \sqrt{x} \cot \sqrt{x}$

33. $y = \tan \sqrt{1 - x}, \qquad y' = -\dfrac{1}{2\sqrt{1 - x}} \sec^2 \sqrt{1 - x} = -\dfrac{\sec^2 \sqrt{1 - x}}{2\sqrt{1 - x}}$

34. $y = 1 - \cos 2x + 2 \cos^2 x$

    $y' = 2 \sin 2x - 4 \cos x \sin x = 2[2 \sin x \cos x] - 4 \sin x \cos x = 0$

35. $y = \dfrac{\sin x}{x^2}, \qquad y' = \dfrac{(x^2)\cos x - (\sin x)(2x)}{x^4} = \dfrac{x \cos x - 2 \sin x}{x^3}$

36. $y = \csc 3x + \cot 3x$

    $y' = -3 \csc 3x \cot 3x - 3 \csc^2 3x = -3 \csc 3x \, (\cot 3x + \csc 3x)$

37. $y = -x \tan x, \qquad y' = -x \sec^2 x - \tan x$

38. $y = \dfrac{\cos (x - 1)}{x - 1}, \qquad y' = \dfrac{-(x - 1) \sin (x - 1) - \cos (x - 1)(1)}{(x - 1)^2}$

    $= -\dfrac{1}{(x - 1)^2}[(x - 1) \sin (x - 1) + \cos (x - 1)]$

39. $y = \dfrac{\sin 4x}{4} + x, \qquad y' = \cos 4x + 1$

40. $y = x \cos x - \sin x, \qquad y' = -x \sin x + \cos x - \cos x = -x \sin x$

41. $f(x) = \sqrt{x^2 + 9}, \qquad f'(x) = \dfrac{1}{2}(x^2 + 9)^{-1/2}(2x) = \dfrac{x}{\sqrt{x^2 + 9}}$

    $f''(x) = \dfrac{\sqrt{x^2 + 9} - (x^2/\sqrt{x^2 + 9})}{x^2 + 9} = \dfrac{x^2 + 9 - x^2}{(x^2 + 9)^{3/2}} = \dfrac{9}{(x^2 + 9)^{3/2}}$

42. $h(x) = x\sqrt{x^2 - 1}$

    $h'(x) = x(\dfrac{1}{2})(x^2 - 1)^{-1/2}(2x) + (x^2 - 1)^{1/2}$

    $= \dfrac{x^2}{\sqrt{x^2 - 1}} + \sqrt{x^2 - 1} = \dfrac{2x^2 - 1}{\sqrt{x^2 - 1}}$

    $h''(x) = \dfrac{4x\sqrt{x^2 - 1} - (2x^2 - 1)x/\sqrt{x^2 - 1}}{x^2 - 1}$

    $= \dfrac{4x^3 - 4x - 2x^3 + x}{(x^2 - 1)\sqrt{x^2 - 1}} = \dfrac{x(2x^2 - 3)}{(x^2 - 1)^{3/2}}$

43. $f(t) = \dfrac{t}{(1 - t)^2}$, $f'(t) = \dfrac{(1 - t)^2 - 2t(1 - t)(-1)}{(1 - t)^4} = \dfrac{t + 1}{(1 - t)^3}$

$f''(t) = \dfrac{(1 - t)^3 - (t + 1)(3)(1 - t)^2(-1)}{(1 - t)^6} = \dfrac{2(t + 2)}{(1 - t)^4}$

44. $h(x) = x^2 + \dfrac{3}{x}$, $h'(x) = 2x - \dfrac{3}{x^2}$, $h''(x) = 2 + \dfrac{6}{x^3}$

45. $g(x) = \dfrac{6x - 5}{x^2 + 1}$, $g'(x) = \dfrac{6(x^2 + 1) - (6x - 5)(2x)}{(x^2 + 1)^2} = \dfrac{2(-3x^2 + 5x + 3)}{(x^2 + 1)^2}$

$g''(x) = 2 \left[\dfrac{(x^2 + 1)^2(-6x + 5) - (-3x^2 + 5x + 3)(2)(x^2 + 1)(2x)}{(x^2 + 1)^4} \right]$

$= \dfrac{2(6x^3 - 15x^2 - 18x + 5)}{(x^2 + 1)^3}$

46. $f(x) = 3x^4 - 6x^3 + 16x^2 - 14x + 21$, $f'(x) = 12x^3 - 18x^2 + 32x - 14$

$f''(x) = 36x^2 - 36x + 32 = 4(9x^2 - 9x + 8)$

47. $f(x) = \cot x$, $f'(x) = -\csc^2 x$, $f''(x) = 2 \csc^2 x \cot x$

48. $y = \sin^2 x$, $y' = 2 \sin x \cos x = \sin 2x$, $y'' = 2 \cos 2x$

49. $y = \dfrac{\cos x}{x}$, $y' = \dfrac{-x \sin x - \cos x}{x^2}$

$y'' = \dfrac{(x^2)[-x \cos x - \sin x + \sin x] - [-x \sin x - \cos x](2x)}{x^4}$

$= \dfrac{-x^2 \cos x + 2x \sin x + 2 \cos x}{x^3}$

50. $g(x) = x \tan x$, $g'(x) = x \sec^2 x + \tan x$

$g''(x) = x[2 \sec x(\sec x \tan x)] + \sec^2 x + \sec^2 x = 2 \sec^2 x(x \tan x + 1)$

51. $x^2 + 3xy + y^3 = 10$, $2x + 3xy' + 3y + 3y^2y' = 0$,

$3(x + y^2)y' = -(2x + 3y)$, $y' = \dfrac{-(2x + 3y)}{3(x + y^2)}$

52. $x^2 + 9y^2 - 4x + 3y + 7 = 0$, $2x + 18yy' - 4 + 3y' = 0$,

$3(6y + 1)y' = 4 - 2x$, $y' = \dfrac{4 - 2x}{3(6y + 1)}$

53. $y\sqrt{x} - x\sqrt{y} = 16$, $\quad y(\frac{1}{2}x^{-1/2}) + x^{1/2}y' - x(\frac{1}{2}y^{-1/2}y') - y^{1/2} = 0$

$(\sqrt{x} - \frac{x}{2\sqrt{y}})\,y' = \sqrt{y} - \frac{y}{2\sqrt{x}}$, $\quad \frac{2\sqrt{xy} - x}{2\sqrt{y}}\,y' = \frac{2\sqrt{xy} - y}{2\sqrt{x}}$,

$y' = \frac{2\sqrt{xy} - y}{2\sqrt{x}}\,\frac{2\sqrt{y}}{2\sqrt{xy} - x} = \frac{2y\sqrt{x} - y\sqrt{y}}{2x\sqrt{y} - x\sqrt{x}}$

54. $y^2 + x^2 - 6y - 2x - 5 = 0$, $\quad 2yy' + 2x - 6y' - 2 = 0$, $\quad y' = \frac{1 - x}{y - 3}$

55. $y^2 - x^2 = 25$, $\quad 2yy' - 2x = 0$, $\quad y' = \frac{x}{y}$

56. $y^2 = x^3 - x^2y + xy - y^2$, $\quad 0 = x^3 - x^2y + xy - 2y^2$,

$0 = 3x^2 - x^2y' - 2xy + xy' + y - 4yy'$

$(x^2 - x + 4y)y' = 3x^2 - 2xy + y$

$y' = \frac{3x^2 - 2xy + y}{x^2 - x + 4y}$

57. $x\sin y = y\cos x$, $\quad (x\cos y)y' + \sin y = -y\sin x + y'\cos x$

$y'(x\cos y - \cos x) = -y\sin x - \sin y$, $\quad y' = \frac{y\sin x + \sin y}{\cos x - x\cos y}$

58. $\cos(x + y) = x$, $\quad -(1 + y')\sin(x + y) = 1$

$-y'\sin(x + y) = 1 + \sin(x + y)$

$y' = -\frac{1 + \sin(x + y)}{\sin(x + y)} = -\csc(x + y) - 1$

59. $y = (x + 3)^3$, $\quad y' = 3(x + 3)^2$ $\quad$ At $(-2, 1)$: $\quad y' = 3$

Tangent line: $\quad y - 1 = 3(x + 2)$, $\quad 3x - y + 7 = 0$

Normal line: $\quad y - 1 = -\frac{1}{3}(x + 2)$, $\quad x + 3y - 1 = 0$

60. $y = (x - 2)^2$, $\quad y' = 2(x - 2)$ $\quad$ At $(2, 0)$: $\quad y' = 0$

Tangent line: $\quad y = 0$ $\qquad\qquad$ Normal line: $\quad x = 2$

61. $x^2 + y^2 = 20$, $2x + 2yy' = 0$, $y' = -x/y$

    At $(2, 4)$: $y' = -1/2$

    Tangent line: $y - 4 = -\dfrac{1}{2}(x - 2)$, $x + 2y - 10 = 0$

    Normal line: $y - 4 = 2(x - 2)$, $2x - y = 0$

62. $x^2 - y^2 = 16$, $2x - 2yy' = 0$, $y' = x/y$ At $(5, 3)$: $y' = 5/3$

    Tangent line: $y - 3 = \dfrac{5}{3}(x - 5)$, $5x - 3y - 16 = 0$

    Normal line: $y - 3 = -\dfrac{3}{5}(x - 5)$, $3x + 5y - 30 = 0$

63. $y = (x - 2)^{2/3}$, $y' = \dfrac{2}{3}(x - 2)^{-1/3} = \dfrac{2}{3\sqrt[3]{x - 2}}$ At $(3, 1)$: $y' = \dfrac{2}{3}$

    Tangent line: $y - 1 = \dfrac{2}{3}(x - 3)$, $2x - 3y - 3 = 0$

    Normal line: $y - 1 = -\dfrac{3}{2}(x - 3)$, $3x + 2y - 11 = 0$

64. $y = \dfrac{2x}{1 - x^2}$, $y' = \dfrac{2(1 - x^2) - 2x(-2x)}{(1 - x^2)^2} = \dfrac{2(x^2 + 1)}{(1 - x^2)^2}$

    At $(0, 0)$: $y' = 2$

    Tangent line: $y = 2x$ Normal line: $y = -\dfrac{1}{2}x$

65. $f(x) = \dfrac{1}{3}x^3 + x^2 - x - 1$, $f'(x) = x^2 + 2x - 1$

    (a) $x^2 + 2x - 1 = -1$, $x(x + 2) = 0$, $(0, -1)$ and $(-2, 7/3)$

    (b) $x^2 + 2x - 1 = 2$, $(x + 3)(x - 1) = 0$, $(-3, 2)$ and $(1, -2/3)$

    (c) $x^2 + 2x - 1 = 0$, $(x + 1)^2 = 2$, $x = -1 \pm \sqrt{2}$

         $\left(-1 + \sqrt{2}, \dfrac{2(1 - 2\sqrt{2})}{3}\right)$ and $\left(-1 - \sqrt{2}, \dfrac{2(1 + 2\sqrt{2})}{3}\right)$

66. $f(x) = x^2 + 1$, $f'(x) = 2x$

    (a) $2x = -1$, $x = -\dfrac{1}{2}$, $\left(-\dfrac{1}{2}, \dfrac{5}{4}\right)$

    (b) $2x = 0$, $x = 0$, $(0, 1)$

    (c) $2x = 1$, $x = \dfrac{1}{2}$, $\left(\dfrac{1}{2}, \dfrac{5}{4}\right)$

67. $f(x) = \dfrac{1}{x^2}$

    (i) $f(x + \Delta x) = \dfrac{1}{(x + \Delta x)^2} = \dfrac{1}{x^2 + 2x\Delta x + (\Delta x)^2}$

    (ii) $f(x + \Delta x) - f(x) = \dfrac{1}{x^2 + 2x\Delta x + (\Delta x)^2} - \dfrac{1}{x^2} = \dfrac{-2x\Delta x - (\Delta x)^2}{x^2[x^2 + 2x\Delta x + (\Delta x)^2]}$

    (iii) $\dfrac{f(x + \Delta x) - f(x)}{\Delta x} = \dfrac{-2x - \Delta x}{x^2[x^2 + 2x\Delta x + (\Delta x)^2]}$

    (iv) $\displaystyle\lim_{\Delta x \to 0} \dfrac{f(x + \Delta x) - f(x)}{\Delta x} = \dfrac{-2x}{x^4} = \dfrac{-2}{x^3}$

68. $f(x) = \dfrac{x + 1}{x - 1}$

    (i) $f(x + \Delta x) = \dfrac{x + \Delta x + 1}{x + \Delta x - 1}$

    (ii) $f(x + \Delta x) - f(x) = \dfrac{x + \Delta x + 1}{x + \Delta x - 1} - \dfrac{x + 1}{x - 1} = \dfrac{-2\Delta x}{(x - 1)(x + \Delta x - 1)}$

    (iii) $\dfrac{f(x + \Delta x) - f(x)}{\Delta x} = \dfrac{-2}{(x - 1)(x + \Delta x - 1)}$

    (iv) $\displaystyle\lim_{\Delta x \to 0} \dfrac{f(x + \Delta x) - f(x)}{\Delta x} = \dfrac{-2}{(x - 1)^2}$

69. $f(x) = \sqrt{x + 2}$

    (i) $f(x + \Delta x) = \sqrt{x + \Delta x + 2}$

    (ii) $f(x + \Delta x) - f(x) = [\sqrt{x + \Delta x + 2} - \sqrt{x + 2}]$

$$\cdot \left[\dfrac{\sqrt{x + \Delta x + 2} + \sqrt{x + 2}}{\sqrt{x + \Delta x + 2} + \sqrt{x + 2}} \right] = \dfrac{\Delta x}{\sqrt{x + \Delta x + 2} + \sqrt{x + 2}}$$

    (iii) $\dfrac{f(x + \Delta x) - f(x)}{\Delta x} = \dfrac{1}{\sqrt{x + \Delta x + 2} + \sqrt{x + 2}}$

    (iv) $\displaystyle\lim_{\Delta x \to 0} \dfrac{f(x + \Delta x) - f(x)}{\Delta x} = \dfrac{1}{2\sqrt{x + 2}}$

70. $f(x) = \dfrac{1}{\sqrt{x}}$

    (i) $f(x + \Delta x) = \dfrac{1}{\sqrt{x + \Delta x}}$

    (ii) $f(x + \Delta x) - f(x) = \dfrac{1}{\sqrt{x + \Delta x}} - \dfrac{1}{\sqrt{x}}$

$$= \left[\dfrac{\sqrt{x} - \sqrt{x + \Delta x}}{\sqrt{x}\,\sqrt{x + \Delta x}}\right]\left[\dfrac{\sqrt{x} + \sqrt{x + \Delta x}}{\sqrt{x} + \sqrt{x + \Delta x}}\right]$$

$$= \dfrac{-\Delta x}{\sqrt{x}\,\sqrt{x + \Delta x}(\sqrt{x} + \sqrt{x + \Delta x})}$$

    (iii) $\dfrac{f(x + \Delta x) - f(x)}{\Delta x} = \dfrac{-1}{\sqrt{x}\,\sqrt{x + \Delta x}(\sqrt{x} + \sqrt{x + \Delta x})}$

    (iv) $\displaystyle\lim_{\Delta x \to 0} \dfrac{f(x + \Delta x) - f(x)}{\Delta x} = \dfrac{-1}{2x^{3/2}}$

71. $y = 2\sin x + 3\cos x$, $y' = 2\cos x - 3\sin x$, $y'' = -2\sin x - 3\cos x$

    $y'' + y = -(2\sin x + 3\cos x) + (2\sin x + 3\cos x) = 0$

72. $y = \dfrac{(10 - \cos x)}{x}$, $xy + \cos x = 10$, $xy' + y - \sin x = 0$,

    $xy' = \sin x - y$, $xy' + y = (\sin x - y) + y = \sin x$

73. $s(t) = t + \dfrac{1}{t + 1}$, $v(t) = s'(t) = 1 - \dfrac{1}{(t + 1)^2}$

    $a(t) = s''(t) = \dfrac{2}{(t + 1)^3}$

74. $s(t) = \dfrac{1}{t^2 + 2t + 1}$ ⟹ $s(t) = (t + 1)^{-2}$

    $v(t) = s'(t) = -\dfrac{2}{(t + 1)^3}$, $a(t) = s''(t) = \dfrac{6}{(t + 1)^4}$

75. $T = 700(t^2 + 4t + 10)^{-1}$, $T' = \dfrac{-1400(t + 2)}{(t^2 + 4t + 10)^2}$

 (a) When $t = 1$, $T' = \dfrac{-1400(1 + 2)}{(1 + 4 + 10)^2} = -18.667$

 (b) When $t = 3$, $T' = \dfrac{-1400(3 + 2)}{(9 + 12 + 10)^2} = -7.284$

 (c) When $t = 5$, $T' = \dfrac{-1400(5 + 2)}{(25 + 20 + 10)^2} = -3.240$

 (d) When $t = 10$, $T' = \dfrac{-1400(10 + 2)}{(100 + 40 + 10)^2} = -0.747$

76. $v = \sqrt{2gh} = \sqrt{2(32)h} = 8\sqrt{h}$, $\dfrac{dv}{dh} = \dfrac{4}{\sqrt{h}}$

 (a) When $h = 9$, $\dfrac{dv}{dh} = \dfrac{4}{3}$ ft/sec   (b) When $h = 4$, $\dfrac{dv}{dh} = 2$ ft/sec

77. Use the equation for a free-falling object from Section 3.2

$$s(t) = -16t^2 + v_0 t + s_0$$

Consider $s_0 = 0$, then $s(t) = -16t^2 + v_0 t$.

This is a parabola opening downward, therefore the vertex will be the

highest point on the graph. Completing the square we have

$$s(t) = -16(t^2 - \frac{v_0}{16} t + \frac{v_0{}^2}{32^2}) + \frac{v_0{}^2}{64}$$

$$= -16(t - \frac{v_0}{32})^2 + \frac{v_0{}^2}{64}$$

Thus the vertex is $(\dfrac{v_0}{32}, \dfrac{v_0{}^2}{64})$ and we let

$\dfrac{v_0{}^2}{64} = 49 \implies v_0 = 56$ ft/sec

78. $s(t) = -16t^2 + 14{,}400 = 0$, $16t^2 = 14{,}400$, $t = 30$ sec

Since 600 mph = 1/6 mi/sec, in 30 seconds the bomb will move

horizontally $(1/6)(30) = 5$ miles

79.(a)

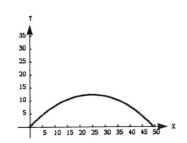

(b)   Total horizontal distance: 50 ft

(c)   Ball reaches its maximum height when x is 25 ft

(d) $y = x - 0.02x^2$, $y' = 1 - 0.04x$

$y'(0) = 1$, $y'(10) = 0.6$, $y'(25) = 0$, $y'(30) = -0.2$, $y'(50) = -1$

(e) $y'(25) = 0$

80.(a)

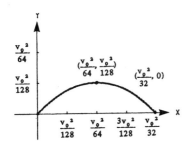

(b) $y = x - \dfrac{32}{v_0^2} x^2 = x(1 - \dfrac{32}{v_0^2} x) = 0$ if $x = 0$ or $x = \dfrac{v_0^2}{32}$

Projectile strikes the ground when $x = \dfrac{v_0^2}{32}$

Projectile reaches its maximum height at $x = \dfrac{v_0^2}{64}$

(c) $y' = 1 - \dfrac{64}{v_0^2} x$, $f'\left(\dfrac{v_0^2}{64}\right) = 0$

81. $y = x - \dfrac{32}{v_0{}^2} x^2 = x(1 - \dfrac{32}{v_0{}^2} x) = 0$ when $x = 0$ and $x = \dfrac{v_0{}^2}{32}$

Therefore the range is $x = \dfrac{v_0{}^2}{32}$

When the initial velocity is doubled the range is $x = \dfrac{(2v_0)^2}{32} = \dfrac{4v_0{}^2}{32}$

or four times the initial range.

From Exercise 80, the maximum height occurs when $x = \dfrac{v_0{}^2}{64}$.

The maximum height is $y(\dfrac{v_0{}^2}{64}) = \dfrac{v_0{}^2}{64} - \dfrac{32}{v_0{}^2}(\dfrac{v_0{}^2}{64})^2 = \dfrac{v_0{}^2}{64} - \dfrac{v_0{}^2}{128} = \dfrac{v_0{}^2}{128}$.

If the initial velocity is doubled the maximum height is

$y\left[\dfrac{(2v_0)^2}{64}\right] = \dfrac{(2v_0)^2}{128} = 4(\dfrac{v_0{}^2}{128})$ or four times the original maximum height.

82. $v_0 = 70$ ft/sec, Range: $x = \dfrac{v_0{}^2}{32} = \dfrac{(70)^2}{32} = 153.125$ ft

Maximum height: $y = \dfrac{v_0{}^2}{128} = \dfrac{(70)^2}{128} = 38.28$ ft

83. $y = \sqrt{x}$, $\dfrac{dy}{dt} = 2$ units/sec

$\dfrac{dy}{dt} = \dfrac{1}{2\sqrt{x}} \dfrac{dx}{dt}$ $\Longrightarrow$ $\dfrac{dx}{dt} = 2\sqrt{x}\, \dfrac{dy}{dt} = 4\sqrt{x}$

(a) When $x = \dfrac{1}{2}$, $\dfrac{dx}{dt} = 2\sqrt{2}$ units/sec (b) When $x = 1$, $\dfrac{dx}{dt} = 4$ units/sec

(c) When $x = 4$, $\dfrac{dx}{dt} = 8$ units/sec

84. $y = \sqrt{x}$, $L^2 = x^2 + y^2$, $\dfrac{dy}{dt} = 2$ units/sec

$L^2 = y^4 + y^2$, $2L\dfrac{dL}{dt} = (4y^3 + 2y)\dfrac{dy}{dt}$

$\dfrac{dL}{dt} = \dfrac{4y^3 + 2y}{2L} \dfrac{dy}{dt}$

$= \dfrac{4y^3 + 2y}{L} = \dfrac{(4x + 2)\sqrt{x}}{L}$

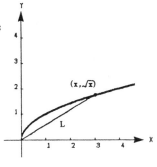

(a)   When $x = \dfrac{1}{2}$, $L = \sqrt{(\dfrac{1}{2})^2 + (\dfrac{1}{\sqrt{2}})^2} = \dfrac{\sqrt{3}}{2}$

$$\dfrac{dL}{dt} = \dfrac{(2 + 2)(1/\sqrt{2})}{\sqrt{3}/2} = \dfrac{8}{\sqrt{6}} \text{ units/sec}$$

(b)   When $x = 1$, $L = \sqrt{(1)^2 + (1)^2} = \sqrt{2}$

$$\dfrac{dL}{dt} = \dfrac{(4 + 2)(1)}{\sqrt{2}} = 3\sqrt{2} \text{ units/sec}$$

(c)   When $x = 4$, $L = \sqrt{(4)^2 + (2)^2} = 2\sqrt{5}$

$$\dfrac{dL}{dt} = \dfrac{(16 + 2)(2)}{2\sqrt{5}} = \dfrac{18}{\sqrt{5}} \text{ units/sec}$$

85. $\dfrac{s}{h} = \dfrac{1/2}{2}$, $s = \dfrac{1}{4}h$, $\dfrac{dV}{dt} = 1$

Width of water when at depth h is $w = 2 + 2(\dfrac{1}{4}h) = \dfrac{4 + h}{2}$

$$V = \dfrac{5}{2}(2 + \dfrac{4 + h}{2})h = \dfrac{5}{4}(8 + h)h$$

$$\dfrac{dV}{dt} = \dfrac{5}{2}(4 + h)\dfrac{dh}{dt}$$

$$\dfrac{dh}{dt} = \dfrac{2 \ dV/dt}{5(4 + h)}$$

When $h = 1$, $\dfrac{dh}{dt} = \dfrac{2}{25}$ ft/min

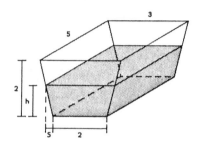

86. $pV^k = C$, $\dfrac{dV}{dt} = 1$, $kpV^{k-1}\dfrac{dV}{dt} + V^k\dfrac{dp}{dt} = 0$, $\dfrac{dp}{dt} = \dfrac{-kp \ dV/dt}{V} = \dfrac{-kp_1}{V_1}$

When $p = p_1$, $V = V_1$ and $\dfrac{dV}{dt} = 1$

87. $\tan \theta = x$, $\dfrac{d\theta}{dt} = 3(2\pi)$

$$\sec^2 \theta (\dfrac{d\theta}{dt}) = \dfrac{dx}{dt}$$

$$\dfrac{dx}{dt} = (\tan^2 \theta + 1)(6\pi) = 6\pi(x^2 + 1)$$

When $x = \dfrac{1}{2}$, $\dfrac{dx}{dt} = 6\pi(\dfrac{1}{4} + 1)$

$$= \dfrac{15\pi}{2} \text{ mi/min} = 450\pi \text{ mph}$$

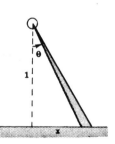

88. $P(t) = 500(1 + \dfrac{4t}{50 + t^2})$

$P'(t) = 500\left[\dfrac{(50 + t^2)(4) - (4t)(2t)}{(50 + t^2)^2}\right] = 2000\left[\dfrac{50 - t^2}{(50 + t^2)^2}\right]$

$P'(2) = 2000\left[\dfrac{50 - 4}{(54)^2}\right] = 31.55$ bacteria/hr

89. $g = \sqrt{x(x + n)} = (x^2 + nx)^{1/2}$

$\dfrac{dg}{dx} = \dfrac{1}{2}(x^2 + nx)^{-1/2}(2x + n) = \dfrac{2x + n}{2\sqrt{x(x + n)}}$

$\dfrac{a}{g} = \dfrac{[x + (x + n)]/2}{\sqrt{x(x + n)}} = \dfrac{2x + n}{2\sqrt{x(x + n)}}$

Thus $\dfrac{dg}{dx} = \dfrac{a}{g}$

90. $y = \dfrac{1}{x}, \qquad y' = -x^{-2} = \dfrac{-1}{x^2}, \qquad y'' = 2 \cdot 1 x^{-3} = \dfrac{2 \cdot 1}{x^3}$

$y''' = -3 \cdot 2 \cdot 1\ x^{-4} = \dfrac{-(3 \cdot 2 \cdot 1)}{x^4}, \qquad y^{(4)} = 4 \cdot 3 \cdot 2 \cdot 1\ x^{-5} = \dfrac{4 \cdot 3 \cdot 2 \cdot 1}{x^5}$

$y^{(n)} = (-1)^n\ n(n - 1)\ \cdots 3 \cdot 2 \cdot 1\ x^{-(n + 1)} = \dfrac{(-1)^n\ n!}{x^{n + 1}}$

91. $f(x) = 4 - |x - 2|$

   (a) continuous at $x = 2$

   (b) Not differentiable at $x = 2$ because of a sharp turn in the graph.

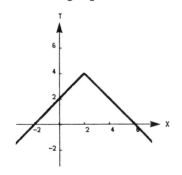

92. $f(x) = \begin{cases} x^2 + 4x + 2, & \text{if } x < -2 \\ 1 - 4x - x^2, & \text{if } x \geq -2 \end{cases}$

   (a) nonremovable discontinuity at $x = -2$

   (b) Not differentiable at $x = -2$ because the function is discontinuous there.

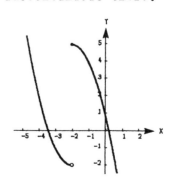

4 Applications of differentiation

4.1
Extrema on an interval

1.(a) Yes   (b) No   2.(a) Yes   (b) No

3.(a) No   (b) Yes   4.(a) No   (b) No

5.(a) Minimum: $(0, -3)$ (b) Minimum: $(0, -3)$
      Maximum: $(2, 1)$

   (c) Maximum: $(2, 1)$ (d) No Extrema

6.(a) Minimum: $(4, 1)$ (b) Maximum: $(1, 4)$
      Maximum: $(1, 4)$

   (c) Minimum: $(4, 1)$ (d) No Extrema

7.(a) Minimum: $(1, -1)$ (b) Maximum: $(3, 3)$
      Maximum: $(-1, 3)$

   (c) Minimum: $(1, -1)$ (d) Minimum: $(1, -1)$
                                           Maximum: $(3, 3)$

8.(a) Minimum: $(-2, 0)$ and $(2, 0)$ (b) Minimum: $(-2, 0)$
      Maximum: $(0, 2)$

   (c) Maximum: $(0, 2)$ (d) Maximum: $(1, \sqrt{3})$

9.(a) No Extrema   (b) Minimum: $(1, 0)$

   (c) Minimum: $(1, 0)$ (d) Maximum: $(3, 2)$
      Maximum: $(3, 2)$

10.(a) Minimum: $(1, -1)$ (b) No Extrema
        Maximum: $(2, 1)$

  (c) Minimum: $(\frac{3}{2}, 0)$ (d) Minimum: $(1, -1)$

       Maximum: $(2, 1)$

11. $f(x) = \dfrac{x^2}{x^2 + 4}$ 12. $f(x) = \cos\dfrac{\pi x}{2}$

      $f'(x) = \dfrac{(x^2 + 4)(2x) - (x^2)(2x)}{(x^2 + 4)^2}$ $f'(x) = -\dfrac{\pi}{2}\sin\dfrac{\pi x}{2}$

      $= \dfrac{8x}{(x^2 + 4)^2}, \quad f'(0) = 0$ $f'(0) = 0$ and $f'(2) = 0$

13. $f(x) = x + \dfrac{32}{x^2}, \quad f'(x) = 1 - \dfrac{64}{x^3}, \quad f'(4) = 0$

14. $f(x) = -3x\sqrt{x + 1},$

      $f'(x) = -3x\left[\dfrac{1}{2}(x + 1)^{-1/2}\right] + \sqrt{x + 1}\,(-3)$

      $= -\dfrac{3}{2}(x + 1)^{-1/2}[x + 2(x + 1)] = -\dfrac{3}{2}(x + 1)^{-1/2}(3x + 2)$

      $f'(-\dfrac{2}{3}) = 0$

15. $f(x) = (x + 2)^{2/3}, \quad f'(x) = \dfrac{2}{3}(x + 2)^{-1/3}, \quad f'(-2)$ is undefined

16.   Using the limit definition of the derivative,

      $\lim\limits_{x \to 0^-} \dfrac{f(x) - f(0)}{x - 0} = \lim\limits_{x \to 0^-} \dfrac{(4 - |x|) - 4}{x} = 1$

      $\lim\limits_{x \to 0^+} \dfrac{f(x) - f(0)}{x - 0} = \lim\limits_{x \to 0^+} \dfrac{(4 - |x|) - 4}{x - 0} = -1$

      $f'(0)$ does not exist, since the one-sided derivatives are not equal.

17. $f(x) = 2(3 - x)$ 18. $f(x) = \dfrac{2x + 5}{3}$

      $f'(x) = -2$ $f'(x) = 2/3$
      Minimum: $(2, 2)$ Minimum: $(0, 5/3)$
      Maximum: $(-1, 8)$ Maximum: $(5, 5)$

19. $f(x) = -x^2 + 3x$
$f'(x) = -2x + 3$
Minimum: $(0, 0)$ and $(3, 0)$
Maximum: $(3/2, 9/4)$

20. $f(x) = x^2 + 2x - 4$
$f'(x) = 2x + 2$
Minimum: $(-1, -5)$
Maximum: $(1, -1)$

21. $f(x) = x^3 - 3x^2$
$f'(x) = 3x^2 - 6x$
Minimum: $(-1, -4)$ and $(2, -4)$
Maximum: $(0, 0)$ and $(3, 0)$

22. $f(x) = x^3 - 12x$
$f'(x) = 3x^2 - 12$
Minimum: $(2, -16)$
Maximum: $(4, 16)$

23. $f(x) = 3x^{2/3} - 2x$
$f'(x) = 2x^{-1/3} - 2$
Minimum: $(0, 0)$
Maximum: $(-1, 5)$

24. $g(t) = \dfrac{t^2}{t^2 + 3}$
$g'(t) = \dfrac{6t}{(t^2 + 3)^2}$
Minimum: $(0, 0)$
Maximum: $(-1, 1/4)$ and $(1, 1/4)$

25. $h(s) = \dfrac{1}{s - 2}$
$h'(s) = \dfrac{-1}{(s - 2)^2}$
Minimum: $(1, -1)$
Maximum: $(0, -1/2)$

26. $h(t) = \dfrac{t}{t - 2}$
$h'(t) = \dfrac{-2}{(t - 2)^2}$
Minimum: $(5, 5/3)$
Maximum: $(3, 3)$

27. $f(x) = \tan x$, f is continuous on $[0, \pi/4]$ but not on $[0, \pi]$.

28. $y = f(x) = \dfrac{1}{x + 1}$, f is continuous on $[0, 2]$ but not on $[-2, 0]$.

29. $f(x) = \dfrac{1}{x^2 + 1}$, $[0, 3]$ $f'(x) = \dfrac{-2x}{(x^2 + 1)^2}$

$f''(x) = \dfrac{-2(1 - 3x^2)}{(x^2 + 1)^3}$ $f'''(x) = \dfrac{24x - 24x^3}{(x^2 + 1)^4}$

Setting $f'''(x) = 0$, we have $x = 0, \pm 1$
$|f''(0)| = 2$ is the maximum value.

30. Using the results of Exercise 29 on the interval $[1/2, 3]$,
$|f''(1)| = 1/2$ is the maximum value.

31. $f(x) = (1 + x^3)^{1/2}$, $[0, 2]$ $\qquad$ $f'(x) = \frac{3}{2}x^2(1 + x^3)^{-1/2}$

$f''(x) = \frac{3}{4}(x^4 + 4x)(1 + x^3)^{-3/2}$ $\qquad$ $f'''(x) = -\frac{3}{8}(x^6 + 20x^3 - 8)(1 + x^3)^{-5/2}$

Setting $f'''(x) = 0$ we have $x^6 + 20x^3 - 8 = 0$

$x^3 = \dfrac{-20 \pm \sqrt{400 - 4(1)(-8)}}{2}$ $\implies$ $x = \sqrt[3]{-10 \pm \sqrt{108}}$

$\left| f''(\sqrt[3]{-10 + \sqrt{108}}) \right| \approx 1.47$ is the maximum value.

32. $f(x) = 3x^5 - 10x^3$, $[0, 1]$ $\qquad$ $f'(x) = 15x^4 - 30x^2$

$f''(x) = 60x^3 - 60x$ $\qquad$ $f'''(x) = 180x^2 - 60 = 60(3x^2 - 1)$

$f'''(x) = 0$ when $x = \dfrac{\sqrt{3}}{3}$ $\qquad$ $\left| f''(\dfrac{\sqrt{3}}{3}) \right| = \dfrac{40\sqrt{3}}{3}$ is the maximum value.

33. $f(x) = 15x^4 - (\dfrac{2x - 1}{2})^6$, $[0, 1]$ $\qquad$ $f'(x) = 60x^3 - 6(\dfrac{2x - 1}{2})^5$

$f''(x) = 180x^2 - 30(\dfrac{2x - 1}{2})^4$ $\qquad$ $f'''(x) = 360x - 120(\dfrac{2x - 1}{2})^3$

$f^{(4)}(x) = 360 - 360(\dfrac{2x - 1}{2})^2$ $\qquad$ $f^{(5)}(x) = -720(\dfrac{2x - 1}{2})$

$f^{(5)}(x) = 0$ when $x = \dfrac{1}{2}$ $\qquad$ $\left| f^{(4)}(\dfrac{1}{2}) \right| = 360$ is the maximum value.

34. $f(x) = x^5 - 5x^4 + 20x^3 + 600$, $[0, 3/2]$

$f'(x) = 5x^4 - 20x^3 + 60x^2$ $\qquad$ $f''(x) = 20x^3 - 60x^2 + 120x$

$f'''(x) = 60x^2 - 120x + 120$ $\qquad$ $f^{(4)}(x) = 120x - 120$

$f^{(5)}(x) = 120$ $\qquad$ $\left| f^{(4)}(0) \right| = 120$ is the maximum value.

35. $f(x) = (x + 1)^{2/3}$, $[0, 2]$ $\qquad$ $f'(x) = \frac{2}{3}(x + 1)^{-1/3}$

$f''(x) = -\frac{2}{9}(x + 1)^{-4/3}$, $\qquad$ $f'''(x) = \frac{8}{27}(x + 1)^{-7/3}$

$f^{(4)}(x) = -\frac{56}{81}(x + 1)^{-10/3}$ $\qquad$ $f^{(5)}(x) = \frac{560}{243}(x + 1)^{-13/3}$

$\left| f^{(4)}(0) \right| = \dfrac{56}{81}$ is the maximum value.

36. $f(x) = \dfrac{1}{x^2 + 1}$, $[-1, 1]$ $f'''(x) = \dfrac{24x - 24x^3}{(x^2 + 1)^4}$ (See Exercise 29.)

$f^{(4)}(x) = \dfrac{24(5x^4 - 10x^2 + 1)}{(x^2 + 1)^5}$ $f^{(5)}(x) = \dfrac{-240x(3x^4 - 10x^2 + 3}{(x^2 + 1)^6}$

$|f^{(4)}(0)| = 24$ is the maximum value.

37. $P = VI - RI^2 = 12I - 0.5I^2$, $0 \leq I \leq 15$

$P = 0$ when $I = 0$, $P = 67.5$ when $I = 15$

$P' = 12 - I = 0$,   Critical number: $I = 12$ amps

When $I = 12$ amps, $P = 27$, the maximum output.

38. $C = 2x + \dfrac{300,000}{x}$, $0 \leq x \leq 300$

$C(0)$ is undefined, $C(300) = 1600$, $C' = 2 - \dfrac{300,000}{x^2} = 0$

$2x^2 = 300,000$, $x^2 = 150,000$, $x = 100\sqrt{15} \approx 387 > 300$

C is minimized when $x = 300$ units.

39. $x = \dfrac{v^2 \sin 2\theta}{32}$, $\dfrac{\pi}{4} \leq \theta \leq \dfrac{3\pi}{4}$, $\dfrac{d\theta}{dt}$ is constant

$\dfrac{dx}{dt} = \dfrac{dx}{d\theta} \dfrac{d\theta}{dt}$ by the Chain Rule $= \dfrac{v^2 \cos 2\theta}{16} \dfrac{d\theta}{dt}$

Thus in the interval $[\pi/4, 3\pi/4]$, $\theta = \pi/4$, $3\pi/4$ indicate minimums for dx/dt and $\theta = \pi/2$ indicates a maximum for dx/dt. This implies that the sprinkler waters longest when $\theta = \pi/4$ and $3\pi/4$, thus the lawn farthest from the sprinkler gets the most water.

4.2
The Mean Value Theorem

1. $f(x) = x^2 - 2x = x(x - 2)$

$f(0) = f(2) = 0$,   f is continuous and differentiable.

$f'(x) = 2x - 2 = 0$ when $x = 1$ Interval: $[0, 2]$

2. $f(x) = x^2 - 3x + 2 = (x - 1)(x - 2)$

$f(1) = f(2) = 0$,   f is continuous and differentiable.

$f'(x) = 2x - 3 = 0$ when $x = 1.5$ Interval: $[1, 2]$

3. $f(x) = x^3 - 6x^2 + 11x - 6 = (x - 1)(x - 2)(x - 3)$

    $f(1) = f(2) = f(3) = 0$,   f is continuous and differentiable.

    $f'(x) = 3x^2 - 12x + 11 = 0$ when $x = \dfrac{12 \pm \sqrt{144 - 132}}{6} = \dfrac{6 \pm \sqrt{3}}{3}$

    In the interval [1, 2]: $f'(\dfrac{6 - \sqrt{3}}{3}) = 0$.

    In the interval [2, 3]: $f'(\dfrac{6 + \sqrt{3}}{3}) = 0$.

4. $f(x) = x(x^2 - x - 2) = x(x + 1)(x - 2)$

    $f(-1) = f(0) = f(2) = 0$,   f is continuous and differentiable.

    $f'(x) = 3x^2 - 2x - 2 = 0$ when $x = \dfrac{1 \pm \sqrt{7}}{3}$

    In the interval [-1, 0]: $f'(\dfrac{1 - \sqrt{7}}{3}) = 0$.

    In the interval [0, 2]: $f'(\dfrac{1 + \sqrt{7}}{3}) = 0$.

5. $f(x) = |x| - 1$ $f(-1) = f(1) = 0$

    However, Rolle's Theorem does not apply since the function is not differentiable at $x = 0$.

6. $f(x) = 3 - |x - 3|$ $f(0) = f(6) = 0$

    However, Rolle's Theorem does not apply since the function is not differentiable at $x = 3$.

7. $f(x) = x^{2/3} - 1$ $f(-1) = f(1) = 0$

    However, Rolle's Theorem does not apply since the function is not differentiable at $x = 0$.

8. $f(x) = x - x^{1/3}$ $\qquad\qquad f(-1) = f(0) = f(1) = 0$

   f is continuous on $[-1, 1]$ and differentiable on $(-1, 0)$ and $(0, 1)$.

   $f'(x) = 1 - \dfrac{1}{3x^{2/3}} = 0$ when $x = \pm\dfrac{\sqrt{3}}{9}$

   In the interval $(-1, 0)$: $f'(-\dfrac{\sqrt{3}}{9}) = 0$

   In the interval $(0, 1)$: $f'(\dfrac{\sqrt{3}}{9}) = 0$

9. $f(x) = \dfrac{x^2 - 2x - 3}{x + 2} = \dfrac{(x + 1)(x - 3)}{x + 2}$

   $f(-1) = f(3) = 0$, $\quad$ f is continuous and differentiable on $[-1, 3]$.

   $f'(x) = \dfrac{x^2 + 4x - 1}{(x + 2)^2} = 0$ when $x = -2 \pm \sqrt{5}$.

   In the interval $[-1, 3]$: $f'(-2 + \sqrt{5}) = 0$

10. $f(x) = (x - 3)(x + 1)^2$

    $f(-1) = f(3) = 0$, $\quad$ f is continuous and differentiable.

    $f'(x) = (x + 1)(3x - 5) = 0$ when $x = -1$ and $x = \dfrac{5}{3}$

    In the interval $[-1, 3]$: $f'(\dfrac{5}{3}) = 0$

11. $f(x) = x^2 - 6x + 10$

    $f(x) \neq 0$ therefore Rolle's Theorem does not apply.

12. $f(x) = \dfrac{x + 1}{x}$

    $f(x) = 0$ only for $x = -1$ therefore Rolle's Theorem does not apply.

13. $f(x) = \sin x$

    $f(n\pi) = 0$ for any integer n, f is continuous and differentiable.

    $f'(x) = \cos x = 0$ when $x = \dfrac{(2n + 1)\pi}{2}$

    In the interval $[n\pi, (n + 1)\pi]$: $f'(\dfrac{(2n + 1)\pi}{2}) = 0$

14. $f(x) = \cos x$

$f(\frac{(2n-1)\pi}{2}) = 0$ for any integer n; f is continuous and differentiable.

$f'(x) = -\sin x = 0$ when $x = n\pi$

In the interval $\left[\frac{(2n-1)\pi}{2}, \frac{(2n+1)\pi}{2}\right]$, $f'(n\pi) = 0$

15. $f(x) = x + \sin x$

$f(x) = 0$ only for $x = 0$ therefore Rolle's Theorem does not apply.

16. $f(x) = 4x - \tan \pi x$

$f(-\frac{1}{4}) = f(0) = f(\frac{1}{4}) = 0$, f is continuous and differentiable on $[-\frac{1}{4}, \frac{1}{4}]$.

$f'(x) = 4 - \pi \sec^2 \pi x = 0$ when $x = \frac{1}{\pi} \operatorname{arcsec}(\pm\frac{2}{\sqrt{\pi}})$

In the interval $[-\frac{1}{4}, 0]$: $f'(\frac{1}{\pi} \operatorname{arcsec}(-\frac{2}{\sqrt{\pi}})) = 0$

In the interval $[0, \frac{1}{4}]$: $f'(\frac{1}{\pi} \operatorname{arcsec} \frac{2}{\sqrt{\pi}}) = 0$

17. $f(x) = \frac{x}{2} - \sin \frac{\pi x}{6}$

$f = (-1) = f(0) = f(1) = 0$, f is continuous and differentiable.

$f'(x) = \frac{1}{2} - \frac{\pi}{6} \cos \frac{\pi x}{6} = 0$ when $x = \pm\frac{6}{\pi} \arccos \frac{3}{\pi}$

In the interval $[-1, 0]$: $f'(-\frac{6}{\pi} \arccos \frac{3}{\pi}) = 0$

In the interval $[0, 1]$: $f'(\frac{6}{\pi} \arccos \frac{3}{\pi}) = 0$

18. $f(x) = \frac{6x}{\pi} - 4 \sin^2 x$

$f(0) = f(\frac{\pi}{6}) = 0$, f is continuous and differentiable.

$f'(x) = \frac{6}{\pi} - 8 \sin x \cos x = \frac{6}{\pi} - 4 \sin 2x = 0$ when $x = \frac{1}{2} \arcsin \frac{3}{2\pi}$

In the interval $[0, \frac{\pi}{6}]$: $f'(\frac{1}{2} \arcsin \frac{3}{2\pi}) = 0$

19. $f(x) = \tan x$ $\qquad\qquad$ $f(n\pi) = 0$ for any integer n.

However, f is not continuous on $[n\pi, (n + 1)\pi]$ therefore Rolle's Theorem does not apply.

20. $f(x) = \sec x$ $\qquad\qquad$ $|f(x)| \geq 1$ thus $f(x) \neq 0$ for any x.

21. $f(x) = x^2$ is continuous and differentiable on $[-2, 1]$.

$$\frac{f(1) - f(-2)}{1 - (-2)} = \frac{1 - 4}{3} = -1$$

$f'(x) = 2x = -1$ when $x = -\frac{1}{2}$ therefore $c = -\frac{1}{2}$.

22. $f(x) = x(x^2 - x - 2)$ is continuous and differentiable on $[-1, 1]$.

$$\frac{f(1) - f(-1)}{1 - (-1)} = -1$$

$f'(x) = 3x^2 - 2x - 2 = -1$, $\qquad (3x + 1)(x - 1) = 0$, $\qquad c = -\frac{1}{3}$

23. $f(x) = x^{2/3}$ is continuous on $[0, 1]$ and differentiable on $(0, 1)$.

$$\frac{f(1) - f(0)}{1 - 0} = 1$$

$f'(x) = \frac{2}{3} x^{-1/3} = 1$, $\qquad x = (\frac{2}{3})^3 = \frac{8}{27}$, $\qquad c = \frac{8}{27}$

24. $f(x) = \frac{x + 1}{x}$ is continuous and differentiable on $[\frac{1}{2}, 2]$.

$$\frac{f(2) - f(1/2)}{2 - (1/2)} = \frac{(3/2) - 3}{3/2} = -1, \quad f'(x) = \frac{-1}{x^2} = -1, \quad x^2 = 1, \quad c = 1$$

25. $f(x) = \frac{x}{x + 1}$ is continuous and differentiable on $[-\frac{1}{2}, 2]$.

$$\frac{f(1) - f(-1/2)}{2 - (-1/2)} = \frac{(2/3) - (-1)}{5/2} = \frac{2}{3}$$

$$f'(x) = \frac{1}{(x + 1)^2} = \frac{2}{3}, \quad (x + 1)^2 = \frac{3}{2}, \quad x = -1 \pm \frac{\sqrt{6}}{2}$$

In the interval $(-\frac{1}{2}, 2)$: $\quad c = -1 + \frac{\sqrt{6}}{2}$

26. $f(x) = \sqrt{x - 2}$ is continuous on $[2, 6]$ and differentiable on $(2, 6)$.

$$\frac{f(6) - f(2)}{6 - 2} = \frac{2 - 0}{4} = \frac{1}{2}, \quad f'(x) = \frac{1}{2\sqrt{x - 2}} = \frac{1}{2}, \quad \sqrt{x - 2} = 1, \quad c = 3$$

27. $f(x) = x^3$ is continuous and differentiable on $[0, 1]$.

$\dfrac{f(1) - f(0)}{1 - 0} = \dfrac{1 - 0}{1} = 1$, $\quad f'(x) = 3x^2 = 1$, $\quad x = \pm \sqrt{3}/3$

In the interval $(0, 1)$: $\quad c = \sqrt{3}/3$

28. $f(x) = \sin x$ is continuous and differentiable on $[0, \pi]$.

$\dfrac{f(\pi) - f(0)}{\pi - 0} = \dfrac{0 - 0}{\pi} = 0$, $\quad f'(x) = \cos x = 0$, $\quad c = \dfrac{\pi}{2}$

29. $f(x) = x - 2\sin x$ is continuous and differentiable on $[-\pi, \pi]$.

$\dfrac{f(\pi) - f(-\pi)}{\pi - (-\pi)} = \dfrac{\pi - (-\pi)}{2\pi} = 1$

$f'(x) = 1 - 2\cos x = 1$, $\quad \cos x = 0$, $\quad c = \pm\dfrac{\pi}{2}$

30. $f(x) = 2\sin x + \sin 2x$ is continuous and differentiable on $[0, \pi]$.

$\dfrac{f(\pi) - f(0)}{\pi - 0} = \dfrac{0 - 0}{\pi} = 0$, $\quad\quad f'(x) = 2\cos x + 2\cos 2x = 0$

$2[\cos x + 2\cos^2 x - 1] = 0$, $\quad 2(2\cos x - 1)(\cos x + 1) = 0$

$\cos x = \dfrac{1}{2}$, $\quad \cos x = -1$, $\quad x = \dfrac{\pi}{3}$, $\quad x = \dfrac{3\pi}{2}$

In the interval $(0, \pi)$: $\quad c = \dfrac{\pi}{3}$

31. $f(t) = -16t^2 + 48t + 32$

(a) $f(1) = f(2) = 64$

(b) $v = f'(t)$ must be 0 at some time in $[1, 2]$

$f'(t) = -32t + 48 = 0$, $\quad t = \dfrac{3}{2}$ seconds

32. $C(x) = 10(\dfrac{1}{x} + \dfrac{x}{x + 3})$

(a) $C(3) = C(6) = \dfrac{25}{3}$

(b) $C'(x) = 10(-\dfrac{1}{x^2} + \dfrac{3}{(x + 3)^2}) = 0$, $\quad \dfrac{3}{x^2 + 6x + 9} = \dfrac{1}{x^2}$

$2x^2 - 6x - 9 = 0$, $\quad x = \dfrac{6 \pm \sqrt{108}}{4} = \dfrac{6 \pm 6\sqrt{3}}{4} = \dfrac{3 \pm 3\sqrt{3}}{2}$

In the interval $[3, 6]$: $\quad c = \dfrac{3 + 3\sqrt{3}}{2} \approx 4.098$

33. $s(t) = -16t^2 + 500$

(a) $V_{avg} = \dfrac{s(3) - s(0)}{3 - 0} = -\dfrac{144}{3} = -48$ ft/sec

(b) $s(t)$ is continuous on $[0, 3]$ and differentiable on $(0, 3)$ therefore the Mean Value Theorem applies.

$v(t) = s'(t) = -32t = -48,$ $\qquad t = \dfrac{3}{2}$ seconds

In the interval $[0, 3]$: $\quad c = \dfrac{3}{2}$

34. $s(t) = 200\left(5 - \dfrac{9}{2 + t}\right)$

(a) $\dfrac{s(12) - s(0)}{12 - 0} = \dfrac{200(5 - 9/14) - 200(5 - 9/2)}{12} = \dfrac{450}{7}$

(b) $s'(t) = 200\left(\dfrac{9}{(2 + t)^2}\right) = \dfrac{450}{7},$ $\qquad \dfrac{1}{(2 + t)^2} = \dfrac{1}{28}$

$2 + t = 2\sqrt{7},$ $\quad t = 2\sqrt{7} - 2 \approx 3.2915$ months

$s'(t)$ is equal to the average value on April 10th (April 9th in a leap year).

4.3
Increasing and decreasing functions and the First Derivative Test

	Increasing on:	Decreasing on:
1. $f(x) = x^2 - 6x + 8,$	$(3, \infty)$	$(-\infty, 3)$
2. $y = -(x + 1)^2,$	$(-\infty, -1)$	$(-1, \infty)$
3. $y = \dfrac{x^3}{4} - 3x,$	$(-\infty, -2)$ and $(2, \infty)$	$(-2, 2)$
4. $f(x) = x^4 - 2x^2,$	$(-1, 0)$ and $(1, \infty)$	$(-\infty, -1)$ and $(0, 1)$
5. $f(x) = \dfrac{1}{x^2},$	$(-\infty, 0)$	$(0, \infty)$
6. $y = \dfrac{x^2}{x + 1},$	$(-\infty, -2)$ and $(0, \infty)$	$(-2, -1)$ and $(-1, 0)$

Complete Solutions Guide to Accompany Calculus

7. $f(x) = -2x^2 + 4x + 3$, $f'(x) = -4x + 4 = 0$, $x = 1$ Critical Number

   Test Intervals: $x < 1$ $x > 1$

   Sign of $f'(x)$: $f' > 0$ $f' < 0$

   Conclusion: Increasing   Decreasing

   Increasing on $(-\infty, 1)$, Decreasing on $(1, \infty)$, Relative Maximum $(1, 5)$

8. $f(x) = x^2 + 8x + 10$, $f'(x) = 2x + 8 = 0$, $x = -4$ Critical Number

   Test Intervals: $x < -4$ $x > -4$

   Sign of $f'(x)$: $f' < 0$ $f' > 0$

   Conclusion: Decreasing   Increasing

   Increasing on $(-4,\infty)$, Decreasing on $(-\infty,-4)$, Relative Minimum $(-4,-6)$

9. $f(x) = x^2 - 6x$, $f'(x) = 2x - 6 = 0$, $x = 3$ Critical Number

   Test Intervals: $x < 3$ $x > 3$

   Sign of $f'(x)$: $f' < 0$ $f' > 0$

   Conclusion: Decreasing   Increasing

   Increasing on $(3, \infty)$, Decreasing on $(-\infty, 3)$, Relative Minimum $(3, -9)$

10. $f(x) = (x - 1)^2 (x + 2)$

    $f'(x) = (x - 1)^2(1) + (x + 2)(2)(x - 1)$

    $= (x - 1)[(x - 1) + 2(x + 2)] = 3(x - 1)(x + 1) = 0$

    $x = -1, 1$ Critical Numbers

    Test Intervals: $x < -1$ $-1 < x < 1$ $x > 1$

    Sign of $f'(x)$: $f' > 0$ $f' < 0$ $f' > 0$

    Conclusion: Increasing   Decreasing   Increasing

    Increasing on $(-\infty, -1)$ and $(1, \infty)$, Decreasing on $(-1, 1)$

    Relative Maximum $(-1, 4)$, Relative Minimum $(1, 0)$

169

11. $f(x) = 2x^3 + 3x^2 - 12x,$ $f'(x) = 6x^2 + 6x - 12 = 6(x + 2)(x - 1) = 0$

   $x = -2, 1$ Critical Numbers

   Test Intervals: $x < -2$ $-2 < x < 1$ $x > 1$

   Sign of $f'(x)$: $f' > 0$ $f' < 0$ $f' > 0$

   Conclusion: Increasing   Decreasing   Increasing

   Increasing on $(-\infty, -2)$ and $(1, \infty)$, Decreasing on $(-2, 1)$

   Relative Maximum $(-2, 20)$, Relative Minimum $(1, -7)$

12. $f(x) = (x - 3)^3,$ $f'(x) = 3(x - 3)^2 = 0,$ $x = 3$ Critical Number

   Test Intervals: $x < 3$ $x > 3$

   Sign of $f'(x)$: $f' > 0$ $f' > 0$

   Conclusion: Increasing   Increasing

   Increasing on $(-\infty, \infty)$, No Relative Extrema

13. $f(x) = x^3 - 6x^2 + 15,$ $f'(x) = 3x^2 - 12x = 3x(x - 4) = 0$

   $x = 0, 4$ Critical Numbers

   Test Intervals: $x < 0$ $0 < x < 4$ $x > 4$

   Sign of $f'(x)$: $f' > 0$ $f' < 0$ $f' > 0$

   Conclusion: Increasing   Decreasing   Increasing

   Increasing on $(-\infty, 0)$ and $(4, \infty)$, Decreasing on $(0, 4)$

   Relative Maximum $(0, 15)$, Relative Minimum $(4, -17)$

14. $f(x) = (x - 1)^{2/3},$ $f'(x) = \dfrac{2}{3(x - 1)^{1/3}},$ $x = 1$ Critical Number

   Test Intervals: $x < 1$ $x > 1$

   Sign of $f'(x)$: $f' < 0$ $f' > 0$

   Conclusion: Decreasing   Increasing

   Increasing on $(1, \infty)$, Decreasing on $(-\infty, 1)$, Relative Minimum $(1, 0)$

15. $f(x) = x^4 - 2x^3$, $f'(x) = 4x^3 - 6x^2 = 2x^2(2x - 3) = 0$

$x = 0, \dfrac{3}{2}$ Critical Numbers

Test Intervals:	$x < 0$	$0 < x < \dfrac{3}{2}$	$x > \dfrac{3}{2}$
Sign of $f'(x)$:	$f' < 0$	$f' < 0$	$f' > 0$
Conclusion:	Decreasing	Decreasing	Increasing

Decreasing on $(-\infty, \dfrac{3}{2})$,   Increasing on $(\dfrac{3}{2}, \infty)$,   Relative Minimum $(\dfrac{3}{2}, -\dfrac{27}{16})$

16. $f(x) = (x - 1)^{1/3}$, $f'(x) = \dfrac{1}{3(x - 1)^{2/3}}$, $x = 1$ Critical Number

Test Intervals:	$x < 1$	$x > 1$
Sign of $f'(x)$:	$f' > 0$	$f' > 0$
Conclusion:	Increasing	Increasing

Increasing on $(-\infty, \infty)$,   No relative extrema

17. $f(x) = x^{1/3} + 1$, $f'(x) = \dfrac{1}{3}x^{-2/3} = \dfrac{1}{3x^{2/3}}$, $x = 0$ Critical Number

Test Intervals:	$x < 0$	$x > 0$
Sign of $f'(x)$:	$f' > 0$	$f' > 0$
Conclusion:	Increasing	Increasing

Increasing on $(-\infty, \infty)$,   No Relative Extrema

18. $f(x) = x^{2/3}(x - 5) = x^{5/3} - 5x^{2/3}$

$f'(x) = \dfrac{5}{3}x^{2/3} - \dfrac{10}{3}x^{-1/3} = \dfrac{5}{3}x^{-1/3}(x - 2)$

$x = 0, 2$ Critical Numbers

Test Intervals:	$x < 0$	$0 < x < 2$	$x > 2$
Sign of $f'(x)$:	$f' > 0$	$f' < 0$	$f' > 0$
Conclusion:	Increasing	Decreasing	Increasing

Increasing on $(-\infty, 0)$ and $(2, \infty)$,   Decreasing on $(0, 2)$

Relative Maximum $(0, 0)$,   Relative Minimum $(2, -3\sqrt[3]{4})$

19. $f(x) = x + \dfrac{1}{x}$, $\qquad f'(x) = 1 - \dfrac{1}{x^2} = \dfrac{x^2 - 1}{x^2}$

$x = -1, 1$ Critical Numbers, $\qquad x = 0$ Discontinuity

Test Intervals:	$x < -1$	$-1 < x < 0$	$0 < x < 1$	$x > 1$
Sign of $f'(x)$:	$f' > 0$	$f' < 0$	$f' < 0$	$f' > 0$
Conclusion:	Increasing	Decreasing	Decreasing	Increasing

Increasing on $(-\infty, -1)$ and $(1, \infty)$, Decreasing on $(-1, 0)$ and $(0, 1)$

Relative Maximum $(-1, -2)$, Relative Minimum $(1, 2)$

20. $f(x) = \dfrac{x}{x + 1}$, $\qquad f'(x) = \dfrac{(x + 1)(1) - (x)(1)}{(x + 1)^2} = \dfrac{1}{(x + 1)^2}$

$x = -1$ Discontinuity

Test Intervals:	$x < -1$	$x > -1$
Sign of $f'(x)$:	$f' > 0$	$f' > 0$
Conclusion:	Increasing	Increasing

Increasing on $(-\infty, -1)$ and $(-1, \infty)$, No Relative Extrema

21. $f(x) = \dfrac{x^2}{x^2 - 9}$, $\qquad f'(x) = \dfrac{(x^2 - 9)(2x) - (x^2)(2x)}{(x^2 - 9)^2} = \dfrac{-18x}{(x^2 - 9)^2}$

$x = 0$ Critical Number, $\qquad x = -3, 3$ Discontinuities

Test Intervals:	$x < -3$	$-3 < x < 0$	$0 < x < 3$	$x > 3$
Sign of $f'(x)$:	$f' > 0$	$f' > 0$	$f' < 0$	$f' < 0$
Conclusion:	Increasing	Increasing	Decreasing	Decreasing

Increasing on $(-\infty, -3)$ and $(-3, 0)$, Decreasing on $(0, 3)$ and $(3, \infty)$

Relative Maximum $(0, 0)$

22. $f(x) = \dfrac{x + 3}{x^2} = \dfrac{1}{x} + \dfrac{3}{x^2}$, $f'(x) = -\dfrac{1}{x^2} - \dfrac{6}{x^3} = \dfrac{-(x + 6)}{x^3}$

   $x = -6$ Critical Number, $x = 0$ Discontinuity

Test Intervals:	$x < -6$	$-6 < x < 0$	$x > 0$
Sign of $f'(x)$:	$f' < 0$	$f' > 0$	$f' < 0$
Conclusion:	Decreasing	Increasing	Decreasing

   Increasing on $(-6, 0)$, Decreasing on $(-\infty, -6)$ and $(0, \infty)$

   Relative Minimum $(-6, -\dfrac{1}{12})$

23. $f(x) = \dfrac{x^5 - 5x}{5}$, $f'(x) = x^4 - 1$, $x = -1, 1$ Critical Numbers

Test Intervals:	$x < -1$	$-1 < x < 1$	$x > 1$
Sign of $f'(x)$:	$f' > 0$	$f' < 0$	$f' > 0$
Conclusion:	Increasing	Decreasing	Increasing

   Increasing on $(-\infty, -1)$ and $(1, \infty)$, Decreasing on $(-1, 1)$

   Relative Maximum $(-1, \dfrac{4}{5})$, Relative Minimum $(1, -\dfrac{4}{5})$

24. $f(x) = x^4 - 32x + 4$, $f'(x) = 4x^3 - 32 = 4(x^3 - 8)$

   $x = 2$ Critical Number

Test Intervals:	$x < 2$	$x > 2$
Sign of $f'(x)$:	$f' < 0$	$f' > 0$
Conclusion:	Decreasing	Increasing

   Increasing on $(2, \infty)$, Decreasing on $(-\infty, 2)$

   Relative Minimum $(2, -44)$

25. $f(x) = \dfrac{x^2 - 2x + 1}{x + 1}$

$f'(x) = \dfrac{(x + 1)(2x - 2) - (x^2 - 2x + 1)(1)}{(x + 1)^2}$

$= \dfrac{x^2 + 2x - 3}{(x + 1)^2} = \dfrac{(x + 3)(x - 1)}{(x + 1)^2}$

$x = -3, 1$ Critical Numbers, $x = -1$ Discontinuity

Test Intervals:	$x < -3$	$-3 < x < -1$	$-1 < x < 1$	$x > 1$
Sign of $f'(x)$:	$f' > 0$	$f' < 0$	$f' < 0$	$f' > 0$
Conclusion:	Increasing	Decreasing	Decreasing	Increasing

Increasing on $(-\infty, -3)$ and $(1, \infty)$, Decreasing on $(-3, -1)$ and $(-1, 1)$

Relative Maximum $(-3, -8)$, Relative Minimum $(1, 0)$

26. $f(x) = \dfrac{x^2 - 3x - 4}{x - 2}$

$f'(x) = \dfrac{(x - 2)(2x - 3) - (x^2 - 3x - 4)(1)}{(x - 2)^2}$

$= \dfrac{x^2 - 4x + 10}{(x - 2)^2},$ $x = 2$ Discontinuity

Test Intervals:	$x < 2$	$x > 2$
Sign of $f'(x)$:	$f' > 0$	$f' > 0$
Conclusion:	Increasing	Increasing

Increasing on $(-\infty, 2)$ and $(2, \infty)$, No Relative Extrema

27. $f(x) = \dfrac{x}{2} + \cos x,$ $0 \le x < 2\pi,$ $f'(x) = \dfrac{1}{2} - \sin x = 0$

$x = \dfrac{\pi}{6}, \dfrac{5\pi}{6}$ Critical Numbers

Test Intervals:	$0 \le x < \dfrac{\pi}{6}$	$\dfrac{\pi}{6} < x < \dfrac{5\pi}{6}$	$\dfrac{5\pi}{6} < x < 2\pi$
Sign of $f'(x)$:	$f' > 0$	$f' < 0$	$f' > 0$
Conclusion:	Increasing	Decreasing	Increasing

Increasing on $(0, \dfrac{\pi}{6})$ and $(\dfrac{5\pi}{6}, 2\pi)$, Decreasing on $(\dfrac{\pi}{6}, \dfrac{5\pi}{6})$

Relative Maximum $(\dfrac{\pi}{6}, \dfrac{\pi + 6\sqrt{3}}{12})$, Relative Minimum $(\dfrac{5\pi}{6}, \dfrac{5\pi - 6\sqrt{3}}{12})$

28. $f(x) = \sin x \cos x = \dfrac{1}{2}\sin 2x \qquad 0 \le x < 2\pi, \qquad f'(x) = \cos 2x = 0$

$x = \dfrac{\pi}{4}, \ \dfrac{3\pi}{4}, \ \dfrac{5\pi}{4}, \ \dfrac{7\pi}{4}$ Critical Numbers

$0 < x < \dfrac{\pi}{4}$,	$\dfrac{\pi}{4} < x < \dfrac{3\pi}{4}$,	$\dfrac{3\pi}{4} < x < \dfrac{5\pi}{4}$,	$\dfrac{5\pi}{4} < x < \dfrac{7\pi}{4}$,	$\dfrac{7\pi}{4} < x < 2\pi$
$f' > 0$	$f' < 0$	$f' > 0$	$f' < 0$	$f' > 0$
Increasing	Decreasing	Increasing	Decreasing	Increasing

Increasing on $(0, \dfrac{\pi}{4})$, $(\dfrac{3\pi}{4}, \dfrac{5\pi}{4})$ and $(\dfrac{7\pi}{4}, 2\pi)$

Decreasing on $(\dfrac{\pi}{4}, \dfrac{3\pi}{4})$ and $(\dfrac{5\pi}{4}, \dfrac{7\pi}{4})$

Relative Maxima $(\dfrac{\pi}{4}, \dfrac{1}{2})$, $(\dfrac{5\pi}{4}, \dfrac{1}{2})$, Relative Minima $(\dfrac{3\pi}{4}, -\dfrac{1}{2})$, $(\dfrac{7\pi}{4}, -\dfrac{1}{2})$

29. $f(x) = \sin^2 x + \sin x, \ 0 \le x < 2\pi$

$f'(x) = 2\sin x \cos x + \cos x = \cos x (2\sin x + 1) = 0$

$x = \dfrac{\pi}{2}, \ \dfrac{7\pi}{6}, \ \dfrac{3\pi}{2}, \ \dfrac{11\pi}{6}$ Critical numbers

$0 \le x < \dfrac{\pi}{2}$,	$\dfrac{\pi}{2} < x < \dfrac{7\pi}{6}$,	$\dfrac{7\pi}{6} < x < \dfrac{3\pi}{2}$,	$\dfrac{3\pi}{2} < x < \dfrac{11\pi}{6}$,	$\dfrac{11\pi}{6} < x < 2\pi$
$f' > 0$	$f' < 0$	$f' > 0$	$f' < 0$	$f' > 0$
Increasing	Decreasing	Increasing	Decreasing	Increasing

Increasing on $(0, \dfrac{\pi}{2})$, $(\dfrac{7\pi}{6}, \dfrac{3\pi}{2})$ and $(\dfrac{11\pi}{6}, 2\pi)$

Decreasing on $(\dfrac{\pi}{2}, \dfrac{7\pi}{6})$ and $(\dfrac{3\pi}{2}, \dfrac{11\pi}{6})$

Relative Minima $(\dfrac{7\pi}{6}, -\dfrac{1}{4})$, $(\dfrac{11\pi}{6}, -\dfrac{1}{4})$, Relative Maxima $(\dfrac{\pi}{2}, 2)$, $(\dfrac{3\pi}{2}, 0)$

30. $f(x) = \dfrac{\cos x}{1 + \sin^2 x}, \quad 0 \le x < 2\pi$

$f'(x) = \dfrac{(1 + \sin^2 x)(-\sin x) - \cos x (2 \sin x \cos x)}{(1 + \sin^2 x)^2}$

$= \dfrac{-\sin x (2 + \cos^2 x)}{(1 + \sin^2 x)^2} = 0$

$x = \pi$ Critical number

Test Intervals: $0 \le x < \pi$ $\pi < x < 2\pi$

Sign of $f'(x)$: $f' < 0$ $f' > 0$

Conclusion: Decreasing   Increasing

Increasing on $(\pi, 2\pi)$, Decreasing on $(0, \pi)$, Relative minimum $(\pi, -1)$

31. $s(t) = 96t - 16t^2 = 16t(6 - t) = 0$ thus $0 \le t \le 6$

$s'(t) = 96 - 32t = 0, \quad t = 3$

Test Intervals: $0 < t < 3$ $3 < t < 6$

Sign of $s'(x)$: $s' > 0$ $s' < 0$

Moving upward when $0 < t < 3$,   Moving downward when $3 < t < 6$

Maximum height: $s(3) = 144$ feet

32. $s(t) = -16t^2 + 64t = 16t(4 - t) = 0$ thus $0 \le t \le 4$

$s'(t) = -32t + 64 = 0, \quad t = 2$

Test Intervals: $0 < t < 2$ $2 < t < 4$

Sign of $s'(x)$: $s' > 0$ $s' < 0$

Moving upward when $0 < t < 2$,   Moving downward when $2 < t < 4$

Maximum height: $s(2) = 64$ feet

33. $v = k(R - r)r^2 = k(Rr^2 - r^3), \quad v' = k(2Rr - 3r^2) = kr(2R - 3r) = 0$

$r = 0$ or $r = \dfrac{2}{3}R, \quad$ Maximum when $r = \dfrac{2}{3}R$

34. $C = \dfrac{3t}{27 + t^3}$, $\quad C' = \dfrac{(27 + t^3)(3) - (3t)(3t^2)}{(27 + t^3)^2} = \dfrac{3(27 - 2t^3)}{(27 + t^3)^2} = 0$

$t = \dfrac{3}{\sqrt[3]{2}} \approx 2.38$ hours

35. $C = 0.29483t + 0.04253t^2 - 0.00035t^3$, $\quad 0 \leq t \leq 120$

$C' = 0.29483 + 0.08506t - 0.00105t^2 = 0$

$t = \dfrac{-0.08506 \pm \sqrt{(0.08506)^2 - 4(-0.00105)(0.29483)}}{2(-0.00105)}$, $\quad t = 84.3388$

Test Intervals: $0 < t < 84.3388$ $\qquad$ $84.3388 < t < 120$

Sign of C': $\qquad$ $C' > 0$ $\qquad\qquad$ $C' < 0$

Increasing when $0 < t < 84.3388$ minutes

Decreasing when $84.3388 < t < 120$ minutes

36. $P = 2.44x - \dfrac{x^2}{20,000} - 5000$, $0 \leq x \leq 35,000$

$P' = 2.44 - \dfrac{x}{10,000} = 0$, $\qquad x = 24,400$

Test Intervals: $0 < x < 24,400$ $\qquad$ $24,400 < x < 35,000$

Sign of P': $\qquad$ $P' > 0$ $\qquad\qquad$ $P' < 0$

Increasing when $0 < x < 24,400$ hamburgers

Decreasing when $24,400 < x < 35,000$ hamburgers

37. $W = 0.033t^2 - 0.3974t + 7.3032$, $\quad 0 \leq t \leq 14$

$W' = 0.066t - 0.3974 = 0$, $\quad t = 6.02$

Test Intervals: $0 < t < 6.02$ $\qquad$ $6.02 < t < 14$

Sign of W': $\qquad$ $W' < 0$ $\qquad\qquad$ $W' > 0$

Increasing when $6.02 < t < 14$ days

Decreasing when $0 < t < 6.02$ days

38. $C = 10(\frac{1}{x} + \frac{x}{x + 3}) = 10(\frac{x^2 + x + 3}{x^2 + 3x})$, $1 \le x$

$C' = 10\left[\frac{(x^2 + 3x)(2x + 1) - (x^2 + x + 3)(2x + 3)}{(x^2 + 3x)^2}\right]$

$\quad = 10\left[\frac{2x^2 - 6x - 9}{(x^2 + 3x)^2}\right] = 0$

$x = \frac{6 \pm \sqrt{36 + 72}}{4} = \frac{6 \pm 6\sqrt{3}}{4} = \frac{3(1 \pm \sqrt{3})}{2}$

(Consider only the positive value for x)

Test Intervals: $1 < x < \frac{3(1 + \sqrt{3})}{2}$ $\qquad x > \frac{3(1 + \sqrt{3})}{2}$

Sign of C': $\qquad$ C' < 0 $\qquad\qquad\qquad$ C' > 0

Increasing on $(\frac{3(1 + \sqrt{3})}{2}, \infty)$, $\quad$ Decreasing on $(1, \frac{3(1 + \sqrt{3})}{2})$

39. $R = \sqrt{0.001T^4 - 4T + 100}$, $\quad R' = \frac{0.004T^3 - 4}{2\sqrt{0.001T^4 - 4T + 100}} = 0$, $\quad T = 10°$

40. $P = \frac{vR_1R_2}{(R_1 + R_2)^2}$, v and R_1 are constant

$\frac{dP}{dR_2} = \frac{(R_1 + R_2)^2(vR_1) - (vR_1R_2)[2(R_1 + R_2)(1)]}{(R_1 + R_2)^4} = \frac{vR_1(R_1 - R_2)}{(R_1 + R_2)^3} = 0$

$R_1 = 0$ or $R_1 = R_2$, $\quad$ Maximum when $R_1 = R_2$

41. $F = 100,000\left[1 + \sin(\frac{2\pi(t - 60)}{365})\right]$, $F' = 100,000\left[\frac{2\pi}{365}\cos(\frac{2\pi(t - 60)}{365})\right] = 0$

$\frac{2\pi(t - 60)}{365} = \frac{\pi}{2}$ or $\frac{2\pi(t - 60)}{365} = \frac{3\pi}{2}$

$t \approx 151$ (May 31), $\quad$ $t \approx 334$ (November 30)

Maximum on May 31, $\quad$ Minimum on November 30

42. $T = 45 - 23\cos\left[\frac{2\pi(t - 32)}{365}\right]$, $\quad T' = 23(\frac{2\pi}{365})\sin\left[\frac{2\pi(t - 32)}{365}\right] = 0$

$\frac{2\pi(t - 32)}{365} = 0$ or $\frac{2\pi(t - 32)}{365} = \pi$

$t = 32$ (February 1), $\quad$ $t = 214.5$ (August 2 and 3)

(a) Warmest Days: August 2 and 3

(b) Coldest Day: February 1

43. $f(x) = ax^3 + bx^2 + cx + d$

   Relative minimum $(0, 0)$, Relative maximum $(2, 2)$

   Since $f(0) = 0 \implies d = 0$, $\quad f'(x) = 3ax^2 + 2bx + c$

   Since $(0, 0)$ is a relative minimum $\implies f'(0) = 0 \implies c = 0$

   Since $(2, 2)$ is a relative maximum $f'(2) = 0 \implies 12a + 4b = 0$

   $f(2) = 2 \implies 8a + 4b = 2$

   Solving this system of equations simultaneously we have

   $a = -\dfrac{1}{2}$ and $b = \dfrac{3}{2}$ therefore $f(x) = -\dfrac{1}{2}x^3 + \dfrac{3}{2}x^2$

44. $f(x) = ax^2 + bx + c$ passes through $(2, 10)$

   Relative maximum $(5, 20)$

   $f'(x) = 2ax + b = 0$, $\quad x = 5$ Critical number

   $x = \dfrac{-b}{2a} = 5 \implies b = -10a$, $\quad f(x) = ax^2 - 10ax + c$

   $f(5) = 25a - 50a + c = 20 \implies -25a + c = 20$

   $f(2) = 4a - 20a + c = 10 \implies \underline{-16a + c = 10}$
   $$-9a \qquad = 10$$

   $a = -\dfrac{10}{9}$, $\qquad b = \dfrac{100}{9}$, $\qquad c = -\dfrac{70}{9}$, $\qquad f(x) = -\dfrac{10}{9}(x^2 - 10x + 7)$

In Exercises 45 - 50, $f'(x) > 0$ on $(-\infty, -4)$, $f'(x) < 0$ on $(-4, 6)$ and $f'(x) > 0$ on $(6, \infty)$.

45. $g(x) = f(x) + 5$, $\qquad g'(x) = f'(x)$, $\qquad g'(0) = f'(0) < 0$

46. $g(x) = 3f(x) - 3$, $\qquad g'(x) = 3f'(x)$, $\qquad g'(-5) = 3f'(-5) > 0$

47. $g(x) = -f(x)$, $\qquad g'(x) = -f'(x)$, $\qquad g'(-6) = -f'(-6) < 0$

48. $g(x) = -f(x)$, $\qquad g'(x) = -f'(x)$, $\qquad g'(0) = -f'(0) > 0$

49. $g(x) = f(x - 10)$, $\qquad g'(x) = f'(x - 10)$, $\qquad g'(0) = f'(-10) > 0$

50. $g(x) = f(x - 10)$, $\qquad g'(x) = f'(x - 10)$, $\qquad g'(8) = f'(-2) < 0$

4.4

Concavity and the Second Derivative Test

1. $f(x) = 4 - x^2$

    (a) f is increasing,
        so f' is positive on $(-\infty, 0)$

    (b) f is decreasing,
        so f' is negative on $(0, \infty)$

    (c) f is not concave upward,
        so f' is never increasing

    (d) f is concave downward,
        so f' is decreasing on $(-\infty, \infty)$

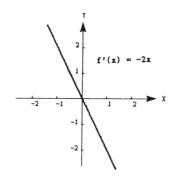

2. $f(x) = \cos \pi x$

    (a) f is increasing,
        so f' is positive on $(1, 2)$

    (b) f is decreasing,
        so f' is negative on $(0, 1)$

    (c) f is concave upward, so f'
        is increasing on $(1/2, 3/2)$

    (d) f is concave downward, so f'
        is decreasing on $(0, 1/2)$
        and $(3/2, 2)$

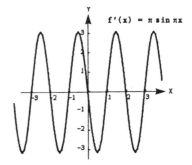

3. $f(x) = -\dfrac{1}{3}x^3 + x^2$

    (a) f is increasing,
        so f' is positive on $(0, 2)$

    (b) f is decreasing, so f' is
        negative on $(-\infty, 0)$ and $(2, \infty)$

    (c) f is concave upward,
        so f' is increasing on $(-\infty, 1)$

    (d) f is concave downward,
        so f' is decreasing on $(1, \infty)$

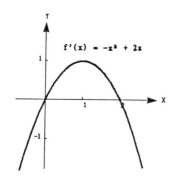

4. $f(x) = \dfrac{3}{2}x^{2/3}$

    (a) f is increasing,
        so f' is positive on $(0, \infty)$

    (b) f is decreasing,
        so f' is negative on $(-\infty, 0)$

    (c) f is not concave upward,
        so f' is never increasing

    (d) f is concave downward except
        at 0, so f' is decreasing on
        $(-\infty, 0)$ and $(0, \infty)$

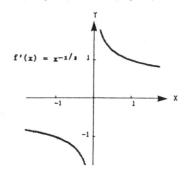

5.

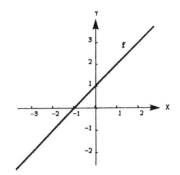

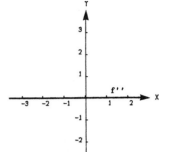

6.

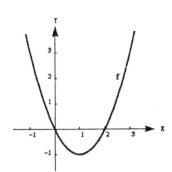

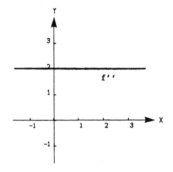

7.

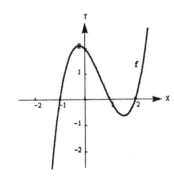

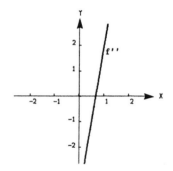

8.

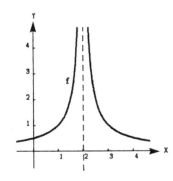

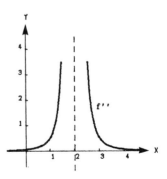

9. $y = x^2 - x - 2$

Concave upward: $(-\infty, \infty)$

10. $f(x) = \dfrac{24}{x^2 + 12}$

Concave upward: $(-\infty, -2)$ and $(2, \infty)$

Concave downward: $(-2, 2)$

11. $y = -x^3 + 3x^2 - 2$

    Concave upward: $(-\infty, 1)$

    Concave downward: $(1, \infty)$

12. $f(x) = \dfrac{x^2 - 1}{2x + 1}$

    Concave upward: $(-\infty, -1/2)$

    Concave downward: $(-1/2, \infty)$

13. $f(x) = \dfrac{x^2 + 1}{x^2 - 1}$

    Concave upward: $(-\infty, -1)$ and $(1, \infty)$

    Concave downward: $(-1, 1)$

14. $y = \dfrac{1}{270}(-3x^5 + 40x^3 + 135x)$

    Concave upward: $(-\infty, -2)$ and $(0, 2)$

    Concave downward: $(-2, 0)$ and $(2, \infty)$

15. $f(x) = 6x - x^2,$ $f'(x) = 6 - 2x,$ $f''(x) = -2$

    Critical number: $x = 3$

    $f''(3) < 0$, therefore $(3, 9)$ is a relative maximum.

16. $f(x) = x^2 + 3x - 8,$ $f'(x) = 2x + 3,$ $f''(x) = 2$

    Critical number: $x = -\dfrac{3}{2}$

    $f''(-\dfrac{3}{2}) > 0$, therefore $(-\dfrac{3}{2}, -\dfrac{41}{4})$ is a relative minimum.

17. $f(x) = (x - 5)^2,$ $f'(x) = 2(x - 5),$ $f''(x) = 2$

    Critical number: $x = 5$

    $f''(5) > 0$, therefore $(5, 0)$ is a relative minimum.

18. $f(x) = -(x - 5)^2,$ $f'(x) = -2(x - 5),$ $f''(x) = -2$

    Critical number: $x = 5$

    $f''(5) < 0$, therefore $(5, 0)$ is a relative maximum.

19. $f(x) = x^3 - 3x^2 + 3,$ $f'(x) = 3x^2 - 6x = 3x(x - 2)$

    $f''(x) = 6x - 6 = 6(x - 1),$ Critical numbers: $x = 0,$ $x = 2$

    $f''(0) = -6$, therefore $(0, 3)$ is a relative maximum.

    $f''(2) = 6$, therefore $(2, -1)$ is a relative minimum.

20. $f(x) = 5 + 3x^2 - x^3$, $\quad$ $f'(x) = 6x - 3x^2 = 3x(2 - x)$

$f''(x) = 6 - 6x = 6(1 - x)$, $\quad$ Critical numbers: $x = 0$, $x = 2$

$f''(0) = 6$, therefore $(0, 5)$ is a relative minimum.

$f''(2) = -6$, therefore $(2, 9)$ is a relative maximum.

21. $f(x) = x^4 - 4x^3 + 2$, $\quad$ $f'(x) = 4x^3 - 12x^2 = 4x^2(x - 3)$

$f''(x) = 12x^2 - 24x = 12x(x - 2)$, $\quad$ Critical numbers: $x = 0$, $x = 3$

However, $f''(0) = 0$, therefore we must use the First Derivative Test. But $f'(x) < 0$ on the intervals $(-\infty, 0)$ and $(0, 3)$, hence $(0, 2)$ is not an extremum. $f''(3) > 0$, therefore $(3, -25)$ is a relative minimum.

22. $f(x) = x^3 - 9x^2 + 27x - 26$, $\quad$ $f'(x) = 3(x - 3)^2$, $\quad$ $f''(x) = 6(x - 3)$

Critical number: $x = 3$

However, $f''(3) = 0$, therefore we must use the First Derivative Test. But $f'(x) \geq 0$ for all x, therefore there are no relative extrema.

23. $f(x) = x^{2/3} - 3$, $\quad$ $f'(x) = \dfrac{2}{3x^{1/3}}$, $\quad$ $f''(x) = \dfrac{-2}{9x^{4/3}}$

Critical number: $x = 0$

However, $f''(0)$ is undefined, therefore we must use the First Derivative Test. Since $f'(x) < 0$ on $(-\infty, 0)$ and $f'(x) > 0$ on $(0, \infty)$, then $(0, -3)$ is a relative minimum.

24. $f(x) = \sqrt{x^2 + 1}$, $\quad$ $f'(x) = \dfrac{x}{\sqrt{x^2 + 1}}$, $\quad$ Critical number: $x = 0$

However, $f''(0) = 0$, therefore we must use the First Derivative Test. Since $f'(x) < 0$ on $(-\infty, 0)$ and $f'(x) > 0$ on $(0, \infty)$, then $(0, 1)$ is a relative minimum.

25. $f(x) = x + \dfrac{4}{x}$, $\quad$ $f'(x) = 1 - \dfrac{4}{x^2}$, $\quad$ $f''(x) = \dfrac{8}{x^3}$, $\quad$ Critical numbers: $x = \pm 2$

$f''(-2) < 0$, therefore $(-2, -4)$ is a relative maximum.

$f''(2) > 0$, therefore $(2, 4)$ is a relative minimum.

26. $f(x) = \dfrac{x}{x - 1}$, $\quad$ $f'(x) = \dfrac{-1}{(x - 1)^2}$

There are no critical numbers and $x = 1$ is not in the domain, therefore there are no relative extrema.

27. $f(x) = x^3 - 12x$

   $f'(x) = 3x^2 - 12$

   $\qquad = 3(x + 2)(x - 2) = 0$

   when $x = \pm 2$

   $f''(x) = 6x = 0$ when $x = 0$

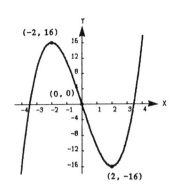

x	f(x)	f'(x)	f''(x)	
-2	16	0	-12	Relative maximum
0	0	-12	0	Point of inflection
2	-16	0	12	Relative minimum

28. $f(x) = x^3 + 1$

   $f'(x) = 3x^2 = 0$ when $x = 0$

   $f''(x) = 6x = 0$ when $x = 0$

   Since $f'(x) \geq 0$ for all x and the
   concavity changes at $x = 0$, then
   $(0, 1)$ is a point of inflection.

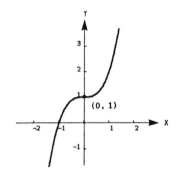

29. $f(x) = x^3 - 6x^2 + 12x - 8$

   $f'(x) = 3x^2 - 12x + 12$

   $\qquad = 3(x - 2)^2 = 0$

   when $x = 2$

   $f''(x) = 6(x - 2) = 0$ when $x = 2$

   Since $f'(x) > 0$ when $x \neq 2$ and the
   concavity changes at $x = 2$, then
   $(2, 0)$ is a point of inflection.

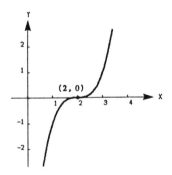

184

30. $f(x) = 2x^3 - 3x^2 - 12x + 8$

$f'(x) = 6x^2 - 6x - 12$

$\qquad = 6(x - 2)(x + 1) = 0$

when $x = -1, 2$

$f''(x) = 12x - 6 = 6(2x - 1) = 0$

when $x = \dfrac{1}{2}$

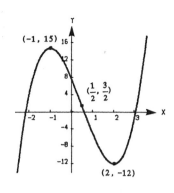

x	f(x)	f'(x)	f''(x)	
−1	15	0	−18	Relative maximum
1/2	3/2	−27/2	0	Point of inflection
2	−12	0	18	Relative minimum

31. $f(x) = \dfrac{x^4}{4} - 2x^2$

$f'(x) = x^3 - 4x$

$\qquad = x(x + 2)(x - 2) = 0$

when $x = 0, \pm 2$

$f''(x) = 3x^2 - 4 = 0$

when $x = \pm \dfrac{2}{\sqrt{3}}$

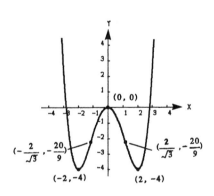

x	f(x)	f'(x)	f''(x)	
−2	−4	0	8	Relative minimum
−2/$\sqrt{3}$	−20/9	−16/3$\sqrt{3}$	0	Point of inflection
0	0	0	−4	Relative maximum
2/$\sqrt{3}$	−20/9	16/3$\sqrt{3}$	0	Point of inflection
2	−4	0	8	Relative minimum

32. $f(x) = 2x^4 - 8x + 3$

$f'(x) = 8x^3 - 8$

$\qquad = 8(x - 1)(x^2 + x + 1) = 0$

when $x = 1$

$f''(x) = 24x^2 = 0$ when $x = 0$

Since $f''(1) > 0$, then $(1, -3)$ is a relative minimum. However, $(0, 3)$ is not a point of inflection since $f''(x) \geq 0$, for all x.

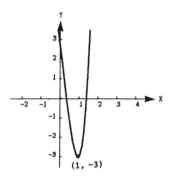

33. $f(x) = x^2 + \dfrac{1}{x^2}$

$f'(x) = 2x - \dfrac{2}{x^3}$

$\qquad = \dfrac{2}{x^3}(x + 1)(x - 1)(x^2 + 1) = 0$

when $x = \pm 1$

$f''(x) = 2 + \dfrac{6}{x^4}$

$\qquad = \dfrac{2}{x^4}(x^4 + 3)$

Since $f''(x) \geq 0$ for all $x \neq 0$, then $(\pm 1, 2)$ are relative minima.

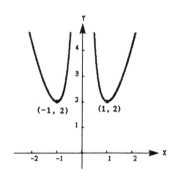

34. $f(x) = \dfrac{x^2}{x^2 - 1}$

$f'(x) = \dfrac{(x^2 - 1)(2x) - x^2(2x)}{(x^2 - 1)^2}$

$\qquad = \dfrac{-2x}{(x^2 - 1)^2} = 0$ when $x = 0$

$f''(x) = \dfrac{(x^2 - 1)^2(-2) + 2x(2)(x^2 - 1)(2x)}{(x^2 - 1)^4}$

$\qquad = \dfrac{2(3x^2 + 1)}{(x^2 - 1)^3}$

Since $f''(0) < 0$, then $(0, 0)$ is a relative maximum. Since $f''(x) \neq 0$ nor is undefined in the domain of f, there are no points of inflection.

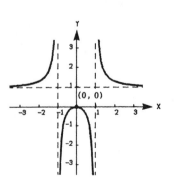

186

35. $f(x) = x\sqrt{x + 3}$ Domain: $[-3, \infty)$

$f'(x) = \dfrac{3(x + 2)}{2\sqrt{x + 3}} = 0$

when $x = -2$

$f''(x) = \dfrac{3(x + 4)}{4(x + 3)^{3/2}}$

Since $f''(-2) > 0$, then

$(-2, -2)$ is a relative minimum.

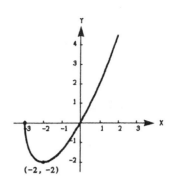

36. $f(x) = x\sqrt{x + 1}$ Domain: $[-1, \infty)$

$f'(x) = (x)\dfrac{1}{2}(x + 1)^{-1/2} + \sqrt{x + 1} = \dfrac{3x + 2}{2\sqrt{x + 1}} = 0$

when $x = -2/3$

$f''(x) = \dfrac{6\sqrt{x + 1} - (3x + 2)(x + 1)^{-1/2}}{4(x + 1)}$

$= \dfrac{3x + 4}{4(x + 1)^{3/2}}$

Since $f''(-\dfrac{2}{3}) > 0$, then $(-\dfrac{2}{3}, -\dfrac{2\sqrt{3}}{9})$

is a relative minimum.

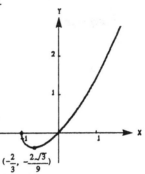

37. $f(x) = \dfrac{x}{x^2 - 4}$

$f'(x) = \dfrac{(x^2 - 4) - x(2x)}{(x^2 - 4)^2} = \dfrac{-(x^2 + 4)}{(x^2 - 4)^2} \neq 0$ for any x in the domain of f.

$f''(x) = \dfrac{(x^2 - 4)^2(-2x) + (x^2 + 4)(2)(x^2 - 4)(2x)}{(x^2 - 4)^4}$

$= \dfrac{2x(x^2 + 12)}{(x^2 - 4)^3} = 0$ when $x = 0$

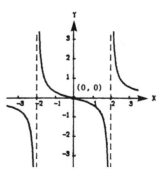

Since $f''(-1) > 0$ and $f''(1) < 0$,

then $(0, 0)$ is a point of inflection.

38. $f(x) = \dfrac{1}{x^2 - x - 2}$

$f'(x) = \dfrac{-(2x - 1)}{(x^2 - x - 2)^2} = 0$ when $x = \dfrac{1}{2}$

$f''(x) = \dfrac{(x^2 - x - 2)^2(-2) + (2x - 1)(2)(x^2 - x - 2)(2x - 1)}{(x^2 - x - 2)^4}$

$\qquad = \dfrac{6(x^2 - x + 1)}{(x^2 - x - 2)^3}$

Since $f''(1/2) < 0$, then $(1/2, -4/9)$ is

a relative maximum. Since $f''(x) \neq 0$

nor is undefined in the domain of f,

there are no points of inflection.

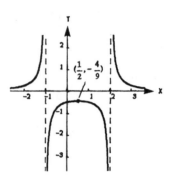

39. $f(x) = \dfrac{x - 2}{x^2 - 4x + 3}$

$f'(x) = \dfrac{(x^2 - 4x + 3) - (x - 2)(2x - 4)}{(x^2 - 4x + 3)^2}$

$\qquad = \dfrac{-x^2 + 4x - 5}{(x^2 - 4x + 3)^2} \neq 0$

$f''(x) = \dfrac{(x^2 - 4x + 3)^2(-2x + 4) - (-x^2 + 4x - 5)(2)(x^2 - 4x + 3)(2x - 4)}{(x^2 - 4x + 3)^4}$

$\qquad = \dfrac{2(x^3 - 6x^2 + 15x - 14)}{(x^2 - 4x + 3)^3} = 0$

when $x = 2$

Since $f''(3/2) > 0$ and $f''(5/2) < 0$,

then $(2, 0)$ is a point of inflection.

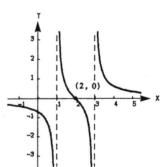

40. $f(x) = \dfrac{x + 1}{x^2 + x + 1}$, $f'(x) = \dfrac{-x(x + 2)}{(x^2 + x + 1)^2} = 0$ when $x = 0, -2$

$f''(x) = \dfrac{2(x^3 + 3x^2 - 1)}{(x^2 + x + 1)^3} = 0*$ when $x \approx 0.5321, -0.6527, -2.8794$

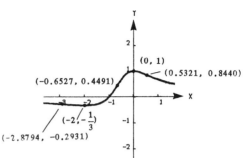

$f''(0) < 0$, therefore $(0, 1)$ is a
relative maximum.
$f''(-2) > 0$, therefore $(-2, -\dfrac{1}{3})$ is a
relative minimum.

To four decimal places:
$(0.5321, 0.8440)$, $(-0.6527, 0.4491)$
and $(-2.8794, -0.2931)$ are points of
inflection.

*We can solve $x^3 + 3x^2 - 1 = 0$ by using an interpolation method on a
programmable calculator or by the Bisection Method written below in
BASIC.

```
10   REM DEFINE THE FUNCTION
20       DEF FNY(X) = X^3 + 3*X^2 - 1
30   REM A = LEFT ENDPOINT, B = RIGHT ENDPOINT
40       INPUT " ENTER LEFT ENDPOINT: ",A
50       INPUT "ENTER RIGHT ENDPOINT: ",B
60   REM DETERMINE MIDPOINT = C
70       C = (A + B)/2
80   REM M = MAXIMUM ERROR
90       M = (B - A)/2
100  REM TEST THE SIZE OF THE ERROR
110      IF M < .00001 GOTO 160
120  REM TEST THE SIGN OF F(C)
130      IF FNY(A)*FNY(C) < 0 THEN B = C: GOTO 70
140      IF FNY(B)*FNY(C) < 0 THEN A = C: GOTO 70
150  REM PRINT THE APPROXIMATION
160      PRINT "A =";A,"C =";C,"B =";B,"MAXIMUM ERROR =";M
170      END
```

The output from the program is:

```
 ENTER LEFT ENDPOINT: -3
ENTER RIGHT ENDPOINT: -2
A =-2.87939   C =-2.87939   B =-2.87938   MAXIMUM ERROR = 7.62939E-06

 ENTER LEFT ENDPOINT: -1
ENTER RIGHT ENDPOINT: 0
A =-.65271   C =-.652702   B =-.652695   MAXIMUM ERROR = 7.62939E-06

 ENTER LEFT ENDPOINT: 0
ENTER RIGHT ENDPOINT: 1
A = .532074   C = .532082   B = .532089   MAXIMUM ERROR = 7.62939E-06
```

41. $f(x) = \sin\left(\dfrac{x}{2}\right)$, $\quad f'(x) = \dfrac{1}{2}\cos\left(\dfrac{x}{2}\right) = 0$ when $x = \pi,\ 3\pi$

$f''(x) = -\dfrac{1}{4}\sin\left(\dfrac{x}{2}\right) = 0$

when $x = 0,\ 2\pi,\ 4\pi$

$f''(\pi) < 0 \implies (\pi,\ 1)$ relative maximum

$f''(3\pi) > 0 \implies (3\pi,\ -1)$ relative minimum

$(2\pi,\ 0)$ point of inflection

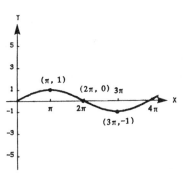

42. $f(x) = 2\csc\left(\dfrac{3x}{2}\right)$, $\quad f'(x) = -3\csc\dfrac{3x}{2}\cot\dfrac{3x}{2} = 0$ when $x = \dfrac{\pi}{3},\ \pi,\ \dfrac{5\pi}{3}$

$f''(x) = \dfrac{9}{2}\left(\csc^3\dfrac{3x}{2} + \csc\dfrac{3x}{2}\cot^2\dfrac{3x}{2}\right) \neq 0$

for any x in the domain of f.

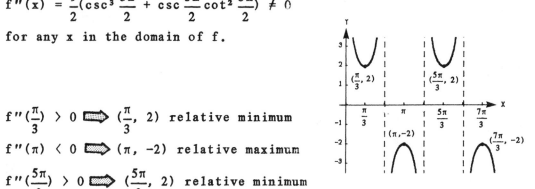

$f''\left(\dfrac{\pi}{3}\right) > 0 \implies \left(\dfrac{\pi}{3},\ 2\right)$ relative minimum

$f''(\pi) < 0 \implies (\pi,\ -2)$ relative maximum

$f''\left(\dfrac{5\pi}{3}\right) > 0 \implies \left(\dfrac{5\pi}{3},\ 2\right)$ relative minimum

43. $f(x) = \sec\left(x - \dfrac{\pi}{2}\right)$, $f'(x) = \sec\left(x - \dfrac{\pi}{2}\right)\tan\left(x - \dfrac{\pi}{2}\right) = 0$

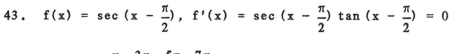

when $x = \dfrac{\pi}{2},\ \dfrac{3\pi}{2},\ \dfrac{5\pi}{2},\ \dfrac{7\pi}{2}$

$f''(x) = \sec^3\left(x - \dfrac{\pi}{2}\right) + \sec\left(x - \dfrac{\pi}{2}\right)\tan^2\left(x - \dfrac{\pi}{2}\right) \neq 0$

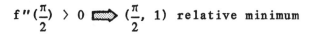

for any x in the domain of f.

$f''\left(\dfrac{\pi}{2}\right) > 0 \implies \left(\dfrac{\pi}{2},\ 1\right)$ relative minimum

$f''\left(\dfrac{3\pi}{2}\right) < 0 \implies \left(\dfrac{3\pi}{2},\ -1\right)$ relative maximum

$f''\left(\dfrac{5\pi}{2}\right) > 0 \implies \left(\dfrac{5\pi}{2},\ 1\right)$ relative minimum

$f''\left(\dfrac{7\pi}{2}\right) < 0 \implies \left(\dfrac{7\pi}{2},\ -1\right)$ relative maximum

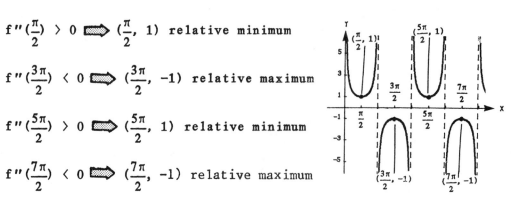

44. $f(x) = \sin x + \cos x$, $f'(x) = \cos x - \sin x = 0$

    when $x = \dfrac{\pi}{4}, \dfrac{5\pi}{4}$

    $f''(x) = -\sin x - \cos x = 0$

    when $x = \dfrac{3\pi}{4}, \dfrac{7\pi}{4}$

    $f''\left(\dfrac{\pi}{4}\right) < 0 \implies \left(\dfrac{\pi}{4}, \sqrt{2}\right)$ relative maximum

    $f''\left(\dfrac{5\pi}{4}\right) > 0 \implies \left(\dfrac{5\pi}{4}, -\sqrt{2}\right)$ relative minimum

    $\left(\dfrac{3\pi}{4}, 0\right)$ point of inflection

    $\left(\dfrac{7\pi}{4}, 0\right)$ point of inflection

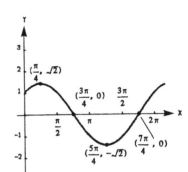

45. $f(x) = 2\sin x + \sin 2x$

    $f'(x) = 2\cos x + 2\cos 2x = 0$

    $2(2\cos^2 x + \cos x - 1) = 0$

    $2(\cos x + 1)(2\cos x - 1) = 0$

    when $\cos x = \dfrac{1}{2}, -1$ or $x = \dfrac{\pi}{3}, \pi, \dfrac{5\pi}{3}$

    $f''(x) = -2\sin x - 4\sin 2x$

    $\qquad = -2\sin x\,(1 + 4\cos x) = 0$

    when $x = 0, 1.823, \pi, 4.460$

    $\left(\dfrac{\pi}{3}, 2.598\right)$ relative maximum

    $\left(\dfrac{5\pi}{3}, -2.598\right)$ relative minimum

    $(0, 0), (1.823, 1.452), (\pi, 0),$
    $(4.46, -1.452)$ points of inflection

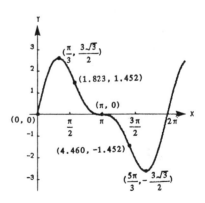

46. $f(x) = 2 \sin x + \cos 2x$

$f'(x) = 2 \cos x - 2 \sin 2x = 2 \cos x(1 - 2 \sin x) = 0$

when $x = \dfrac{\pi}{6}, \dfrac{\pi}{2}, \dfrac{5\pi}{6}, \dfrac{3\pi}{2}$

$f''(x) = -2 \sin x - 4 \cos 2x = 4 \sin^2 x - \sin x - 2 = 0$

when $x = 1.0030, 2.1386, 3.7765, 5.6483$

$\left(\dfrac{\pi}{6}, \dfrac{3}{2}\right)$, $\left(\dfrac{5\pi}{6}, \dfrac{3}{2}\right)$ relative maxima

$\left(\dfrac{\pi}{2}, 1\right)$, $\left(\dfrac{3\pi}{2}, -3\right)$ relative minima

$(1.0030, 1.2646)$, $(2.1386, 1.2646)$,
$(3.7765, -0.8897)$, $(5.6483, -0.8897)$
points of inflection

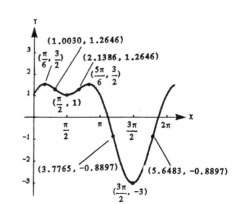

47. $f(x) = x - \sin x$

$f'(x) = 1 - \cos x \geq 0$,

therefore f is nondecreasing

$f''(x) = \sin x = 0$

when $x = 0, \pi, 2\pi, 3\pi, 4\pi$

$(0, 0)$, (π, π), $(2\pi, 2\pi)$, $(3\pi, 3\pi)$,
$(4\pi, 4\pi)$ points of inflection

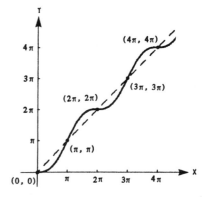

48. $f(x) = \cos x - x$

$f'(x) = -\sin x - 1 \leq 0$

therefore f is non-increasing

$f''(x) = -\cos x = 0$

when $x = \dfrac{\pi}{2}, \dfrac{3\pi}{2}, \dfrac{5\pi}{2}, \dfrac{7\pi}{2}$

$\left(\dfrac{\pi}{2}, -\dfrac{\pi}{2}\right)$, $\left(\dfrac{3\pi}{2}, -\dfrac{3\pi}{2}\right)$, $\left(\dfrac{5\pi}{2}, -\dfrac{5\pi}{2}\right)$,

$\left(\dfrac{7\pi}{2}, \dfrac{-7\pi}{2}\right)$ points of inflection

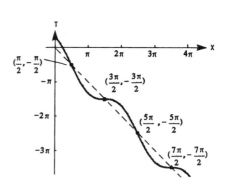

49.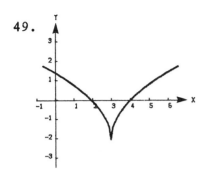

50.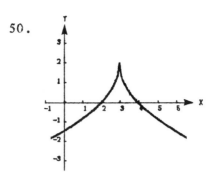

51. $f(x) = ax^3 + bx^2 + cx + d$,   Relative maximum $(3, 3)$

Relative minimum $(5, 1)$,   Point of inflection $(4, 2)$

$f'(x) = 3ax^2 + 2bx + c$, $f''(x) = 6ax + 2b$

$\left.\begin{array}{l} f(3) = 27a + 9b + 3c + d = 3 \\ f(5) = 125a + 25b + 5c + d = 1 \end{array}\right\}$ $98a + 16b + 2c = -2$ ⟹ $49a + 8b + c = -1$

$f'(3) = 27a + 6b + c = 0$, $f''(4) = 24a + 2b = 0$

$\begin{array}{ll} 49a + 8b + c = -1 & \qquad 24a + 2b \quad\;\; = 0 \\ \underline{27a + 6b + c = \;\; 0} & \qquad \underline{22a + 2b \qquad = -1} \\ 22a + 2b \qquad = -1 & \qquad 2a \qquad\qquad = 1 \end{array}$

$a = \dfrac{1}{2}$, $b = -6$, $c = \dfrac{45}{2}$, $d = -24$, $f(x) = \dfrac{1}{2}x^3 - 6x^2 + \dfrac{45}{2}x - 24$

52. $f(x) = ax^3 + bx^2 + cx + d$,   Relative maximum $(2, 4)$

Relative minimum $(4, 2)$,   Point of inflection $(3, 3)$

$f'(x) = 3ax^2 + 2bx + c$, $f''(x) = 6ax + 2b$

$\left.\begin{array}{l} f(2) = 8a + 4b + 2c + d = 4 \\ \\ f(4) = 64a + 16b + 4c + d = 2 \end{array}\right\}$ $56a + 12b + 2c = -2$ ⟹ $28a + 6b + c = -1$

$f'(2) = 12a + 4b + c = 0$, $f'(4) = 48a + 8b + c = 0$

$f''(3) = 18a + 2b = 0$

$\begin{array}{ll} 28a + 6b + c = -1 & \qquad 18a + 2b = \;\; 0 \\ \underline{12a + 4b + c = \;\; 0} & \qquad \underline{16a + 2b = -1} \\ 16a + 2b \qquad = -1 & \qquad 2a \qquad\;\; = \;\; 1 \end{array}$

$a = \dfrac{1}{2}$, $b = -\dfrac{9}{2}$, $c = 12$, $d = -6$, $f(x) = \dfrac{1}{2}x^3 - \dfrac{9}{2}x^2 + 12x - 6$

53. $f(x) = x(x - 6)^2 = x^3 - 12x^2 + 36x$

    $f'(x) = 3x^2 - 24x + 36 = 3(x - 2)(x - 6) = 0$

    Relative extrema $(2, 32)$ and $(6, 0)$

    Point of inflection $(4, 16)$ is midway between the relative extrema of f.

54. Assume the zeros of f are all real. Then express the function as

    $$f(x) = a(x - r_1)(x - r_2)(x - r_3)$$

    where r_1, r_2, and r_3 are the zeros of f.

    From the note below Example 3 in Section 3.4, we have

    $f'(x) = a[(x - r_1)(x - r_2) + (x - r_1)(x - r_3) + (x - r_2)(x - r_3)]$

    $f''(x) = a[(x - r_1) + (x - r_2) + (x - r_1) + (x - r_3) + (x - r_2) + (x - r_3)]$

    $\qquad = a[6x - 2(r_1 + r_2 + r_3)]$

    Consequently, $f''(x) = 0$ if

    $x = \dfrac{2(r_1 + r_2 + r_3)}{6} = \dfrac{r_1 + r_2 + r_3}{3}$ = (average of r_1, r_2, and r_3)

55. $D = 2x^4 - 5Lx^3 + 3L^2x^2$

    $D' = 8x^3 - 15Lx^2 + 6L^2x = x(8x^2 - 15Lx + 6L^2) = 0$

    $x = 0$ or $x = \dfrac{15L \pm \sqrt{33}L}{16} = (\dfrac{15 \pm \sqrt{33}}{16})L$

    By the Second Derivative Test the deflection is maximum when

    $x = (\dfrac{15 - \sqrt{33}}{16})L \approx 0.578L$ feet

56. $E = T(x^2 + a^2)^{-3/2}$, $\qquad E' = -\dfrac{3}{2}T(2x)(x^2 + a^2)^{-5/2} = 0$, $\qquad x = 0$

    By the First Derivative Test, E is maximized when $x = 0$.

57. $C = 0.5x^2 + 15x + 5000$

    $\overline{C} = \dfrac{C}{x} = 0.5x + 15 + \dfrac{5000}{x}$, $\quad \overline{C}$ = average cost per unit

    $\overline{C}' = 0.5 - \dfrac{5000}{x^2} = 0$, $\qquad x = 100$

    By the First Derivative Test, $\overline{C}$ is maximized when $x = 100$ units.

58. $C = 2x + \dfrac{300,000}{x}$

    $C' = 2 - \dfrac{300,000}{x^2} = 0, \qquad x = 100\sqrt{15} \approx 387$

    By the First Derivative Test, C is maximized when $x \approx 387$ units.

59. $f(x) = x \sin \left(\dfrac{1}{x}\right)$

    $f'(x) = x \left[-\dfrac{1}{x^2} \cos \left(\dfrac{1}{x}\right) \right] + \sin \left(\dfrac{1}{x}\right) = -\dfrac{1}{x} \cos \left(\dfrac{1}{x}\right) + \sin \left(\dfrac{1}{x}\right)$

    $f''(x) = -\dfrac{1}{x} \left[\dfrac{1}{x^2} \sin \left(\dfrac{1}{x}\right) \right] + \dfrac{1}{x^2} \cos \left(\dfrac{1}{x}\right) - \dfrac{1}{x^2} \cos \left(\dfrac{1}{x}\right) = -\dfrac{1}{x^3} \sin \left(\dfrac{1}{x}\right) = 0$

    $x = \dfrac{1}{\pi}, \qquad$ Point of inflection $\left(\dfrac{1}{\pi}, 0\right)$

    When $x > \dfrac{1}{\pi}$, $f'' < 0$ so the graph is concave down.

60. $v = -1200\pi \sin \theta, \quad \dfrac{d\theta}{dt} = (200)(2\pi) = 400\pi$ radians/min

    $\dfrac{dv}{dt} = -1200\pi \cos \theta \dfrac{d\theta}{dt} = -1200\pi(400\pi) \cos \theta = 0$

    when $\theta = \dfrac{(2n + 1)\pi}{2}$ where n is any integer. By the First

    Derivative Test v (velocity) is maximum when $\theta = \dfrac{3\pi}{2} + 2n\pi$.

    The speed $|v|$ is maximum when $\theta = \dfrac{\pi}{2} + 2n\pi$ and $\dfrac{3\pi}{2} + 2n\pi$.

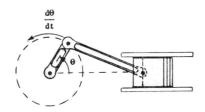

4.5
Limits at infinity

1. (h)   2. (c)   3. (e)   4. (a)

5. (d)   6. (g)   7. (b)   8. (f)

9. $\displaystyle\lim_{x \to \infty} \frac{2x - 1}{3x + 2} = \lim_{x \to \infty} \frac{2 - 1/x}{3 + 2/x} = \frac{2 - 0}{3 + 0} = \frac{2}{3}$

10. $\displaystyle\lim_{x \to \infty} \frac{5x^3 + 1}{10x^3 - 3x^2 + 7} = \lim_{x \to \infty} \frac{5 + (1/x^3)}{10 - (3/x) + (7/x^3)} = \frac{1}{2}$

11. $\displaystyle\lim_{x \to \infty} \frac{x}{x^2 - 1} = \lim_{x \to \infty} \frac{1/x}{1 - (1/x^2)} = \frac{0}{1} = 0$

12. $\displaystyle\lim_{x \to \infty} \frac{2x^{10} - 1}{10x^{11} - 3} = \lim_{x \to \infty} \frac{(2/x) - (1/x^{11})}{10 - (3/x^{11})} = \frac{0}{10} = 0$

13. $\displaystyle\lim_{x \to -\infty} \frac{5x^2}{x + 3} = \lim_{x \to -\infty} \frac{5}{(1/x) + (3/x^2)}$,   Limit does not exist

14. $\displaystyle\lim_{x \to \infty} \frac{x^3 - 2x^2 + 3x + 1}{x^2 - 3x + 2} = \lim_{x \to \infty} \frac{1 - 2/x + 3/x^2 + 1/x^3}{1/x - 3/x^2 + 2/x^3}$

    Limit does not exist

15. $\displaystyle\lim_{x \to \infty} \left(2x - \frac{1}{x^2}\right) = \lim_{x \to \infty} \frac{2x^3 - 1}{x^2}$,   Limit does not exist

16. $\displaystyle\lim_{x \to \infty} \frac{1}{(x + 3)^2} = 0$

17. $\displaystyle\lim_{x \to -\infty} \left(\frac{2x}{x - 1} + \frac{3x}{x + 1}\right) = \lim_{x \to -\infty} \left(\frac{2}{1 - (1/x)} + \frac{3}{1 + (1/x)}\right) = 2 + 3 = 5$

18. $\displaystyle\lim_{x \to \infty} \left(\frac{2x^2}{x - 1} + \frac{3x}{x + 1}\right) = \lim_{x \to \infty} \left(\frac{2}{(1/x) - (1/x^2)} + \frac{3}{1 + (1/x)}\right) = \infty + 3$

    Limit does not exist

19. $\displaystyle\lim_{x \to -\infty} (x + \sqrt{x^2 + 3}) = \lim_{x \to -\infty} \left[(x + \sqrt{x^2 + 3}) \cdot \frac{x - \sqrt{x^2 + 3}}{x - \sqrt{x^2 + 3}} \right]$

$\displaystyle = \lim_{x \to -\infty} \frac{-3}{x - \sqrt{x^2 + 3}} = 0$

20. $\displaystyle\lim_{x \to \infty} (2x - \sqrt{4x^2 + 1}) = \lim_{x \to \infty} \left[(2x - \sqrt{4x^2 + 1}) \cdot \frac{2x + \sqrt{4x^2 + 1}}{2x + \sqrt{4x^2 + 1}} \right]$

$\displaystyle = \lim_{x \to \infty} \frac{-1}{2x + \sqrt{4x^2 + 1}} = 0$

21. $\displaystyle\lim_{x \to \infty} (x - \sqrt{x^2 + x}) = \lim_{x \to \infty} \left[(x - \sqrt{x^2 + x}) \cdot \frac{x + \sqrt{x^2 + x}}{x + \sqrt{x^2 + x}} \right]$

$\displaystyle = \lim_{x \to \infty} \frac{-x}{x + \sqrt{x^2 + x}} = \lim_{x \to \infty} \frac{-1}{1 + \sqrt{1 + (1/x)}} = -\frac{1}{2}$

22. $\displaystyle\lim_{x \to -\infty} (3x + \sqrt{9x^2 - x}) = \lim_{x \to -\infty} \left[(3x + \sqrt{9x^2 - x}) \cdot \frac{3x - \sqrt{9x^2 - x}}{3x - \sqrt{9x^2 - x}} \right]$

$\displaystyle = \lim_{x \to -\infty} \frac{x}{3x - \sqrt{9x^2 - x}}$

$\displaystyle = \lim_{x \to -\infty} \frac{1}{3 - \dfrac{\sqrt{9x^2 - x}}{-\sqrt{x^2}}} \qquad (\text{for } x < 0 \text{ we have } x = -\sqrt{x^2})$

$\displaystyle = \lim_{x \to -\infty} \frac{1}{3 + \sqrt{9 - (1/x)}} = \frac{1}{6}$

23. $\displaystyle\lim_{x \to -\infty} \frac{x}{\sqrt{x^2 - x}} = \lim_{x \to -\infty} \frac{1}{\dfrac{\sqrt{x^2 - x}}{-\sqrt{x^2}}} \qquad (\text{for } x < 0 \text{ we have } x = -\sqrt{x^2})$

$\displaystyle = \lim_{x \to -\infty} \frac{-1}{\sqrt{1 - (1/x)}} = -1$

24. $\displaystyle\lim_{x \to \infty} \frac{x}{\sqrt{x^2 + 1}} = \lim_{x \to \infty} \frac{1}{\sqrt{1 + (1/x^2)}} = 1$

25. $\displaystyle\lim_{x \to \infty} \frac{2x + 1}{\sqrt{x^2 - x}} = \lim_{x \to \infty} \frac{2 + (1/x)}{\sqrt{1 - (1/x)}} = 2$

26. $\lim\limits_{x \to -\infty} \dfrac{-3x + 1}{\sqrt{x^2 + x}} = \lim\limits_{x \to -\infty} \dfrac{-3 + (1/x)}{\dfrac{\sqrt{x^2 + x}}{-\sqrt{x^2}}}$ (for $x < 0$ we have $-\sqrt{x^2} = x$)

$= \lim\limits_{x \to -\infty} \dfrac{3 - (1/x)}{\sqrt{1 + (1/x)}} = 3$

27. $\lim\limits_{x \to \infty} \dfrac{x^2 - x}{\sqrt{x^4 + x}} = \lim\limits_{x \to \infty} \dfrac{1 - (1/x)}{\sqrt{1 + (1/x^3)}} = 1$

28. $\lim\limits_{x \to \infty} \dfrac{2x}{\sqrt{4x^2 + 1}} = \lim\limits_{x \to \infty} \dfrac{2}{\sqrt{4 + 1/x^2}} = 1$

29. Since $-\dfrac{1}{x} \leq \dfrac{\sin(2x)}{x} \leq \dfrac{1}{x}$ for all $x \neq 0$, we have by the Squeeze Theorem,

$\lim\limits_{x \to \infty} -\dfrac{1}{x} \leq \lim\limits_{x \to \infty} \dfrac{\sin(2x)}{x} \leq \lim\limits_{x \to \infty} \dfrac{1}{x}$

$0 \leq \lim\limits_{x \to \infty} \dfrac{\sin(2x)}{x} \leq 0$, therefore $\lim\limits_{x \to \infty} \dfrac{\sin(2x)}{x} = 0$

30. $\lim\limits_{x \to \infty} \dfrac{x - \cos x}{x} = \lim\limits_{x \to \infty} \left(1 - \dfrac{\cos x}{x}\right) = 1 - 0 = 1$

Note: $\lim\limits_{x \to \infty} \dfrac{\cos x}{x} = 0$ by the Squeeze Theorem since $-\dfrac{1}{x} \leq \dfrac{\cos x}{x} \leq \dfrac{1}{x}$

31. $\lim\limits_{x \to \infty} \dfrac{1}{2x + \sin x} = 0$

32. $\lim\limits_{x \to \infty} \sin\dfrac{1}{x} = \sin 0 = 0$

33. $\lim\limits_{x \to \infty} x \sin\dfrac{1}{x} = \lim\limits_{t \to 0} \dfrac{\sin t}{t} = 1$ (Let $x = \dfrac{1}{t}$)

34. $\lim\limits_{x \to \infty} x \tan\dfrac{1}{x} = \lim\limits_{t \to 0} \dfrac{\tan t}{t} = \lim\limits_{t \to 0} \left[\dfrac{\sin t}{t} \cdot \dfrac{1}{\cos t}\right] = (1)(1) = 1$

35. $y = \dfrac{2 + x}{1 - x}$

Intercepts: $x = -2$, $y = 2$
Symmetry: none
Asymptotes:

$$\lim_{x \to -\infty} \frac{2 + x}{1 - x} = -1 = \lim_{x \to \infty} \frac{2 + x}{1 - x}$$

$y = -1$

Discontinuity: $x = 1$

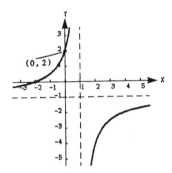

36. $y = \dfrac{x - 3}{x - 2}$

Intercepts: $x = 3$, $y = 3/2$
Symmetry: none
Asymptotes:

$$\lim_{x \to -\infty} \frac{x - 3}{x - 2} = 1 = \lim_{x \to \infty} \frac{x - 3}{x - 2}$$

$y = 1$

Discontinuity: $x = 2$

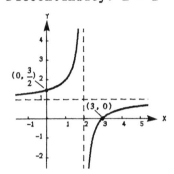

37. $y = \dfrac{x^2}{x^2 + 9}$

Intercepts: $x = 0$, $y = 0$
Symmetry: y-axis
Asymptotes:

$$\lim_{x \to -\infty} \frac{x^2}{x^2 + 9} = 1 = \lim_{x \to \infty} \frac{x^2}{x^2 + 9}$$

$y = 1$

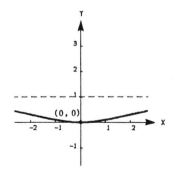

38. $y = \dfrac{x^2}{x^2 - 9}$

Intercepts: $x = 0$, $y = 0$
Symmetry: y-axis
Asymptotes:

$$\lim_{x \to -\infty} \frac{x^2}{x^2 - 9} = 1 = \lim_{x \to \infty} \frac{x^2}{x^2 - 9}$$

$y = 1$

Discontinuities: $x = \pm 3$

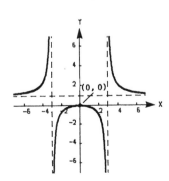

39. $xy^2 = 4$, Domain: $x > 0$

Intercepts: none
Symmetry: x-axis
Asymptote:

$$\lim_{x \to \infty} \frac{2}{\sqrt{x}} = 0 = \lim_{x \to \infty} -\frac{2}{\sqrt{x}}$$

$y = 0$

Discontinuity: $x = 0$

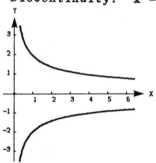

40. $x^2y = 4$

Intercepts: none
Symmetry: y-axis
Asymptote:

$$\lim_{x \to -\infty} \frac{4}{x^2} = 0 = \lim_{x \to \infty} \frac{4}{x^2}$$

$y = 0$

Discontinuity: $x = 0$

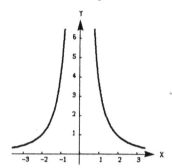

41. $y = \dfrac{2x}{1 - x}$

Intercepts: $x = 0$, $y = 0$
Symmetry: none
Asymptote:

$$\lim_{x \to -\infty} \frac{2x}{1 - x} = -2 = \lim_{x \to \infty} \frac{2x}{1 - x}$$

$y = -2$

Discontinuity: $x = 1$

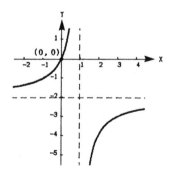

42. $y = \dfrac{2x}{1 - x^2}$

Intercepts: $x = 0$, $y = 0$
Symmetry: origin
Asymptotes:

$$\lim_{x \to -\infty} \frac{2x}{1 - x^2} = 0 = \lim_{x \to \infty} \frac{2x}{1 - x^2}$$

$y = 0$

Discontinuities: $x = \pm 1$

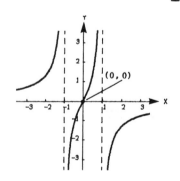

43. $y = 2 - \dfrac{3}{x^2}$

Intercept: $x = \pm \sqrt{3/2}$
Symmetry: y-axis
Asymptote:

$$\lim_{x \to -\infty} \left(2 - \dfrac{3}{x^2}\right) = 2 = \lim_{x \to \infty} \left(2 - \dfrac{3}{x^2}\right)$$

$y = 2$

Discontinuity: $x = 0$

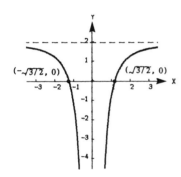

44. $y = 1 + \dfrac{1}{x}$

Intercept: $x = -1$
Symmetry: none
Asymptote:

$$\lim_{x \to -\infty} \left(1 + \dfrac{1}{x}\right) = 1 = \lim_{x \to \infty} \left(1 + \dfrac{1}{x}\right)$$

$y = 1$

Discontinuity: $x = 0$

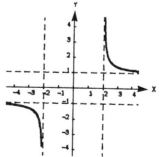

45. $y = \dfrac{x^3}{\sqrt{x^2 - 4}}$, Domain: $|x| > 2$

Intercepts: none
Symmetry: origin
Asymptotes: none

Vertical: $x = \pm 2$

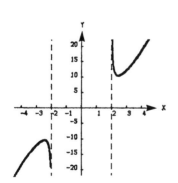

46. $y = \dfrac{x}{\sqrt{x^2 - 4}}$, Domain: $|x| > 2$

Intercepts: none
Symmetry: origin
Asymptote: $\lim_{x \to \infty} \dfrac{x}{\sqrt{x^2 - 4}} = 1$

$$\lim_{x \to -\infty} \dfrac{x}{\sqrt{x^2 - 4}} = -1$$

$y = 1,\ y = -1$
Vertical: $x = \pm 2$

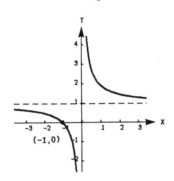

47.

x	1	10	10^2	10^4	10^6
f(x)	2.000	0.348	0.101	0.010	0.001

$$\lim_{x \to \infty} \frac{x + 1}{x\sqrt{x}} = 0$$

48.

x	1	10	10^2	10^3	10^6
f(x)	1	0.513	0.501	0.500	0.500

$$\lim_{x \to \infty} (x - \sqrt{x(x - 1)}) = \frac{1}{2}$$

49.

x	1	10	10^2	10^3	10^6
f(x)	0.479	0.500	0.500	0.500	0.500

$$\lim_{x \to \infty} x \sin\left(\frac{1}{2x}\right) = \frac{1}{2}$$

50.

x	1	10	10^2	10^4	10^6
f(x)	1.0	5.1	50.1	5000.1	500000.2

$$\lim_{x \to \infty} [x^2 - x\sqrt{x(x - 1)}] = \infty, \text{ Limit does not exist}$$

51. $C = 0.5x + 500$, $\bar{C} = \dfrac{C}{x}$

$\bar{C} = 0.5 + \dfrac{500}{x}$, $\quad \lim_{x \to \infty} \left(0.5 + \dfrac{500}{x}\right) = 0.5$

52. $\lim_{v \to c} \dfrac{m_0}{\sqrt{1 - (v^2/c^2)}} = \dfrac{m_0}{0} \to \infty$, $\quad$ Limit does not exist

53. $\lim_{v_1/v_2 \to \infty} 100[1 - (v_1/v_2)(c)] = 100[1 - 0] = 100\%$

4.6
A summary of curve sketching

1. $y = x^3 - 3x^2 + 3$

    $y' = 3x^2 - 6x = 3x(x - 2) = 0$
    when $x = 0, 2$

    $y'' = 6x - 6 = 6(x - 1) = 0$ when $x = 1$

    $y'' < 0$ when $x = 0$, therefore $(0, 3)$ is a relative maximum.

    $y'' > 0$ when $x = 2$, therefore $(2, -1)$ is a relative minimum.

    $(1, 1)$ is a point of inflection.

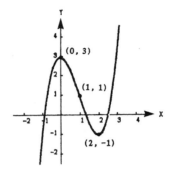

2. $y = -\frac{1}{3}(x^3 - 3x + 2)$

    $y' = -x^2 + 1 = 0$ when $x = \pm 1$

    $y'' = -2x = 0$ when $x = 0$

    $y'' > 0$ when $x = -1$, therefore $(-1, -\frac{4}{3})$ is a relative minimum.

    $y'' < 0$ when $x = 1$, therefore $(1, 0)$ is a relative maximum.

    $(0, -\frac{2}{3})$ is a point of inflection.

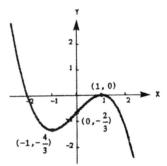

3. $y = 2 - x - x^3$

    $y' = -1 - 3x^2 < 0$ for all x

    $y'' = -6x = 0$ when $x = 0$

    $(0, 2)$ is a point of inflection.

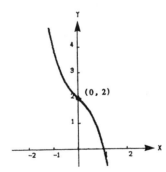

4. $y = x^3 + 3x^2 + 3x + 2$

   $y' = 3x^2 + 6x + 3 = 3(x^2 + 2x + 1)$

   $\quad = 3(x + 1)^2 > 0$ if $x \neq -1$

   $y'' = 6(x + 1) = 0$ when $x = -1$

   $(-1, 1)$ is a point of inflection.

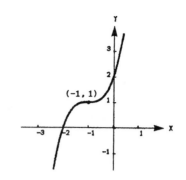

5. $y = 3x^3 - 9x + 1$

   $y' = 9x^2 - 9 = 9(x + 1)(x - 1) = 0$
   when $x = \pm 1$

   $y'' = 18x = 0$ when $x = 0$

   $y'' < 0$ when $x = -1$, therefore $(-1, 7)$
   is a relative maximum.

   $y'' > 0$ when $x = 1$, therefore $(1, -5)$
   is a relative minimum.

   $(0, 1)$ is a point of inflection.

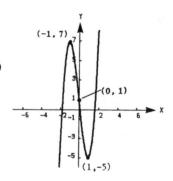

6. $y = (x + 1)(x - 2)(x - 5)$

   $y' = (x - 2)(x - 5) + (x + 1)(x - 5) + (x + 1)(x - 2)$

   $\quad = 3(x^2 - 4x + 1) = 0$

   when $x = 2 \pm \sqrt{3}$

   $y'' = 6(x - 2) = 0$ when $x = 2$

   $y'' < 0$ when $x = 2 - \sqrt{3}$, therefore
   $(2 - \sqrt{3}, 6\sqrt{3})$ is a relative maximum.

   $y'' > 0$ when $x = 2 + \sqrt{3}$, therefore
   $(2 + \sqrt{3}, -6\sqrt{3})$ is a relative minimum.

   $(2, 0)$ is a point of inflection.

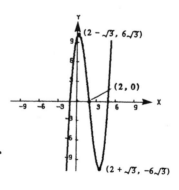

7. $y = -x^3 + 3x^2 + 9x - 2$

    $y' = -3x^2 + 6x + 9 = -3(x + 1)(x - 3) = 0$
    when $x = -1, 3$

    $y'' = -6x + 6 = -6(x - 1) = 0$
    when $x = 1$

    $y'' > 0$ when $x = -1$, therefore $(-1, -7)$
    is a relative minimum.

    $y'' < 0$ when $x = 3$, therefore $(3, 25)$
    is a relative maximum.

    $(1, 9)$ is a point of inflection.

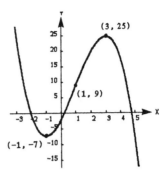

8. $y = \dfrac{1}{3}(x - 1)^3 + 2$

    $y' = (x - 1)^2 > 0$ when $x \neq 1$

    $y'' = 2(x - 1) = 0$ when $x = 1$

    $(1, 2)$ is a point of inflection.

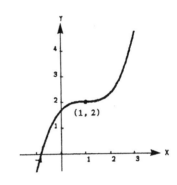

9. $y = 3x^4 + 4x^3 = x^3(3x + 4)$

    $y' = 12x^3 + 12x^2 = 12x^2(x + 1) = 0$
    when $x = 0, -1$

    $y'' = 36x^2 + 24x = 12x(3x + 2) = 0$

    when $x = 0, -\dfrac{2}{3}$

    $y'' > 0$ when $x = -1$, therefore $(-1, -1)$
    is a relative minimum.

    $(0, 0)$ is a point of inflection.

    $(-\dfrac{2}{3}, -\dfrac{16}{27})$ is a point of inflection.

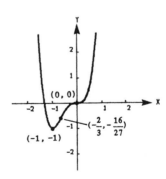

10. $y = 3x^4 - 6x^2$

$y' = 12x^3 - 12x = 12x(x^2 - 1) = 0$ when $x = 0, \pm 1$

$y'' = 36x^2 - 12 = 12(3x^2 - 1) = 0$ when $x = \pm \dfrac{\sqrt{3}}{3}$

$y'' > 0$ when $x = -1$, therefore $(-1, -3)$ is a relative minimum.

$y'' < 0$ when $x = 0$, therefore $(0, 0)$ is a relative maximum.

$y'' > 0$ when $x = 1$, therefore $(1, -3)$ is a relative minimum.

$(-\dfrac{\sqrt{3}}{3}, -\dfrac{5}{3})$ is a point of inflection.

$(\dfrac{\sqrt{3}}{3}, -\dfrac{5}{3})$ is a point of inflection.

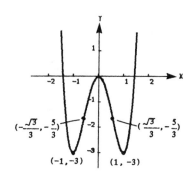

11. $y = x^4 - 4x^3 + 16x$

$y' = 4x^3 - 12x^2 + 16 = 4(x + 1)(x - 2)^2 = 0$ when $x = -1, 2$

$y'' = 12x^2 - 24x = 12x(x - 2) = 0$ when $x = 0, 2$

$y'' > 0$ when $x = -1$, therefore $(-1, -11)$ is a relative minimum.

$(0, 0)$ is a point of inflection.

$(2, 16)$ is a point of inflection.

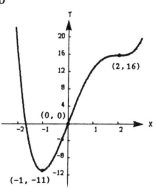

12. $y = x^4 - 8x^3 + 18x^2 - 16x + 5$

$y' = 4x^3 - 24x^2 + 36x - 16 = 4(x - 4)(x - 1)^2 = 0$ when $x = 1, 4$

$y'' = 12x^2 - 48x + 36$
$= 12(x - 1)(x - 3) = 0$
when $x = 1, 3$

$y'' > 0$ when $x = 4$, therefore $(4, -27)$ is a relative minimum.

$(1, 0)$ is a point of inflection.

$(3, -16)$ is a point of inflection.

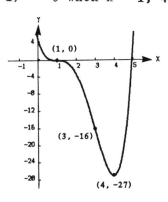

13. $y = x^4 - 4x^3 + 16x - 16$

$y' = 4x^3 - 12x^2 + 16$
$\quad = 4(x + 1)(x - 2)^2 = 0$
when $x = -1, 2$

$y'' = 12x^2 - 24x = 12x(x - 2) = 0$
when $x = 0, 2$

$y'' > 0$ when $x = -1$, therefore $(-1, -27)$
is a relative minimum.

$(0, -16)$ is a point of inflection.

$(2, 0)$ is a point of inflection.

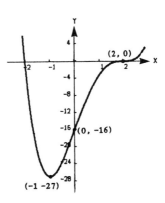

14. $y = x^5 + 1$

$y' = 5x^4 > 0$ when $x \neq 0$

$y'' = 20x^3 = 0$ when $x = 0$

$(0, 1)$ is a point of inflection.

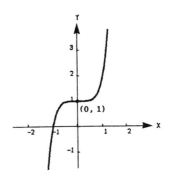

15. $y = x^5 - 5x$

$y' = 5x^4 - 5 = 5(x + 1)(x - 1)(x^2 + 1) = 0$ when $x = \pm 1$

$y'' = 20x^3 = 0$ when $x = 0$

$y'' < 0$ when $x = -1$, therefore $(-1, 4)$
is a relative maximum.

$y'' > 0$ when $x = 1$, therefore $(1, -4)$
is a relative minimum.

$(0, 0)$ is a point of inflection.

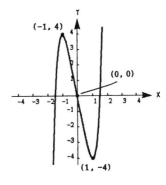

16. $y = (x - 1)^5$

$y' = 5(x - 1)^4 = 0$ when $x = 1$

$y'' = 20(x - 1)^3 = 0$ when $x = 1$

(1, 0) is a point of inflection.

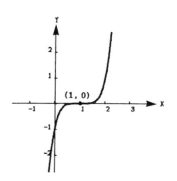

17. $y = |2x - 3|$

$y' = \dfrac{2(2x - 3)}{|2x - 3|}$ undefined at $x = \dfrac{3}{2}$

$y' = -2$ when $x < 3/2$

$y' = 2$ when $x > 3/2$

Therefore (3/2, 0) is a relative minimum.

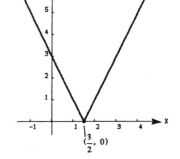

18. $y = |x^2 - 6x + 5|$

$y' = \dfrac{2(x - 3)(x^2 - 6x + 5)}{|x^2 - 6x + 5|}$

$= \dfrac{2(x - 3)(x - 5)(x - 1)}{|(x - 5)(x - 1)|} = 0$

when $x = 3$ and undefined when $x = 1, 5$

$y' < 0$ in the intervals $(-\infty, 1)$ and $(3, 5)$
$y' > 0$ in the intervals $(1, 3)$ and $(5, \infty)$

Therefore (1, 0) is a relative minimum,
(3, 4) is a relative maximum, and
(5, 0) is a relative minimum.

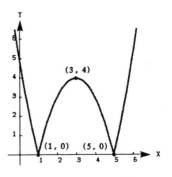

19. $y = \dfrac{x^2}{x^2 + 3}$

$y' = \dfrac{6x}{(x^2 + 3)^2} = 0$ when $x = 0$

$y'' = \dfrac{6(x^2 + 3)^2 - 6x(2)(x^2 + 3)(2x)}{(x^2 + 3)^4}$

$\quad = \dfrac{18(1 + x)(1 - x)}{(x^2 + 3)^3} = 0$ when $x = \pm 1$

$y'' > 0$ when $x = 0$, therefore $(0, 0)$
is a relative minimum.

$(\pm 1, \dfrac{1}{4})$ are points of inflection.

Horizontal asymptote: $y = 1$

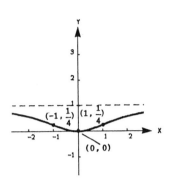

20. $y = \dfrac{x}{x^2 + 1}$

$y' = \dfrac{1 - x^2}{(x^2 + 1)^2} = \dfrac{(1 - x)(1 + x)}{(x^2 + 1)^2} = 0$ when $x = \pm 1$

$y'' = -\dfrac{2x(3 - x^2)}{(x^2 + 1)^3} = 0$ when $x = 0, \pm \sqrt{3}$

$y'' > 0$ when $x = -1$, therefore $(-1, -\dfrac{1}{2})$
is a relative minimum.

$y'' < 0$ when $x = 1$, therefore $(1, \dfrac{1}{2})$
is a relative maximum.

$(-\sqrt{3}, -\dfrac{\sqrt{3}}{4})$ is a point of inflection.

$(0, 0)$ is a point of inflection.

$(\sqrt{3}, \dfrac{\sqrt{3}}{4})$ is a point of inflection.

Horizontal asymptote: $y = 0$

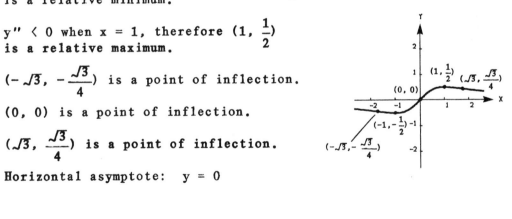

21. $y = 3x^{2/3} - 2x$

$y' = 2x^{-1/3} - 2 = \dfrac{2(1 - x^{1/3})}{x^{1/3}} = 0$

when $x = 1$ and undefined when $x = 0$

$y'' = \dfrac{-2}{3x^{4/3}} < 0$ when $x \neq 0$

Therefore $(1, 1)$ is a relative maximum.

$y' < 0$ when $x = -1$ and $y' > 0$ when $x = \dfrac{1}{2}$

Therefore $(0, 0)$ is a relative minimum.

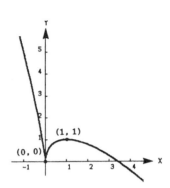

22. $y = 3x^{2/3} - x^2$

$y' = \dfrac{2}{x^{1/3}} - 2x = \dfrac{2(1 - x^{4/3})}{x^{1/3}} = 0$

when $x = \pm 1$ and undefined when $x = 0$.

$y'' = -2(\dfrac{1}{3x^{4/3}} + 1) < 0$ when $x \neq 0$

Therefore $(\pm 1, 2)$ are maxima.

$y' < 0$ when $x = -\dfrac{1}{2}$ and $y' > 0$ when $x = \dfrac{1}{2}$

Therefore $(0, 0)$ is a relative minimum.

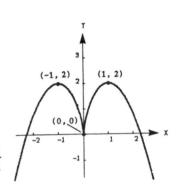

23. $y = \dfrac{x}{\sqrt{x^2 + 7}}$

$y' = \dfrac{7}{(x^2 + 7)^{3/2}} > 0$ for all x

$y'' = \dfrac{-21x}{(x^2 + 7)^{5/2}} = 0$ when $x = 0$

$(0, 0)$ is a point of inflection.

Horizontal asymptotes: $y = 1$, $y = -1$

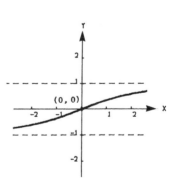

24. $f(x) = \dfrac{4x}{\sqrt{x^2 + 15}}$

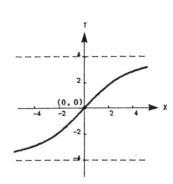

$f'(x) = \dfrac{60}{(x^2 + 15)^{3/2}} > 0$ for all x

$f''(x) = \dfrac{-180x}{(x^2 + 15)^{5/2}} = 0$ when $x = 0$

$(0, 0)$ is a point of inflection.

Horizontal asymptotes: $y = 4$, $y = -4$

25. $y = \sin x - \dfrac{1}{18}\sin 3x$, $0 \le x \le 2\pi$

$y' = \cos x - \dfrac{1}{6}\cos 3x = 0$, when $x = \dfrac{\pi}{2}, \dfrac{3\pi}{2}$

$y'' = -\sin x + \dfrac{1}{2}\sin 3x = 0$ when $x = 0, \dfrac{\pi}{6}, \dfrac{5\pi}{6}, \pi, \dfrac{7\pi}{6}, \dfrac{11\pi}{6}$

Relative maximum: $\left(\dfrac{\pi}{2}, \dfrac{19}{18}\right)$

Relative minimum: $\left(\dfrac{3\pi}{2}, -\dfrac{19}{18}\right)$

Inflection points: $\left(\dfrac{\pi}{6}, \dfrac{4}{9}\right)$, $\left(\dfrac{5\pi}{6}, \dfrac{4}{9}\right)$, $(\pi, 0)$, $\left(\dfrac{7\pi}{6}, -\dfrac{4}{9}\right)$, $\left(\dfrac{11\pi}{6}, -\dfrac{4}{9}\right)$

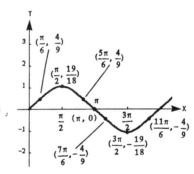

26. $y = \cos x - \dfrac{1}{2}\cos 2x$, $0 \le x \le 2\pi$

$y' = -\sin x + \sin 2x = -\sin x(1 - 2\cos x) = 0$ when $x = 0, \pi, \dfrac{\pi}{3}, \dfrac{5\pi}{3}$

$y'' = -\cos x + 2\cos 2x = -\cos x + 2(2\cos^2 x - 1) = 4\cos^2 x - \cos x - 2 = 0$

When $\cos x = \dfrac{1 \pm \sqrt{33}}{8} = 0.8431, -0.5931$

Therefore $x = 0.5678$ or 5.7154,

$x = 2.2057$ or 4.0775

Relative maxima: $\left(\dfrac{\pi}{3}, \dfrac{3}{4}\right)$, $\left(\dfrac{5\pi}{3}, \dfrac{3}{4}\right)$

Relative minima: $\left(0, \dfrac{1}{2}\right)$, $\left(\pi, -\dfrac{3}{2}\right)$

Inflection points: $(0.5678, 0.6323)$, $(2.2057, -0.4449)$, $(5.7154, 0.6323)$, $(4.0775, -0.4449)$

27. $y = 2x - \tan x$, $-\dfrac{\pi}{2} < x < \dfrac{\pi}{2}$

$y' = 2 - \sec^2 x = 0$ when $x = \pm\dfrac{\pi}{4}$

$y'' = -2\sec^2 x \tan x = 0$ when $x = 0$

Relative maximum: $\left(\dfrac{\pi}{4}, \dfrac{\pi}{2} - 1\right)$

Relative minimum: $\left(-\dfrac{\pi}{4}, 1 - \dfrac{\pi}{2}\right)$

Inflection point: $(0, 0)$

Vertical asymptotes: $x = -\dfrac{\pi}{2}$, $x = \dfrac{\pi}{2}$

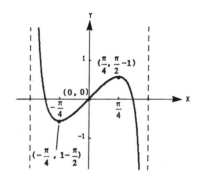

28. $y = 2x + \cot x$, $0 < x < \pi$

$y' = 2 - \csc^2 x = 0$ when $x = \dfrac{\pi}{4}, \dfrac{3\pi}{4}$

$y'' = 2\csc^2 x \cot x = 0$ when $x = \dfrac{\pi}{2}$

Relative maximum: $\left(\dfrac{3\pi}{4}, \dfrac{3\pi}{2} - 1\right)$

Relative minimum: $\left(\dfrac{\pi}{4}, \dfrac{\pi}{2} + 1\right)$

Inflection point: $\left(\dfrac{\pi}{2}, \pi\right)$

Vertical asymptotes: $x = 0$, $x = \pi$

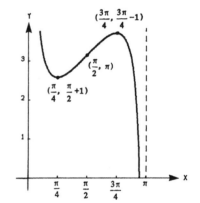

29. $y = \dfrac{1}{x - 2} - 3$

$y' = -\dfrac{1}{(x - 2)^2} < 0$ when $x \neq 2$

No relative extrema.

Intercepts: $\left(\dfrac{7}{3}, 0\right)$, $\left(0, -\dfrac{7}{2}\right)$

Vertical asymptote: $x = 2$

Horizontal asymptote: $y = -3$

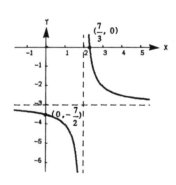

30. $y = \dfrac{x^2 + 1}{x^2 - 2}$

$y' = \dfrac{-6x}{(x^2 - 2)^2} = 0$ when $x = 0$

$y'' = \dfrac{6(3x^2 + 2)}{(x^2 - 2)^3}$

$y'' < 0$ when $x = 0$, therefore $(0, -\dfrac{1}{2})$ is a relative maximum.

Intercept: $(0, -1/2)$

Vertical asymptotes: $x = \pm \sqrt{2}$

Horizontal asymptote: $y = 1$

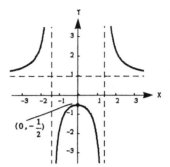

31. $y = \dfrac{2x}{x^2 - 1}$

$y' = \dfrac{-2(x^2 + 1)}{(x^2 - 1)^2} < 0$ if $x \neq 1$

$y'' = \dfrac{4x(x^2 + 3)}{(x^2 - 1)^3} = 0$ if $x = 0$

$(0, 0)$ is a point of inflection.

Intercept: $(0, 0)$

Vertical asymptotes: $x = \pm 1$

Horizontal asymptote: $y = 0$

Symmetry with respect to the origin

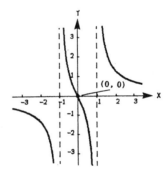

32. $y = \dfrac{x^2 - 6x + 12}{x - 4} = (x - 2) + \dfrac{4}{x - 4}$

$y' = 1 - \dfrac{4}{(x - 4)^2} = \dfrac{(x - 2)(x - 6)}{(x - 4)^2} = 0$ when $x = 2, 6$

$y'' = \dfrac{8}{(x - 4)^3}$

$y'' < 0$ when $x = 2$, therefore $(2, -2)$ is a relative maximum.

$y'' > 0$ when $x = 6$, therefore $(6, 6)$ is a relative minimum.

Vertical asymptote: $x = 4$

Slant asymptote: $y = x - 2$

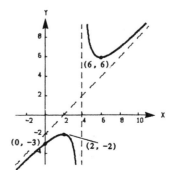

33. $y = x\sqrt{4 - x}$, Domain $(-\infty, 4]$

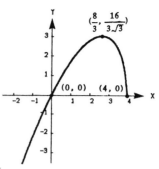

$y' = \dfrac{8 - 3x}{2\sqrt{4 - x}} = 0$ when $x = \dfrac{8}{3}$ and

undefined when $x = 4$

$y'' = \dfrac{3x - 16}{4(4 - x)^{3/2}} > 0$ when $x \neq 4$

therefore $(\dfrac{8}{3}, \dfrac{16}{3\sqrt{3}})$ is a relative maximum.

34. $y = x\sqrt{4 - x^2}$, Domain $[-2, 2]$

$y' = \dfrac{4 - 2x^2}{\sqrt{4 - x^2}} = 0$ when $x = \pm\sqrt{2}$ and

undefined when $x = \pm 2$

$y'' = \dfrac{2x(x^2 - 6)}{(4 - x^2)^{3/2}} = 0$

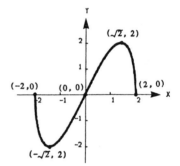

when $x = 0, \pm\sqrt{6}$

$y'' > 0$ when $x = -\sqrt{2}$, therefore $(-\sqrt{2}, -2)$ is a relative minimum.

$y'' < 0$ when $x = \sqrt{2}$, therefore $(\sqrt{2}, 2)$ is a relative maximum.

$(0, 0)$ is the only point of inflection in the domain.

35. $y = \dfrac{x + 2}{x} = 1 + \dfrac{2}{x}$

$y' = \dfrac{-2}{x^2} < 0$ when $x \neq 0$

$y'' = \dfrac{4}{x^3} \neq 0$

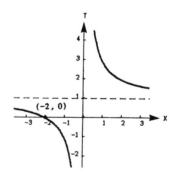

Intercept: $(-2, 0)$

Vertical asymptote: $x = 0$

Horizontal asymptote: $y = 1$

36. $y = x + \dfrac{32}{x^2}$

$y' = 1 - \dfrac{64}{x^3} = \dfrac{(x - 4)(x^2 + 4x + 16)}{x^3} = 0$ when $x = 4$

$y'' = \dfrac{192}{x^4} > 0$ if $x \neq 0$

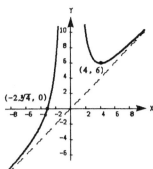

Therefore $(4, 6)$ is a relative minimum.

Intercept: $(-2\sqrt[3]{4}, 0)$

Vertical asymptote: $x = 0$

Slant asymptote: $y = x$

37. $f(x) = \dfrac{x^2 + 1}{x} = x + \dfrac{1}{x}$

$f'(x) = 1 - \dfrac{1}{x^2} = 0$ when $x = \pm 1$

$f''(x) = \dfrac{2}{x^3} \neq 0$

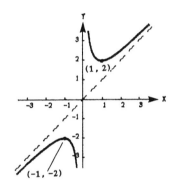

Relative maximum: $(-1, -2)$

Relative minimum: $(1, 2)$

Vertical asymptote: $y = 0$

Slant Asymptote: $y = x$

38. $f(x) = \dfrac{x^3}{x^2 - 1} = x + \dfrac{x}{x^2 - 1}$

$f'(x) = \dfrac{x^2(x^2 - 3)}{(x^2 - 1)^2} = 0$ when $x = 0, \pm\sqrt{3}$

$f''(x) = \dfrac{2x(x^2 + 3)}{(x^2 - 1)^3} = 0$ when $x = 0$

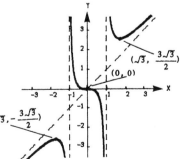

Relative maximum: $\left(-\sqrt{3}, -\dfrac{3\sqrt{3}}{2}\right)$

Relative minimum: $\left(\sqrt{3}, \dfrac{3\sqrt{3}}{2}\right)$

Inflection point: $(0, 0)$

Vertical asymptotes: $x = 1, x = -1$

Slant asymptote: $y = x$

215

39. $y = \dfrac{x^3}{2x^2 - 8} = \dfrac{x}{2} + \dfrac{2x}{x^2 - 4}$

$y' = \dfrac{x^2(x^2 - 12)}{2(x^2 - 4)^2} = 0$ when $x = 0, \ \pm 2\sqrt{3}$

$y'' = \dfrac{4x(x^2 + 12)}{(x^2 - 4)^3} = 0$ when $x = 0$

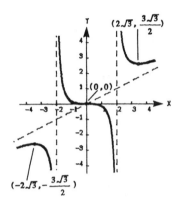

Relative maximum: $(-2\sqrt{3}, \ -\dfrac{3\sqrt{3}}{2})$

Relative minimum: $(2\sqrt{3}, \ \dfrac{3\sqrt{3}}{2})$

Inflection point: $(0, 0)$

Vertical asymptotes: $x = 2, -2$

Slant asymptote: $y = \dfrac{x}{2}$

40. $y = \dfrac{2x^2 - 5x + 5}{x - 2} = 2x - 1 + \dfrac{3}{x - 2}$

$y' = 2 - \dfrac{3}{(x - 2)^2} = \dfrac{2x^2 - 8x + 5}{(x - 2)^2} = 0$ when $x = \dfrac{4 \pm \sqrt{6}}{2}$

$y'' = \dfrac{6}{(x - 2)^3} \neq 0$

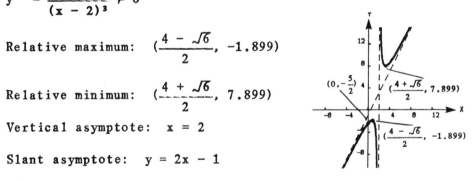

Relative maximum: $(\dfrac{4 - \sqrt{6}}{2}, \ -1.899)$

Relative minimum: $(\dfrac{4 + \sqrt{6}}{2}, \ 7.899)$

Vertical asymptote: $x = 2$

Slant asymptote: $y = 2x - 1$

In Exercises 41 - 46: $f(x) = ax^3 + bx^2 + cx + d$
$f'(x) = 3ax^2 + 2bx + c$
$f''(x) = 6ax + 2b$

$f'(x) = 0$ when $x = \dfrac{-2b \pm \sqrt{4b^2 - 12ac}}{6a} = \dfrac{-b \pm \sqrt{b^2 - 3ac}}{3a}$

Furthermore $\lim\limits_{x \to \infty} f(x) = \infty$ if and only if $a > 0$ and

$\lim\limits_{x \to \infty} f(x) = -\infty$ if and only if $a < 0$.

41. Since $\lim\limits_{x \to \infty} f(x) = -\infty$, $a < 0$

Also $f'(x) < 0$ for all x.
Therefore the discriminant is negative.

$b^2 - 3ac < 0$ ⟹ $b^2 < 3ac$

42. Since $\lim\limits_{x \to \infty} f(x) = \infty$, $a > 0$

Also $f'(x) > 0$ for all x.
Therefore the discriminant is negative.

$b^2 - 3ac < 0$ ⟹ $b^2 < 3ac$

43. Since $\lim\limits_{x \to \infty} f(x) = -\infty$, $a < 0$

Also since there is only one critical point the discriminant is zero.

$b^2 - 3ac = 0$ ⟹ $b^2 = 3ac$

44. Since $\lim\limits_{x \to \infty} f(x) = \infty$, $a > 0$

Also since there is only one critical point the discriminant is zero.

$b^2 - 3ac = 0$ ⟹ $b^2 = 3ac$

45. Since $\lim\limits_{x \to \infty} f(x) = -\infty$, $a < 0$

Also since there are two critical points the discriminant is positive.

$b^2 - 3ac > 0$ ⟹ $b^2 > 3ac$

46. Since $\lim\limits_{x \to \infty} f(x) = \infty$, $a > 0$

Also since there are two critical points the discriminant is positive.

$b^2 - 3ac > 0$ ⟹ $b^2 > 3ac$

47. $f'(x) = -x(x - 1)(x - 3) = -x^2 + 4x - 3$

Relative minimum when $x = 1$, Relative maximum when $x = 3$

$f(x) = -\dfrac{x^3}{3} + 2x^2 - 3x + C$, C any constant

Let $C = 0$

$f(x) = -\dfrac{x^3}{3} + 2x^2 - 3x$

$f(1) = -\dfrac{4}{3}$, $f(3) = 0$

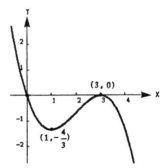

48. $f' > 0$ for $x > 0$ ⇒ $f(x)$ is increasing

    $f' < 0$ for $x < 0$ ⇒ $f(x)$ is decreasing

    f' is undefined when $x = 0$

    One such function is $f(x) = -\dfrac{1}{x^2}$

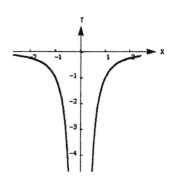

49. $f'(x) = 2$

    $f(x) = 2x + C$, C any constant

    If $C = 0$, $f(x) = 2x$

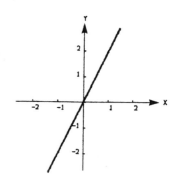

50. f' is linear ⇒ f is quadratic

    $f' < 0$ when $x < 0$ ⇒ f is decreasing
    $f' > 0$ when $x > 0$ ⇒ f is increasing
    Therefore $(0, f(0))$ is a relative
    minimum.

    $f(x) = ax^2 + b$, $a > 0$

    If $a = 2$ and $b = 0$, then $f(x) = 2x^2$

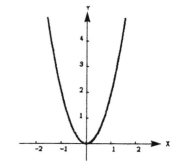

51. $f''(x) = 2 > 0$ ⇒ f is concave up
    $f'(x) = 2x + C_1$

    $f(x) = x^2 + C_1 x + C_2$

    Let $C_1 = C_2 = 0$, then $f(x) = x^2$

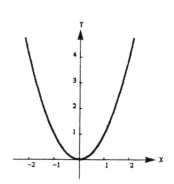

52. f'' is linear $\implies$ f is cubic

$$f(x) = ax^3 + bx^2 + cx + d$$

$f'' < 0$ when $x < 0$ $\implies$ concave down
$f'' > 0$ when $x > 0$ $\implies$ concave up
Therefore $(0, f(0))$ is a point of inflection.

Since f is concave up to the right and concave down to the left $\implies$ $a > 0$

Let $f(x) = x^3$

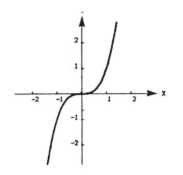

4.7
Optimization problems

1. Let x be the length and y the width of the rectangle.

$$2x + 2y = 100, \qquad y = 50 - x, \qquad A = xy = x(50 - x)$$

$\dfrac{dA}{dx} = 50 - 2x = 0$ when $x = 25$, $\dfrac{d^2A}{dx^2} = -2$

A is maximum when $x = y = 25$.

2. Let x be the positive number. Then the indicated sum is $s = x + \dfrac{1}{x}$

$\dfrac{ds}{dx} = 1 - \dfrac{1}{x^2} = 0$ when $x = 1$, $\qquad \dfrac{d^2s}{dx^2} = \dfrac{2}{x^3} > 0$ for $x > 0$

The sum is a minimum when $x = 1$.

3. Let x and y be two numbers such that $x + 2y = 24$

$P = xy = (24 - 2y)y$, $\qquad \dfrac{dP}{dy} = 24 - 4y = 0$ when $y = 6$, $\qquad \dfrac{d^2P}{dy^2} = -4$

P is maximum when $y = 6$ and $x = 12$.

4. Let x and y be two numbers such that $x - y = 50$

$P = xy = x(x - 50)$, $\qquad \dfrac{dP}{dx} = 2x - 50 = 0$ when $x = 25$, $\qquad \dfrac{d^2P}{dx^2} = 2$

P is minimum when $x = 25$ and $y = -25$.

5. Let x and y be two positive numbers such that x + y = 110

$$P = xy = x(110 - x) = 110x - x^2$$

$$\frac{dP}{dx} = 110 - 2x = 0 \text{ when } x = 55, \qquad \frac{d^2P}{dx^2} = -2$$

P is a maximum when x = y = 55.

6. Let x and y be two positive numbers such that x + 2y = 100

$$P = xy = y(100 - 2y) = 100y - 2y^2$$

$$\frac{dP}{dy} = 100 - 4y = 0 \text{ when } y = 25, \qquad \frac{d^2P}{dy^2} = -4$$

P is a maximum when x = 50 and y = 25.

7. Let x and y be two positive numbers such that xy = 192

$$S = x + y = x + \frac{192}{x}$$

$$\frac{dS}{dx} = 1 - \frac{192}{x^2} = 0 \text{ when } x = \sqrt{192}, \qquad \frac{d^2S}{dx^2} = \frac{384}{x^3} > 0 \text{ when } x > 0$$

S is a minimum when x = y = $\sqrt{192}$.

8. Let x and y be two positive numbers such that xy = 192

$$S = x + 3y = 3y + \frac{192}{y}$$

$$\frac{dS}{dy} = 3 - \frac{192}{y^2} = 0 \text{ when } y = 8, \qquad \frac{d^2S}{dy^2} = \frac{384}{y^3} > 0 \text{ when } y > 0$$

S is a minimum when y = 8 and x = 24.

9. From the figure we see that the perimeter is 4x + 3y = 200

$$A = 2xy = 2x(\frac{200 - 4x}{3}) = \frac{8}{3}(50x - x^2)$$

$$\frac{dA}{dx} = \frac{8}{3}(50 - 2x) = 0 \text{ when } x = 25$$

$$\frac{d^2A}{dx^2} = -\frac{16}{3}$$

A is a maximum when x = 25 and y = $\frac{100}{3}$.

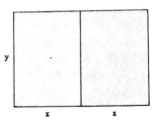

10. From the figure we see that $xy = 180{,}000$

    $S = x + 2y = (x + \dfrac{360{,}000}{x})$ where

    S is the length of fence needed.

    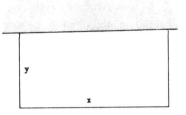

    $\dfrac{dS}{dx} = 1 - \dfrac{360{,}000}{x^2} = 0$ when $x = 600$

    $\dfrac{d^2S}{dx^2} = \dfrac{720{,}000}{x^3} > 0$ when $x > 0$

    S is a minimum when $x = 600$ and $y = 300$.

11. From the figure we see that $V = x(12 - 2x)^2$

    $\dfrac{dV}{dx} = 2x(12 - 2x)(-2) + (12 - 2x)^2 = 12(6 - x)(2 - x) = 0$ when $x = 2, 6$

    $\dfrac{d^2V}{dx^2} = 12(2x - 8)$, therefore

    when $x = 6$, $\dfrac{d^2V}{dx^2} = 48 > 0$ and

    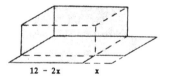

    when $x = 2$, $\dfrac{d^2V}{dx^2} = -48 < 0$

    $V = 128$ is maximum when $x = 2$.

12.(a) $V = x(s - 2x)^2$

    $\dfrac{dV}{dx} = 2x(s - 2x)(-2) + (s - 2x)^2 = (s - 2x)(s - 6x) = 0$ when $x = \dfrac{s}{2}, \dfrac{s}{6}$

    $\dfrac{d^2V}{dx^2} = 24x - 8s$, therefore

    when $x = \dfrac{s}{2}$, $\dfrac{d^2V}{dx^2} > 0$ and when $x = \dfrac{s}{6}$, $\dfrac{d^2V}{dx^2} < 0$

    $V = \dfrac{2s^3}{27}$ is maximum when $x = \dfrac{s}{6}$.

    (b) If the length is doubled:

    $V = \dfrac{2}{27}(2s)^3 = 8(\dfrac{2}{27} s^3)$ Volume is increased by a factor of 8.

13. From the figure we see that $V = x(3 - 2x)(2 - 2x)$

$$\frac{dV}{dx} = (3 - 2x)(2 - 2x) + x(2 - 2x)(-2) + x(3 - 2x)(-2)$$

$$\frac{dV}{dx} = 12x^2 - 20x + 6 = 0 \text{ when } x = \frac{5 \pm \sqrt{7}}{6}$$

$$\frac{d^2V}{dx^2} = 24x - 20 < 0 \text{ when } x = \frac{5 - \sqrt{7}}{6}$$

The dimensions $(\frac{5 - \sqrt{7}}{6}$ ft) by

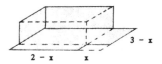

$(\frac{1 + \sqrt{7}}{3}$ ft) by $(\frac{4 + \sqrt{7}}{3}$ ft) or

(0.392 ft) by (1.215 ft) by (2.215 ft)
yield a maximum volume.

14. From the figure we see that $S = x^2 + 3xy$

$$x^2y = \frac{250}{3}, \qquad y = \frac{250}{3x^2}, \qquad S = x^2 + \frac{250}{x}$$

$$\frac{dS}{dx} = 2x - \frac{250}{x^2} = \frac{2x^3 - 250}{x^2} = 0 \text{ when } x = 5$$

$$\frac{d^2S}{dx^2} = 2 + \frac{500}{x^3} > 0 \text{ when } x > 0$$

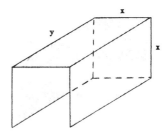

S is a minimum when $x = 5$ and $y = 10/3$.

15. From the figure we see that $A = xy$ and since $y = \frac{6 - x}{2}$

$$A = x(\frac{6 - x}{2}) = \frac{1}{2}(6x - x^2)$$

$$\frac{dA}{dx} = \frac{1}{2}(6 - 2x) = 0 \text{ when } x = 3$$

$$\frac{d^2A}{dx^2} = -1$$

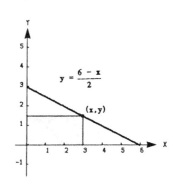

A is a maximum when $x = 3$ and $y = 3/2$.

16. $A = \frac{1}{2}(2\sqrt{r^2 - x^2})(x + r) = \sqrt{r^2 - x^2}(x + r)$

$\frac{dA}{dx} = \sqrt{r^2 - x^2} + (x + r)(\frac{1}{2})(r^2 - x^2)^{-1/2}(-2x)$

$= \frac{r^2 - rx - 2x^2}{\sqrt{r^2 - x^2}} = \frac{(r - 2x)(r + x)}{\sqrt{r^2 - x^2}} = 0$ when $x = \frac{r}{2}$

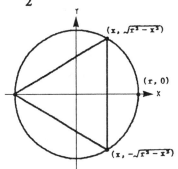

By the First Derivative Test, A is a
maximum at the critical point $x = r/2$.
The vertices are then at $(-r, 0)$, $(\frac{r}{2}, \frac{\sqrt{3}r}{2})$
and $(\frac{r}{2}, \frac{-\sqrt{3}r}{2})$, and the triangle is
equilateral with sides of length $\sqrt{3}r$.

17. $P = 2x + 2\pi r = 2x + 2\pi(\frac{y}{2})$, $P = 2x + \pi y = 200$

$y = \frac{200 - 2x}{\pi}$

$A = xy = \frac{2x}{\pi}(100 - x) = \frac{2}{\pi}(100x - x^2)$

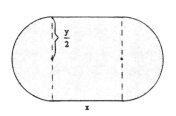

$\frac{dA}{dx} = \frac{2}{\pi}(100 - 2x) = 0$
when $x = 50$ and $y = \frac{100}{\pi}$

18. From the graph we see that $xy = 30 \implies y = \frac{30}{x}$ and $A = (x + 2)(\frac{30}{x} + 4)$

$\frac{dA}{dx} = (x + 2)(\frac{-30}{x^2}) + (\frac{30}{x} + 4)$

$= \frac{4(x^2 - 15)}{x^2} = 0$ when $x = \sqrt{15}$

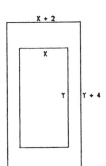

$y = \frac{30}{\sqrt{15}} = 2\sqrt{15}$

By the First Derivative Test the
dimensions $(x + 2)$ by $(y + 4)$ or
$(2 + \sqrt{15})$ by $2(2 + \sqrt{15})$ or approximately
(5.873) by (11.746) yield a maximum area.

19. $A = \dfrac{xy}{2} = \dfrac{1}{2}x(3 + \dfrac{6}{x-2}) = \dfrac{3x^2}{2(x-2)}$

$\dfrac{dA}{dx} = \dfrac{3x(x-4)}{2(x-2)^2} = 0$

when $x = 0$, $x = 4$

By the First Derivative Test A is a
minimum when $x = 4$ and $y = 6$.
The vertices of the triangle are
$(0, 0)$, $(4, 0)$, and $(0, 6)$.

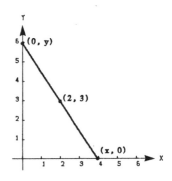

20. $y = 2 + \dfrac{2}{x-1}$

Find $x > 0$ so that the square of the distance is minimum.

$L = d^2 = (x-0)^2 + \left[0 - (2 + \dfrac{2}{x-1})\right]^2 = x^2 + 4 + \dfrac{8}{x-1} + \dfrac{4}{(x-1)^2}$

$\dfrac{dL}{dx} = 2x - \dfrac{8}{(x-1)^2} - \dfrac{8}{(x-1)^3}$

$\quad = \dfrac{2[x(x-1)^3 - 4(x-1) - 4]}{(x-1)^3} = 0$

when $x = 1 + \sqrt[3]{4}$

By the First Derivative Test, the
vertices of the triangle with a minimum
hypotenuse are $(0, 0)$, $(1 + \sqrt[3]{4}, 0)$, and
$(0, 2 + \sqrt[3]{2})$, or approximately
$(0, 0)$, $(2.587, 0)$, and $(0, 3.260)$.

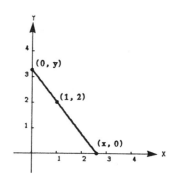

21. From the figure we see that $A = 2x\sqrt{25 - x^2}$

$\dfrac{dA}{dx} = 2x(\dfrac{1}{2})(\dfrac{-2x}{\sqrt{25 - x^2}}) + 2\sqrt{25 - x^2}$

$\quad = 2(\dfrac{25 - 2x^2}{\sqrt{25 - x^2}}) = 0$

when $x = y = \dfrac{5\sqrt{2}}{2} \approx 3.54$

By the First Derivative Test the
inscribed rectangle of maximum area
has vertices

$(\pm\dfrac{5\sqrt{2}}{2}, 0)$ and $(\pm\dfrac{5\sqrt{2}}{2}, \dfrac{5\sqrt{2}}{2})$

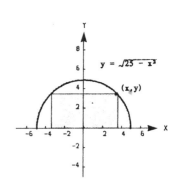

22. From the figure we see that $A = 2x\sqrt{r^2 - x^2}$

$$\frac{dA}{dx} = \frac{2(r^2 - 2x^2)}{\sqrt{r^2 - x^2}} = 0$$

when $x = \dfrac{\sqrt{2}\,r}{2}$

By the First Derivative Test A is maximum when the rectangle has

dimensions $\sqrt{2}\,r$ by $\dfrac{\sqrt{2}\,r}{2}$.

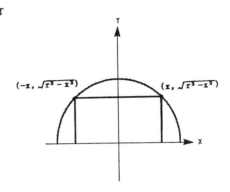

23. $d = \sqrt{(x - 4)^2 + (\sqrt{x} - 0)^2}$

$d' = \dfrac{2(x - 4) + 1}{\sqrt{(x - 4)^2 + x}} = 0$ when $x = \dfrac{7}{2}$

By the First Derivative Test the point

nearest to $(4, 0)$ is $(\dfrac{7}{2}, \sqrt{7/2})$.

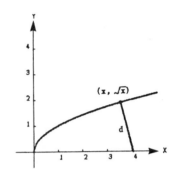

24. $d = \sqrt{(x - 2)^2 + [y - (1/2)]^2} = \sqrt{(x - 2)^2 + [x^2 - (1/2)]^2}$

$d' = \dfrac{2(x^3 - 1)}{\sqrt{(x - 2)^2 + [x^2 - (1/2)]^2}} = 0$

$2(x^3 - 1) = 0$ when $x = 1$ and $y = 1$

By the First Derivative Test the point nearest to $(2, \dfrac{1}{2})$ is $(1, 1)$.

25. $V = \pi r^2 h = 21.66$ cubic inches or $h = \dfrac{21.66}{\pi r^2}$

$s = 2\pi r^2 + 2\pi rh = 2(\pi r^2 + \dfrac{21.66}{r})$

$\dfrac{ds}{dr} = 2(2\pi r - \dfrac{21.66}{r^2}) = 0$ when $r \approx 1.51$ inches

$h = \dfrac{21.66}{\pi(1.51)^2} \approx 3.02$ inches $= 2r$

By the First Derivative Test this will yield the minimum surface area.

26. From the graph we see that $A = \frac{1}{2}(2r + 2x)\sqrt{r^2 - x^2} = (r + x)\sqrt{r^2 - x^2}$

$\frac{dA}{dx} = (r + x)(\frac{1}{2})(r^2 - x^2)^{-1/2}(-2x) + \sqrt{r^2 - x^2}$

$\qquad = \frac{-x(r + x)}{\sqrt{r^2 - x^2}} + \sqrt{r^2 - x^2}$

$\qquad = \frac{r^2 - rx - 2x^2}{\sqrt{r^2 - x^2}}$

$\qquad = \frac{(r - 2x)(r + x)}{\sqrt{r^2 - x^2}} = 0$ when $x = \frac{r}{2}$

By the First Derivative Test A will
be a maximum when the trapezoid bases
are r and 2r, and the altitude is $\sqrt{3}r/2$.

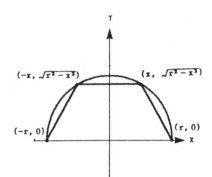

27. Let x be the sides of the square ends and y the length of the package.

$P = 4x + y = 108, \qquad V = x^2y = x^2(108 - 4x) = 108x^2 - 4x^3$

$\frac{dV}{dx} = 216x - 12x^2 = 12x(18 - x) = 0$ when $x = 18$

$\frac{d^2V}{dx^2} = 216 - 24x = -216$ when $x = 18$

The volume is maximum when $x = 18$ inches and
$y = 108 - 4(18) = 36$ inches.

28. From the graph we see that $V = \pi r^2 x$, $x + 2\pi r = 108 \implies x = 108 - 2\pi r$

$V = \pi r^2(108 - 2\pi r) = \pi(108r^2 - 2\pi r^3)$

$\frac{dV}{dr} = \pi(216r - 6\pi r^2)$

$\qquad = 6\pi r(36 - \pi r) = 0$

$r = 0, \frac{36}{\pi}$ when $x = 36$

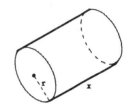

Volume is maximum when $x = 36$ inches

and $r = \frac{36}{\pi} \approx 11.459$ inches.

29. From the figure we see that $V = \frac{1}{3}\pi x^2 h = \frac{1}{3}\pi x^2(r + \sqrt{r^2 - x^2})$

$$\frac{dV}{dx} = \frac{1}{3}\pi\left[\frac{-x^3}{\sqrt{r^2 - x^2}} + 2x(r + \sqrt{r^2 - x^2})\right]$$

$$= \frac{\pi x}{3\sqrt{r^2 - x^2}}(2r^2 + 2r\sqrt{r^2 - x^2} - 3x^2) = 0$$

when $x = 0$, $\dfrac{2\sqrt{2}\,r}{3}$

By the First Derivative Test the volume

is a maximum when $x = \dfrac{2\sqrt{2}\,r}{3}$ and

$h = r + \sqrt{r^2 - x^2} = \dfrac{4r}{3}$. Thus the

maximum volume is $V = \dfrac{1}{3}\pi\,\dfrac{8r^2}{9}\left(\dfrac{4r}{3}\right) = \dfrac{32\pi r^3}{81}$.

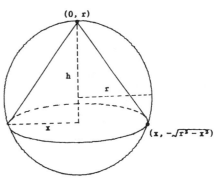

30. From the figure we see that $V = \pi r^2 h = \pi x^2(2\sqrt{r^2 - x^2}) = 2\pi x^2 \sqrt{r^2 - x^2}$

$$\frac{dV}{dx} = 2\pi\left[x^2\left(\frac{1}{2}\right)(r^2 - x^2)^{-1/2}(-2x) + 2x\sqrt{r^2 - x^2}\right]$$

$$= \frac{2\pi x}{\sqrt{r^2 - x^2}}(2r^2 - 3x^2) = 0$$

when $x = 0$ and $x^2 = \dfrac{2r^2}{3}$ ⟹ $x = \dfrac{\sqrt{6}\,r}{3}$

By the First Derivative Test the volume

is a maximum when $x = \dfrac{\sqrt{6}\,r}{3}$ and $h = \dfrac{2r}{\sqrt{3}}$.

Thus the maximum volume is

$V = \pi\,\dfrac{2}{3}r^2\left(\dfrac{2r}{\sqrt{3}}\right) = \dfrac{4\pi r^3}{3\sqrt{3}}$

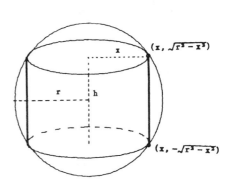

31. From the figure we have $A = 4\pi r^2 + 2\pi rh$

and since $\dfrac{4}{3}\pi r^3 + \pi r^2 h = 12 \implies h = \dfrac{12 - 4\pi r^3/3}{\pi r^2}$

$A = 4\pi r^2 + 2\pi r\left[\dfrac{12 - 4\pi r^3/3}{\pi r^2}\right] = \dfrac{4}{3}\pi r^2 + \dfrac{24}{r}$

$\dfrac{dA}{dr} = \dfrac{8}{3}\pi r - \dfrac{24}{r^2} = 0$

when $r = \sqrt[3]{9/\pi} \approx 1.42$ inches

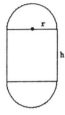

By the First Derivative Test this value
for r yields a minimum surface area.
This radius yields a height of zero.
The resulting solid is a sphere of
radius 1.42 inches.

32. Let $s = x + y = x + \sqrt{r^2 - x^2}$

$\dfrac{ds}{dx} = 1 + \dfrac{-x}{\sqrt{r^2 - x^2}} = 0$ when $x = \sqrt{r^2 - x^2} = y$

$\dfrac{ds}{dx} > 0$ when $x < \sqrt{r^2 - x^2}$, $\dfrac{ds}{dx} < 0$ when $x > \sqrt{r^2 - x^2}$

Therefore s is maximum if $x = y$.

33. Let x be the length of a side of the square and y the length of a side
of the triangle.

$4x + 3y = 10$, $A = x^2 + \dfrac{1}{2}y\left(\dfrac{\sqrt{3}}{2}y\right) = \dfrac{(10 - 3y)^2}{16} + \dfrac{\sqrt{3}}{4}y^2$

$\dfrac{dA}{dy} = \dfrac{1}{8}(10 - 3y)(-3) + \dfrac{\sqrt{3}}{2}y = 0$, $-30 + 9y + 4\sqrt{3}\,y = 0$, $y = \dfrac{30}{9 + 4\sqrt{3}}$

$\dfrac{d^2A}{dy^2} = \dfrac{9 + 4\sqrt{3}}{8} > 0$ therefore A is minimum when $y = \dfrac{30}{9 + 4\sqrt{3}}$

and $x = \dfrac{10\sqrt{3}}{9 + 4\sqrt{3}}$.

34. Let x be the length of a side of the square and r the radius of the circle. $4x + 2\pi r = 16$, $x = 4 - \frac{1}{2}\pi r$, $A = x^2 + \pi r^2 = (4 - \frac{1}{2}\pi r)^2 + \pi r^2$

$$\frac{dA}{dr} = 2(4 - \frac{1}{2}\pi r)(-\frac{1}{2}\pi) + 2\pi r$$

$$-8\pi + (\pi^2 + 4\pi)r = 0 \text{ when } r = \frac{8}{4 + \pi}$$

$$\frac{d^2A}{dr^2} = \frac{\pi^2}{2} > 0 \text{ therefore } A \text{ is a minimum when } r = \frac{8}{4 + \pi} \text{ and } x = \frac{16}{4 + \pi}.$$

35. Let x be the length of the sides of the isosceles triangle (the hypotenuse will be $\sqrt{2}\,x$) and r the radius of the circle.

$$(2x + \sqrt{2}x) + 2\pi r = 10, \quad x = \frac{10 - 2\pi r}{2 + \sqrt{2}}$$

$$A = \frac{1}{2}x^2 + \pi r^2 = \frac{1}{2}(\frac{10 - 2\pi r}{2 + \sqrt{2}})^2 + \pi r^2 = \frac{1}{3 + 2\sqrt{2}}(25 - 10\pi r + \pi^2 r^2) + \pi r^2$$

$$\frac{dA}{dr} = \frac{1}{3 + 2\sqrt{2}}(-10\pi + 2\pi^2 r) + 2\pi r = 0 \implies r = \frac{5}{\pi + 3 + 2\sqrt{2}}$$

  (a) Minimum: $2\pi r = 2\pi(\dfrac{5}{\pi + 3 + 2\sqrt{2}}) = \dfrac{10\pi}{\pi + 3 + 2\sqrt{2}} \approx 3.5$ feet

  (b) Maximum: $2\pi r = 10$ feet (endpoint extrema)

36.(a) Let x be the side of the triangle and y the side of the square.

$$A = \frac{3}{4}(\cot\frac{\pi}{3})x^2 + \frac{4}{4}(\cot\frac{\pi}{4})y^2 \text{ where } 3x + 4y = 20$$

$$= \frac{\sqrt{3}}{4}x^2 + (5 - \frac{3}{4}x)^2, \quad 0 \le x \le \frac{20}{3}$$

$$\frac{dA}{dx} = \frac{\sqrt{3}}{2}x + 2(5 - \frac{3}{4}x)(-\frac{3}{4}) = 0, \quad x = \frac{60}{4\sqrt{3} + 9}$$

When $x = 0$, $A = 25$, when $x = \dfrac{60}{4\sqrt{3} + 9}$, $A \approx 10.847$, and

when $x = \dfrac{20}{3}$, $A \approx 19.245$.

Area is maximum when all 20 feet are used on the square.

36.(b)   Let x be the side of the square and y the side of the pentagon.

$$A = \frac{4}{4}(\cot \frac{\pi}{4})x^2 + \frac{5}{4}(\cot \frac{\pi}{5})y^2 \text{ where } 4x + 5y = 20$$

$$= x^2 + 1.7204774(4 - \frac{4}{5}x)^2, \qquad 0 \le x \le 5$$

$$A' = 2x - 2.75276384(4 - \frac{4}{5}x) = 0, \qquad x \approx 2.62$$

When x = 0, A ≈ 27.528, when x ≈ 2.62, A ≈ 13.102, and
when x = 5, A ≈ 25.

Area is maximum when all 20 feet are used on the pentagon.

(c)   Let x be the side of the pentagon and y the side of the hexagon.

$$A = \frac{5}{4}(\cot \frac{\pi}{5})x^2 + \frac{6}{4}(\cot \frac{\pi}{6})y^2 \text{ where } 5x + 6y = 20$$

$$= \frac{5}{4}(\cot \frac{\pi}{5})x^2 + \frac{3}{2}(\sqrt{3})(\frac{20 - 5x}{6})^2, \qquad 0 \le x \le 4$$

$$A' = \frac{5}{2}(\cot \frac{\pi}{5})x + 3\sqrt{3}(-\frac{5}{6})(\frac{20 - 5x}{6}) = 0, \qquad x \approx 2.0475$$

When x = 0, A ≈ 28.868, when x ≈ 2.0475, A ≈ 14.091, and
when x = 4, A ≈ 27.528.

Area is maximum when all 20 feet are used on the hexagon.

(d)   Let x be the side of the hexagon and r the radius of the circle.

$$A = \frac{6}{4}(\cot \frac{\pi}{6})x^2 + \pi r^2 \text{ where } 6x + 2\pi r = 20$$

$$= \frac{3\sqrt{3}}{2}x^2 + \pi(\frac{10}{\pi} - \frac{3x}{\pi})^2, \qquad 0 \le x \le \frac{10}{3}$$

$$A' = 3\sqrt{3}x - 6(\frac{10}{\pi} - \frac{3x}{\pi}) = 0, \qquad x \approx 1.748$$

When x = 0, A ≈ 31.831, when x ≈ 1.748, A ≈ 15.138, and
when x = 10/3, A ≈ 28.868.

Area is maximum when all 20 feet are used on the circle.

37. Let S be the strength and k the constant of proportionality.

    $S = kwh^2$ Given: $h^2 + w^2 = 24^2$, $h^2 = 24^2 - w^2$

    $S = kw(576 - w^2) = k(576w - w^3)$

    $\dfrac{dS}{dw} = k(576 - 3w^2) = 0$, when $w = 8\sqrt{3}$ and $h = 8\sqrt{6}$

    $\dfrac{d^2S}{dw^2} = -6kw < 0$ for $w > 0$ thus these values yield a maximum.

38. Let F be the illumination at point P which is x units from source 1.

    $F = \dfrac{kI_1}{x^2} + \dfrac{kI_2}{(d - x)^2}$, $\dfrac{dF}{dx} = \dfrac{-2kI_1}{x^3} + \dfrac{2kI_2}{(d - x)^3} = 0$ when

    $\dfrac{2kI_1}{x^3} = \dfrac{2kI_2}{(d - x)^3}$, $\dfrac{\sqrt[3]{I_1}}{\sqrt[3]{I_2}} = \dfrac{x}{d - x}$, $x = \dfrac{d\,\sqrt[3]{I_1}}{\sqrt[3]{I_1} + \sqrt[3]{I_2}}$

    $\dfrac{d^2F}{dx^2} = \dfrac{6kI_1}{x^4} + \dfrac{6kI_2}{(d - x)^4} > 0$, when $x > 0$, thus this is the minimum point.

39. $S = \sqrt{x^2 + 4}$, $L = \sqrt{1 + (3 - x)^2} = \sqrt{x^2 - 6x + 10}$

    $T = \dfrac{\sqrt{x^2 + 4}}{2} + \dfrac{\sqrt{x^2 - 6x + 10}}{4}$

    $\dfrac{dT}{dx} = \dfrac{x}{2\sqrt{x^2 + 4}} + \dfrac{x - 3}{4\sqrt{x^2 - 6x + 10}} = 0$

    $\dfrac{x^2}{x^2 + 4} = \dfrac{9 - 6x + x^2}{4(x^2 - 6x + 10)}$

    $x^4 - 6x^3 + 9x^2 + 8x - 12 = 0$

    We can find the roots of this equation in the interval [0, 3] with the bisection method (see Section 4.4, Exercise 40). This equation has only one root in this interval ($x = 1$). Testing at this value and at the endpoints we see that $x = 1$ yields the minimum time. Thus the man should row to a point 1 mile from the nearest point on the coast.

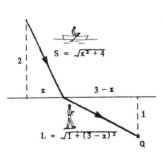

40. $T = \dfrac{\sqrt{x^2 + 4}}{4} + \dfrac{\sqrt{x^2 - 6x + 10}}{4}$

$\dfrac{dT}{dx} = \dfrac{1}{4}\left[\dfrac{x}{\sqrt{x^2 + 4}} + \dfrac{x - 3}{\sqrt{x^2 - 6x + 10}}\right] = 0$ when $\dfrac{x^2}{(x^2 + 4)} = \dfrac{(x - 3)^2}{x^2 - 6x + 10}$

Thus $x^2 - 8x + 12 = 0 \implies x = 2, 6$

In the interval $[0, 3]$ we have $x = 2$ miles. Testing at $x = 2$ and at the endpoints we see that $x = 2$ yields a minimum time.

41. $T = \dfrac{\sqrt{x^2 + 4}}{v_1} + \dfrac{\sqrt{x^2 - 6x + 10}}{v_2}$

$\dfrac{dT}{dx} = \dfrac{x}{v_1\sqrt{x^2 + 4}} + \dfrac{x - 3}{v_2\sqrt{x^2 - 6x + 10}} = 0$

Since $\dfrac{x}{\sqrt{x^2 + 4}} = \sin\theta_1$ and

$\dfrac{x - 3}{\sqrt{x^2 - 6x + 10}} = -\sin\theta_2$ we have

$\dfrac{\sin\theta_1}{v_1} - \dfrac{\sin\theta_2}{v_2} = 0 \implies \dfrac{\sin\theta_1}{v_1} = \dfrac{\sin\theta_2}{v_2}$

Since $\dfrac{d^2T}{dx^2} = \dfrac{4}{v_1(x^2 + 4)^{3/2}} + \dfrac{1}{v_2(x^2 - 6x + 10)^{3/2}} > 0$
this condition yields a minimum time.

42. $T = \dfrac{\sqrt{x^2 + d_1^2}}{v_1} + \dfrac{\sqrt{d_2^2 + (a - x)^2}}{v_2}$

$\dfrac{dT}{dx} = \dfrac{x}{v_1\sqrt{x^2 + d_1^2}} + \dfrac{x - a}{v_2\sqrt{d_2^2 + (a - x)^2}} = 0$

Since $\dfrac{x}{\sqrt{x^2 + d_1^2}} = \sin\theta_1$ and $\dfrac{x - a}{\sqrt{d_2^2 + (a - x)^2}} = -\sin\theta_2$

we have $\dfrac{\sin\theta_1}{v_1} - \dfrac{\sin\theta_2}{v_2} = 0 \implies \dfrac{\sin\theta_1}{v_1} = \dfrac{\sin\theta_2}{v_2}$

Since $\dfrac{d^2T}{dx^2} = \dfrac{d_1^2}{v_1(x^2 + d_1^2)^{3/2}} + \dfrac{d_2^2}{v_2[d_2^2 + (a - x)^2]^{3/2}} > 0$
this condition yields a minimum time.

43. $F \cos\theta = k(W - F\sin\theta)$, $F = \dfrac{kW}{\cos\theta + k\sin\theta}$

$\dfrac{dF}{d\theta} = \dfrac{-kW(k\cos\theta - \sin\theta)}{(\cos\theta + k\sin\theta)^2} = 0$

$k\cos\theta = \sin\theta \implies k = \tan\theta \implies \theta = \arctan k$

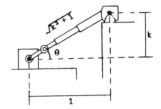

Since $\cos\theta + k\sin\theta = \dfrac{1}{\sqrt{k^2 + 1}} + \dfrac{k^2}{\sqrt{k^2 + 1}}$
$= \sqrt{k^2 + 1}$

the minimum force is

$F = \dfrac{kW}{\cos\theta + k\sin\theta} = \dfrac{kW}{\sqrt{k^2 + 1}}$

4.8
Business and economics applications

1. $R = 900x - 0.1x^2$

    $\dfrac{dR}{dx} = 900 - 0.2x = 0$ when $x = 4500$ maximum, by the First

    Derivative Test.

2. $R = 600x^2 - 0.02x^3$

    $\dfrac{dR}{dx} = 1200x - 0.06x^2 = 6x(200 - 0.01x) = 0$ when $x = 0, 20,000$

    The maximum occurs when $x = 20,000$ by the First Derivative Test.

3. $R = \dfrac{1,000,000x}{0.02x^2 + 1800}$

    $\dfrac{dR}{dx} = 1,000,000 \dfrac{0.02x^2 + 1800 - x(0.04x)}{(0.02x^2 + 1800)^2} = 0$

    $1800 - 0.02x^2 = 0$ when $x = 300$ maximum by the First Derivative Test.

4. $R = 30x^{2/3} - 2x$, $\dfrac{dR}{dx} = 20x^{-1/3} - 2 = \dfrac{20}{x^{1/3}} - 2 = 0$ when $= 1000$ maximum,

    by the First Derivative Test.

5. $\overline{C} = 0.125x + 20 + \dfrac{5000}{x}$, $\dfrac{d\overline{C}}{dx} = 0.125 - \dfrac{5000}{x^2} = 0$ when $x = 200$

    By the First Derivative Test, $x = 200$ yields the minimum average cost.

6. $\bar{C} = 0.001x^2 - 5 + \dfrac{250}{x}, \qquad \dfrac{d\bar{C}}{dx} = 0.002x - \dfrac{250}{x^2} = 0$ when $x = 50$

By the First Derivative Test, $x = 50$ yields the minimum average cost.

7. $\bar{C} = 3000 - x(300 - x)^{1/2}$

$\dfrac{d\bar{C}}{dx} = -x(\dfrac{1}{2})(300 - x)^{-1/2}(-1) - (300 - x)^{1/2} = 0$ when $x = 200$

By the First Derivative Test, $x = 200$ yields the minimum average cost.

8. $\bar{C} = \dfrac{2x^2 - x + 5000}{x^2 + 2500}$

$\dfrac{d\bar{C}}{dx} = \dfrac{(x^2 + 2500)(4x - 1) - (2x^2 - x + 5000)(2x)}{(x^2 + 2500)^2}$

$= \dfrac{x^2 - 2500}{(x^2 + 2500)^2} = 0$ when $x = 50$

By the First Derivative Test, $x = 50$ yields the minimum average cost.

9. $C = 100 + 30x, \qquad p = 90 - x$

$P = xp - C = 90x - x^2 - 30x - 100 = -x^2 + 60x - 100$

$\dfrac{dP}{dx} = -2x + 60 = 0$ when $x = 30$ so $p = 60$

By the First Derivative Test, $x = 30$ is a maximum.

10. $C = 2400x + 5200, \qquad p = \dfrac{2}{5}(15{,}000 - x^2)$

$P = \dfrac{2}{5}(15{,}000x - x^3) - 2400x - 5200 = -\dfrac{2x^3}{5} + 3600x - 5200$

$\dfrac{dP}{dx} = -\dfrac{6x^2}{5} + 3600 = 0$ when $x \approx 55$ so $p \approx 4800$

By the First Derivative Test, $x \approx 55$ is a maximum.

11. $C = 4000 - 40x + 0.02x^2, \qquad p = 50 - 0.01x$

$P = 50x - 0.01x^2 - 4000 + 40x - 0.02x^2 = -0.03x^2 + 90x - 4000$

$\dfrac{dP}{dx} = -0.06x + 90 = 0$ when $x = 1500$ so $p = 35$

By the First Derivative Test, $x = 1500$ is a maximum.

12. $C = 35x + 2\sqrt{x - 1}, \qquad p = 40 - \sqrt{x - 1}$

$P = 40x - x\sqrt{x - 1} - 35x - 2\sqrt{x - 1} = 5x - (x + 2)\sqrt{x - 1}$

$\dfrac{dP}{dx} = 5 - \left[(x + 2)(\dfrac{1}{2})(x - 1)^{-1/2} + (x - 1)^{1/2} \right] = 5 - \left[\dfrac{x + 2 + 2(x - 1)}{2\sqrt{x - 1}} \right]$

$\qquad = 5 - \dfrac{3x}{2\sqrt{x - 1}} = \dfrac{10\sqrt{x - 1} - 3x}{2\sqrt{x - 1}} = 0$ when $x = 10$ so $p = 37$

By the First Derivative Test, $x = 10$ is a maximum.

13. $C = 800 - 10x + \dfrac{1}{4}x^2, \qquad \dfrac{dC}{dx} = -10 + \dfrac{1}{2}x = 0$ when $x = 20$

C is minimum when $x = 20$, since $\dfrac{d^2C}{dx^2} = \dfrac{1}{2} > 0$.

14. $P = 230 + 20s - \dfrac{1}{2}s^2, \qquad \dfrac{dP}{ds} = 20 - s = 0$ when $s = 20$

P is maximum when advertising is \$2,000, since $\dfrac{d^2P}{dx^2} = -1 < 0$.

15. Let x be the number of units purchased, p be the price per unit, and P be the total profit.

$p = 90 - (0.10)(x - 100) = 100 - 0.1x$

Since each radio costs the manufacturer \$60, the profit per radio is $p - 60$ and the total profit is $P = x(p - 60) = 40x - 0.1x^2$.

$\dfrac{dP}{dx} = 40 - 0.2x = 0$ when $x = \dfrac{40}{0.2} = 200$ (maximum, since $\dfrac{d^2P}{dx^2} = -0.2 < 0$)

Manufacturer's largest ordering size for maximum profit is 200 units.

16. Let x be the number of vacant apartments, and p be the rent per unit.

$p = 270 + 15x, \qquad P = \text{Revenue} - \text{Cost}$

$P = (\text{number of apartments rented})(\text{rent/unit}) - \$18(\text{no. of units rented})$

$\quad = (50 - x)(270 + 15x) - 18(50 - x) = (50 - x)(252 + 15x)$

$\dfrac{dP}{dx} = (50 - x)(15) + (252 + 15x)(-1)$

$\qquad = 498 - 30x = 0$ when $x = 16.6$, (maximum since $\dfrac{d^2P}{dx^2} = -30 < 0$)

Maximum profit when 17 apartments are vacant and the rent is \$270 + \$255 = \$525.

17. Let T be the total cost

$$T = 8(5280) \sqrt{x^2 + (1/4)} + 6(5280)(6 - x)$$

$$= 2(5280) [4 \sqrt{x^2 + (1/4)} + 18 - 3x]$$

$$\frac{dT}{dx} = 2(5280) \left[\frac{4(2x)}{2\sqrt{x^2 + (1/4)}} - 3 \right] = 2(5280) \left[\frac{4x - 3\sqrt{x^2 + (1/4)}}{\sqrt{x^2 + (1/4)}} \right] = 0$$

when $4x - 3\sqrt{x^2 + (1/4)} = 0$, $x = \dfrac{3}{2\sqrt{7}} \approx 0.57$ miles

By the First Derivative Test $x \approx 0.57$ is a minimum.

18. Let K be the cost per mile for laying pipe on land and T be the total cost. $T = 2K \sqrt{x^2 + 1} + K(2 - x) = K(2 \sqrt{x^2 + 1} + 2 - x)$

$$\frac{dT}{dx} = K \left[\frac{2x}{\sqrt{x^2 + 1}} - 1 \right] = K \left[\frac{2x - \sqrt{x^2 + 1}}{\sqrt{x^2 + 1}} \right] = 0$$

when $2x - \sqrt{x^2 + 1} = 0$, $x = \dfrac{1}{\sqrt{3}} \approx 0.58$ miles

By the First Derivative Test $x = 1/\sqrt{3}$ is a minimum.

19. Let d be the amount deposited in the bank, i be interest rate paid by bank, and P be profit.

$P = (0.12)d - id$, $d = ki^2$ (since d is proportional to i^2)

$P = (0.12)(ki^2) - i(ki^2) = k(0.12i^2 - i^3)$

$\dfrac{dP}{di} = k(0.24i - 3i^2) = 0$ when $i = 0.24/3 = 0.08$

$\dfrac{d^2P}{di^2} = 0.24 - 6i < 0$ when $i = 0.08$ so the profit is a maximum when $i = 8\%$.

20. Let $\overline{C} = \dfrac{C}{x} = \dfrac{f(x)}{x}$ = average cost.

$\dfrac{d\overline{C}}{dx} = \dfrac{xf'(x) - f(x)}{x^2} = 0 \implies xf'(x) - f(x) = 0$ when $f'(x) = \dfrac{f(x)}{x}$

marginal cost = average cost

This condition will yield a minimum if it exists.

21. $C = 2x^2 + 5x + 18$

   (a) Average cost $= \dfrac{C}{x} = \overline{C} = 2x + 5 + \dfrac{18}{x}$

   $\dfrac{d\overline{C}}{dx} = 2 - \dfrac{18}{x^2} = 0$ when $x = 3$, $\quad \overline{C}(3) = 6 + 5 + 6 = 17$

   By the First Derivative Test, $x = 3$ is a minimum.

   (b) Marginal cost $= \dfrac{dC}{dx} = 4x + 5$, $\quad$ at $x = 3$: $\dfrac{dC}{dx} = 17 = \overline{C}(3)$

22. $C = x^3 - 6x^2 + 13x$

   (a) Average cost $= \dfrac{C}{x} = \overline{C} = x^2 - 6x + 13$, $\quad \dfrac{d\overline{C}}{dx} = 2x - 6 = 0$ when $x = 3$

   (b) $\dfrac{dC}{dx} = 3x^2 - 12x + 13$ when $x = 3$, $\dfrac{dC}{dx} = 27 - 36 + 13 = 4$ (marginal cost)

   $\overline{C} = 9 - 18 + 13 = 4$ (average cost)

23. $Q = 6.9 + \cos\left[\dfrac{\pi(2t - 1)}{12}\right]$, $\quad Q' = -\dfrac{\pi}{6} \sin\left[\dfrac{\pi(2t - 1)}{12}\right] = 0$

   $t = \dfrac{1}{2}$ $\quad$ or $\quad$ $t = \dfrac{13}{2}$

   (a) By the First Derivative Test, maximum when $t = \dfrac{1}{2}$ or January 15th

   $Q(\dfrac{1}{2}) = 7.9$ quads/month

   (b) By the First Derivative Test, minimum when $t = \dfrac{13}{2}$ or July 15th

   $Q(\dfrac{13}{2}) = 5.9$ quads/month

24. $S = 164.68 + 0.56t + 13.60 \cos (\frac{\pi t}{3})$

$S' = 0.56 - 13.60(\frac{\pi}{3}) \sin (\frac{\pi t}{3}) = 0, \qquad \sin (\frac{\pi t}{3}) = 0.03932$

$\frac{\pi t}{3} = 0.03933 + n\pi, \ n = 0, 1, \cdots, 7 \implies t = \frac{3}{\pi}(0.03933 + n\pi)$

(a) Applying the First Derivative Test to each test interval, we have

Relative Maxima	Relative Minima
(0.0376, 178.29)	(3.0376, 152.79)
(6.0376, 181.65)	(9.0376, 156.15)
(12.0376, 185.01)	(15.0376, 159.51)
(18.0376, 188.37)	(21.0376, 162.87)

(b) $S(31.5) = 182.32$ billion kilowatt hours

25. $Q = 6.9 + \frac{2}{3} \sin \frac{\pi(2t + 5)}{12} - \frac{1}{3} \cos \frac{\pi(2t + 5)}{6}$

$Q' = \frac{\pi}{9} \cos \frac{\pi(2t + 5)}{12} + \frac{\pi}{9} \sin \frac{\pi(2t + 5)}{6}$

Q' is zero when $\cos \frac{\pi(2t + 5)}{12} + \sin \frac{\pi(2t + 5)}{6} = 0$

$\cos \frac{\pi(2t + 5)}{12} + 2 \sin \frac{\pi(2t + 5)}{12} \cos \frac{\pi(2t + 5)}{12} = 0$

$\cos \frac{\pi(2t + 5)}{12} \left[1 + 2 \sin \frac{\pi(2t + 5)}{12} \right] = 0, \qquad \cos \frac{\pi(2t + 5)}{12} = 0$

$\frac{\pi(2t + 5)}{12} = \frac{\pi}{2}, \frac{3\pi}{2} \implies t = \frac{1}{2}, \frac{13}{2}$

or $\sin \frac{\pi(2t + 5)}{12} = -\frac{1}{2} \implies \frac{\pi(2t + 5)}{12} = \frac{7\pi}{6}, \frac{11\pi}{6} \implies t = \frac{9}{2}, \frac{17}{2}$

Critical Points: $(\frac{1}{2}, 7.9)$, $(\frac{13}{2}, 6.566)$, $(\frac{9}{2}, 6.4)$, $(\frac{17}{2}, 6.4)$

Applying the First Derivative Test on these intervals, we have

(a) Highest consumption: $t = \frac{1}{2}$, January 15
     Rate: 7.9 quads/month

(b) Lowest consumption: $t = \frac{9}{2}$, May 15, $t = \frac{17}{2}$, September 15
     Rate: 6.4 quads/month

26.(a) $P = R - C = xp - C = x(100 - \frac{1}{2}x^2) - (4x + 375)$

$\frac{dP}{dx} = x(-x) + (100 - \frac{1}{2}x^2) - 4 = -\frac{3}{2}x^2 + 96 = 0$ when $x = 8$

$p = 100 - \frac{1}{2}(64) = \$68$ (maximum, by First Derivative Test)

(b) $\bar{C}(x) = \frac{C}{x} = 4 + \frac{375}{x}$, $\quad \bar{C}(8) = 4 + \frac{375}{8} \approx \$50.88$

27.(a) $P = R - C = x(100 - \frac{1}{2}x^2) - (26.5x + 37.5)$

$\frac{dP}{dx} = x(-x) + (100 - \frac{1}{2}x^2) - 26.5 = -\frac{3}{2}x^2 + 73.5 = 0$ when $x = 7$

$p = 100 - \frac{1}{2}(49) = \$75.50$ (maximum, by First Derivative Test)

(b) $\bar{C}(x) = \frac{C}{x} = 26.5 + (\frac{37.5}{x})$, $\quad \bar{C}(7) = 26.5 + \frac{37.5}{7} \approx \$31.86$

28. For the linear demand function the rate of change (slope) is $m = -\frac{25}{5}$.

The demand x is therefore: $x = 800 - \frac{25}{5}(p - 25) = 925 - 5p$

$R = xp = -5p^2 + 925p$, $\quad \frac{dR}{dp} = -10p + 925 = 0$ when $p = \$92.50$

29. The demand function is $x = 20 - 2p^2$.

(a) If $p = 2$ and $x = 12$ and p decreases by 5%, then
$p = 2 - (0.05)(2) = 1.9$ and $x = 20 - 2(1.9)^2 = 12.78$.

The percentage increase in x is $\frac{12.78 - 12}{12} = 6.5\%$

(b) The exact elasticity of demand at (2, 12) is

$\eta = \frac{p/x}{(dp/dx)} = (\frac{p}{x})(-4p) = (\frac{2}{12})[-4(2)] = -\frac{4}{3}$

(c) $R = xp = (20 - 2p^2)p = 20p - 2p^3$

$\frac{dR}{dp} = 20 - 6p^2 = 0$, $\quad 6p^2 = 20$, $\quad p = \sqrt{10/3} \approx \$1.83$

$x = 20 - 2(\sqrt{10/3})^2 = \frac{40}{3}$ (maximum, since $\frac{d^2R}{dp^2} = -12p < 0$ when $p > 0$)

(d) If $x = \frac{40}{3}$ and $p = \sqrt{10/3}$, $|\eta| = \left|\frac{p}{x}(-4p)\right| = \left|\frac{\sqrt{10/3}}{40/3}(-4\sqrt{10/3})\right| = 1$

30. $p^3 + x^3 = 9$, $3p^2 \dfrac{dp}{dx} + 3x^2 = 0$, $\dfrac{dp}{dx} = -\dfrac{x^2}{p^2}$

(a) $\eta = \dfrac{p/x}{dp/dx} = \dfrac{p/x}{-x^2/p^2} = -\dfrac{p^3}{x^3} = -\dfrac{9 - x^3}{x^3}$

   When $x = 2$, $\eta = -\dfrac{9 - 8}{8} = -\dfrac{1}{8}$

(b) $R = xp = x\sqrt[3]{9 - x^3}$

   $\dfrac{dR}{dx} = \dfrac{-x^3}{\sqrt[3]{(9 - x^3)^2}} + \sqrt[3]{9 - x^3} = \dfrac{9 - 2x^3}{\sqrt[3]{(9 - x^3)^2}} = 0$ when $x = \sqrt[3]{9/2}$

   $p = \sqrt[3]{9 - (9/2)} = \sqrt[3]{9/2}$

(c) When $x = \sqrt[3]{9/2}$, $|\eta| = \left|\dfrac{9 - x^3}{x^3}\right| = \dfrac{9 - (9/2)}{9/2} = 1$

31. $p = (16 - x)^{1/2}$, $0 \le x \le 16$

   $R = xp = x\sqrt{16 - x}$, $\dfrac{dR}{dx} = \dfrac{-x}{2\sqrt{16 - x}} + \sqrt{16 - x} = \dfrac{32 - 3x}{2\sqrt{16 - x}} = 0$

   when $x = \dfrac{32}{3}$ and $p = \sqrt{16 - (32/3)} = \dfrac{4\sqrt{3}}{3}$

   (maximum by the First Derivative Test)

32. $x = \dfrac{a}{p^m}$, $p^m = \dfrac{a}{x}$, $mp^{m-1} \dfrac{dp}{dx} = \dfrac{-a}{x^2}$, $\dfrac{dp}{dx} = -\dfrac{-a}{mp^{m-1}x^2}$

   $\eta = \dfrac{p/x}{dp/dx} = \dfrac{p}{x}(-\dfrac{mp^{m-1}x^2}{a}) = -\dfrac{1}{a}mp^m x = -m$ since $p^m = \dfrac{a}{x}$

4.9
Newton's Method

1. $f(x) = x^2 - 3$, $f'(x) = 2x$, $x_1 = 1.7$

n	x_n	$f(x_n)$	$f'(x_n)$	$f(x_n)/f'(x_n)$	$x_n - [f(x_n)/f'(x_n)]$
1	1.700	−0.110	3.400	−0.032	1.732

2. $f(x) = 3x^3 - 2$, $f'(x) = 9x^2$, $x_1 = 1$

n	x_n	$f(x_n)$	$f'(x_n)$	$f(x_n)/f'(x_n)$	$x_n - [f(x_n)/f'(x_n)]$
1	1.000	1.000	9.000	0.111	0.889

3. $f(x) = \sin x$, $\quad f'(x) = \cos x$, $\quad x_1 = 3$

n	x_n	$f(x_n)$	$f'(x_n)$	$\dfrac{f(x_n)}{f'(x_n)}$	$x_n - \dfrac{f(x_n)}{f'(x_n)}$
1	3	0.141	−0.990	−0.143	3.143

4. $f(x) = \tan x$, $\quad f'(x) = \sec^2 x$, $\quad x_1 = 0.1$

n	x_n	$f(x_n)$	$f'(x_n)$	$\dfrac{f(x_n)}{f'(x_n)}$	$x_n - \dfrac{f(x_n)}{f'(x_n)}$
1	0.1	0.100	1.010	0.099	0.001

5. $f(x) = x^3 + x - 1$, $\quad f'(x) = 3x^2 + 1$

n	x_n	$f(x_n)$	$f'(x_n)$	$\dfrac{f(x_n)}{f'(x_n)}$	$x_n - \dfrac{f(x_n)}{f'(x_n)}$
1	0.5000	−0.3750	1.7500	−0.2143	0.7143
2	0.7143	0.0787	2.5306	0.0311	0.6832
3	0.6832	0.0021	2.4002	0.0009	0.6823

Approximation to the zero: $x = 0.682$

6. $f(x) = x^5 + x - 1$, $\quad f'(x) = 5x^4 + 1$

n	x_n	$f(x_n)$	$f'(x_n)$	$\dfrac{f(x_n)}{f'(x_n)}$	$x_n - \dfrac{f(x_n)}{f'(x_n)}$
1	0.5000	−0.4687	1.3125	−0.3571	0.8571
2	0.8571	0.3198	3.6983	0.0865	0.7707
3	0.7707	0.0426	2.7640	0.0154	0.7553
4	0.7553	0.0011	2.6272	0.0004	0.7549

Approximation to the zero: $x = 0.755$

7. $f(x) = 3\sqrt{x - 1} - x$, $\quad f'(x) = \dfrac{3}{2\sqrt{x - 1}} - 1$

n	x_n	$f(x_n)$	$f'(x_n)$	$\dfrac{f(x_n)}{f'(x_n)}$	$x_n - \dfrac{f(x_n)}{f'(x_n)}$
1	1.2000	0.1416	2.3541	0.0602	1.1398
2	1.1398	−0.0180	3.0113	−0.0060	1.1458
3	1.1458	−0.0003	2.9283	−0.0001	1.1459

Approximation to the zero: $x = 1.146$

8. $f(x) = x^3 - 3.9x^2 + 4.79x - 1.881,$ $f'(x) = 3x^2 - 7.8x + 4.79$

n	x_n	$f(x_n)$	$f'(x_n)$	$\dfrac{f(x_n)}{f'(x_n)}$	$x_n - \dfrac{f(x_n)}{f'(x_n)}$
1	0.5000	−0.3360	1.6400	−0.2049	0.7049
2	0.7049	−0.0921	0.7824	−0.1177	0.8226
3	0.8226	−0.0231	0.4037	−0.0572	0.8799
4	0.8799	−0.0045	0.2494	−0.0181	0.8980
5	0.8980	−0.0004	0.2048	−0.0019	0.8999
6	0.8999	0.0000	0.2002	0.0000	0.8999

Approximation to the zero: $x = 0.900$

n	x_n	$f(x_n)$	$f'(x_n)$	$\dfrac{f(x_n)}{f'(x_n)}$	$x_n - \dfrac{f(x_n)}{f'(x_n)}$
1	1.1	0.001	−0.16	−0.0062	1.1062
2	1.1062	0.000	−0.1673	0.0000	1.1062

Approximation to the zero: $x = 1.106$

n	x_n	$f(x_n)$	$f'(x_n)$	$\dfrac{f(x_n)}{f'(x_n)}$	$x_n - \dfrac{f(x_n)}{f'(x_n)}$
1	1.9	0.0010	0.8000	0.0012	1.8988
2	1.8988	0.0000	0.7957	0.0000	1.8988

Approximation to the zero: $x = 1.899$

9. $f(x) = x^4 - 10x^2 - 11,$ $f'(x) = 4x^3 - 20x$

n	x_n	$f(x_n)$	$f'(x_n)$	$\dfrac{f(x_n)}{f'(x_n)}$	$x_n - \dfrac{f(x_n)}{f'(x_n)}$
1	3.5000	16.5625	101.5000	0.1632	3.3368
2	3.3368	1.6307	81.8748	0.0199	3.3169
3	3.3169	0.0224	79.6298	0.0003	3.3166

Approximation to the zero: $x = 3.317$

10. $f(x) = x^3 + 3,$ $f'(x) = 3x^2$

n	x_n	$f(x_n)$	$f'(x_n)$	$\dfrac{f(x_n)}{f'(x_n)}$	$x_n - \dfrac{f(x_n)}{f'(x_n)}$
1	−1.5000	−0.3750	6.7500	−0.0556	−1.4444
2	−1.4444	−0.0137	6.2589	−0.0022	−1.4422
3	−1.4422	−0.0000	6.2398	0.0000	−1.4422

Approximation to the zero: $x = -1.442$

11. $f(x) = x + \sin(x + 1), \qquad f'(x) = 1 + \cos(x + 1)$

n	x_n	$f(x_n)$	$f'(x_n)$	$\dfrac{f(x_n)}{f'(x_n)}$	$x_n - \dfrac{f(x_n)}{f'(x_n)}$
1	-0.5	-0.0206	1.8776	-0.0110	-0.4890
2	-0.4890	0.0000	1.8722	0.0000	-0.4890

Approximation to the zero: $x = -0.489$

12. $f(x) = x^3 - \cos x, \qquad f'(x) = 3x^2 + \sin x$

n	x_n	$f(x_n)$	$f'(x_n)$	$\dfrac{f(x_n)}{f'(x_n)}$	$x_n - \dfrac{f(x_n)}{f'(x_n)}$
1	0.9	0.1074	3.2133	0.0334	0.8666
2	0.8666	0.0034	3.0151	0.0011	0.8655
3	0.8655	0.0001	3.0087	0.0000	0.8655

Approximation to the zero: $x = 0.866$

13. $f(x) = 2x + 1 - \sqrt{x + 4}, \qquad f'(x) = 2 - \dfrac{1}{2\sqrt{x + 4}}$

n	x_n	$f(x_n)$	$f'(x_n)$	$\dfrac{f(x_n)}{f'(x_n)}$	$x_n - \dfrac{f(x_n)}{f'(x_n)}$
1	0.6	0.0552	1.7669	0.0312	0.5688
2	0.5688	0.0001	1.7661	0.0001	0.5687
3	0.5687	-0.0001	1.7661	-0.0001	0.5688

Approximation to the zero: $x = 0.569$

14. $f(x) = 3 - x - \dfrac{1}{x^2 + 1}, \qquad f'(x) = -1 + \dfrac{2x}{(x^2 + 1)^2}$

n	x_n	$f(x_n)$	$f'(x_n)$	$\dfrac{f(x_n)}{f'(x_n)}$	$x_n - \dfrac{f(x_n)}{f'(x_n)}$
1	2.9	-0.0063	-0.9345	0.0067	2.8933
2	2.8933	0.0000	-0.9341	0.0000	2.8933

Approximation to the zero: $x = 2.893$

15. $f(x) = x - \tan x, \qquad f'(x) = 1 - \sec^2 x$

n	x_n	$f(x_n)$	$f'(x_n)$	$\dfrac{f(x_n)}{f'(x_n)}$	$x_n - \dfrac{f(x_n)}{f'(x_n)}$
1	4.5	−0.1373	−21.5048	0.0064	4.4936
2	4.4936	−0.0038	−20.2271	0.0002	4.4934

Approximation to the zero: $x = 4.493$

16. $f(x) = x^2 - \cos x, \qquad f'(x) = 2x + \sin x$

n	x_n	$f(x_n)$	$f'(x_n)$	$\dfrac{f(x_n)}{f'(x_n)}$	$x_n - \dfrac{f(x_n)}{f'(x_n)}$
1	0.8	−0.0567	2.3174	−0.0245	0.8245
2	0.8245	0.0009	2.3832	0.0004	0.8241

Approximation to the zero: $x = 0.824$

Since $f(x) = x^2$ and $g(x) = \cos x$ are both symmetric with respect to the y-axis, the other root is $x = -0.824$.

17. $y = 2x^3 - 6x^2 + 6x - 1, \qquad y' = 6x^2 - 12x + 6, \qquad x_1 = 1$

n	x_n	$f(x_n)$	$f'(x_n)$
1	1	1	0

$f'(x_1) = 0$
therefore method fails

18. $y = 4x^3 - 12x^2 + 12x - 3, \qquad y' = 12x^2 - 24x + 12, \qquad x_1 = 3/2$

n	x_n	$f(x_n)$	$f'(x_n)$	$\dfrac{f(x_n)}{f'(x_n)}$	$x_n - \dfrac{f(x_n)}{f'(x_n)}$
1	3/2	3/2	3	1/2	1
2	1	1	0		

$f'(x_2) = 0$ therefore the method fails

19. $y = -x^3 + 3x^2 - x + 1, \qquad y' = -3x^2 + 6x - 1, \qquad x_1 = 1$

n	x_n	$f(x_n)$	$f'(x_n)$	$\dfrac{f(x_n)}{f'(x_n)}$	$x_n - \dfrac{f(x_n)}{f'(x_n)}$
1	1	2	2	1	0
2	0	1	−1	−1	1
3	1				

Fails to converge

20. $f(x) = 2 \sin x + \cos 2x$, $f'(x) = 2 \cos x - 2 \sin 2x$, $x_1 = \dfrac{3\pi}{2}$

n	x_n	$f(x_n)$	$f'(x_n)$
1	$\dfrac{3\pi}{2}$	-3	0

Fails because $f'(x_1) = 0$

21. $f(x) = x^n - a_n = 0$, $f'(x) = nx^{n-1}$

$$x_{i+1} = x_i - \frac{x_i^n - a}{nx_i^{n-1}} = \frac{nx_i^n - x_i^n + a}{nx_i^{n-1}} = \frac{(n-1)x_i^n + a}{nx_i^{n-1}}$$

22. $x_{i+1} = \dfrac{2x_i^3 + 7}{3x_i^2}$

$\sqrt[3]{7} \approx 1.913$

i	1	2	3	4
x_i	2.0000	1.9167	1.9129	1.9129

23. $x_{i+1} = \dfrac{3x_i^4 + 6}{4x_i^3}$

$\sqrt[4]{6} \approx 1.565$

i	1	2	3	4
x_i	1.5000	1.5694	1.5651	1.5651

24. $x_{i+1} = \dfrac{x_i^2 + 5}{2x_i}$

i	1	2	3	4	5	6	7
x_i	2	2.25	2.3611	2.3937	2.2412	2.2361	2.2361

$\sqrt{5} \approx 2.236$

25. $f(x) = 1 + \cos x$ $f'(x) = -\sin x$

n	x_n	$f(x_n)$	$f'(x_n)$	$\dfrac{f(x_n)}{f'(x_n)}$	$x_n - \dfrac{f(x_n)}{f'(x_n)}$
1	3	0.0100	−0.1411	−0.0709	3.0709
2	3.0709	0.0025	−0.0706	−0.0354	3.1063
3	3.1063	0.0006	−0.0353	−0.0177	3.1239
4	3.1239	0.0002	−0.0177	−0.0088	3.1328
5	3.1328	0.0000	−0.0088	−0.0044	3.1372
6	3.1372	0.0000	−0.0044	−0.0022	3.1394
7	3.1394	0.0000	−0.0022	−0.0011	3.1405
8	3.1405	0.0000	−0.0011	−0.0005	3.1410

Approximation to the zero: 3.141

26. $f(x) = \tan x \qquad f'(x) = \sec^2 x$

n	x_n	$f(x_n)$	$f'(x_n)$	$\dfrac{f(x_n)}{f'(x_n)}$	$x_n - \dfrac{f(x_n)}{f'(x_n)}$
1	3	−0.1425	1.0203	−0.1397	3.1397
2	3.1397	−0.0019	1.0000	−0.0019	3.1416
3	3.1416	0.0000	1.0000	0.0000	3.1416

Approximation to the zero: 3.142

27. $f(x) = x \cos x, \qquad f'(x) = -x \sin x + \cos x = 0$

Letting $F(x) = f'(x)$ we can use Newton's Method as follows:

$[F'(x) = -2 \sin x - x \cos x]$

n	x_n	$F(x_n)$	$F'(x_n)$	$\dfrac{F(x_n)}{F'(x_n)}$	$x_n - \dfrac{F(x_n)}{F'(x_n)}$
1	0.9000	−0.0834	−2.1261	0.0392	0.8608
2	0.8608	−0.0009	−2.0778	0.0004	0.8603

Approximation to the critical number: 0.860

28. $f(x) = x \sin x, \qquad f'(x) = x \cos x + \sin x = 0$

The simple root is $x = 0$. Letting $F(x) = f'(x)$ we can use Newton's Method as follows:

$[F'(x) = 2 \cos x - x \sin x]$

n	x_n	$F(x_n)$	$F'(x_n)$	$\dfrac{F(x_n)}{F'(x_n)}$	$x_n - \dfrac{F(x_n)}{F'(x_n)}$
1	2.0000	0.0770	−2.6509	−0.0290	2.0291
2	2.0291	−0.0008	−2.7045	0.0003	2.0288

Approximation to the critical number: 2.029

29.

$f(x) = x \cos x, \qquad 0 \le x \le \pi$

30. 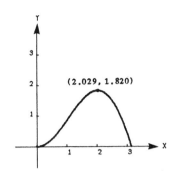 $f(x) = x \sin x, \qquad 0 \leq x \leq \pi$

(2.029, 1.820)

31. $y = f(x) = 4 - x^2, \qquad (1, 0)$

$d = \sqrt{(x - 1)^2 + (y - 0)^2} = \sqrt{(x - 1)^2 + (4 - x^2)^2} = \sqrt{x^4 - 7x^2 - 2x + 17}$

d is minimized when $d_i = x^4 - 7x^2 - 2x + 17$ is a minimum

$g(x) = d'_i = 4x^3 - 14x^2 - 2, \qquad g'(x) = 12x^2 - 14$

n	x_n	$g(x_n)$	$g'(x_n)$	$\dfrac{g(x_n)}{g'(x_n)}$	$x_n - \dfrac{g(x_n)}{g'(x_n)}$
1	2	2	34	0.0588	1.9412
2	1.9412	0.0830	31.2191	0.0026	1.9386
3	1.9386	0.0020	31.0980	0.0001	1.9385

$x \approx 1.939$

Point closest to $(1, 0)$
is $\approx (1.939, 0.240)$

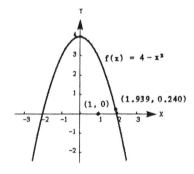

$f(x) = 4 - x^2$

(1.939, 0.240)

(1, 0)

32. $y = f(x) = x^2, \qquad (4, -3)$

$d = \sqrt{(x - 4)^2 + (y + 3)^2} = \sqrt{(x - 4)^2 + (x^2 + 3)^2} = \sqrt{x^4 + 7x^2 - 8x + 25}$

d is minimum when $d_i = x^4 + 7x^2 - 8x + 25$ is minimum

$g(x) = d_i = 4x^3 + 14x - 8 = 2(2x^3 + 7x - 4) = 0, \qquad g'(x) = 12x^2 + 14$

32. (continued)

n	x_n	$g(x_n)$	$g'(x_n)$	$\dfrac{g(x_n)}{g'(x_n)}$	$x_n - \dfrac{g(x_n)}{g'(x_n)}$
1	0.5	-0.5000	17	-0.0294	0.5294
2	0.5294	0.0051	17.3632	0.0003	0.5291
3	0.5291	-0.0001	17.3594	0.0000	0.5291

$x \approx 0.529$

Point closest to $(4, -3)$
is $(0.529, 0.280)$

33. Minimize: $T = \dfrac{\text{Distance Rowed}}{\text{Rate Rowed}} + \dfrac{\text{Distance Walked}}{\text{Rate Walked}}$

$$T = \frac{\sqrt{x^2 + 4}}{3} + \frac{\sqrt{x^2 - 6x + 10}}{4}$$

$$T' = \frac{x}{3\sqrt{x^2 + 4}} + \frac{x - 3}{4\sqrt{x^2 - 6x + 10}} = 0$$

$$4x\sqrt{x^2 - 6x + 10} = -3(x - 3)\sqrt{x^2 + 4}$$

$$16x^2(x^2 - 6x + 10) = 9(x - 3)^2(x^2 + 4)$$

$$7x^4 - 42x^3 + 43x^2 + 216x - 324 = 0$$

Let $f(x) = 7x^4 - 42x^3 + 43x^2 + 216x - 324$

$f'(x) = 28x^3 - 126x^2 + 86x + 216$

Since $f(1) = -100$ and $f(2) = 56$, the solution is in the interval $(1, 2)$.

n	x_n	$f(x_n)$	$f'(x_n)$	$\dfrac{f(x_n)}{f'(x_n)}$	$x_n - \dfrac{f(x_n)}{f'(x_n)}$
1	1.7000	19.5887	135.6240	0.1444	1.5556
2	1.5556	-1.0531	150.2820	-0.0070	1.5626
3	1.5626	-0.0025	149.5620	-0.0000	1.5626

Approximation: $x \approx 1.563$ miles

34. $\bar{C} = \dfrac{C}{x} = \dfrac{800}{x} + 0.4 + 0.02x + 0.0001x^2$

$\bar{C}' = -\dfrac{800}{x^2} + 0.02 + 0.0002x = 0$

$f(x) = 0.0002x^3 + 0.02x^2 - 800 = 0$

$f'(x) = 0.0006x^2 + 0.04x$

n	x_n	$f(x_n)$	$f'(x_n)$	$\dfrac{f(x_n)}{f'(x_n)}$	$x_n - \dfrac{f(x_n)}{f'(x_n)}$
1	130	−22.5997	15.3400	−1.4733	131.4733
2	131.4733	0.2125	15.6301	0.0136	131.4597
3	131.4597	−0.0006	15.6274	0.0000	131.4597

Approximation: $x \approx 131$ units

35. Maximize: $C = \dfrac{3t^2 + t}{50 + t^3}$, $\qquad C' = -\dfrac{3t^4 - 2t^3 + 300t + 50}{(50 + t^3)^2} = 0$

Let $f(x) = 3t^4 + 2t^3 - 300t - 50$, $\qquad f'(x) = 12t^3 + 6t^2 - 300$

Since $f(4) = -354$ and $f(5) = 575$, the solution is in the interval $(4, 5)$.

n	x_n	$f(x_n)$	$f'(x_n)$	$\dfrac{f(x_n)}{f'(x_n)}$	$x_n - \dfrac{f(x_n)}{f'(x_n)}$
1	4.5000	12.4374	915.0000	0.0136	4.4864
2	4.4864	0.0720	904.3880	0.0001	4.4863

Approximation: $t \approx 4.486$ hours

36. $C = 100\left(\dfrac{200}{x^2} + \dfrac{x}{x + 30}\right)$, $\qquad 1 \le x$

$C' = 100\left(-\dfrac{400}{x^3} + \dfrac{30}{(x + 30)^2}\right) = 0$

$f(x) = 10(3x^3 - 40x^2 - 2400x - 36000)$, $\qquad f'(x) = 10(9x^2 - 80x - 2400)$

n	x_n	$f(x_n)$	$f'(x_n)$	$\dfrac{f(x_n)}{f'(x_n)}$	$x_n - \dfrac{f(x_n)}{f'(x_n)}$
1	40	−4000	8800	−0.4545	40.4545
2	40.4545	66.3229	9092.76	0.0073	40.4472
3	40.4472	0.0397	9088.03	0.0000	40.4472

Approximation: $x \approx 40$ units

4.10
Differentials

1. $y = 3x^2 - 4$, $\quad dy = 6x\, dx$

2. $y = 2x^{3/2}$, $\quad dy = 3\sqrt{x}\, dx$

3. $y = \dfrac{x + 1}{2x - 1}$, $\quad dy = \dfrac{-3}{(2x - 1)^2}\, dx$

4. $y = \sqrt{x^2 - 4}$, $\quad dy = \dfrac{x}{\sqrt{x^2 - 4}}\, dx$

5. $y = x\sqrt{1 - x^2}$, $\quad dy = \left[x\, \dfrac{-x}{\sqrt{1 - x^2}} + \sqrt{1 - x^2} \right] dx = \dfrac{1 - 2x^2}{\sqrt{1 - x^2}}\, dx$

6. $y = \sqrt{x} + \dfrac{1}{\sqrt{x}}$, $\quad dy = \left[\dfrac{1}{2\sqrt{x}} - \dfrac{1}{2x\sqrt{x}} \right] dx = \dfrac{x - 1}{2x\sqrt{x}}\, dx$

7. $y = \dfrac{\sec^2 x}{x^2 + 1}$, $\quad dy = \left[\dfrac{(x^2 + 1)\, 2\sec^2 x \tan x - \sec^2 x(2x)}{(x^2 + 1)^2} \right] dx$

$$= \left[\dfrac{2\sec^2 x [x^2 \tan x + \tan x - x]}{(x^2 + 1)^2} \right] dx$$

8. $y = x \sin x$, $\quad dy = [x \cos x + \sin x]\, dx$

9. $y = \dfrac{1}{3} \cos \left(\dfrac{6\pi x - 1}{2} \right)$, $\quad dy = -\pi \sin \left(\dfrac{6\pi x - 1}{2} \right) dx$

10. $y = x - \tan^2 x$, $\quad dy = [1 - 2 \tan x \sec^2 x]\, dx$

11. $y = x^2$, $\quad dy = 2x\, dx$, $\quad x = 2$

dx = Δx	dy	Δy	Δy − dy	dy/Δy
1.0000	4.0000	5.0000	1.0000	0.8000
0.5000	2.0000	2.2500	0.2500	0.8889
0.1000	0.4000	0.4100	0.0100	0.9756
0.0100	0.0400	0.0401	0.0001	0.9975
0.0010	0.0040	0.0040	0.0000	1.0000

12. $y = \dfrac{1}{x^2}$, $\quad dy = \dfrac{-2}{x^3}\, dx$, $\quad x = 2$

Δx = dx	dy	Δy	Δy − dy	dy/Δy
1.0000	−0.2500	−0.1389	0.1111	1.8000
0.5000	−0.1250	−0.0900	0.0350	1.3889
0.1000	−0.0250	−0.0232	0.0018	1.0756
0.0100	−0.0025	−0.0025	0.0000	1.0075
0.0010	−0.0003	−0.0003	0.0000	1.0009

13. $y = x^5$, $\qquad dy = 5x^4\ dx$, $\qquad x = 2$

$\Delta x = dx$	dy	Δy	$\Delta y - dy$	$dy/\Delta y$
1.0000	80.0000	211.0000	131.0000	0.3791
0.5000	40.0000	65.6562	25.6562	0.6092
0.1000	8.0000	8.8410	0.8410	0.9049
0.0100	0.8000	0.8080	0.0080	0.9901
0.0010	0.0800	0.0801	0.0001	0.9990

14. $y = \sqrt{x}$, $\qquad dy = \dfrac{dx}{2\sqrt{x}}$, $\qquad x = 2$

$\Delta x = dx$	dy	Δy	$\Delta y - dy$	$dy/\Delta y$
1.0000	0.3536	0.3178	−0.0357	1.1124
0.5000	0.1768	0.1669	−0.0099	1.0590
0.1000	0.0354	0.0349	−0.0004	1.0124
0.0100	0.0035	0.0035	0.0000	1.0012
0.0010	0.0004	0.0004	0.0000	0.9999

15. $y = \sin x$, $\qquad dy = \cos x\ dx$, $\qquad x = 2$

$\Delta x = dx$	dy	Δy	$\Delta y - dy$	$dy/\Delta y$
1.0000	−0.4161	−0.7682	−0.3520	0.5417
0.5000	−0.2081	−0.3108	−0.1028	0.6694
0.1000	−0.0416	−0.0461	−0.0045	0.9029
0.0100	−0.0042	−0.0042	0.0000	0.9892
0.0010	−0.0004	−0.0004	0.0000	0.9990

16. $y = \csc x$, $\qquad dy = -\csc x \cot x\ dx$, $\qquad x = 2$

$\Delta x = dx$	dy	Δy	$\Delta y - dy$	$dy/\Delta y$
1.0000	0.5033	5.9864	5.4831	0.0841
0.5000	0.2517	0.5712	0.3195	0.4406
0.1000	0.0503	0.0587	0.0084	0.8572
0.0100	0.0050	0.0051	0.0001	0.9846
0.0010	0.0005	0.0005	0.0000	0.9986

17. $A(x) = x^2$

    (a) $dA = 2x\,dx = 2x\Delta x$ (b)

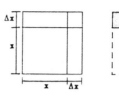 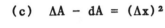

           $\Delta A = (x + \Delta x)^2 - x^2$

                $= 2x\Delta x + (\Delta x)^2$

    (c) $\Delta A - dA = (\Delta x)^2$

18. $A = x^2,\qquad x = 12,\qquad \Delta x = dx = \pm\dfrac{1}{64}$

    $dA = 2x\,dx,\qquad \Delta A \approx dA = 2(12)(\pm\dfrac{1}{64}) = \pm\dfrac{3}{8}$ square inches

19. $A = \pi r^2 = \pi(\dfrac{x}{2})^2,\qquad x = 28,\qquad \Delta x = dx = \pm\dfrac{1}{4}$

    $\Delta A \approx dA = \dfrac{\pi x}{2}\,dx = \pi(\dfrac{28}{2})(\pm\dfrac{1}{4}) = \pm\dfrac{7\pi}{2}$ square inches

20. $x = 12$ inches, $\Delta x = dx = 0.03$ inch

    (a) $V = x^3,\qquad dV = 3x^2\,dx = 3(12)^2(\pm 0.03) = \pm 12.96$ cubic inches

    (b) $S = 6x^2,\qquad dS = 12x\,dx = 12(12)(\pm 0.03) = \pm 4.32$ square inches

21. $r = 6$ inches, $\Delta r = dr = 0.02$ inch

    (a) $V = \dfrac{4}{3}\pi r^3,\qquad dV = 4\pi r^2\,dr = 4\pi(6)^2(\pm 0.02) = \pm 2.88\pi$ cubic inches

    (b) $S = 4\pi r^2,\qquad dS = 8\pi r\,dr = 8\pi(6)(\pm 0.02) = \pm 0.96\pi$ square inches

    (c) relative error $= \dfrac{dV}{V} = \dfrac{4\pi r^2\,dr}{(4/3)\pi r^3} = \dfrac{3}{r}\,dr = \dfrac{3}{6}(0.02) = 0.01 = 1\%$

            relative error $= \dfrac{dS}{S} = \dfrac{8\pi r\,dr}{4\pi r^2} = \dfrac{2}{r}\,dr = \dfrac{2(0.02)}{6} = 0.0067 = \dfrac{2}{3}\%$

22. $P = (500x - x^2) - (\dfrac{1}{2}x^2 - 77x + 3000),\qquad x$ changes from 115 to 120

    $dP = (500 - 2x - x + 77)\,dx = (577 - 3x)\,dx$

       $= [577 - 3(115)](120 - 115) = 1160$

    Approximate percentage change $= \dfrac{dP}{P}(100) = \dfrac{1160}{43517.50}(100) = 2.7\%$

23.(a) $T = 2\pi\sqrt{L/g}$, $dT = \dfrac{\pi}{g\sqrt{L/g}}\ dL$

relative error $= \dfrac{dT}{T} = \dfrac{(\pi dL)/(g\sqrt{L/g})}{2\pi\sqrt{L/g}}$

$= \dfrac{dL}{2L} = \dfrac{1}{2}(\text{relative error in L}) = \dfrac{1}{2}(0.005) = 0.0025$

percentage error $= \dfrac{dT}{T}(100) = 0.25\% = \dfrac{1}{4}\%$

(b) $(0.0025)(3600)(24) = 216$ seconds $= 3.6$ minutes

24. $T = 2\pi\sqrt{L/g}$, $T^2 = 4\pi^2(L/g)$, $g = (4\pi^2 L)/(T^2)$

$dg = 4\pi^2 L \dfrac{(-2)}{T^3}\ dT$

relative error in g $= \dfrac{dg}{g} = \dfrac{(-8\pi^2 L\ dT)/(T^3)}{(4\pi^2 L)/(T^2)} = -2\dfrac{dT}{T}$

$= -2(\text{relative error in period}) = -2(0.001) = -0.002$

percentage error in g $= \left|\dfrac{dg}{g}\right|(100) = \dfrac{1}{5}\%$

25. $r = \dfrac{v_0{}^2}{32}(\sin 2\theta)$, $v_0 = 2200$ feet/second

θ changes from 10° to 11°, $dr = \dfrac{(2200)^2}{16}(\cos 2\theta)d\theta$

$\theta = 10(\dfrac{\pi}{180})$, $d\theta = (11 - 10)\dfrac{\pi}{180}$

$\Delta r \approx dr = \dfrac{(2200)^2}{16}\cos(\dfrac{20\pi}{180})(\dfrac{\pi}{180}) \approx 4961$ feet

26.   From the figure we see that

h = 50 tan θ,   θ = 71.5°

dh = 50 sec² θ dθ

$\dfrac{dh}{h} = \dfrac{\sec^2 71.5}{\tan 71.5}\ d\theta \le \pm 0.06$

dθ ≤ ±0.018°

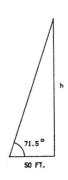

27.   Since Δy --> dy as Δx --> 0 then Δy - dy --> 0

Review Exercises for Chapter 4

1. $f(x) = 4x - x^2 = x(4 - x)$

    $f'(x) = 4 - 2x = 0$ when $x = 2$

    $f''(x) = -2$, therefore
    $(2, 4)$ is a relative maximum.

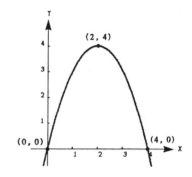

2. $f(x) = 4x^3 - x^4 = x^3(4 - x)$

    $f'(x) = 12x^2 - 4x^3 = 4x^2(3 - x) = 0$
    when $x = 0, 3$

    $f''(x) = 24x - 12x^2 = 12x(2 - x) = 0$
    when $x = 0, 2$

    $f''(3) < 0$, therefore $(3, 27)$
    is a relative maximum.

    $(0, 0)$, $(2, 16)$ are points of inflection.

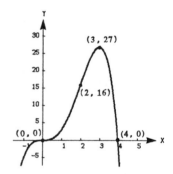

3. $f(x) = x\sqrt{16 - x^2}$ Domain: $[-4, 4]$

    $f'(x) = \dfrac{16 - 2x^2}{\sqrt{16 - x^2}} = 0$ when $x = \pm 2\sqrt{2}$ and undefined when $x = \pm 4$.

    $f''(x) = \dfrac{2x(x^2 - 24)}{(16 - x^2)^{3/2}}$

    $f''(-2\sqrt{2}) > 0$, therefore $(-2\sqrt{2}, -8)$
    is a relative minimum.

    $f''(2\sqrt{2}) < 0$, therefore $(2\sqrt{2}, 8)$
    is a relative maximum.

    $(0, 0)$ is a point of inflection.
    Intercepts: $(-4, 0)$, $(0, 0)$, $(4, 0)$
    Symmetry with respect to origin.

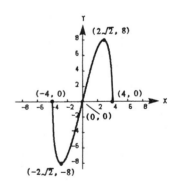

4. $f(x) = x + \dfrac{4}{x^2}$

$f'(x) = 1 - \dfrac{8}{x^3} = \dfrac{x^3 - 8}{x^3} = 0$ when $x = 2$

$f''(x) = \dfrac{24}{x^4} > 0$

$f''(2) > 0$, therefore $(2, 3)$
is a relative minimum.

Intercept: $(-\sqrt[3]{4}, 0)$
Vertical asymptote: $x = 0$
Slant asymptote: $y = x$

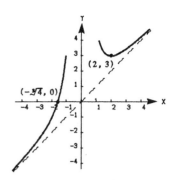

5. $f(x) = \dfrac{x + 1}{x - 1}$

$f'(x) = \dfrac{-2}{(x - 1)^2} < 0$ if $x \neq 1$

$f''(x) = \dfrac{4}{(x - 1)^3}$

Horizontal asymptote: $y = 1$
Vertical asymptote: $x = 1$
Intercepts: $(-1, 0)$, $(0, -1)$

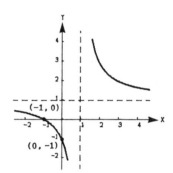

6. $f(x) = x^2 + \dfrac{1}{x} = \dfrac{x^3 + 1}{x}$

$f'(x) = 2x - \dfrac{1}{x^2} = \dfrac{2x^3 - 1}{x^2} = 0$

when $x = 1/\sqrt[3]{2}$

$f''(x) = 2 + \dfrac{2}{x^3} = \dfrac{2(x^3 + 1)}{x^3} = 0$ when $x = -1$

$f''(1/\sqrt[3]{2}) > 0$, therefore $(1/\sqrt[3]{2}, 3/\sqrt[3]{4})$
is a relative minimum.

$(-1, 0)$ is a point of inflection.
Intercept: $(-1, 0)$
Vertical asymptote: $x = 0$

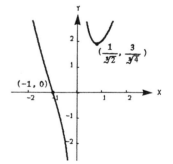

7. $f(x) = x^3 + x + \dfrac{4}{x}$

$f'(x) = 3x^2 + 1 - \dfrac{4}{x^2} = \dfrac{3x^4 + x - 4}{x^2} = 0$ when $x = \pm 1$

$f''(x) = 6x + \dfrac{8}{x^3} = \dfrac{6x^4 + 8}{x^3} \neq 0$

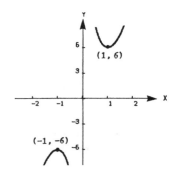

$f''(-1) < 0$, therefore $(-1, -6)$
is a relative maximum.

$f''(1) > 0$, therefore $(1, 6)$
is a relative minimum.

Symmetric with respect to origin
Vertical asymptote: $x = 0$

8. $f(x) = x^3 (x + 1)$

$f'(x) = x^3 + 3x^2(x + 1) = x^2(4x + 3) = 0$ when $x = 0, -3/4$

$f''(x) = 12x^2 + 6x = 6x(2x + 1) = 0$
when $x = 0, -1/2$

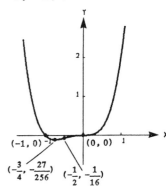

$f''(-3/4) > 0$, therefore $(-3/4, -27/256)$
is a relative minimum.

$(-1/2, -1/16)$, $(0, 0)$ are points
of inflection.
Intercepts: $(-1, 0)$, $(0, 0)$

9. $f(x) = (x - 1)^3 (x - 3)^2$

$f'(x) = (x - 1)^2 (x - 3)(5x - 11) = 0$ when $x = 1, \dfrac{11}{5}, 3$

$f''(x) = 4(x - 1)(5x^2 - 22x + 23) = 0$ when $x = 1, \dfrac{11 \pm \sqrt{6}}{5}$

$f''(3) > 0$, therefore $(3, 0)$ is a relative minimum.

$f''\left(\dfrac{11}{5}\right) < 0$, therefore $\left(\dfrac{11}{5}, \dfrac{3456}{3125}\right)$
is a relative maximum.

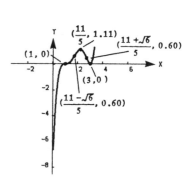

$(1, 0)$, $\left(\dfrac{11 - \sqrt{6}}{5}, 0.60\right)$,

$\left(\dfrac{11 + \sqrt{6}}{5}, 0.46\right)$ are points of inflection.

Intercepts: $(0, -9)$, $(1, 0)$, $(3, 0)$

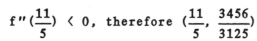

10. $f(x) = (x - 3)(x + 2)^3$

$f'(x) = (x - 3)(3)(x + 2)^2 + (x + 2)^3 = (4x - 7)(x + 2)^2 = 0$
when $x = -2, 7/4$

$f''(x) = (4x - 7)(2)(x + 2) + (x + 2)^2(4) = 6(2x - 1)(x + 2) = 0$
when $x = -2, 1/2$

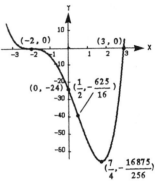

$f''(\frac{7}{4}) > 0$, therefore $(\frac{7}{4}, -\frac{16875}{256})$
is a relative minimum.

$(-2, 0)$, $(\frac{1}{2}, -\frac{625}{16})$ are points

of inflection.

Intercepts: $(-2, 0)$, $(0, -24)$, $(3, 0)$

11. $f(x) = (5 - x)^3$

$f'(x) = -3(5 - x)^2 < 0$ if $x \neq 5$

$f''(x) = 6(5 - x) = 0$ if $x = 5$

$(5, 0)$ is a point of inflection.

Intercepts: $(5, 0)$, $(0, 125)$

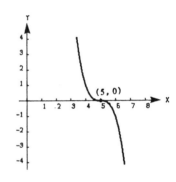

12. $f(x) = (x^2 - 4)^2$

$f'(x) = 4x(x^2 - 4) = 0$ when $x = 0, \pm 2$

$f''(x) = 4(3x^2 - 4) = 0$ when $x = \pm \frac{2\sqrt{3}}{3}$

$f''(0) < 0$, therefore $(0, 16)$
is a relative maximum.

$f''(\pm 2) > 0$, therefore $(\pm 2, 0)$
are relative minima.

$(\pm \frac{2\sqrt{3}}{3}, \frac{64}{9})$ are points of inflection.

Symmetry with respect to y-axis.
Intercepts: $(-2, 0)$, $(0, 16)$, $(2, 0)$

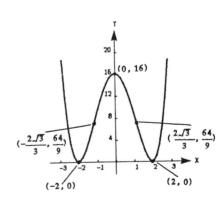

13. $f(x) = x^{1/3}(x + 3)^{2/3}$

$f'(x) = \dfrac{x + 1}{(x + 3)^{1/3}x^{2/3}} = 0$ when $x = -1$ and undefined when $x = -3, 0$

$f''(x) = \dfrac{-2}{x^{5/3}(x + 3)^{4/3}}$ is undefined when $x = 0, -3$

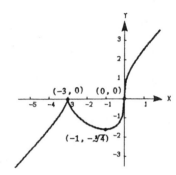

$f'(-4) > 0$ and $f'(-2) < 0$, therefore $(-3, 0)$ is a relative maximum.

$f''(-1) > 0$, therefore $(-1, -\sqrt[3]{4})$ is a relative minimum.

$f''(-1) > 0$ and $f''(1) < 0$, therefore $(0, 0)$ is a point of inflection.

Intercepts: $(-3, 0)$, $(0, 0)$

14. $f(x) = (x - 2)^{1/3}(x + 1)^{2/3}$

Graph of exercise 13 translated two units to the right (x replaced by $x - 2$).

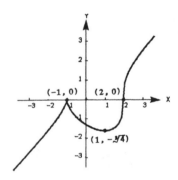

15. $f(x) = x^3 + \dfrac{243}{x} = \dfrac{x^4 + 243}{x}$

$f'(x) = 3x^2 - \dfrac{243}{x^2} = \dfrac{3(x^4 - 81)}{x^2} = 0$ when $x = \pm 3$

$f''(x) = 6x + \dfrac{2(243)}{x^3} = \dfrac{6(x^4 + 81)}{x^3}$

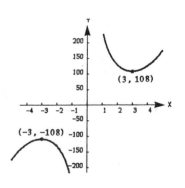

$f''(-3) < 0$, therefore $(-3, -108)$ is a relative maximum.

$f''(3) > 0$, therefore $(3, 108)$ is a relative minimum.

Symmetric to the origin.

16. $f(x) = \dfrac{2x}{1 + x^2}$

$f'(x) = \dfrac{2(1 - x)(1 + x)}{(1 + x^2)^2} = 0$ when $x = \pm 1$

$f''(x) = \dfrac{-2x(3 - x^2)}{(1 + x^2)^3} = 0$ when $x = 0, \pm\sqrt{3}$

$f''(1) < 0$, therefore $(1, 1)$ is a relative maximum.

$f''(-1) > 0$, therefore $(-1, -1)$ is a relative minimum.

$(-\sqrt{3}, -\dfrac{\sqrt{3}}{2})$, $(0, 0)$, $(\sqrt{3}, \dfrac{\sqrt{3}}{2})$

are points of inflection.
Intercept: $(0, 0)$
Symmetric to the origin.
Horizontal asymptote: $y = 0$

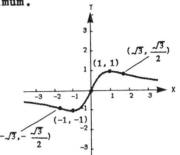

17. $f(x) = \dfrac{4}{1 + x^2}$

$f'(x) = \dfrac{-8x}{(1 + x^2)^2} = 0$ when $x = 0$

$f''(x) = \dfrac{-8(1 - 3x^2)}{(1 + x^2)^3} = 0$ when $x = \pm\dfrac{\sqrt{3}}{3}$

$f''(0) < 0$, therefore $(0, 4)$ is a relative maximum.

$(\pm\dfrac{\sqrt{3}}{3}, 3)$ are points of inflection.

Intercept: $(0, 4)$
Horizontal asymptote: $y = 0$
Symmetric to the y-axis.

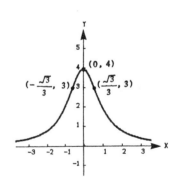

18. $f(x) = \dfrac{x^2}{1 + x^4}$

$f'(x) = \dfrac{(1 + x^4)(2x) - x^2(4x^3)}{(1 + x^4)^2} = \dfrac{2x(1 - x)(1 + x)(1 + x^2)}{(1 + x^4)^2} = 0$

when $x = 0, \pm 1$

$f''(x) = \dfrac{(1 + x^4)^2(2 - 10x^4) - (2x - 2x^5)(2)(1 + x^4)(4x^3)}{(1 + x^4)^4}$

$= \dfrac{2(1 - 12x^4 + 3x^8)}{(1 + x^4)^3} = 0$ when $x = \pm\sqrt[4]{\dfrac{6 \pm \sqrt{33}}{3}}$

18. (continued)

f"(±1) < 0, therefore (±1, 1/2) are relative maxima.

f"(0) > 0, therefore (0, 0) is a relative minimum.

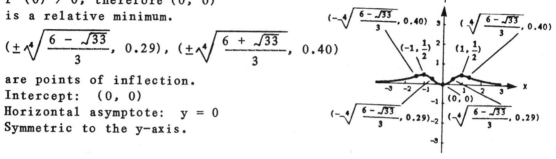

$(\pm\sqrt[4]{\dfrac{6 - \sqrt{33}}{3}}, 0.29)$, $(\pm\sqrt[4]{\dfrac{6 + \sqrt{33}}{3}}, 0.40)$

are points of inflection.
Intercept: (0, 0)
Horizontal asymptote: y = 0
Symmetric to the y-axis.

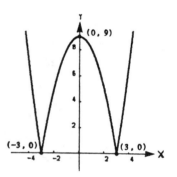

19. $f(x) = |x^2 - 9|$

$f'(x) = \dfrac{2x(x^2 - 9)}{|x^2 - 9|} = 0$ when x = 0 and is undefined when x = ±3.

$f''(x) = \dfrac{2(x^2 - 9)}{|x^2 - 9|}$ is undefined at x = ±3.

f"(0) < 0, therefore (0, 9) is a relative maximum.
(±3, 0) are relative minima.
(±3, 0) are points of inflection.
Intercepts: (±3, 0), (0, 9)
Symmetric to the y axis.

20. $f(x) = |9 - x^2| = |x^2 - 9|$ Same as exercise 19.

21. $f(x) = |x^3 - 3x^2 + 2x| = |x(x - 2)(x - 1)|$

$f'(x) = \dfrac{(3x^2 - 6x + 2)(x^3 - 3x^2 + 2x)}{|x^3 - 3x^2 + 2x|} = 0$ when $x = \dfrac{3 \pm \sqrt{3}}{3}$ and

undefined when x = 0, 2, 1

$f''(x) = \dfrac{(6x - 6)(x^3 - 3x^2 + 2x)}{|x^3 - 3x^2 + 2x|}$

$f''\left[\dfrac{3 \pm \sqrt{3}}{3}\right] < 0$, therefore

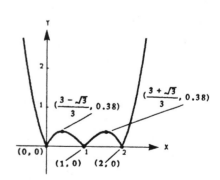

$(\dfrac{3 \pm \sqrt{3}}{3}, 0.38)$ are relative maxima.

By the First Derivative Test (1, 0), (0, 0) and (2, 0) are relative minima.
Intercepts: (0, 0), (1, 0), (2, 0)

22. $f(x) = |x - 1| + |x - 3|$

$$= \begin{cases} -2x + 4 & x \le 1 \\ 2, & 1 < x \le 3 \\ 2x - 4, & x > 3 \end{cases}$$

Intercept: $(0, 4)$

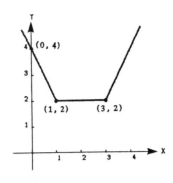

23. $f(x) = \dfrac{1}{|x - 1|}$

$$= \begin{cases} \dfrac{1}{x - 1} & x > 1 \\ \dfrac{1}{1 - x} & x < 1 \end{cases}$$

Vertical asymptote: $x = 1$
Horizontal asymptote: $y = 0$
Intercept: $(0, 1)$

24. $f(x) = \dfrac{x - 1}{1 + 3x^2}$

$f'(x) = \dfrac{1 + 6x - 3x^2}{(1 + 3x^2)^2} = 0$ when $x = \dfrac{3 \pm 2\sqrt{3}}{3}$

By the First Derivative Test $\left(\dfrac{3 - 2\sqrt{3}}{3}, \dfrac{-3 - 2\sqrt{3}}{6}\right)$ is a relative

minimum and $\left(\dfrac{3 + 2\sqrt{3}}{3}, \dfrac{2\sqrt{3} - 3}{6}\right)$ is a relative maximum.

$f''(x) = \dfrac{6(1 - 3x - 9x^2 + 3x^3)}{(1 + 3x^2)^3}$

Using Newton's Method, we find that
$f''(x) = 0$ when $x = -0.484, 0.210, 3.274$.

$(-0.484, -0.872), (0.210, -0.698),$
$(3.274, 0.069)$ are points of inflection.

Intercepts: $(1, 0), (0, -1)$
Horizontal asymptote: $y = 0$

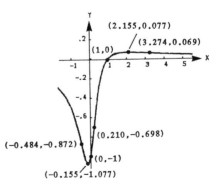

25. $f(x) = x + \cos x, \quad 0 \le x \le 2\pi$

$f'(x) = 1 - \sin x \ge 0$, f is increasing

$f''(x) = -\cos x = 0$ when $x = \dfrac{\pi}{2}, \dfrac{3\pi}{2}$

$(\dfrac{\pi}{2}, \dfrac{\pi}{2})$ and $(\dfrac{3\pi}{2}, \dfrac{3\pi}{2})$ are points of inflection.

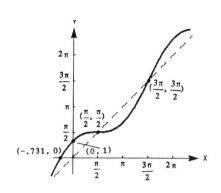

26. $f(x) = \dfrac{1}{\pi}(2 \sin \pi x - \sin 2\pi x), \quad -1 \le x \le 1$

$f'(x) = 2(\cos \pi x - \cos 2\pi x) = -2(2\cos \pi x + 1)(\cos \pi x - 1) = 0$

Critical numbers: $x = \pm\dfrac{2}{3}, 0$

By the First Derivative Test $(-\dfrac{2}{3}, \dfrac{-3\sqrt{3}}{2\pi})$ is relative minimum

and $(\dfrac{2}{3}, \dfrac{3\sqrt{3}}{2\pi})$ is a relative maximum.

$f''(x) = 2\pi(-\sin \pi x + 2 \sin 2\pi x)$

$= 2\pi \sin \pi x(-1 + 4\cos \pi x) = 0$

when $x = 0, \pm1, \pm0.420$

$(-0.420, -0.462), (0.420, 0.462), (\pm1, 0),$
$(0, 0)$ are points of inflection.
Symmetric to origin.
Intercepts: $(-1, 0), (0, 0), (1, 0)$

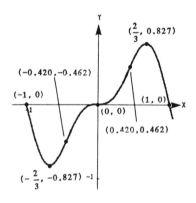

27.(a) $x^2 + 4y^2 - 2x - 16y + 13 = 0, \quad \dfrac{(x-1)^2}{4} + \dfrac{(y-2)^2}{1} = 1$

Therefore the maximum and minimum are at the endpoints of the minor axis: (1, 3) and (1, 1)

(b) $2x + 8yy' - 2 - 16y' = 0, \quad 8(y-2)y' = 2(1-x), \quad y' = \dfrac{1-x}{4(y-2)}$

The ellipse has horizontal tangents when $x = 1$.

$1 + 4y^2 - 2 - 16y + 13 = 0, \quad 4y^2 - 16y + 12 = 0$

$(y-3)(y-1) = 0$

Maximum: (1, 3), Minimum: (1, 1)

28. (a) n even

    (b) n odd. If n is odd, then n = 2k + 1 and $f(x) = x^{2k+1}$,

    $f'(x) = (2k + 1)x^{2k}$, $f''(x) = 2k(2k + 1)x^{2k-1}$. Hence $2k - 1$

    is odd and thus f" changes sign as x increases through x = 0.

29. $f(x) = \dfrac{2x + 3}{3x + 2}$, $1 \leq x \leq 5$

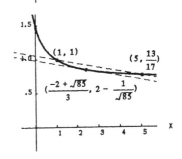

$\dfrac{f(b) - f(a)}{b - a} = \dfrac{(13/17) - 1}{5 - 1} = -\dfrac{1}{17}$

$f'(x) = \dfrac{-5}{(3x + 2)^2} = -\dfrac{1}{17}$

$(3x + 2)^2 = 85$

$x = \dfrac{-2 \pm \sqrt{85}}{3}$, $c = \dfrac{-2 + \sqrt{85}}{3}$

30. $f(x) = \dfrac{1}{x}$, $1 \leq x \leq 4$

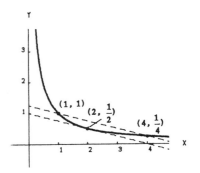

$\dfrac{f(b) - f(a)}{b - a} = \dfrac{(1/4) - 1}{4 - 1} = \dfrac{-3/4}{3} = -\dfrac{1}{4}$

$f'(x) = \dfrac{-1}{x^2} = -\dfrac{1}{4}$

$x = \pm 2$, $c = 2$

31. $f(x) = x^{2/3}$, $1 \leq x \leq 8$

    $\dfrac{f(b) - f(a)}{b - a} = \dfrac{4 - 1}{8 - 1} = \dfrac{3}{7}$

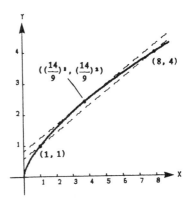

$f'(x) = \dfrac{2}{3}x^{-1/3} = \dfrac{3}{7}$

$x = (\dfrac{14}{9})^3$, $c = (\dfrac{14}{9})^3$

32. $f(x) = |x^2 - 9|$, $\qquad 0 \leq x \leq 2$

$\dfrac{f(b) - f(a)}{b - a} = \dfrac{5 - 9}{2 - 0} = -2$

On the interval [0, 2]:

$f(x) = 9 - x^2$, $f'(x) = -2x$ $\implies$ $-2x = -2$

$x = 1$, $\qquad c = 1$

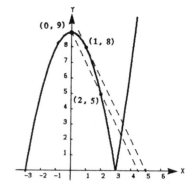

33. $f(x) = x - \cos x$, $\qquad -\pi/2 \leq x \leq \pi/2$

$\dfrac{f(b) - f(a)}{b - a} = \dfrac{(\pi/2) - (-\pi/2)}{(\pi/2) - (-\pi/2)} = 1$

$f'(x) = 1 + \sin x = 1$

$x = 0$, $\qquad c = 0$

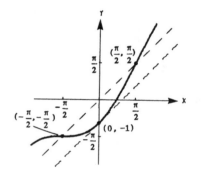

34. $f(x) = \sqrt{x} - 2x$, $\qquad 0 \leq x \leq 4$

$\dfrac{f(b) - f(a)}{b - a} = \dfrac{-6 - 0}{4 - 0} = -\dfrac{3}{2}$

$f'(x) = \dfrac{1}{2\sqrt{x}} - 2 = -\dfrac{3}{2}$

$x = 1$, $\qquad c = 1$

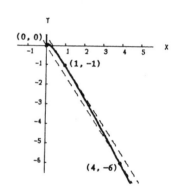

35. No, the function is discontinuous on [-2, 1].

36. $f(x) = 3 - |x - 4|$, $\quad f(1) = 3 - |1 - 4| = 0$, $\qquad f(7) = 3 - |7 - 4| = 0$

$f'(x) = -\dfrac{x - 4}{|x - 4|} \neq 0$ on [1, 7]

Rolle's Theorem requires that the function be differentiable in (1, 7), but this function is not differentiable at $x = 4$.

37. $f(x) = Ax^2 + Bx + C$

$$\frac{f(x_2) - f(x_1)}{x_2 - x_1} = \frac{A(x_2{}^2 - x_1{}^2) + B(x_2 - x_1)}{x_2 - x_1} = A(x_1 + x_2) + B$$

$f'(x) = 2Ax + B = A(x_1 + x_2) + B, \quad 2Ax = A(x_1 + x_2)$

$x = \dfrac{x_1 + x_2}{2} = $ midpoint of $[x_1, x_2]$

38. $f(x) = 2x^2 - 3x + 1, \qquad \dfrac{f(b) - f(a)}{b - a} = \dfrac{21 - 1}{4 - 0} = 5$

$f'(x) = 4x - 3 = 5, \qquad x = 2 = $ midpoint of $[0, 4]$.

39. Let $t = 0$ at noon

$L = d^2 = (100 - 12t)^2 + (-10t)^2 = 10{,}000 - 2400t + 244t^2$

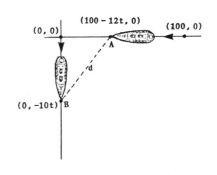

$\dfrac{dL}{dt} = -2400 + 488t = 0$

when $t = \dfrac{300}{61} \approx 4.92$ hr

ship A at $(40.98, 0)$, ship B at $(0, -49.18)$

$d^2 = 10{,}000 - 2400t + 244t^2$

$\quad = 4098.36$ when $t = 4.92 \approx$ 4:55 PM

$d \approx 64$ miles

40. $C = \dfrac{1}{4}x^2 + 62x + 125, \qquad p = 75 - \dfrac{1}{3}x, \qquad R = xp$

(a) $P = R - C = x(75 - \dfrac{1}{3}x) - (\dfrac{1}{4}x^2 + 62x + 125)$

$\dfrac{dP}{dx} = -\dfrac{1}{3}x + 75 - \dfrac{1}{3}x - \dfrac{1}{2}x - 62 = -\dfrac{7}{6}x + 13 = 0$

when $x = 78/7$ (11 units)

(b) $\overline{C}(x) = \dfrac{C}{x} = \dfrac{1}{4}x + 62 + \dfrac{125}{x}$ (average cost)

$\dfrac{d\overline{C}}{dx} = \dfrac{1}{4} - \dfrac{125}{x^2} = 0, \qquad x^2 - 500 = 0$ when $x = 10\sqrt{5}$ (22 units)

(c) $\eta = \dfrac{p/x}{dp/dx} = \dfrac{(75/x) - (1/3)}{-1/3} = 1 - \dfrac{225}{x} = \dfrac{x - 225}{x}$

41. $p = 36 - 4x,$ $C = 2x^2 + 6,$ $R = xp$

$P = R - C = x(36 - 4x) - (2x^2 + 6) = -6(x^2 - 6x + 1)$

$\dfrac{dP}{dx} = -6(2x - 6) = 0$ when $x = 3$ and $P = \$48$

42. Ellipse: $\dfrac{x^2}{144} + \dfrac{y^2}{16} = 1$

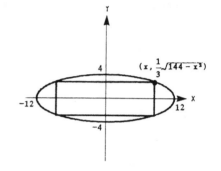

$A = (2x)\left(\dfrac{2}{3}\sqrt{144 - x^2}\right) = \dfrac{4}{3}x\sqrt{144 - x^2}$

$\dfrac{dA}{dx} = \dfrac{4}{3}\left[\dfrac{-x^2}{\sqrt{144 - x^2}} + \sqrt{144 - x^2}\right]$

$\quad = \dfrac{4}{3}\left[\dfrac{144 - 2x^2}{\sqrt{144 - x^2}}\right] = 0$

when $x = \sqrt{72}$ or $x = 6\sqrt{2}$

The dimensions of the rectangle are
$12\sqrt{2}$ by $4\sqrt{2}$.

43. We have points $(0, y),$ $(x, 0),$ and $(1, 8)$ thus

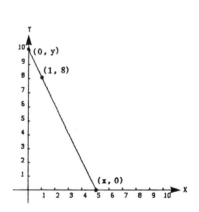

$m = \dfrac{y - 8}{1} = \dfrac{8}{x - 1}$ or $y = \dfrac{8x}{x - 1}$

$L^2 = x^2 + \left(\dfrac{8x}{x - 1}\right)^2$

$2LL' = 2x + 128\left(\dfrac{x}{x - 1}\right)\left(\dfrac{(x - 1) - x}{(x - 1)^2}\right) = 0$

$x - \dfrac{64x}{(x - 1)^3} = 0,$ $x[(x - 1)^3 - 64] = 0$

when $x = 0,\ 5$ (minimum)

Vertices of the triangle are
$(0, 0),$ $(5, 0),$ and $(0, 10)$.

44. We have points $(0, y)$, $(x, 0)$, and $(4, 5)$ thus

$$m = \frac{y - 5}{0 - 4} = \frac{5 - 0}{4 - x} \quad \text{or} \quad y = \frac{5x}{x - 4}$$

$$L^2 = x^2 + \left(\frac{5x}{x - 4}\right)^2, \qquad 2LL' = 2x + 50\left(\frac{x}{x - 4}\right)\left(\frac{x - 4 - x}{(x - 4)^2}\right) = 0$$

$$x - \frac{100x}{(x - 4)^3} = 0$$

$$x[(x - 4)^3 - 100] = 0 \quad \text{when} \quad x = 0 \quad \text{or}$$

$$x = 4 + \sqrt[3]{100}$$

$$L = \sqrt{x^2 + \frac{25x^2}{(x - 4)^2}} = \frac{x}{x - 4}\sqrt{(x - 4)^2 + 25}$$

$$= \frac{\sqrt[3]{100} + 4}{\sqrt[3]{100}}\sqrt{100^{2/3} + 25} = 12.7 \text{ feet}$$

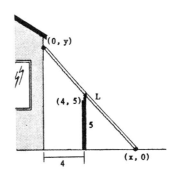

45. Let x be the number of people over 80 who go.

$$R = (\text{number who go})(\text{price/person}) = (80 + x)(8 - 0.05x)$$

$$\frac{dR}{dx} = (80 + x)(-0.05) + (8 - 0.05x) = 4 - 0.10x = 0 \quad \text{when} \quad x = 40$$

Revenue will be maximum when 120 people go.

46. Width of rectangle: $a - \frac{a - b}{b}x$, height of rectangle: $\frac{c}{b}x$ (see figure)

$$A = (\text{width})(\text{height}) = \left(a - \frac{a - b}{b}x - x\right)\left(\frac{c}{b}x\right) = \left(a - \frac{a}{b}x\right)\frac{c}{b}x$$

$$\frac{dA}{dx} = \left(a - \frac{a}{b}x\right)\frac{c}{b} + \left(\frac{c}{b}x\right)\left(-\frac{a}{b}\right)$$

$$= \frac{ac}{b} - \frac{2ac}{b^2}x = 0 \quad \text{when} \quad x = \frac{b}{2}$$

$$A\left(\frac{b}{2}\right) = \left(a - \frac{a}{b}\frac{b}{2}\right)\left(\frac{c}{b}\frac{b}{2}\right) = \left(\frac{a}{2}\right)\left(\frac{c}{2}\right)$$

$$= \frac{1}{4}ac = \frac{1}{2}\left(\frac{1}{2}ac\right) = \frac{1}{2}(\text{area of triangle})$$

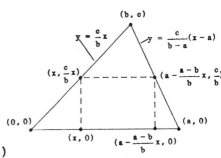

47. We see from the figure that

A = (average of bases)(height)

$$= (\frac{x + s}{2})\frac{\sqrt{3s^2 + 2sx - x^2}}{2}$$

$$\frac{dA}{dx} = \frac{1}{4}\left[\frac{(s - x)(s + x)}{\sqrt{3s^2 + 2sx - x^2}} + \sqrt{3s^2 + 2sx - x^2}\right]$$

$$= \frac{2(2s - x)(s + x)}{4\sqrt{3s^2 + 2sx - x^2}} = 0 \text{ when } x = 2s, -s$$

A is a maximum when x = 2s.

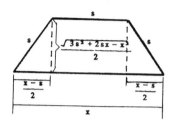

48. The cost C is proportional to $s^{3/2}$,   C = 50 when s = 25
    Fixed costs are $100

$$C = ks^{3/2} \text{ where } 50 = k(25)^{3/2} \implies k = \frac{2}{5}$$

$$\overline{C} = \frac{(2/5)s^{3/2} + 100}{s} = \frac{2}{5}s^{1/2} + \frac{100}{s},$$

$$\overline{C}' = \frac{1}{5s^{1/2}} - \frac{100}{s^2} = 0 \text{ when } s^{3/2} = 500$$

$$s = 500^{2/3} \approx 63 \text{ mi/hr}$$

49. $m = \frac{y - 6}{0 - 4} = \frac{6 - 0}{4 - x}$, $LL' = x[(x - 4)^3 - 144] = 0$

L = 14.05 feet   (See solution to Exercise 44.)

50. We can form a right triangle with vertices (0, y), (0, 0), and (x, 0).
    Choosing a point (a, b) on the hypotenuse (assuming the triangle is in
    the first quadrant), the slope is:

$$m = \frac{b - y}{0 - a} = \frac{b - 0}{a - x} \implies y = \frac{-bx}{a - x}$$

$$L^2 = x^2 + y^2 = x^2 + \left[\frac{-bx}{a - x}\right]^2, \qquad 2LL' = 2x + 2\left[\frac{-bx}{a - x} \frac{-ab}{(a - x)^2}\right]$$

$$\frac{2x[(a - x)^3 + ab^2]}{(a - x)^2} = 0 \text{ when } x = 0, \ a + \sqrt[3]{ab^2}$$

Choosing the nonzero value we have $y = b + \sqrt[3]{a^2b}$.

$$L = \sqrt{(a + \sqrt[3]{ab^2})^2 + (b + \sqrt[3]{a^2b})^2}$$

$$= (a^2 + 3a^{4/3}b^{2/3} + 3a^{2/3}b^{4/3} + b^2)^{1/2} = (a^{2/3} + b^{2/3})^{3/2} \text{ feet}$$

51. We see from the graph that $\csc \theta = \dfrac{L_1}{6}$ or $L_1 = 6 \csc \theta$

$\csc(\dfrac{\pi}{2} - \theta) = \dfrac{L_2}{9}$ or $L_2 = 9 \csc(\dfrac{\pi}{2} - \theta)$

$L = 6 \csc \theta + 9 \csc(\dfrac{\pi}{2} - \theta) = 6 \csc \theta + 9 \sec \theta$

$\dfrac{dL}{d\theta} = -6 \csc \theta \cot \theta + 9 \sec \theta \tan \theta = 0$

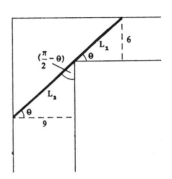

$\tan^3 \theta = \dfrac{2}{3} \implies \tan \theta = \dfrac{\sqrt[3]{2}}{\sqrt[3]{3}}$

$\theta = \arctan \sqrt[3]{2/3}$

$L = 6 \dfrac{(3^{2/3} + 2^{2/3})^{1/2}}{2^{1/3}} + 9 \dfrac{(3^{2/3} + 2^{2/3})^{1/2}}{3^{1/3}}$

$\quad = 3(3^{2/3} + 2^{2/3})^{3/2}$ feet

52. Using Exercise 51 as a guide we have $L_1 = a \csc \theta$ and $L_2 = b \sec \theta$

Then $\dfrac{dL}{d\theta} = -a \csc \theta \cot \theta + b \sec \theta \tan \theta = 0$ when $\theta = \arctan \sqrt[3]{a/b}$

$L = a \dfrac{(a^{2/3} + b^{2/3})^{1/2}}{a^{1/3}} + b \dfrac{(a^{2/3} + b^{2/3})^{1/2}}{b^{1/3}}$

$\quad = (a^{2/3} + b^{2/3})^{3/2}$ (This matches the result of Exercise 50.)

53. $y = \dfrac{1}{3} \cos(12t) - \dfrac{1}{4} \sin(12t)$, $v = y' = -4 \sin(12t) - 3 \cos(12t)$

(a)   when $t = \dfrac{\pi}{8}$, $y = \dfrac{1}{4}$ inches and $v = y' = 4$ inches/second

(b) $y' = -4 \sin(12t) - 3 \cos(12t) = 0$ when $\dfrac{\sin(12t)}{\cos(12t)} = -3/4$

     $\tan(12t) = -3/4$ therefore $\sin(12t) = 3/5$ and $\cos(12t) = -4/5$

     and the maximum displacement is $y = -5/12$ inch

(c)   Period: $\dfrac{2\pi}{12} = \dfrac{\pi}{6}$,   Frequency: $\dfrac{1}{\pi/6} = \dfrac{6}{\pi}$

54. $y = A \sin (\sqrt{k/m}\, t) + B \cos (\sqrt{k/m}\, t)$

$y' = A\sqrt{k/m} \cos (\sqrt{k/m}\, t) - B\sqrt{k/m} \sin (\sqrt{k/m}\, t) = 0$

when $\dfrac{\sin \sqrt{k/m}\, t}{\cos \sqrt{k/m}\, t} = \dfrac{A}{B}$ or $\tan (\sqrt{k/m}\, t) = \dfrac{A}{B}$

Therefore $\sin (\sqrt{k/m}\, t) = \dfrac{A}{\sqrt{A^2 + B^2}}$, $\cos (\sqrt{k/m}\, t) = \dfrac{B}{\sqrt{A^2 + B^2}}$

Therefore when $v = y' = 0$, $y = A(\dfrac{A}{\sqrt{A^2 + B^2}}) + B(\dfrac{B}{\sqrt{A^2 + B^2}}) = \sqrt{A^2 + B^2}$

Period $= \dfrac{2\pi}{\sqrt{k/m}}$,   Frequency $= \dfrac{1}{2\pi/\sqrt{k/m}} = \dfrac{1}{2\pi} \sqrt{k/m}$

Thus the frequency is changed as the square root of k. Furthermore, the frequency is inversely proportional to the square root of the mass.

55. $C = (\dfrac{Q}{x}) s + (\dfrac{x}{2}) r$

$\dfrac{dC}{dx} = -\dfrac{Qs}{x^2} + \dfrac{r}{2} = 0$, $\dfrac{Qs}{x^2} = \dfrac{r}{2}$, $x^2 = \dfrac{2Qs}{r}$, $x = \sqrt{\dfrac{2Qs}{r}}$

56. $p = 600 - 3x$, $C = 0.3x^2 + 6x + 600$, $P = xp - C - xt$

$P = -3.3x^2 + (594 - t)x - 600$

$\dfrac{dP}{dx} = -6.6x + 594 - t = 0$, $x = \dfrac{594 - t}{6.6}$

(a) $t = 5$, $x = 89$, $P = 25681.70$
(b) $t = 10$, $x = 88$, $P = 25236.80$
(c) $t = 20$, $x = 87$, $P = 24360.30$

57. $f(x) = x^3 - 3x - 1$, $[-1, 0]$

$f'(x) = 3x^2 - 3$

$x_{n+1} = x_n - \dfrac{x_n^3 - 3x_n - 1}{3x_n^2 - 3}$

$x \approx -0.347$

n	x_n	$f(x_n)$
1	−0.5000	0.3750
2	−0.3333	−0.0370
3	−0.3472	−0.0002
4	−0.3473	0.0000

58. Find the zeros to $f(x) = x^4 - x - 3$, $\quad$ f changes sign in $[-2, -1]$

$f'(x) = 4x^3 - 1$

$x_{n+1} = x_n - \dfrac{x_n^4 - x_n - 3}{4x_n^3 - 1}$

On the interval $[-2, -1]$:
$x \approx -1.164$

n	x_n	$f(x_n)$
1	-1.2000	0.2736
2	-1.1654	0.0101
3	-1.1640	0.0000

f changes sign in $[1, 2]$

n	x_n	$f(x_n)$
1	1.2000	-2.1264
2	1.5597	1.3578
3	1.4639	0.1285
4	1.4528	0.0016
5	1.4526	0.0000

On the interval $[1, 2]$:
$x \approx 1.453$

59. $f(x) = x^3 + 2x + 1$

$f'(x) = 3x^2 + 2$

$x_{n+1} = x_n - \dfrac{x_n^3 + 2x_n + 1}{3x_n^2 + 2}$

On the interval $[-1, 0]$:
$x \approx -0.453$

f changes sign in $[-1, 0]$

n	x_n	$f(x_n)$
1	-0.5000	-0.1250
2	-0.4545	-0.0030
3	-0.4534	0.0000

60. $s = 4\pi r^2$, $\quad ds = 8\pi r\, dr = 8\pi(9)(0.025) \approx 1.8\pi$ square inches

$V = \dfrac{4}{3}\pi r^3$, $\quad dV = 4\pi r^2\, dr = 4\pi(9)^2(0.025) \approx 8.1\pi$ cubic inches

61. $s = 6x^2$, $\quad ds = 12x\, dx$,

% error $= \dfrac{ds}{s}(100) = \dfrac{12x\, dx}{6x^2}(100) = 2\left[\dfrac{dx}{x}(100)\right] = 2(1\%) = 2\%$

$V = x^3$, $\quad dV = 3x^2\, dx$

% error $= \dfrac{dV}{V}(100) = \dfrac{3x^2\, dx}{x^3}(100) = 3\left[\dfrac{dx}{x}(100)\right] = 3(1\%) = 3\%$

62. $p = 75 - \dfrac{1}{4}x$, $\quad \Delta p = p(8) - p(7) = \left(75 - \dfrac{8}{4}\right) - \left(75 - \dfrac{7}{4}\right) = -\dfrac{1}{4}$

$dp = -\dfrac{1}{4}\, dx = -\dfrac{1}{4}(1) = -\dfrac{1}{4}$

5 Integration

Antiderivatives and indefinite integration

Given	Rewrite	Integrate	Simplify
1. $\int \sqrt[3]{x}\, dx$	$\int x^{1/3}\, dx$	$\dfrac{x^{4/3}}{4/3} + C$	$\dfrac{3}{4} x^{4/3} + C$
2. $\int \dfrac{1}{x^2}\, dx$	$\int x^{-2}\, dx$	$\dfrac{x^{-1}}{-1} + C$	$-\dfrac{1}{x} + C$
3. $\int \dfrac{1}{x\sqrt{x}}\, dx$	$\int x^{-3/2}\, dx$	$\dfrac{x^{-1/2}}{-1/2} + C$	$-\dfrac{2}{\sqrt{x}} + C$
4. $\int x(x^2 + 3)\, dx$	$\int (x^3 + 3x)\, dx$	$\dfrac{x^4}{4} + 3\left(\dfrac{x^2}{2}\right) + C$	$\dfrac{1}{4} x^4 + \dfrac{3}{2} x^2 + C$
5. $\int \dfrac{1}{2x^3}\, dx$	$\dfrac{1}{2} \int x^{-3}\, dx$	$\dfrac{1}{2}\left(\dfrac{x^{-2}}{-2}\right) + C$	$-\dfrac{1}{4x^2} + C$
6. $\int \dfrac{1}{(2x)^3}\, dx$	$\dfrac{1}{8} \int x^{-3}\, dx$	$\dfrac{1}{8}\left(\dfrac{x^{-2}}{-2}\right) + C$	$-\dfrac{1}{16x^2} + C$

7. $\int (x^3 + 2)\, dx = \dfrac{1}{4} x^4 + 2x + C$, Check: $\dfrac{d}{dx}[\dfrac{1}{4} x^4 + 2x + C] = x^3 + 2$

8. $\int (x^2 - 2x + 3)\, dx = \dfrac{1}{3} x^3 - x^2 + 3x + C$

    Check: $\dfrac{d}{dx}[\dfrac{1}{3} x^3 - x^2 + 3x + C] = x^2 - 2x + 3$

9. $\int (x^{3/2} + 2x + 1)\, dx = \dfrac{2}{5} x^{5/2} + x^2 + x + C$

   Check: $\dfrac{d}{dx}[\dfrac{2}{5} x^{5/2} + x^2 + x + C] = x^{3/2} + 2x + 1$

10. $\int (\sqrt{x} + \dfrac{1}{2\sqrt{x}})\, dx = \int (x^{1/2} + \dfrac{1}{2} x^{-1/2})\, dx$

    $\qquad\qquad\qquad = \dfrac{x^{3/2}}{3/2} + \dfrac{1}{2}(\dfrac{x^{1/2}}{1/2}) + C = \dfrac{2}{3} x^{3/2} + x^{1/2} + C$

    Check: $\dfrac{d}{dx}[\dfrac{2}{3} x^{3/2} + x^{1/2} + C] = x^{1/2} + \dfrac{1}{2} x^{-1/2} = \sqrt{x} + \dfrac{1}{2\sqrt{x}}$

11. $\int \sqrt[3]{x^2}\, dx = \int x^{2/3}\, dx = \dfrac{x^{5/3}}{5/3} + C = \dfrac{3}{5} x^{5/3} + C$

    Check: $\dfrac{d}{dx}[\dfrac{3}{5} x^{5/3} + C] = x^{2/3} = \sqrt[3]{x^2}$

12. $\int (\sqrt[4]{x^3} + 1)\, dx = \int (x^{3/4} + 1)\, dx = \dfrac{4}{7} x^{7/4} + x + C$

    Check: $\dfrac{d}{dx}[\dfrac{4}{7} x^{7/4} + x + C] = x^{3/4} + 1 = \sqrt[4]{x^3} + 1$

13. $\int \dfrac{1}{x^3}\, dx = \int x^{-3}\, dx = \dfrac{x^{-2}}{-2} + C = -\dfrac{1}{2x^2} + C,$ Check: $\dfrac{d}{dx}[-\dfrac{1}{2x^2} + C] = \dfrac{1}{x^3}$

14. $\int \dfrac{1}{x^4}\, dx = \int x^{-4}\, dx = \dfrac{x^{-3}}{-3} + C = -\dfrac{1}{3x^3} + C,$ Check: $\dfrac{d}{dx}[-\dfrac{1}{3x^3} + C] = \dfrac{1}{x^4}$

15. $\int \dfrac{1}{4x^2}\, dx = \dfrac{1}{4} \int x^{-2}\, dx = \dfrac{1}{4}(\dfrac{x^{-1}}{-1}) + C = -\dfrac{1}{4x} + C$

    Check: $\dfrac{d}{dx}[-\dfrac{1}{4x} + C] = \dfrac{1}{4x^2}$

16. $\int (2x + x^{-1/2})\, dx = x^2 + 2x^{1/2} + C$

    Check: $\dfrac{d}{dx}[x^2 + 2x^{1/2} + C] = 2x + x^{-1/2}$

17. $\int \dfrac{x^2 + x + 1}{\sqrt{x}}\, dx = \int (x^{3/2} + x^{1/2} + x^{-1/2})\, dx$

    $\qquad\qquad\qquad = \dfrac{2}{5} x^{5/2} + \dfrac{2}{3} x^{3/2} + 2x^{1/2} + C = \dfrac{2}{15} x^{1/2}(3x^2 + 5x + 15) + C$

    Check: $\dfrac{d}{dx}[\dfrac{2}{5} x^{5/2} + \dfrac{2}{3} x^{3/2} + 2x^{1/2} + C] = x^{3/2} + x^{1/2} + x^{-1/2} = \dfrac{x^2 + x + 1}{\sqrt{x}}$

18. $\displaystyle\int \frac{x^2 + 1}{x^2}\, dx = \int (1 + x^{-2})\, dx = x + \frac{x^{-1}}{-1} + C = x - \frac{1}{x} + C$

Check: $\dfrac{d}{dx}[x - \dfrac{1}{x} + C] = 1 + \dfrac{1}{x^2} = \dfrac{x^2 + 1}{x^2}$

19. $\displaystyle\int (x + 1)(3x - 2)\, dx = \int (3x^2 + x - 2)\, dx = x^3 + \frac{1}{2}x^2 - 2x + C$

Check: $\dfrac{d}{dx}[x^3 + \dfrac{1}{2}x^2 - 2x + C] = 3x^2 + x - 2 = (x + 1)(3x - 2)$

20. $\displaystyle\int (2t^2 - 1)^2\, dt = \int (4t^4 - 4t^2 + 1)\, dt = \frac{4}{5}t^5 - \frac{4}{3}t^3 + t + C$

Check: $\dfrac{d}{dt}[\dfrac{4}{5}t^5 - \dfrac{4}{3}t^3 + t + C] = 4t^4 - 4t^2 + 1 = (2t^2 - 1)^2$

21. $\displaystyle\int \frac{t^2 + 2}{t^2}\, dt = \int (1 + 2t^{-2})\, dt = t + 2\left(\frac{t^{-1}}{-1}\right) + C = t - \frac{2}{t} + C$

Check: $\dfrac{d}{dt}[t - \dfrac{2}{t} + C] = 1 + \dfrac{2}{t^2} = \dfrac{t^2 + 2}{t^2}$

22. $\displaystyle\int (1 - 2y + 3y^2)\, dy = y - y^2 + y^3 + C$

Check: $\dfrac{d}{dy}[y - y^2 + y^3 + C] = 1 - 2y + 3y^2$

23. $\displaystyle\int y^2 \sqrt{y}\, dy = \int y^{5/2}\, dy = \frac{2}{7}y^{7/2} + C$

Check: $\dfrac{d}{dy}[\dfrac{2}{7}y^{7/2} + C] = y^{5/2} = y^2\sqrt{y}$

24. $\displaystyle\int (1 + 3t)t^2\, dt = \int (t^2 + 3t^3)\, dt = \frac{1}{3}t^3 + \frac{3}{4}t^4 + C$

Check: $\dfrac{d}{dt}[\dfrac{1}{3}t^3 + \dfrac{3}{4}t^4 + C] = t^2 + 3t^3 = (1 + 3t)t^2$

25. $\displaystyle\int dx = \int 1\, dx = x + C$, Check: $\dfrac{d}{dx}[x + C] = 1$

26. $\displaystyle\int 3\, dt = 3t + C$, Check: $\dfrac{d}{dt}[3t + C] = 3$

27. $\displaystyle\int (2\sin x + 3\cos x)\, dx = -2\cos x + 3\sin x + C$

Check: $\dfrac{d}{dx}[-2\cos x + 3\sin x + C] = 2\sin x + 3\cos x$

28. $\displaystyle\int (t^2 - \sin t)\ dt = \frac{1}{3}t^3 + \cos t + C$

   Check: $\dfrac{d}{dt}[\frac{1}{3}t^3 + \cos t + C] = t^2 - \sin t$

29. $\displaystyle\int (1 - \csc t \cot t)\ dt = t + \csc t + C$

   Check: $\dfrac{d}{dt}[t + \csc t + C] = 1 - \csc t \cot t$

30. $\displaystyle\int (\theta^2 + \sec^2 \theta)\ d\theta = \frac{1}{3}\theta^3 + \tan \theta + C$

   Check: $\dfrac{d}{d\theta}[\frac{1}{3}\theta^3 + \tan \theta + C] = \theta^2 + \sec^2 \theta$

31. $\displaystyle\int (\sec^2 \theta - \sin \theta)\ d\theta = \tan \theta + \cos \theta + C$

   Check: $\dfrac{d}{d\theta}[\tan \theta + \cos \theta + C] = \sec^2 \theta - \sin \theta$

32. $\displaystyle\int \sec y(\tan y - \sec y)\ dy = \int (\sec y \tan y - \sec^2 y)\ dy = \sec y - \tan y + C$

   Check: $\dfrac{d}{dy}[\sec y - \tan y + C] = \sec y \tan y - \sec^2 y = \sec y(\tan y - \sec y)$

33. $\displaystyle\int (\tan^2 y + 1)\ dy = \int \sec^2 y\ dy = \tan y + C$

   Check: $\dfrac{d}{dy}[\tan y + C] = \sec^2 y = \tan^2 y + 1$

34. $\displaystyle\int \frac{\sin x}{1 - \sin^2 x}\ dx = \int \frac{\sin x}{\cos^2 x}\ dx = \int \left(\frac{1}{\cos x}\right)\left(\frac{\sin x}{\cos x}\right)\ dx$

   $$= \int \sec x \tan x\ dx = \sec x + C$$

   Check: $\dfrac{d}{dx}[\sec x + C] = \sec x \tan x = \dfrac{\sin x}{\cos^2 x} = \dfrac{\sin x}{1 - \sin^2 x}$

35. $\dfrac{dy}{dx} = 2x - 1,\qquad (1,\ 1),\qquad y = \displaystyle\int (2x - 1)\ dx = x^2 - x + C$

   $1 = (1)^2 - (1) + C \implies C = 1,\qquad y = x^2 - x + 1$

36. $\dfrac{dy}{dx} = 2(x - 1) = 2x - 2,\qquad (3,\ 2),\qquad y = \displaystyle\int 2(x - 1)\ dx = x^2 - 2x + C$

   $2 = (3)^2 - 2(3) + C \implies C = -1,\qquad y = x^2 - 2x - 1$

37. $\dfrac{dy}{dx} = \cos x$, $(0, 4)$, $y = \displaystyle\int \cos x \, dx = \sin x + C$

$4 = \sin 0 + C \implies C = 4$, $y = \sin x + 4$

38. $\dfrac{dy}{dx} = -\dfrac{1}{x^2} = -x^{-2}$, $(1, 3)$, $y = \displaystyle\int -x^{-2} \, dx = \dfrac{1}{x} + C$

$3 = \dfrac{1}{1} + C \implies C = 2$, $y = \dfrac{1}{x} + 2$

39. $f''(x) = 2$, $f'(2) = 5$, $f(2) = 10$

$f'(x) = \displaystyle\int 2 \, dx = 2x + C_1$

$f'(2) = 4 + C_1 = 5 \implies C_1 = 1$, $f'(x) = 2x + 1$

$f(x) = \displaystyle\int (2x + 1) \, dx = x^2 + x + C_2$, $f(2) = 6 + C_2 = 10 \implies C_2 = 4$

$f(x) = x^2 + x + 4$

40. $f''(x) = x^2$, $f'(0) = 6$, $f(0) = 3$

$f'(x) = \displaystyle\int x^2 \, dx = \dfrac{1}{3}x^3 + C_1$

$f'(0) = 0 + C_1 = 6 \implies C_1 = 6$, $f'(x) = \dfrac{1}{3}x^3 + 6$

$f(x) = \displaystyle\int (\dfrac{1}{3}x^3 + 6) \, dx = \dfrac{1}{12}x^4 + 6x + C_2$

$f(0) = 0 + 0 + C_2 = 3 \implies C_2 = 3$, $f(x) = \dfrac{1}{12}x^4 + 6x + 3$

41. $f''(x) = x^{-3/2}$, $f'(4) = 2$, $f(0) = 0$

$f'(x) = \displaystyle\int x^{-3/2} \, dx = -2x^{-1/2} + C_1 = -\dfrac{2}{\sqrt{x}} + C_1$

$f'(4) = -\dfrac{2}{2} + C_1 = 2 \implies C_1 = 3$, $f'(x) = -\dfrac{2}{\sqrt{x}} + 3$

$f(x) = \displaystyle\int (-2x^{-1/2} + 3) \, dx = -4x^{1/2} + 3x + C_2$

$f(0) = 0 + 0 + C_2 = 0 \implies C_2 = 0$, $f(x) = -4x^{1/2} + 3x = -4\sqrt{x} + 3x$

42. $f''(x) = x^{-3/2}, \qquad f'(1) = 2, \qquad f(9) = -4$

$f'(x) = \displaystyle\int x^{-3/2}\ dx = -2x^{-1/2} + C_1 = -\dfrac{2}{\sqrt{x}} + C_1$

$f'(1) = -2 + C_1 = 2 \implies C_1 = 4, \qquad f'(x) = -\dfrac{2}{\sqrt{x}} + 4$

$f(x) = \displaystyle\int (-2x^{-1/2} + 4)\ dx = -4x^{1/2} + 4x + C_2 = -4\sqrt{x} + 4x + C_2$

$f(9) = -12 + 36 + C_2 = -4 \implies C_2 = -28$

$f(x) = -4\sqrt{x} + 4x - 28 = -4(\sqrt{x} - x + 7)$

43. $s_0 = 1600, \qquad v_0 = 0, \qquad v(t) = \displaystyle\int -32\ dt = -32t + C_1$

$v(0) = C_1 = 0 \implies v(t) = -32t, \qquad s(t) = \displaystyle\int -32t\ dt = -16t^2 + C_2$

$s(0) = C_2 = 1600 \implies s(t) = -16t^2 + 1600$

$-16t^2 + 1600 = 0, \qquad t = 10 \text{ seconds}$

44. $s_0 = 0, \qquad v_0 = 60$

$v(t) = \displaystyle\int -32\ dt = -32t + C_1, \qquad v(0) = C_1 = 60, \qquad v(t) = -32t + 60$

$s(t) = \displaystyle\int (-32t + 60)\ dt = -16t^2 + 60t + C_2, \qquad s(0) = C_2 = 0$

$s(t) = -16t^2 + 60t$

$s'(t) = v(t) = -32t + 60 = 0 \text{ when } t = 1.875 \quad (s''(t) = -32 < 0)$

Maximum height at $s(1.875) = -56.25 + 112.50 = 56.25$ feet

45. $v(t) = \displaystyle\int -32\ dt = -32t + C_1, \qquad v_0 = C_1$

$s(t) = \displaystyle\int -32t + v_0 = -16t^2 + v_0 t + C_2$

$s(0) = C_2 = 0 \implies s(t) = -16t^2 + v_0 t$

$s'(t) = -32t + v_0 = 0 \text{ when } t = \dfrac{v_0}{32} = \text{time to reach maximum height}$

$s\left(\dfrac{v_0}{32}\right) = -16\left(\dfrac{v_0}{32}\right)^2 + v_0\left(\dfrac{v_0}{32}\right) = 550, \qquad -\dfrac{v_0^2}{64} + \dfrac{v_0^2}{32} = 550$

$v_0^2 = 35200, \qquad v_0 \approx 187.617 \text{ ft/sec}$

46. $s''(t) = a(t) = -32 \text{ ft/sec}^2$, $s'(0) = v_0$, $s(0) = s_0$

$s'(t) = v(t) = \int -32 \, dt = -32t + C_1$, $s'(0) = 0 + C_1 = v_0$ $\Longrightarrow$ $C_1 = v_0$

$s'(t) = -32t + v_0$, $s(t) = \int (-32t + v_0)dt = -16t^2 + v_0 t + C_2$

$s(0) = 0 + 0 + C_2 = s_0$ $\Longrightarrow$ $C_2 = s_0$, $s(t) = -16t^2 + v_0 t + s_0$

47. $v_0 = 16 \text{ ft/sec}$, $s_0 = 64 \text{ ft}$

(a) $s(t) = -16t^2 + 16t + 64 = 0$, $-16(t^2 - t - 4) = 0$, $t = \dfrac{1 \pm \sqrt{17}}{2}$

Choosing the positive value, $t = \dfrac{1 + \sqrt{17}}{2} \approx 2.562$ seconds

(b) $v(t) = s'(t) = -32t + 16$

$v\left(\dfrac{1 + \sqrt{17}}{2}\right) = -32\left(\dfrac{1 + \sqrt{17}}{2}\right) + 16 = -16\sqrt{17} \approx -65.970 \text{ ft/sec}$

48. $a(t) = k$, $v(t) = kt$, $s(t) = \dfrac{k}{2}t^2$ since $v(0) = s(0) = 0$

At the time of lift-off $kt = 160$ and $\dfrac{k}{2}t^2 = 0.7$

Since $\dfrac{k}{2}t^2 = 0.7$, $t = \sqrt{\dfrac{1.4}{k}}$ and $k\sqrt{\dfrac{1.4}{k}} = 160$

$1.4k = 160^2$ $\Longrightarrow$ $k = \dfrac{160^2}{1.4} \approx 18285.714 \text{ mi/hr}^2 \approx 7.45 \text{ ft/sec}^2$

49. $v(0) = 15 \text{ mph} = 22 \text{ ft/sec}$, $v(13) = 50 \text{ mph} = 220/3 \text{ ft/sec}$, $s(0) = 0$

(a) $a(t) = a$ (constant acceleration)

$v(t) = at + C$

$v(0) = C = 22 \text{ ft/sec}$

$v(t) = at + 22$

$v(13) = 13a + 22 = \dfrac{220}{3}$

when $a = \dfrac{154}{39} \approx 3.95 \text{ ft/sec}^2$

(b) $v(t) = \dfrac{154}{39} + 22$

$s(t) = \dfrac{77}{39}^2 + 22t$ since $s(0) = 0$

$s(13) = \dfrac{77}{39}(13)^2 + 22(13)$

$= \dfrac{1859}{3} \text{ ft} \approx 619.67 \text{ ft}$

50. $v(0) = 45$ mph $= 66$ ft/sec,　　30 mph $= 44$ ft/sec,　　15 mph $= 22$ ft/sec

$a(t) = -a$,　　$v(t) = -at + 66$,　　$s(t) = -\dfrac{a}{2}t^2 + 66t$ (Let $s(0) = 0$)

$v(t) = 0$ after car moves 132 ft,　$-at + 66 = 0$ when $t = 66/a$

$s(\dfrac{66}{a}) = -\dfrac{a}{2}(\dfrac{66}{a})^2 + 66(\dfrac{66}{a}) = 132$ when $a = \dfrac{33}{2} = 16.5$

$a(t) = -16.5$,　　$v(t) = -16.5t + 66$,　　$s(t) = -8.25t^2 + 66t$

(a)　$-16.5t + 66 = 44$,　　$t = \dfrac{22}{16.5}$,　　$s(\dfrac{22}{16.5}) \approx 73.33$

(b)　$-16.5t + 66 = 22$,　　$t = \dfrac{44}{16.5}$,　　$s(\dfrac{44}{16.5}) \approx 117.33$

(c)　

51. Truck:　$v(t) = 30$,　　$s(t) = 30t$,　(Let $s(0) = 0$)

Automobile:　$a(t) = 6$,　　$v(t) = 6t$　($v(0) = 0$)
　　　　　　　　$s(t) = 3t^2$　($s(0) = 0$)

At the point where the automobile overtakes the truck:

$30t = 3t^2$,　　$0 = 3t^2 - 30t$,　　$0 = 3t(t - 10)$ when $t = 10$ sec

(a)　$s(10) = 3(10)^2 = 300$ ft

(b)　$v(10) = 6(10) = 60$ ft/sec ≈ 41 mph

52. $a(t) = a$ (since acceleration is constant),　　$v(t) = at$　($v(0) = 0$)

$s(t) = \dfrac{at^2}{2}$ (Let $s(0) = 0$),　　$s(4) = \dfrac{a(4)^2}{2} = 8a = 100$,　　$a = 12.5$ cm/sec^2

53. Let d be the distance traversed and a the uniform acceleration. Furthermore, note that $v(0) = 0$ and let $s(0) = 0$.

$a(t) = a$,　　$v(t) = at$,　　$s(t) = \dfrac{at^2}{2} = d$ when $t = \sqrt{\dfrac{2d}{a}}$

The highest speed is $v(\sqrt{\dfrac{2d}{a}}) = a\sqrt{\dfrac{2d}{a}} = \sqrt{2ad}$ and the mean speed

is $\dfrac{\sqrt{2ad}}{2} = \sqrt{\dfrac{ad}{2}}$. The time necessary to traverse the distance

at the mean speed must satisfy the equation $\sqrt{\dfrac{ad}{2}}\, t = d$ or $t = \sqrt{\dfrac{2d}{a}}$.

This is the same time as under uniform acceleration.

54. If $C'(x) = k$ then $C(x) = kx + C_1$ where k and C_1 are constants.

    Thus the cost function is linear.

55. $\dfrac{dC}{dx} = 2x - 12,$ $\quad C(0) = 50$

    $C(x) = x^2 - 12x + C,$ $\quad C(0) = 0 + C = 50 \implies C = 50$

    $C(x) = x^2 - 12x + 50,$ and thus $\overline{C}(x) = \dfrac{C(x)}{x} = x - 12 + \dfrac{50}{x}$

56. $\dfrac{dR}{dx} = 100 - 5x,$ $\quad R = 100x - \dfrac{5}{2}x^2 + C$

    ($C = 0$, if no units are sold then no revenue is generated.)

    $R = 100x - \dfrac{5}{2}x^2,$ $\quad R = xp = x(100 - \dfrac{5}{2}x),$ thus $p = 100 - \dfrac{5}{2}x$

57. $\dfrac{dR}{dx} = 10 - 6x - 2x^2,$ $\quad R = 10x - 3x^2 - \dfrac{2}{3}x^3 + C$

    ($C = 0$, if no units are sold then no revenue is generated.)

    $R = xp = x(10 - 3x - \dfrac{2}{3}x^2),$ thus $p = 10 - 3x - \dfrac{2}{3}x^2$

5.2
Integration by substitution

$\displaystyle\int f(g(x))g'(x)\,dx$	$u = g(x)$	$du = g'(x)\,dx$
1. $\displaystyle\int (5x^2 + 1)^2(10x)\,dx$	$5x^2 + 1$	$10x\,dx$
2. $\displaystyle\int x^2\sqrt{x^3 + 1}\,dx$	$x^3 + 1$	$3x^2\,dx$
3. $\displaystyle\int \dfrac{x}{\sqrt{x^2 + 1}}\,dx$	$x^2 + 1$	$2x\,dx$
4. $\displaystyle\int \sec 2x \tan 2x\,dx$	$2x$	$2\,dx$
5. $\displaystyle\int \tan^2 x \sec^2 x\,dx$	$\tan x$	$\sec^2 x\,dx$
6. $\displaystyle\int \dfrac{\cos x}{\sin^2 x}\,dx$	$\sin x$	$\cos x\,dx$

7. $\displaystyle\int (1 + 2x)^4\ 2\ dx = \frac{(1 + 2x)^5}{5} + C$

Check: $\dfrac{d}{dx}[\dfrac{(1 + 2x)^5}{5} + C] = 2(1 + 2x)^4$

8. $\displaystyle\int (x^2 - 1)^3\ 2x\ dx = \frac{(x^2 - 1)^4}{4} + C$

Check: $\dfrac{d}{dx}[\dfrac{(x^2 - 1)^4}{4} + C] = 2x(x^2 - 1)^3$

9. $\displaystyle\int x^2(x^3 - 1)^4\ dx = \frac{1}{3}\int (x^3 - 1)^4(3x^2)\ dx = \frac{1}{3}\left[\frac{(x^3 - 1)^5}{5}\right] + C$

$= \dfrac{(x^3 - 1)^5}{15} + C,$ Check: $\dfrac{d}{dx}[\dfrac{(x^3 - 1)^5}{15} + C] = x^2(x^3 - 1)^4$

10. $\displaystyle\int x(1 - 2x^2)^3\ dx = -\frac{1}{4}\int (1 - 2x^2)^3(-4x)dx = -\frac{1}{4}\left[\frac{(1 - 2x^2)^4}{4}\right] + C$

$= -\dfrac{(1 - 2x^2)^4}{16} + C,$ Check: $\dfrac{d}{dx}[-\dfrac{(1 - 2x^2)^4}{16} + C] = x(1 - 2x^2)^3$

11. $\displaystyle\int x(x^2 - 1)^7\ dx = \frac{1}{2}\int (x^2 - 1)^7(2x)dx = \frac{1}{2}\left[\frac{(x^2 - 1)^8}{8}\right] + C$

$= \dfrac{(x^2 - 1)^8}{16} + C,$ Check: $\dfrac{d}{dx}[\dfrac{(x^2 - 1)^8}{16} + C] = x(x^2 - 1)^7$

12. $\displaystyle\int \frac{x^2}{(x^3 - 1)^2}\ dx = \frac{1}{3}\int (x^3 - 1)^{-2}(3x^2)dx = \frac{1}{3}\left[\frac{(x^3 - 1)^{-1}}{-1}\right] + C$

$= -\dfrac{1}{3(x^3 - 1)} + C,$ Check: $\dfrac{d}{dx}[-\dfrac{1}{3(x^3 - 1)} + C] = \dfrac{x^2}{(x^3 - 1)^2}$

13. $\displaystyle\int \frac{4x}{\sqrt{1 + x^2}}\ dx = 2\int (1 + x^2)^{-1/2}(2x)\ dx = 2\left[\frac{(1 + x^2)^{1/2}}{1/2}\right] + C$

$= 4\sqrt{1 + x^2} + C,$ Check: $\dfrac{d}{dx}[4\sqrt{1 + x^2} + C] = \dfrac{4x}{\sqrt{1 + x^2}}$

14. $\displaystyle\int \frac{6x}{(1 + x^2)^3}\ dx = 3\int (1 + x^2)^{-3}(2x)\ dx = 3\left[\frac{(1 + x^2)^{-2}}{-2}\right] + C$

$= -\dfrac{3}{2(1 + x^2)^2} + C,$ Check: $\dfrac{d}{dx}[-\dfrac{3}{2(1 + x^2)^2} + C] = \dfrac{6x}{(1 + x^2)^3}$

15. $\displaystyle\int 5x\sqrt[3]{1+x^2}\,dx = \frac{5}{2}\int (1+x^2)^{1/3}(2x)\,dx = \frac{5}{2}\left[\frac{(1+x^2)^{4/3}}{4/3}\right]+C$

$= \frac{15}{8}(1+x^2)^{4/3}+C,\quad \text{Check: } \frac{d}{dx}[\frac{15}{8}(1+x^2)^{4/3}+C]=5x\sqrt[3]{1+x^2}$

16. $\displaystyle\int 3(x-3)^{5/2}\,dx = 3\int (x-3)^{5/2}\,dx = 3\left[\frac{(x-3)^{7/2}}{7/2}\right]+C$

$= \frac{6}{7}(x-3)^{7/2}+C,\quad \text{Check: } \frac{d}{dx}[\frac{6}{7}(x-3)^{7/2}+C]=3(x-3)^{5/2}$

17. $\displaystyle\int \frac{-3}{\sqrt{2x+3}}\,dx = -\frac{3}{2}\int (2x+3)^{-1/2}\,2\,dx = -\frac{3}{2}\left[\frac{(2x+3)^{1/2}}{1/2}\right]+C$

$= -3\sqrt{2x+3}+C,\quad \text{Check: } \frac{d}{dx}[-3\sqrt{2x+3}+C]=-\frac{3}{\sqrt{2x+3}}$

18. $\displaystyle\int \frac{4x+6}{(x^2+3x+7)^3}\,dx = 2\int (x^2+3x+7)^{-3}(2x+3)\,dx$

$= 2\left[\frac{(x^2+3x+7)^{-2}}{-2}\right]+C = -\frac{1}{(x^2+3x+7)^2}+C$

Check: $\frac{d}{dx}[-\frac{1}{(x^2+3x+7)^2}+C]=\frac{4x+6}{(x^2+3x+7)^3}$

19. $\displaystyle\int \frac{x+1}{(x^2+2x-3)^2}\,dx = \frac{1}{2}\int (x^2+2x-3)^{-2}(2x+2)\,dx$

$= \frac{1}{2}\left[\frac{(x^2+2x-3)^{-1}}{-1}\right]+C = -\frac{1}{2(x^2+2x-3)}+C$

Check: $\frac{d}{dx}[-\frac{1}{2(x^2+2x-3)}+C]=\frac{x+1}{(x^2+2x-3)^2}$

20. $\displaystyle\int u^3\sqrt{u^4+2}\,du = \frac{1}{4}\int (u^4+2)^{1/2}(4u^3)\,du = \frac{1}{4}\left[\frac{(u^4+2)^{3/2}}{3/2}\right]+C$

$= \frac{1}{6}(u^4+2)^{3/2}+C,\quad \text{Check: } \frac{d}{dx}[\frac{1}{6}(u^4+2)^{3/2}+C]=u^3\sqrt{u^4+2}$

21. $\displaystyle\int \frac{1}{\sqrt{x}(1+\sqrt{x})^2}\,dx = 2\int (1+\sqrt{x})^{-2}(\frac{1}{2\sqrt{x}})\,dx = 2\left[\frac{(1+\sqrt{x})^{-1}}{-1}\right]+C$

$= -\frac{2}{1+\sqrt{x}}+C,\quad \text{Check: } \frac{d}{dx}[-\frac{2}{1+\sqrt{x}}+C]=\frac{1}{\sqrt{x}(1+\sqrt{x})^2}$

22. $\displaystyle\int (1 + \frac{1}{t})^3(\frac{1}{t^2})\, dt = -\int (1 + \frac{1}{t})^3(-\frac{1}{t^2})\, dt = -\frac{(1 + 1/t)^4}{4} + C$

    Check: $\dfrac{d}{dt}[-\dfrac{(1 + 1/t)^4}{4} + C] = \dfrac{1}{t^2}(1 + \dfrac{1}{t})^3$

23. $\displaystyle\int \frac{x^2}{(1 + x^3)^2}\, dx = \frac{1}{3}\int (1 + x^3)^{-2}(3x^2)\, dx = \frac{1}{3}\left[\frac{(1 + x^3)^{-1}}{-1}\right] + C$

    $= -\dfrac{1}{3(1 + x^3)} + C$, Check: $\dfrac{d}{dx}[-\dfrac{1}{3(1 + x^3)} + C] = \dfrac{x^2}{(1 + x^3)^2}$

24. $\displaystyle\int \frac{x^2}{\sqrt{1 + x^3}}\, dx = \frac{1}{3}\int (1 + x^3)^{-1/2}(3x^2)\, dx = \frac{1}{3}\left[\frac{(1 + x^3)^{1/2}}{1/2}\right] + C$

    $= \dfrac{2}{3}\sqrt{1 + x^3} + C$, Check: $\dfrac{d}{dx}[\dfrac{2}{3}\sqrt{1 + x^3} + C] = \dfrac{x^2}{\sqrt{1 + x^3}}$

25. $\displaystyle\int \frac{x^3}{\sqrt{1 + x^4}}\, dx = \frac{1}{4}\int (1 + x^4)^{-1/2}(4x^3)\, dx = \frac{1}{4}\left[\frac{(1 + x^4)^{1/2}}{1/2}\right] + C$

    $= \dfrac{1}{2}\sqrt{1 + x^4} + C$, Check: $\dfrac{d}{dx}[\dfrac{1}{2}\sqrt{1 + x^4} + C] = \dfrac{x^3}{\sqrt{1 + x^4}}$

26. $\displaystyle\int \frac{t + 2t^2}{\sqrt{t}}\, dt = \int (t^{1/2} + 2t^{3/2})\, dt = \frac{2}{3}t^{3/2} + \frac{4}{5}t^{5/2} + C$

    $= \dfrac{2}{15}t^{3/2}(5 + 6t) + C$

    Check: $\dfrac{d}{dt}[\dfrac{2}{15}t^{3/2}(5 + 6t) + C] = t^{1/2} + 2t^{3/2} = \dfrac{t + 2t^2}{\sqrt{t}}$

27. $\displaystyle\int \frac{1}{2\sqrt{x}}\, dx = \frac{1}{2}\int x^{-1/2}\, dx = \frac{1}{2}\left[\frac{x^{1/2}}{1/2}\right] + C = \sqrt{x} + C$

    Check: $\dfrac{d}{dx}[\sqrt{x} + C] = \dfrac{1}{2\sqrt{x}}$

28. $\displaystyle\int \frac{1}{(3x)^2}\, dx = \frac{1}{9}\int x^{-2}\, dx = \frac{1}{9}(\frac{x^{-1}}{-1}) + C = -\frac{1}{9x} + C$

    Check: $\dfrac{d}{dx}[-\dfrac{1}{9x} + C] = \dfrac{1}{9x^2} = \dfrac{1}{(3x)^2}$

29. $\displaystyle\int \frac{1}{\sqrt{2x}}\, dx = \frac{1}{2}\int (2x)^{-1/2}\,2\, dx = \frac{1}{2}\left[\frac{(2x)^{1/2}}{1/2}\right] + C = \sqrt{2x} + C$

    Check: $\dfrac{d}{dx}[\sqrt{2x} + C] = \dfrac{1}{\sqrt{2x}}$

30. $\int \dfrac{1}{3x^2}\ dx = \dfrac{1}{3} \int x^{-2}\ dx = \dfrac{1}{3}\left(\dfrac{x^{-1}}{-1}\right) + C = -\dfrac{1}{3x} + C$

Check: $\dfrac{d}{dx}\left[-\dfrac{1}{3x} + C\right] = \dfrac{1}{3x^2}$

31. $\int \dfrac{x^2 + 3x + 7}{\sqrt{x}}\ dx = \int (x^{3/2} + 3x^{1/2} + 7x^{-1/2})\ dx$

$= \dfrac{2}{5}x^{5/2} + 2x^{3/2} + 14x^{1/2} + C = \dfrac{2}{5}\sqrt{x}(x^2 + 5x + 35) + C$

Check: $\dfrac{d}{dx}\left[\dfrac{2}{5}\sqrt{x}(x^2 + 5x + 35) + C\right] = \dfrac{2}{5}\sqrt{x}(2x + 5) + \dfrac{x^2 + 5x + 35}{5\sqrt{x}}$

$= \dfrac{5x^2 + 15x + 35}{5\sqrt{x}} = \dfrac{x^2 + 3x + 7}{\sqrt{x}}$

32. $\int \dfrac{x^{5/2} + 5x^{1/2}}{x^{5/2}}\ dx = \int (1 + 5x^{-2})\ dx = x + 5\left(\dfrac{x^{-1}}{-1}\right) + C = x - \dfrac{5}{x} + C$

Check: $\dfrac{d}{dx}\left[x - \dfrac{5}{x} + C\right] = 1 + \dfrac{5}{x^2} = 1 + 5x^{-2} = \dfrac{x^{5/2} + 5x^{1/2}}{x^{5/2}}$

33. $\int t^2\left(t - \dfrac{2}{t}\right)\ dt = \int (t^3 - 2t)\ dt = \dfrac{1}{4}t^4 - t^2 + C$

Check: $\dfrac{d}{dt}\left[\dfrac{1}{4}t^4 - t^2 + C\right] = t^3 - 2t = t^2\left(t - \dfrac{2}{t}\right)$

34. $\int \left(\dfrac{t^3}{3} + \dfrac{1}{4t^2}\right)\ dt = \int \left(\dfrac{1}{3}t^3 + \dfrac{1}{4}t^{-2}\right) dt = \dfrac{1}{3}\left(\dfrac{t^4}{4}\right) + \dfrac{1}{4}\left(\dfrac{t^{-1}}{-1}\right) + C$

$= \dfrac{1}{12}t^4 - \dfrac{1}{4t} + C,\quad$ Check: $\dfrac{d}{dt}\left[\dfrac{1}{12}t^4 - \dfrac{1}{4t} + C\right] = \dfrac{1}{3}t^3 + \dfrac{1}{4t^2}$

35. $\int (9 - y)\ \sqrt{y}\ dy = \int (9y^{1/2} - y^{3/2})\ dy = 9\left(\dfrac{2}{3}y^{3/2}\right) - \dfrac{2}{5}y^{5/2} + C$

$= \dfrac{2}{5}y^{3/2}(15 - y) + C$

Check: $\dfrac{d}{dy}\left[\dfrac{2}{5}y^{3/2}(15 - y) + C\right] = 9y^{1/2} - y^{3/2} = (9 - y)\sqrt{y}$

36. $\int 2\pi y(8 - y^{3/2})\ dy = 2\pi \int (8y - y^{5/2})\ dy = 2\pi\left(4y^2 - \dfrac{2}{7}y^{7/2}\right) + C$

$= \dfrac{4\pi y^2}{7}(14 - y^{3/2}) + C$

Check: $\dfrac{d}{dy}\left[\dfrac{4\pi y^2}{7}(14 - y^{3/2}) + C\right] = 16\pi y - 2\pi y^{5/2} = 2\pi y(8 - y^{3/2})$

37. $\displaystyle\int (2x - 1)^2 \, dx = \frac{1}{2}\int (2x - 1)^2 \, 2 \, dx = \frac{1}{6}(2x - 1)^3 + C_1$

$$= \frac{4}{3}x^3 - 2x^2 + x - \frac{1}{6} + C_1$$

$\displaystyle\int (2x - 1)^2 \, dx = \int (4x^2 - 4x + 1) \, dx = \frac{4}{3}x^3 - 2x^2 + x + C_2$

They differ by a constant: $C_2 = C_1 - \dfrac{1}{6}$

38. $\displaystyle\int x(x^2 - 1)^2 \, dx = \frac{1}{2}\int (x^2 - 1)^2 \, 2x \, dx = \frac{1}{6}(x^2 - 1)^3 + C_1$

$$= \frac{1}{6}x^6 - \frac{1}{2}x^4 + \frac{1}{2}x^2 - \frac{1}{6} + C_1$$

$\displaystyle\int x(x^2 - 1)^2 \, dx = \int (x^5 - 2x^3 + x) \, dx = \frac{1}{6}x^6 - \frac{1}{2}x^4 + \frac{1}{2}x^2 + C_2$

They differ by a constant: $C_2 = C_1 - \dfrac{1}{6}$

39. $u = \sqrt{x - 3}, \qquad x = u^2 + 3, \qquad dx = 2u \, du$

$\displaystyle\int x\sqrt{x - 3} \, dx = \int (u^2 + 3)u(2u \, du) = 2\int (u^4 + 3u^2) \, du = 2\left[\frac{u^5}{5} + u^3\right] + C$

$$= \frac{2u^3}{5}(u^2 + 5) + C = \frac{2}{5}(x - 3)^{3/2}(x + 2) + C$$

40. $u = \sqrt{2x + 1}, \qquad x = (u^2 - 1)/2, \qquad dx = u \, du$

$\displaystyle\int x\sqrt{2x + 1} \, dx = \frac{1}{2}\int (u^2 - 1)u(u \, du) = \frac{1}{2}\int (u^4 - u^2) \, du = \frac{1}{2}\left[\frac{u^5}{5} - \frac{u^3}{3}\right] + C$

$$= \frac{u^3}{30}(3u^2 - 5) + C = \frac{(2x + 1)^{3/2}}{30}(6x - 2) + C = \frac{(2x + 1)^{3/2}}{15}(3x - 1) + C$$

41. $u = \sqrt{1 - x}, \qquad x = 1 - u^2, \qquad dx = -2u \, du$

$\displaystyle\int x^2\sqrt{1 - x} \, dx = \int (1 - u^2)^2 u(-2u \, du) = -2\int (u^2 - 2u^4 + u^6) \, du$

$$= -2\left[\frac{u^3}{3} - \frac{2u^5}{5} + \frac{u^7}{7}\right] + C = \frac{-2u^3}{105}[35 - 42u^2 + 15u^4] + C$$

$$= \frac{-2}{105}(1 - x)^{3/2}(15x^2 + 12x + 8) + C$$

42. $u = \sqrt{x + 3}, \quad x = u^2 - 3, \quad dx = 2u\,du$

$$\int \frac{2x - 1}{\sqrt{x + 3}}\,dx = \int \frac{2u^2 - 7}{u}(2u)\,du = 2\left[\frac{2}{3}u^3 - 7u\right] + C = \frac{2u}{3}(2u^2 - 21) + C$$

$$= \frac{2}{3}\sqrt{x + 3}(2x - 15) + C$$

43. $u = \sqrt{2x - 1}, \quad x = (u^2 + 1)/2, \quad dx = u\,du$

$$\int \frac{x^2 - 1}{\sqrt{2x - 1}}\,dx = \frac{1}{4}\int \frac{u^4 + 2u^2 - 3}{u}u\,du = \frac{1}{4}\left[\frac{1}{5}u^5 + \frac{2}{3}u^3 - 3u\right] + C$$

$$= \frac{u}{60}[3u^4 + 10u^2 - 45] + C = \frac{\sqrt{2x - 1}}{15}(3x^2 + 2x - 13) + C$$

44. $u = \sqrt{x + 2}, \quad x = u^2 - 2, \quad dx = 2u\,du$

$$\int x^3\sqrt{x + 2}\,dx = \int (u^2 - 2)^3 u(2u\,du) = 2\int (u^8 - 6u^6 + 12u^4 - 8u^2)\,du$$

$$= 2\left[\frac{u^9}{9} - \frac{6}{7}u^7 + \frac{12}{5}u^5 - \frac{8}{3}u^3\right] + C$$

$$= \frac{2}{315}(x + 2)^{3/2}[35(x + 2)^3 - 270(x + 2)^2 + 756(x + 2) - 840] + C$$

$$= \frac{2}{315}(x + 2)^{3/2}(35x^3 - 60x^2 + 96x - 128) + C$$

45. $u = \sqrt{x + 1}, \quad x = u^2 - 1, \quad dx = 2u\,du$

$$\int \frac{-x}{(x + 1) - \sqrt{x + 1}}\,dx = \int \frac{-(u^2 - 1)}{u^2 - u}(2u)\,du = -2\int \frac{u^2 - 1}{u - 1}\,du$$

$$= -2\int (u + 1)\,du = -2\left(\frac{u^2}{2} + u\right) + C = -u^2 - 2u + C$$

$$= -[(x + 1) + 2\sqrt{x + 1}] + C = -(x + 2\sqrt{x + 1}) + C_1$$

46. $u = \sqrt[3]{t + 1}, \quad t = u^3 - 1, \quad dt = 3u^2\,du$

$$\int t\sqrt[3]{t + 1}\,dt = 3\int (u^3 - 1)u^3\,du = 3\int (u^6 - u^3)\,du = 3\left[\frac{u^7}{7} - \frac{u^4}{4}\right] + C$$

$$= \frac{3u^4}{28}(4u^3 - 7) + C = \frac{3}{28}(t + 1)^{4/3}(4t - 3) + C$$

47. $u = \sqrt{2x + 1}$, $\qquad x = (u^2 - 1)/2$, $\qquad dx = u\ du$

$$\int \frac{x}{\sqrt{2x + 1}}\ dx = \int (\frac{u^2 - 1}{2})u^{-1}\ u\ du = \frac{1}{2} \int (u^2 - 1)\ du$$

$$= \frac{1}{2}(\frac{u^3}{3} - u) + C = \frac{u}{6}(u^2 - 3) + C$$

$$= \frac{1}{6}\sqrt{2x + 1}(2x + 1 - 3) + C = \frac{1}{3}\sqrt{2x + 1}(x - 1) + C$$

48. $u = \sqrt{2 - x}$, $\qquad x = 2 - u^2$, $\qquad dx = -2u\ du$

$$\int (x + 1)\sqrt{2 - x}\ dx = \int (3 - u^2)(u)(-2u)\ du = -2 \int (3u^2 - u^4)\ du$$

$$= -2(u^3 - \frac{u^5}{5}) + C = -\frac{2u^3}{5}(5 - u^2) + C$$

$$= -\frac{2}{5}(2 - x)^{3/2}[5 - (2 - x)] + C = -\frac{2}{5}(2 - x)^{3/2}(x + 3) + C$$

49. $\displaystyle\int \sin 2x\ dx = \frac{1}{2} \int (\sin 2x)(2)\ dx = -\frac{1}{2}\cos 2x + C$

50. $\displaystyle\int x \sin x^2\ dx = \frac{1}{2} \int (\sin x^2)(2x)\ dx = -\frac{1}{2}\cos x^2 + C$

51. $\displaystyle\int x \cos x^2\ dx = \frac{1}{2} \int (\cos x^2)(2x)\ dx = \frac{1}{2}\sin x^2 + C$

52. $\displaystyle\int \cos 6x\ dx = \frac{1}{6} \int (\cos 6x)(6)\ dx = \frac{1}{6}\sin 6x + C$

53. $\displaystyle\int \sec^2 (\frac{x}{2})\ dx = 2 \int \sec^2 (\frac{x}{2})(\frac{1}{2})\ dx = 2 \tan (\frac{x}{2}) + C$

54. $\displaystyle\int \csc^2 (\frac{x}{2})\ dx = 2 \int \csc^2 (\frac{x}{2})(\frac{1}{2})\ dx = -2 \cot (\frac{x}{2}) + C$

55. $\displaystyle\int \sec (1 - x) \tan (1 - x)\ dx = - \int [\sec (1 - x)\ \tan (1 - x)](-1)\ dx$

$$= -\sec (1 - x) + C$$

56. $\displaystyle\int \sin 2x \cos 2x \, dx = \frac{1}{2}\int (\sin 2x)(2\cos 2x) \, dx$

OR:
$$= \frac{1}{2}\frac{(\sin 2x)^2}{2} + C = \frac{1}{4}(\sin 2x)^2 + C$$

$\displaystyle\int \sin 2x \cos 2x \, dx = -\frac{1}{2}\int (\cos 2x)(-2\sin 2x) \, dx$

$$= -\frac{1}{2}\frac{(\cos 2x)^2}{2} + C = -\frac{1}{4}(\cos 2x)^2 + C$$

57. $\displaystyle\int \frac{\csc^2 x}{\cot^3 x} \, dx = -\int (\cot x)^{-3}(-\csc^2 x) \, dx = -\frac{(\cot x)^{-2}}{-2} + C = \frac{1}{2\cot^2 x} + C$

$$= \frac{1}{2}\tan^2 x + C = \frac{1}{2}(\sec^2 x - 1) + C = \frac{1}{2}\sec^2 x + C_1$$

58. $\displaystyle\int \csc 2x \cot 2x \, dx = \frac{1}{2}\int (\csc 2x \cot 2x)(2) \, dx = -\frac{1}{2}\csc 2x + C$

59. $\displaystyle\int \cot^2 x \, dx = \int (\csc^2 x - 1) \, dx = -\cot x - x + C$

60. $\displaystyle\int \frac{\sin x}{\cos^2 x} \, dx = \int \left(\frac{1}{\cos x}\right)\left(\frac{\sin x}{\cos x}\right) \, dx = \int \sec x \tan x \, dx = \sec x + C$

61. $\displaystyle\int \tan^4 x \sec^2 x \, dx = \frac{\tan^5 x}{5} + C = \frac{1}{5}\tan^5 x + C$

62. $\displaystyle\int \sqrt{\cot x} \csc^2 x \, dx = -\int (\cot x)^{1/2}(-\csc^2 x) \, dx = -\frac{2}{3}(\cot x)^{3/2} + C$

63.(a) $\displaystyle\int \sin x \cos x \, dx = \frac{(\sin^2 x)}{2} + C_1 \qquad u = \sin x \qquad du = \cos x \, dx$

(b) $\displaystyle\int \sin x \cos x \, dx = -\int \cos x(-\sin x) \, dx \qquad u = \cos x \qquad du = -\sin x \, dx$

$$= -\frac{(\cos^2 x)}{2} + C_2$$

(c) $-\dfrac{1}{2}(\cos^2 x) + C_2 = -\dfrac{1}{2}(1 - \sin^2 x) + C_2 = \dfrac{1}{2}\sin^2 x - \dfrac{1}{2} + C_2$

$$= \frac{1}{2}\sin^2 x + C_1$$

They differ by a constant $C_1 = C_2 - \dfrac{1}{2}$

64.(a) $\displaystyle\int \sqrt{x^2 + 1}\ dx$ cannot be evaluated by the General Power Rule.

   (b) $\displaystyle\int x\,\sqrt{x^2 + 1}\ dx$,   let $u = x^2 + 1$ and $du = 2x\ dx$.

   (c) $\displaystyle\int \frac{x}{\sqrt{x^2 + 1}}\ dx$,   let $u = x^2 + 1$ and $du = 2x\ dx$.

65.(a) $\displaystyle\int \sin^2 x \cos x\ dx$,   let $u = \sin x$ and $du = \cos x\ dx$.

   (b) $\displaystyle\int \sin^2 x \cos^2 x\ dx$ cannot be evaluated by the General Power Rule
       without using a trigonometric identity.

   (c) $\displaystyle\int \sqrt{\sin x}\ \cos x\ dx$,   let $u = \sin x$ and $du = \cos x\ dx$.

66.(a) $\displaystyle\int \sin x \cos^3 x\ dx$,   let $u = \cos x$ and $du = -\sin x\ dx$.

   (b) $\displaystyle\int \sin^3 x \cos^3 x\ dx$ cannot be evaluated by the General Power Rule
       without using a trigonometric identity.

   (c) $\displaystyle\int \sin x \sqrt{\cos x}\ dx$,   let $u = \cos x$ and $du = -\sin x\ dx$.

67. $f'(x) = x\sqrt{1 - x^2}$, $(0, 4/3)$

   $\displaystyle f(x) = \int x\sqrt{1 - x^2}\ dx = -\frac{1}{2}\int (1 - x^2)^{1/2}(-2x)\ dx = -\frac{1}{3}(1 - x^2)^{3/2} + C$

   $f(0) = -\dfrac{1}{3} + C = \dfrac{4}{3} \implies C = \dfrac{5}{3}$

   $f(x) = -\dfrac{1}{3}(1 - x^2)^{3/2} + \dfrac{5}{3} = -\dfrac{1}{3}[(1 - x^2)^{3/2} - 5]$

68. $f'(x) = x\sqrt{1 - x^2}$, $(0, 7/3)$

   $\displaystyle f(x) = \int x\sqrt{1 - x^2}\ dx = -\frac{1}{3}(1 - x^2)^{3/2} + C$ (See Exercise 67)

   $f(0) = -\dfrac{1}{3} + C = \dfrac{7}{3} \implies C = \dfrac{8}{3}$

   $f(x) = -\dfrac{1}{3}(1 - x^2)^{3/2} + \dfrac{8}{3} = -\dfrac{1}{3}[(1 - x^2)^{3/2} - 8]$

69. $\dfrac{dW}{dt} = \dfrac{12}{\sqrt{16t + 9}}$

   (a) $W = \displaystyle\int \dfrac{12}{\sqrt{16t + 9}}\, dt = \dfrac{12}{16}\displaystyle\int (16t + 9)^{-1/2}\, 16\, dt$

                $= \dfrac{3}{4}\left[\dfrac{(16t + 9)^{1/2}}{1/2}\right] + C = \dfrac{3}{2}\sqrt{16t + 9} + C$

            $W(0) = \dfrac{9}{2} + C = 0 \implies C = -\dfrac{9}{2}$

            $W(t) = \dfrac{3}{2}\sqrt{16t + 9} - \dfrac{9}{2} = \dfrac{3}{2}[\sqrt{16t + 9} - 3]$

   (b) $W(100) = \dfrac{3}{2}[\sqrt{1609} - 3] \approx 55.67 \text{ lb.}$

70. $\dfrac{dC}{dx} = \dfrac{4}{\sqrt{x + 1}}$

   (a) $C = \displaystyle\int \dfrac{4}{\sqrt{x + 1}}\, dx$ (b)

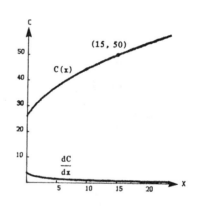

           $= 4\displaystyle\int (x + 1)^{-1/2}\, dx$

           $= 8\sqrt{x + 1} + C_1$

      $C(15) = 32 + C_1 = 50$

                $C_1 = 18$

      $C(x) = 8\sqrt{x + 1} + 18$

71. $\dfrac{dC}{dx} = \dfrac{12}{\sqrt[3]{12x + 1}}$

   (a) $C(x) = \displaystyle\int \dfrac{12}{\sqrt[3]{12x + 1}}\, dx$ (b)

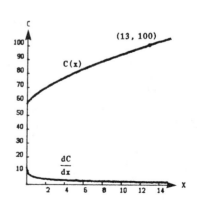

           $= \displaystyle\int (12x + 1)^{-1/3}\, 12\, dx$

           $= \dfrac{3}{2}(12x + 1)^{2/3} + C_1$

      $C(13) = 43.65 + C_1 = 100$

                $C_1 = 56.35$

      $C(x) = \dfrac{3}{2}(12x + 1)^{2/3} + 56.35$

5.3

Sigma notation and the limit of a sequence

1. $\displaystyle\sum_{i=1}^{5} (2i + 1) = 2\sum_{i=1}^{5} i + \sum_{i=1}^{5} 1 = 2(1 + 2 + 3 + 4 + 5) + 5 = 35$

2. $\displaystyle\sum_{i=1}^{6} 2i = 2 + 4 + 6 + 8 + 10 + 12 = 42$

3. $\displaystyle\sum_{k=0}^{4} \frac{1}{k^2 + 1} = 1 + \frac{1}{2} + \frac{1}{5} + \frac{1}{10} + \frac{1}{17} = \frac{158}{85}$

4. $\displaystyle\sum_{j=3}^{5} \frac{1}{j} = \frac{1}{3} + \frac{1}{4} + \frac{1}{5} = \frac{47}{60}$

5. $\displaystyle\sum_{k=1}^{4} c = c + c + c + c = 4c$

6. $\displaystyle\sum_{n=1}^{10} \frac{3}{n + 1} = 3(\frac{1}{2} + \frac{1}{3} + \frac{1}{4} + \frac{1}{5} + \frac{1}{6} + \frac{1}{7} + \frac{1}{8} + \frac{1}{9} + \frac{1}{10} + \frac{1}{11}) = \frac{55991}{9240}$

7. $\displaystyle\sum_{i=1}^{4} [(i - 1)^2 + (i + 1)^3] = (0 + 8) + (1 + 27) + (4 + 64) + (9 + 125) = 238$

8. $\displaystyle\sum_{k=2}^{5} (k + 1)(k - 3) = (3)(-1) + (4)(0) + (5)(1) + (6)(2) = 14$

9. $\displaystyle\sum_{i=1}^{9} \frac{1}{3i}$

10. $\displaystyle\sum_{i=1}^{15} \frac{5}{1 + i}$

11. $\displaystyle\sum_{j=1}^{8} [2(\frac{j}{8}) + 3]$

12. $\displaystyle\sum_{j=1}^{4} [1 - (\frac{j}{4})^2]$

13. $\displaystyle\sum_{k=1}^{6} [(\frac{k}{6})^2 + 2](\frac{1}{6}) = \frac{1}{6}\sum_{k=1}^{6} [(\frac{k}{6})^2 + 2]$

14. $\displaystyle\sum_{k=1}^{n} [(\frac{k}{n})^2 + 2](\frac{1}{n}) = \frac{1}{n}\sum_{k=1}^{n} [(\frac{k}{n})^2 + 2]$

15. $\displaystyle\frac{2}{n}\sum_{i=1}^{n} [(\frac{2i}{n})^3 - (\frac{2i}{n})]$

16. $\displaystyle\frac{2}{n}\sum_{i=1}^{n} [1 - (\frac{2i}{n} - 1)^2]$

17. $\dfrac{3}{n} \sum\limits_{i=1}^{n} [2(1 + \dfrac{3i}{n})^2]$

18. $\dfrac{1}{n} \sum\limits_{i=1}^{n-1} \sqrt{1 - (\dfrac{i}{n})^2}$

19. $\sum\limits_{i=1}^{20} 2i = 2 \sum\limits_{i=1}^{20} i = 2 \left[\dfrac{20(21)}{2} \right] = 420$

20. $\sum\limits_{i=1}^{10} i(i^2 + 1) = \sum\limits_{i=1}^{10} i^3 + \sum\limits_{i=1}^{10} i = \left[\dfrac{10(11)}{2} \right]^2 + \dfrac{10(11)}{2} = 3080$

21. $\sum\limits_{i=1}^{20} (i - 1)^2 = \sum\limits_{i=1}^{19} i^2 = \left[\dfrac{19(20)(39)}{6} \right] = 2470$

22. $\sum\limits_{i=1}^{15} (2i - 3) = 2 \sum\limits_{i=1}^{15} i - 3(15) = 2 \left[\dfrac{15(16)}{2} \right] - 45 = 195$

23. $\sum\limits_{i=1}^{15} \dfrac{1}{n^3}(i - 1)^2 = \dfrac{1}{n^3} \sum\limits_{i=1}^{14} i^2 = \dfrac{1}{n^3} \left[\dfrac{14(15)(29)}{6} \right] = \dfrac{1015}{n^3}$

24. $\sum\limits_{i=1}^{10} (i^2 - 1) = \sum\limits_{i=1}^{10} i^2 - \sum\limits_{i=1}^{10} 1 = \left[\dfrac{10(11)(21)}{6} \right] - 10 = 375$

25. $\lim\limits_{n \to \infty} \left[(\dfrac{4}{3n^3})(2n^3 + 3n^2 + n) \right] = \lim\limits_{n \to \infty} \left[\dfrac{8}{3} + \dfrac{4}{n} + \dfrac{4}{3n^2} \right] = \dfrac{8}{3}$

26. $\lim\limits_{n \to \infty} (\dfrac{8}{3} + \dfrac{4}{n} + \dfrac{4}{3n^2}) = \dfrac{8}{3}$

27. $\lim\limits_{n \to \infty} \left[(\dfrac{81}{n^4}) \dfrac{n^2(n+1)^2}{4} \right] = \dfrac{81}{4} \lim\limits_{n \to \infty} \left[\dfrac{n^4 + 2n^3 + n^2}{n^4} \right] = \dfrac{81}{4}(1) = \dfrac{81}{4}$

28. $\lim\limits_{n \to \infty} \left[(\dfrac{64}{n^3}) \dfrac{n(n+1)(2n+1)}{6} \right] = \dfrac{64}{6} \lim\limits_{n \to \infty} \left[\dfrac{2n^3 + 3n^2 + n}{n^3} \right] = \dfrac{64}{6}(2) = \dfrac{64}{3}$

29. $\lim\limits_{n \to \infty} \left[(\dfrac{18}{n^2}) \dfrac{n(n+1)}{2} \right] = \dfrac{18}{2} \lim\limits_{n \to \infty} \left[\dfrac{n^2 + n}{n^2} \right] = \dfrac{18}{2}(1) = 9$

30. $\lim\limits_{n \to \infty} \left[(\dfrac{1}{n^2}) \dfrac{n(n+1)}{2} \right] = \dfrac{1}{2} \lim\limits_{n \to \infty} \left[\dfrac{n^2 + n}{n^2} \right] = \dfrac{1}{2}(1) = \dfrac{1}{2}$

31. $\displaystyle\lim_{n \to \infty} \sum_{i=1}^{n} \frac{1}{n^3}(i-1)^2 = \lim_{n \to \infty} \frac{1}{n^3} \sum_{i=1}^{n-1} i^2$

$\displaystyle = \lim_{n \to \infty} \frac{1}{n^3}\left[\frac{(n-1)(n)(2n-1)}{6}\right]$

$\displaystyle = \lim_{n \to \infty} \frac{1}{6}\left[\frac{2n^3 - 3n^2 + n}{n^3}\right] = \lim_{n \to \infty}\left[\frac{1}{6}\left(\frac{2 - (3/n) + (1/n^2)}{1}\right)\right] = \frac{1}{3}$

32. $\displaystyle\lim_{n \to \infty} \sum_{i=1}^{n} (1 + \frac{2i}{n})^2(\frac{2}{n}) = \lim_{n \to \infty} \frac{2}{n^3} \sum_{i=1}^{n} (n + 2i)^2$

$\displaystyle = \lim_{n \to \infty} \frac{2}{n^3}\left[\sum_{i=1}^{n} n^2 + 4n \sum_{i=1}^{n} i + 4 \sum_{i=1}^{n} i^2\right]$

$\displaystyle = \lim_{n \to \infty} \frac{2}{n^3}\left[n^3 + (4n)(\frac{n(n+1)}{2}) + \frac{4(n)(n+1)(2n+1)}{6}\right]$

$\displaystyle = 2 \lim_{n \to \infty}\left[1 + 2 + \frac{2}{n} + \frac{4}{3} + \frac{2}{n} + \frac{2}{3n^2}\right] = 2(1 + 2 + \frac{4}{3}) = \frac{26}{3}$

33. $\displaystyle\lim_{n \to \infty} \sum_{i=1}^{n} \frac{16i}{n^2} = \lim_{n \to \infty} \frac{16}{n^2} \sum_{i=1}^{n} i = \lim_{n \to \infty} \frac{16}{n^2}(\frac{n(n+1)}{2})$

$\displaystyle = \lim_{n \to \infty}\left[8(\frac{n^2 + n}{n^2})\right] = 8 \lim_{n \to \infty} (1 + \frac{1}{n}) = 8$

34. $\displaystyle\lim_{n \to \infty} \sum_{i=1}^{n} (\frac{2i}{n})(\frac{2}{n}) = \lim_{n \to \infty} \frac{4}{n^2} \sum_{i=1}^{n} i = \lim_{n \to \infty} \frac{4}{n^2}(\frac{n(n+1)}{2})$

$\displaystyle = \lim_{n \to \infty} \frac{4}{2}(1 + \frac{1}{n}) = 2$

35. $\lim\limits_{n \to \infty} \sum\limits_{i=1}^{n} (1 + \frac{2i}{n})^3 (\frac{2}{n}) = 2 \lim\limits_{n \to \infty} \frac{1}{n^4} \sum\limits_{i=1}^{n} (n + 2i)^3$

$= 2 \lim\limits_{n \to \infty} \frac{1}{n^4} \sum\limits_{i=1}^{n} (n^3 + 6n^2 i + 12ni^2 + 8i^3)$

$= 2 \lim\limits_{n \to \infty} \frac{1}{n^4} \left[n^4 + 6n^2(\frac{n(n+1)}{2}) + 12n(\frac{n(n+1)(2n+1)}{6}) + 8(\frac{n(n+1)}{2})^2 \right]$

$= 2 \lim\limits_{n \to \infty} (1 + 3 + \frac{3}{n} + 4 + \frac{6}{n} + \frac{2}{n^2} + 2 + \frac{4}{n} + \frac{2}{n^2})$

$= 2 \lim\limits_{n \to \infty} (10 + \frac{13}{n} + \frac{4}{n^2}) = 20$

36. $\lim\limits_{n \to \infty} \sum\limits_{i=1}^{n} (1 + \frac{i}{n})(\frac{2}{n}) = 2 \lim\limits_{n \to \infty} \frac{1}{n} \left[\sum\limits_{i=1}^{n} 1 + \frac{1}{n} \sum\limits_{i=1}^{n} i \right]$

$= 2 \lim\limits_{n \to \infty} \frac{1}{n} \left[n + \frac{1}{n}(\frac{n(n+1)}{2}) \right] = 2 \lim\limits_{n \to \infty} \left[1 + \frac{n^2 + n}{2n^2} \right] = 2(1 + \frac{1}{2}) = 3$

37. $N = \frac{10(5 + 3t)}{1 + 0.04t}$, $\quad N(5) \approx 167$, $\quad N(10) \approx 250$, $\quad N(25) \approx 400$

$\lim\limits_{t \to \infty} \frac{10(5 + 3t)}{1 + 0.04t} = \lim\limits_{t \to \infty} \frac{30t + 50}{0.04t + 1} = \frac{30}{0.04} = 750$

38. $\lim\limits_{n \to \infty} \frac{b + \theta a(n - 1)}{1 + \theta(n - 1)} = \frac{\theta a}{\theta} = a$

39. $P = \frac{0.5 + 0.9(n - 1)}{1 + 0.9(n - 1)}$

n	1	2	3	4	5	6	7	8	9	10
P	0.500	0.737	0.821	0.865	0.891	0.909	0.922	0.932	0.939	0.945

$\lim\limits_{n \to \infty} \frac{0.5 + 0.9(n - 1)}{1 + 0.9(n - 1)} = \frac{0.9}{0.9} = 1$

5.4
Area

1. $S(\Delta) = \sqrt{\frac{1}{4}}(\frac{1}{4}) + \sqrt{\frac{1}{2}}(\frac{1}{4}) + \sqrt{\frac{3}{4}}(\frac{1}{4}) + \sqrt{1}(\frac{1}{4}) = \dfrac{3 + \sqrt{2} + \sqrt{3}}{8} \approx 0.768$

   $s(\Delta) = 0(\frac{1}{4}) + \sqrt{\frac{1}{4}}(\frac{1}{4}) + \sqrt{\frac{1}{2}}(\frac{1}{4}) + \sqrt{\frac{3}{4}}(\frac{1}{4}) = \dfrac{1 + \sqrt{2} + \sqrt{3}}{8} \approx 0.518$

2. $S(\Delta) = (\sqrt{\frac{1}{4}} + 1)\frac{1}{4} + (\sqrt{\frac{1}{2}} + 1)\frac{1}{4} + (\sqrt{\frac{3}{4}} + 1)\frac{1}{4} + (\sqrt{1} + 1)\frac{1}{4}$

   $\quad + (\sqrt{\frac{5}{4}} + 1)\frac{1}{4} + (\sqrt{\frac{3}{2}} + 1)\frac{1}{4} + (\sqrt{\frac{7}{4}} + 1)\frac{1}{4} + (\sqrt{2} + 1)\frac{1}{4}$

   $\quad = \frac{1}{4}(8 + \frac{1}{2} + \frac{\sqrt{2}}{2} + \frac{\sqrt{3}}{2} + 1 + \frac{\sqrt{5}}{2} + \frac{\sqrt{6}}{2} + \frac{\sqrt{7}}{2} + \sqrt{2}) \approx 4.038$

   $s(\Delta) = (0 + 1)\frac{1}{4} + (\sqrt{\frac{1}{4}} + 1)\frac{1}{4} + (\sqrt{\frac{1}{2}} + 1)\frac{1}{4} + \ldots + (\sqrt{\frac{7}{4}} + 1)\frac{1}{4} \approx 3.685$

3. $S(\Delta) = 1(\frac{1}{5}) + \frac{1}{6/5}(\frac{1}{5}) + \frac{1}{7/5}(\frac{1}{5}) + \frac{1}{8/5}(\frac{1}{5}) + \frac{1}{9/5}(\frac{1}{5})$

   $\quad = \frac{1}{5} + \frac{1}{6} + \frac{1}{7} + \frac{1}{8} + \frac{1}{9} \approx 0.746$

   $s(\Delta) = \frac{1}{6/5}(\frac{1}{5}) + \frac{1}{7/5}(\frac{1}{5}) + \frac{1}{8/5}(\frac{1}{5}) + \frac{1}{9/5}(\frac{1}{5}) + \frac{1}{2}(\frac{1}{5})$

   $\quad = \frac{1}{6} + \frac{1}{7} + \frac{1}{8} + \frac{1}{9} + \frac{1}{10} \approx 0.646$

4. $S(\Delta) = \frac{1}{4 - 2}(\frac{1}{2}) + \frac{1}{4.5 - 2}(\frac{1}{2}) + \frac{1}{5 - 2}(\frac{1}{2}) + \frac{1}{5.5 - 2}(\frac{1}{2})$

   $\quad = \frac{1}{2}(\frac{1}{2} + \frac{2}{5} + \frac{1}{3} + \frac{2}{7}) \approx 0.760$

   $s(\Delta) = \frac{1}{4.5 - 2}(\frac{1}{2}) + \frac{1}{5 - 2}(\frac{1}{2}) + \frac{1}{5.5 - 2}(\frac{1}{2}) + \frac{1}{6 - 2}(\frac{1}{2})$

   $\quad = \frac{1}{2}(\frac{2}{5} + \frac{1}{3} + \frac{2}{7} + \frac{1}{4}) \approx 0.635$

5. $S(\Delta) = 1(\frac{1}{5}) + \sqrt{1 - (\frac{1}{5})^2}\,(\frac{1}{5}) + \sqrt{1 - (\frac{2}{5})^2}\,(\frac{1}{5})$

$$+ \sqrt{1 - (\frac{3}{5})^2}\,(\frac{1}{5}) + \sqrt{1 - (\frac{4}{5})^2}\,(\frac{1}{5})$$

$$= \frac{1}{5}\left[1 + \frac{\sqrt{24}}{5} + \frac{\sqrt{21}}{5} + \frac{\sqrt{16}}{5} + \frac{\sqrt{9}}{5}\right] \approx 0.859$$

$s(\Delta) = \sqrt{1 - (\frac{1}{5})^2}\,(\frac{1}{5}) + \sqrt{1 - (\frac{2}{5})^2}\,(\frac{1}{5}) + \sqrt{1 - (\frac{3}{5})^2}\,(\frac{1}{5})$

$$+ \sqrt{1 - (\frac{4}{5})^2}\,(\frac{1}{5}) + 0 \approx 0.659$$

6. $S(\Delta) = \sqrt{\frac{1}{4} + 1}\,(\frac{1}{4}) + \sqrt{\frac{1}{2} + 1}\,(\frac{1}{4}) + \sqrt{\frac{3}{4} + 1}\,(\frac{1}{4}) + \sqrt{2}\,(\frac{1}{4})$

$$= \frac{\sqrt{5} + \sqrt{6} + \sqrt{7} + \sqrt{8}}{8} \approx 1.270$$

$s(\Delta) = 1(\frac{1}{4}) + \sqrt{\frac{1}{4} + 1}\,(\frac{1}{4}) + \sqrt{\frac{1}{2} + 1}\,(\frac{1}{4}) + \sqrt{\frac{3}{4} + 1}\,(\frac{1}{4})$

$$= \frac{2 + \sqrt{5} + \sqrt{6} + \sqrt{7}}{8} \approx 1.166$$

7.(a)

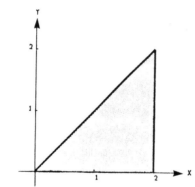

(b) $\Delta x = \dfrac{2 - 0}{n} = \dfrac{2}{n}$

Endpoints $0 < 1(\frac{2}{n}) < 2(\frac{2}{n}) < \cdots < (n-1)(\frac{2}{n}) < n(\frac{2}{n}) = 2$

(c) Since $y = x$ is increasing $f(m_i) = f(x_{i-1})$ on $[x_{i-1}, x_i]$

$$s(n) = \sum_{i=1}^{n} f(x_{i-1})\,\Delta x = \sum_{i=1}^{n} f(\frac{2i-2}{n})(\frac{2}{n}) = \sum_{i=1}^{n}\left[(i-1)(\frac{2}{n})\right](\frac{2}{n})$$

7.(d) $f(M_i) = f(x_i)$ on $[x_{i-1}, x_i]$

$$S(n) = \sum_{i=1}^{n} f(x_i)\Delta x = \sum_{i=1}^{n} f\left(\frac{2i}{n}\right) \frac{2}{n} = \sum_{i=1}^{n} \left[i\left(\frac{2}{n}\right)\right]\left(\frac{2}{n}\right)$$

(e)

n	5	10	50	100
s(n)	1.6	1.8	1.96	1.98
S(n)	2.4	2.2	2.04	2.02

(f) $$\lim_{n \to \infty} \sum_{i=1}^{n} \left[(i-1)\left(\frac{2}{n}\right)\right]\left(\frac{2}{n}\right) = \lim_{n \to \infty} \frac{4}{n^2} \sum_{i=1}^{n}(i-1)$$

$$= \lim_{n \to \infty} \frac{4}{n^2}\left[\frac{n(n+1)}{2} - n\right] = \lim_{n \to \infty} \left[\frac{2(n+1)}{n} - \frac{4}{n}\right] = 2$$

$$\lim_{n \to \infty} \sum_{i=1}^{n} \left[i\left(\frac{2}{n}\right)\right]\left(\frac{2}{n}\right) = \lim_{n \to \infty} \frac{4}{n^2} \sum_{i=1}^{n} i = \lim_{n \to \infty} \left(\frac{4}{n^2}\right)\frac{n(n+1)}{2}$$

$$= \lim_{n \to \infty} \frac{2(n+1)}{n} = 2$$

8.(a)

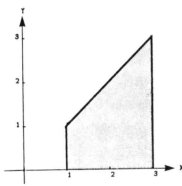

(b) $$\Delta x = \frac{3-1}{n} = \frac{2}{n}$$

Endpoints $1 < 1 + \dfrac{2}{n} < 1 + \dfrac{4}{n} < \ldots < 1 + \dfrac{2n}{n} = 3$

$$1 < 1 + 1\left(\frac{2}{n}\right) < 1 + 2\left(\frac{2}{n}\right) < \ldots < 1 + (n-1)\left(\frac{2}{n}\right) < 1 + n\left(\frac{2}{n}\right)$$

8.(c) Since $y = x$ is increasing $f(m_i) = f(x_{i-1})$ on $[x_{i-1}, x_i]$

$$s(n) = \sum_{i=1}^{n} f(x_{i-1}) \Delta x = \sum_{i=1}^{n} f(1 + (i-1)(\frac{2}{n}))(\frac{2}{n})$$

$$= \sum_{i=1}^{n} \left[1 + (i-1)(\frac{2}{n}) \right] (\frac{2}{n})$$

(d) $f(M_i) = f(x_i)$ on $[x_{i-1}, x_i]$

$$S(n) = \sum_{i=1}^{n} f(x_i)\Delta x = \sum_{i=1}^{n} f(1 + i(\frac{2}{n}))(\frac{2}{n}) = \sum_{i=1}^{n} \left[1 + i(\frac{2}{n}) \right] (\frac{2}{n})$$

(e)

n	5	10	50	100
s(n)	3.6	3.8	3.96	3.98
S(n)	4.4	4.2	4.04	4.02

(f)

$$\lim_{n \to \infty} \sum_{i=1}^{n} \left[1 + (i-1)(\frac{2}{n}) \right] (\frac{2}{n}) = \lim_{n \to \infty} (\frac{2}{n}) \left[n + \frac{2}{n} (\frac{n(n+1)}{2} - n) \right]$$

$$= \lim_{n \to \infty} \left[2 + \frac{2n+2}{n} - \frac{4}{n} \right] = \lim_{n \to \infty} \left[4 - \frac{2}{n} \right] = 4$$

$$\lim_{n \to \infty} \sum_{i=1}^{n} \left[1 + i(\frac{2}{n}) \right] (\frac{2}{n}) = \lim_{n \to \infty} \frac{2}{n} \left[n + (\frac{2}{n})\frac{n(n+1)}{2} \right]$$

$$= \lim_{n \to \infty} \left[2 + \frac{2(n+1)}{n} \right] = \lim_{n \to \infty} \left[4 + \frac{2}{n} \right] = 4$$

9. $y = -2x + 3$ on $[0, 1]$ (note: $\Delta x = \dfrac{1 - 0}{n} = \dfrac{1}{n}$)

$$S(n) = \sum_{i=1}^{n} f(\tfrac{i}{n})(\tfrac{1}{n}) = \sum_{i=1}^{n} \left[-2(\tfrac{i}{n}) + 3 \right] (\tfrac{1}{n})$$

$$= 3 - \frac{2}{n^2} \sum_{i=1}^{n} i = 3 - \frac{2(n + 1)n}{2n^2} = 2 - \frac{1}{n}$$

$$\text{Area} = \lim_{n \to \infty} S(n) = 2$$

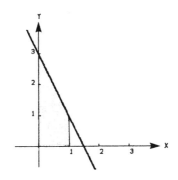

10. $y = 1 - x^2$ on $[-1, 1]$
First find the area of the region over the interval $[0, 1]$:

$$S(n) = \sum_{i=1}^{n} f(\tfrac{i}{n})(\tfrac{1}{n}) \quad \text{(note: } \Delta x = \tfrac{1}{n})$$

$$= \sum_{i=1}^{n} \left[1 - (\tfrac{i}{n})^2 \right] \frac{1}{n} = 1 - \frac{1}{n^3} \sum_{i=1}^{n} i^2 = 1 - \frac{n(n + 1)(2n + 1)}{6n^3}$$

$$= 1 - \frac{1}{6} (2 + \frac{3}{n} + \frac{1}{n^2})$$

$$\frac{1}{2} \text{ Area} = \lim_{n \to \infty} S(n) = 1 - \frac{1}{3} = \frac{2}{3} \quad \text{thus Area} = \frac{4}{3}$$

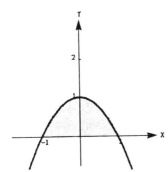

11. $y = x^2 + 2$ on $[0, 1]$ (note: $\Delta x = \dfrac{1}{n}$)

$$S(n) = \sum_{i=1}^{n} f(\frac{i}{n})(\frac{1}{n}) = \sum_{i=1}^{n} \left[(\frac{i}{n})^2 + 2 \right] (\frac{1}{n}) = \left[\frac{1}{n^3} \sum_{i=1}^{n} i^2 \right] + 2$$

$$= \frac{n(n + 1)(2n + 1)}{6n^3} + 2 = \frac{1}{6}(2 + \frac{3}{n} + \frac{1}{n^2}) + 2$$

$$\text{Area} = \lim_{n \to \infty} S(n) = \frac{7}{3}$$

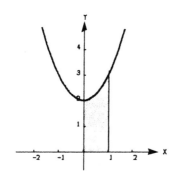

12. $y = 3x - 4$ on $[2, 5]$ (note: $\Delta x = \dfrac{5 - 2}{n} = \dfrac{3}{n}$)

$$S(n) = \sum_{i=1}^{n} f(2 + \frac{3i}{n})(\frac{3}{n}) = \sum_{i=1}^{n} \left[3(2 + \frac{3i}{n}) - 4 \right](\frac{3}{n})$$

$$= 18 + 3(\frac{3}{n})^2 \sum_{i=1}^{n} i - 12 = 6 + \frac{27}{n^2}(\frac{(n + 1)n}{2}) = 6 + \frac{27}{2}(1 + \frac{1}{n})$$

$$\text{Area} = \lim_{n \to \infty} S(n) = 6 + \frac{27}{2} = \frac{39}{2}$$

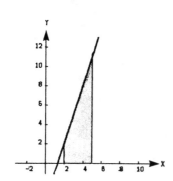

13. $y = 2x^2$ on $[1, 3]$ (note: $\Delta x = \dfrac{3-1}{n} = \dfrac{2}{n}$)

$$S(n) = \sum_{i=1}^{n} f\left(1 + \frac{2i}{n}\right)\left(\frac{2}{n}\right) = \sum_{i=1}^{n} \left[2\left(1 + \frac{2i}{n}\right)^2\right]\left(\frac{2}{n}\right) = \frac{4}{n} \sum_{i=1}^{n} \left(1 + \frac{4i}{n} + \frac{4i^2}{n^2}\right)$$

$$= \frac{4}{n}\left[n + \frac{4}{n}\left(\frac{n(n+1)}{2}\right) + \left(\frac{4}{n^2}\right)\frac{n(n+1)(2n+1)}{6}\right] = \frac{52}{3} + \frac{16}{n} + \frac{8}{3n^2}$$

$$\text{Area} = \lim_{n \to \infty} S(n) = \frac{52}{3}$$

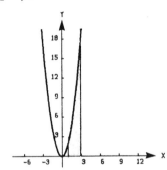

14. $y = 2x - x^3$ on $[0, 1]$ (note: $\Delta x = \dfrac{1-0}{n} = \dfrac{1}{n}$)

Since y both increases and decreases on $[0, 1]$, $T(n)$ is neither an upper nor a lower sum.

$$T(n) = \sum_{i=1}^{n} f\left(\frac{i}{n}\right)\left(\frac{1}{n}\right) = \sum_{i=1}^{n}\left[2\left(\frac{i}{n}\right) - \left(\frac{i}{n}\right)^3\right]\left(\frac{1}{n}\right)$$

$$= \frac{2}{n^2}\sum_{i=1}^{n} i - \frac{1}{n^4}\sum_{i=1}^{n} i^3$$

$$= \frac{n(n+1)}{n^2} - \frac{1}{n^4}\left[\frac{n(n+1)}{2}\right]^2$$

$$= 1 - \frac{1}{n} - \frac{1}{4} - \frac{2}{4n} - \frac{1}{4n^2}$$

$$\text{Area} = \lim_{n \to \infty} T(n) = 1 - \frac{1}{4} = \frac{3}{4}$$

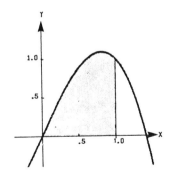

15. $y = 1 - x^3$ on $[0, 1]$ (note: $\Delta x = \dfrac{1 - 0}{n} = \dfrac{1}{n}$)

$$S(n) = \sum_{i=1}^{n} f(\tfrac{i}{n})(\tfrac{1}{n}) = \sum_{i=1}^{n}\left[1 - (\tfrac{i}{n})^3\right](\tfrac{1}{n}) = 1 - \frac{1}{n^4}\sum_{i=1}^{n} i^3$$

$$= 1 - \frac{1}{n^4}\left[\frac{n(n+1)}{2}\right]^2 = 1 - \frac{1}{4} - \frac{1}{2n} - \frac{1}{4n^2}$$

$$\text{Area} = \lim_{n \to \infty} S(n) = 1 - \frac{1}{4} = \frac{3}{4}$$

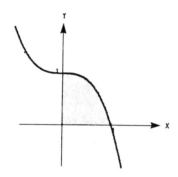

16. $y = x^2 - x^3$ on $[0, 1]$ (note: $\Delta x = \dfrac{1 - 0}{n} = \dfrac{1}{n}$)

Again, $T(n)$ is neither an upper nor lower sum.

$$T(n) = \sum_{i=1}^{n} f(\tfrac{i}{n})(\tfrac{1}{n}) = \sum_{i=1}^{n}\left[(\tfrac{i}{n})^2 - (\tfrac{i}{n})^3\right](\tfrac{1}{n}) = \frac{1}{n^3}\sum_{i=1}^{n} i^2 - \frac{1}{n^4}\sum_{i=1}^{n} i^3$$

$$= \frac{1}{n^3}\left[\frac{n(n+1)(2n+1)}{6}\right] - \frac{1}{n^4}\left[\frac{n(n+1)}{2}\right]^2$$

$$= \frac{1}{6}(2 + \frac{3}{n} + \frac{1}{n^2}) - \frac{1}{4}(1 + \frac{2}{n} + \frac{1}{n^2})$$

$$\text{Area} = \lim_{n \to \infty} T(n) = \frac{1}{3} - \frac{1}{4} = \frac{1}{12}$$

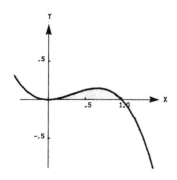

17. $y = x^2 - x^3$ on $[-1, 0]$ (note: $\Delta x = \dfrac{0 - (-1)}{n} = \dfrac{1}{n}$)

$$S(n) = \sum_{i=1}^{n} f(-1 + \frac{i}{n})(\frac{1}{n})$$

$$= \sum_{i=1}^{n} \left[(-1 + \frac{i}{n})^2 - (-1 + \frac{i}{n})^3 \right] (\frac{1}{n})$$

$$= \sum_{i=1}^{n} \left[2 - \frac{5i}{n} + \frac{4i^2}{n^2} - \frac{i^3}{n^3} \right] (\frac{1}{n})$$

$$= 2 - \frac{5}{n^2} \sum_{i=1}^{n} i + \frac{4}{n^3} \sum_{i=1}^{n} i^2 - \frac{1}{n^4} \sum_{i=1}^{n} i^3$$

$$= 2 - \frac{5}{2} - \frac{5}{2n} + \frac{4}{3} + \frac{2}{n} + \frac{2}{3n^3} - \frac{1}{4} - \frac{1}{2n} - \frac{1}{4n^2}$$

$$\text{Area} = \lim_{n \to \infty} S(n) = 2 - \frac{5}{2} + \frac{4}{3} - \frac{1}{4} = \frac{7}{12}$$

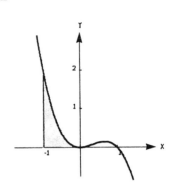

18. $y = 2x^2 - x + 1$ on $[0, 2]$ (note: $\Delta x = \dfrac{2 - 0}{n} = \dfrac{2}{n}$)

$$S(n) = \sum_{i=1}^{n} f(\frac{2i}{n})(\frac{2}{n}) = \sum_{i=1}^{n} \left[2(\frac{2i}{n})^2 - (\frac{2i}{n}) + 1 \right] (\frac{2}{n})$$

$$= \frac{16}{n^3} \sum_{i=1}^{n} i^2 - \frac{4}{n^2} \sum_{i=1}^{n} i + \frac{2}{n} \sum_{i=1}^{n} 1$$

$$= \frac{16}{n^3} \left[\frac{n(n + 1)(2n + 1)}{6} \right] - \frac{4}{n^2} \left[\frac{n(n + 1)}{2} \right] + 2$$

$$= \frac{8}{3}(2 + \frac{3}{n} + \frac{1}{n^2}) - 2(1 + \frac{1}{n}) + 2$$

$$\text{Area} = \lim_{n \to \infty} S(n) = \frac{16}{3} - 2 + 2 = \frac{16}{3}$$

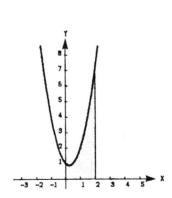

19. $f(y) = 3y$ on $[0, 2]$ (note: $\Delta y = \dfrac{2 - 0}{n} = \dfrac{2}{n}$)

$$s(n) = \sum_{i=1}^{n} f(m_i)\Delta y = \sum_{i=1}^{n} f\left(\frac{2i}{n}\right)\left(\frac{2}{n}\right)$$

$$= \sum_{i=1}^{n} 3\left(\frac{2i}{n}\right)\left(\frac{2}{n}\right) = \frac{12}{n^2}\sum_{i=1}^{n} i$$

$$= \left(\frac{12}{n^2}\right)\frac{n(n + 1)}{2} = \frac{6(n + 1)}{n} = 6 + \frac{6}{n}$$

$$\text{Area} = \lim_{n \to \infty} s(n) = \lim_{n \to \infty}\left(6 + \frac{6}{n}\right) = 6$$

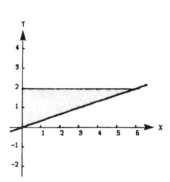

20. $f(y) = y^2$ on $[0, 3]$ (note: $\Delta y = \dfrac{3 - 0}{n} = \dfrac{3}{n}$)

$$S(n) = \sum_{i=1}^{n} f\left(\frac{3i}{n}\right)\left(\frac{3}{n}\right) = \sum_{i=1}^{n}\left(\frac{3i}{n}\right)^2\left(\frac{3}{n}\right)$$

$$= \frac{27}{n^3}\sum_{i=1}^{n} i^2 = \frac{27}{n^3}\cdot\frac{n(n + 1)(2n + 1)}{6}$$

$$= \frac{9}{n^2}\left(\frac{2n^2 + 3n + 1}{2}\right) = 9 + \frac{27}{2n} + \frac{9}{2n^2}$$

$$\text{Area} = \lim_{n \to \infty} S(n) = \lim_{n \to \infty}\left(9 + \frac{27}{2n} + \frac{9}{2n^2}\right) = 9$$

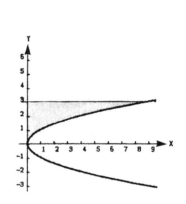

21. $f(x) = x^2 + 3$, $0 \le x \le 2$, $n = 4$

$$\Delta x = \frac{1}{2}, \qquad c_1 = \frac{1}{4}, \qquad c_2 = \frac{3}{4}, \qquad c_3 = \frac{5}{4}, \qquad c_4 = \frac{7}{4}$$

$$\text{Area} \approx \sum_{i=1}^{n} f(c_i)\Delta x = \sum_{i=1}^{4} [c_i^2 + 3]\left(\frac{1}{2}\right)$$

$$= \frac{1}{2}\left[\left(\frac{1}{16} + 3\right) + \left(\frac{9}{16} + 3\right) + \left(\frac{25}{16} + 3\right) + \left(\frac{49}{16} + 3\right)\right] = \frac{69}{8}$$

22. $f(x) = x^2 + 4x$, $\quad 0 \leq x \leq 4$, $\quad n = 4$

$\Delta x = 1$, $\quad c_1 = \frac{1}{2}$, $\quad c_2 = \frac{3}{2}$, $\quad c_3 = \frac{5}{2}$, $\quad c_4 = \frac{7}{2}$

Area $\approx \sum\limits_{i=1}^{n} f(c_i)\Delta x = \sum\limits_{i=1}^{4} [c_i^2 + 4c_i](1)$

$\quad = \left[(\frac{1}{4} + 2) + (\frac{9}{4} + 6) + (\frac{25}{4} + 10) + (\frac{49}{4} + 14) \right] = 53$

23. $f(x) = \tan x$, $\quad 0 \leq x \leq \frac{\pi}{4}$, $\quad n = 4$

$\Delta x = \frac{\pi}{16}$, $\quad c_1 = \frac{\pi}{32}$, $\quad c_2 = \frac{3\pi}{32}$, $\quad c_3 = \frac{5\pi}{32}$, $\quad c_4 = \frac{7\pi}{32}$

Area $\approx \sum\limits_{i=1}^{n} f(c_i)\Delta x = \sum\limits_{i=1}^{4} (\tan c_i)(\frac{\pi}{16})$

$\quad = \frac{\pi}{16} \left[\tan \frac{\pi}{32} + \tan \frac{3\pi}{32} + \tan \frac{5\pi}{32} + \tan \frac{7\pi}{32} \right] \approx 0.345$

24. $f(x) = \sec x$, $\quad 0 \leq x \leq \frac{\pi}{4}$, $\quad n = 4$

$\Delta x = \frac{\pi}{16}$, $\quad c_1 = \frac{\pi}{32}$, $\quad c_2 = \frac{3\pi}{32}$, $\quad c_3 = \frac{5\pi}{32}$, $\quad c_4 = \frac{7\pi}{32}$

Area $\approx \sum\limits_{i=1}^{n} f(c_i)\Delta x = \sum\limits_{i=1}^{4} (\sec c_i)(\frac{\pi}{16})$

$\quad = \frac{\pi}{16} \left[\sec \frac{\pi}{32} + \sec \frac{3\pi}{32} + \sec \frac{5\pi}{32} + \sec \frac{7\pi}{32} \right] \approx 0.879$

5.5
Riemann sums and the definite integral

1. $\int_{0}^{5} 3 \, dx$

2. $\int_{0}^{2} (4 - 2x) \, dx$

3. $\int_{-4}^{4} (4 - |x|) \, dx$

4. $\int_{0}^{2} x^2 \, dx$

5. $\int_{-2}^{2} (4 - x^2) \, dx$

6. $\int_{0}^{2} (y - 2)^2 \, dy$

7. $\displaystyle\int_0^2 y^3\ dy$

8. $\displaystyle\int_0^{\pi/4} \tan x\ dx$

9. $\displaystyle\int_0^{\pi} \sin x\ dx$

10. $\displaystyle\int_{-1}^{1} \frac{1}{x^2 + 1}\ dx$

11. Rectangle, $A = bh = 3(4)$

$A = \displaystyle\int_0^3 4\ dx = 12$

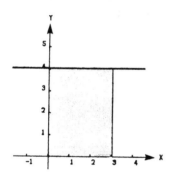

12. Rectangle, $A = bh = 2(4)(a)$

$A = \displaystyle\int_{-a}^{a} 4\ dx = 8a$

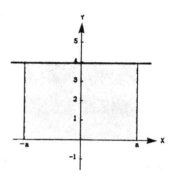

13. Triangle, $A = \dfrac{1}{2}bh = \dfrac{1}{2}(4)(4)$

$A = \displaystyle\int_0^4 x\ dx = 8$

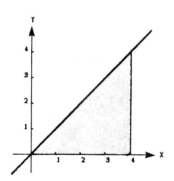

14. Triangle, $A = \dfrac{1}{2}bh = \dfrac{1}{2}(4)(2)$

$A = \displaystyle\int_0^4 \frac{x}{2}\ dx = 4$

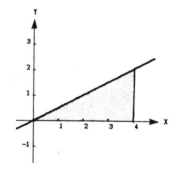

15. Trapezoid, $A = \dfrac{b_1 + b_2}{2}h = 2\left(\dfrac{5+9}{2}\right)$ 16. Triangle, $A = \dfrac{1}{2}bh = \dfrac{1}{2}(5)(5)$

$A = \displaystyle\int_0^2 (2x + 5)\,dx = 14$ $A = \displaystyle\int_0^5 (5 - x)\,dx = \dfrac{25}{2}$

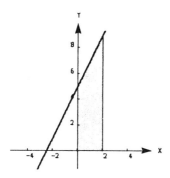   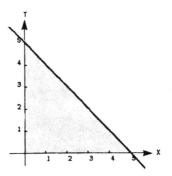

17. Triangle, $A = \dfrac{1}{2}bh = \dfrac{1}{2}(2)(1)$ 18. Triangle, $A = \dfrac{1}{2}bh = \dfrac{1}{2}(2a)a$

$A = \displaystyle\int_{-1}^1 (1 - |x|)\,dx = 1$ $A = \displaystyle\int_{-a}^a (a - |x|)\,dx = a^2$

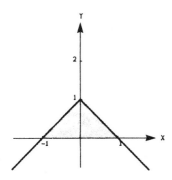   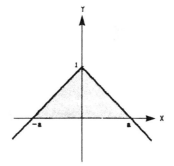

19. Semi-circle, $A = \dfrac{1}{2}\pi r^2 = \dfrac{1}{2}\pi(3)^2$ 20. Semi-circle, $A = \dfrac{1}{2}\pi r^2$

$A = \displaystyle\int_{-3}^3 \sqrt{9 - x^2}\,dx = \dfrac{9\pi}{2}$ $A = \displaystyle\int_{-r}^r \sqrt{r^2 - x^2}\,dx = \dfrac{1}{2}\pi r^2$

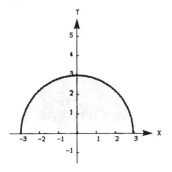   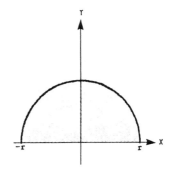

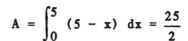

21.(a) $\displaystyle\int_0^7 f(x)\ dx = \int_0^5 f(x)\ dx + \int_5^7 f(x)\ dx = 10 + 3 = 13$

(b) $\displaystyle\int_5^0 f(x)\ dx = -\int_0^5 f(x)\ dx = -10$

(c) $\displaystyle\int_5^5 f(x)\ dx = 0$

(d) $\displaystyle\int_0^5 3f(x)\ dx = 3\int_0^5 f(x)\ dx = 3(10) = 30$

22.(a) $\displaystyle\int_0^6 f(x)\ dx = \int_0^3 f(x)\ dx + \int_3^6 f(x)\ dx = 4 + (-1) = 3$

(b) $\displaystyle\int_6^3 f(x)\ dx = -\int_3^6 f(x)\ dx = -(-1) = 1$

(c) $\displaystyle\int_4^4 f(x)\ dx = 0$

(d) $\displaystyle\int_3^6 -5f(x)\ dx = -5\int_3^6 f(x)\ dx = -5(-1) = 5$

23.(a) $\displaystyle\int_2^6 [f(x) + g(x)]\ dx = \int_2^6 f(x)\ dx + \int_2^6 g(x)\ dx = 10 + (-2) = 8$

(b) $\displaystyle\int_2^6 [g(x) - f(x)]\ dx = \int_2^6 g(x)\ dx - \int_2^6 f(x)\ dx = -2 - 10 = -12$

(c) $\displaystyle\int_2^6 [2f(x) - 3g(x)]\ dx = 2\int_2^6 f(x)\ dx - 3\int_2^6 g(x)\ dx = 2(10) - 3(-2) = 26$

(d) $\displaystyle\int_2^6 3f(x)\ dx = 3\int_2^6 f(x)\ dx = 3(10) = 30$

24.(a) $\displaystyle\int_{-1}^0 f(x)\ dx = \int_{-1}^1 f(x)\ dx - \int_0^1 f(x)\ dx = 0 - 5 = -5$

(b) $\displaystyle\int_0^1 f(x)\ dx - \int_{-1}^0 f(x)\ dx = 5 - (-5) = 10$

(c) $\displaystyle\int_{-1}^1 3f(x)\ dx = 3\int_{-1}^1 f(x)\ dx = 3(0) = 0$

(d) $\displaystyle\int_0^1 3f(x)\ dx = 3\int_0^1 f(x)\ dx = 3(5) = 15$

25. $y = 6$ on $[4, 10]$ (note: $\Delta x = \dfrac{10 - 4}{n} = \dfrac{6}{n}$, $\Delta x \to 0$ as $n \to \infty$)

$$S(n) = \sum_{i=1}^{n} f\left(4 + \frac{6i}{n}\right)\left(\frac{6}{n}\right) = \sum_{i=1}^{n} 6\left(\frac{6}{n}\right) = \sum_{i=1}^{n} \frac{36}{n} = 36$$

$$\int_{4}^{10} 6 \, dx = \lim_{\Delta x \to 0} S(n) = \lim_{n \to \infty} 36 = 36$$

26. $y = x$ on $[-2, 3]$ (note: $\Delta x = \dfrac{3 - (-2)}{n} = \dfrac{5}{n}$, $\Delta x \to 0$ as $n \to \infty$)

$$S(n) = \sum_{i=1}^{n} f\left(-2 + \frac{5i}{n}\right)\left(\frac{5}{n}\right) = \sum_{i=1}^{n} \left(-2 + \frac{5i}{n}\right)\left(\frac{5}{n}\right) = -10 + \frac{25}{n^2} \sum_{i=1}^{n} i$$

$$= -10 + \left(\frac{25}{n^2}\right)\frac{n(n + 1)}{2} = -10 + \frac{25}{2}\left(1 + \frac{1}{n}\right) = \frac{5}{2} + \frac{25}{2n}$$

$$\int_{-2}^{3} x \, dx = \lim_{\Delta x \to 0} S(n) = \lim_{n \to \infty} \left(\frac{5}{2} + \frac{25}{2n}\right) = \frac{5}{2}$$

27. $y = x^3$ on $[-1, 1]$ (note: $\Delta x = \dfrac{1 - (-1)}{n} = \dfrac{2}{n}$, $\Delta x \to 0$ as $n \to \infty$)

$$S(n) = \sum_{i=1}^{n} f\left(-1 + \frac{2i}{n}\right)\left(\frac{2}{n}\right) = \sum_{i=1}^{n} \left(-1 + \frac{2i}{n}\right)^3\left(\frac{2}{n}\right)$$

$$= \sum_{i=1}^{n} \left[-1 + \frac{6i}{n} - \frac{12i^2}{n^2} + \frac{8i^3}{n^3}\right]\left(\frac{2}{n}\right)$$

$$= -2 + \frac{12}{n^2} \sum_{i=1}^{n} i - \frac{24}{n^3} \sum_{i=1}^{n} i^2 + \frac{16}{n^4} \sum_{i=1}^{n} i^3$$

$$= -2 + 6\left(1 + \frac{1}{n}\right) - 4\left(2 + \frac{3}{n} + \frac{1}{n^2}\right) + 4\left(1 + \frac{2}{n} + \frac{1}{n^2}\right) = \frac{2}{n}$$

$$\int_{-1}^{1} x^3 \, dx = \lim_{\Delta x \to 0} S(n) = \lim_{n \to \infty} \frac{2}{n} = 0$$

28. $y = x^3$ on $[0, 1]$ (note: $\Delta x = \dfrac{1 - 0}{n} = \dfrac{1}{n}$, $\Delta x \to 0$ as $n \to \infty$)

$$S(n) = \sum_{i=1}^{n} f\left(\frac{i}{n}\right)\left(\frac{1}{n}\right) = \sum_{i=1}^{n} \left(\frac{i}{n}\right)^3\left(\frac{1}{n}\right) = \frac{1}{n^4} \sum_{i=1}^{n} i^3$$

$$= \frac{1}{n^4}\left[\frac{n(n + 1)}{2}\right]^2 = \frac{1}{4}\left(1 + \frac{2}{n} + \frac{1}{n^2}\right)$$

$$\int_{0}^{1} x^3 \, dx = \lim_{\Delta x \to 0} S(n) = \frac{1}{4} \lim_{n \to \infty} \left(1 + \frac{2}{n} + \frac{1}{n^2}\right) = \frac{1}{4}$$

29. $y = x^2 + 1$ on $[1, 2]$ (note: $\Delta x = \dfrac{2 - 1}{n} = \dfrac{1}{n}$, $\Delta x \to 0$ as $n \to \infty$)

$$S(n) = \sum_{i=1}^{n} f(1 + \frac{i}{n})(\frac{1}{n}) = \sum_{i=1}^{n}\left[(1 + \frac{i}{n})^2 + 1\right](\frac{1}{n}) = \sum_{i=1}^{n}\left[1 + \frac{2i}{n} + \frac{i^2}{n^2} + 1\right](\frac{1}{n})$$

$$= 2 + \frac{2}{n^2}\sum_{i=1}^{n} i + \frac{1}{n^3}\sum_{i=1}^{n} i^2 = 2 + (1 + \frac{1}{n}) + \frac{1}{6}(2 + \frac{3}{n} + \frac{1}{n^2})$$

$$= \frac{10}{3} + \frac{3}{2n} + \frac{1}{6n^2}$$

$$\int_{1}^{2} (x^2 + 1)\, dx = \lim_{\Delta x \to 0} S(n) = \lim_{n \to \infty} (\frac{10}{3} + \frac{3}{2n} + \frac{1}{6n^2}) = \frac{10}{3}$$

30. $y = 4x^2$ on $[1, 2]$ (note: $\Delta x = \dfrac{2 - 1}{n} = \dfrac{1}{n}$, $\Delta x \to 0$ as $n \to \infty$)

$$S(n) = \sum_{i=1}^{n} f(1 + \frac{i}{n})(\frac{1}{n}) = \sum_{i=1}^{n}\left[4(1 + \frac{i}{n})^2\right](\frac{1}{n}) = 4\sum_{i=1}^{n}\left[1 + \frac{2i}{n} + \frac{i^2}{n^2}\right](\frac{1}{n})$$

$$= 4\left[1 + \frac{1}{n^2}\sum_{i=1}^{n} i + \frac{1}{n^3}\sum_{i=1}^{n} i^2\right]$$

$$= 4\left[1 + (\frac{2}{n^2})\frac{n(n + 1)}{2} + (\frac{1}{n^3})\frac{n(n + 1)(2n + 1)}{6}\right]$$

$$= 4\left[1 + (1 + \frac{1}{n}) + \frac{1}{6}(2 + \frac{3}{n} + \frac{1}{n^2})\right] = 4(\frac{7}{3} + \frac{3}{2n} + \frac{1}{6n^2})$$

$$\int_{1}^{2} 4x^2\, dx = \lim_{\Delta x \to 0} S(n) = 4\lim_{n \to \infty}(\frac{7}{3} + \frac{3}{2n} + \frac{1}{6n^2}) = \frac{28}{3}$$

31. $f(x) = \sqrt{x}$, $y = 0$, $x = 0$, $x = 2$, $c_i = \dfrac{2i^2}{n^2}$

$$\Delta x_i = \frac{2i^2}{n^2} - \frac{2(i - 1)^2}{n^2} = \frac{2(2i - 1)}{n^2}$$

$$\sum_{i=1}^{n} f(c_i)\Delta x_i = \sum_{i=1}^{n} \sqrt{\frac{2i^2}{n^2}}\left[\frac{2(2i - 1)}{n^2}\right] = \frac{2\sqrt{2}}{n^3}\sum_{i=1}^{n}(2i^2 - i)$$

$$= \frac{2\sqrt{2}}{n^3}\left[2(\frac{n(n + 1)(2n + 1)}{6}) - \frac{n(n + 1)}{2}\right]$$

$$= \frac{2\sqrt{2}}{n^3}\left[\frac{4n^3 + 3n^2 - n}{6}\right] = \sqrt{2}\left[\frac{4}{3} + \frac{1}{n} - \frac{2}{n^2}\right]$$

$$\lim_{n \to \infty}\sum_{i=1}^{n} f(c_i)\Delta x_i = \lim_{n \to \infty}\sqrt{2}\left[\frac{4}{3} + \frac{1}{n} - \frac{2}{n^2}\right] = \frac{4\sqrt{2}}{3}$$

32. $f(x) = \sqrt[3]{x}$, $\quad y = 0$, $\quad x = 0$, $\quad x = 1$, $\quad c_i = \dfrac{i^3}{n^3}$

$\Delta x_i = \dfrac{i^3}{n^3} - \dfrac{(i-1)^3}{n^3} = \dfrac{3i^2 - 3i + 1}{n^3}$

$\displaystyle\sum_{i=1}^{n} f(c_i)\Delta x_i = \sum_{i=1}^{n} \sqrt[3]{\dfrac{i^3}{n^3}} \left[\dfrac{3i^2 - 3i + 1}{n^3}\right] = \dfrac{1}{n^4} \sum_{i=1}^{n} (3i^3 - 3i^2 + i)$

$\qquad = \dfrac{1}{n^4}\left[3(\dfrac{n^2(n+1)^2}{4}) - 3(\dfrac{n(n+1)(2n+1)}{6}) + \dfrac{n(n+1)}{2}\right]$

$\qquad = \dfrac{1}{n^4}\left[\dfrac{3n^4 + 6n^3 + 3n^2}{4} - \dfrac{2n^3 + 3n^2 + n}{2} + \dfrac{n^2 + n}{2}\right]$

$\qquad = \dfrac{1}{n^4}\left[\dfrac{3n^4}{4} + \dfrac{n^3}{2} - \dfrac{n^2}{4}\right] = \left[\dfrac{3}{4} + \dfrac{1}{2n} - \dfrac{1}{4n^2}\right]$

$\displaystyle\lim_{n \to \infty} \sum_{i=1}^{n} f(c_i)\Delta x_i = \lim_{n \to \infty}\left[\dfrac{3}{4} + \dfrac{1}{2n} - \dfrac{1}{4n^2}\right] = \dfrac{3}{4}$

5.6
The Fundamental Theorem of Calculus

1. $\displaystyle\int_{0}^{1} 2x \, dx = x^2 \Big]_{0}^{1} = 1 - 0 = 1$

2. $\displaystyle\int_{2}^{7} 3 \, dv = 3v \Big]_{2}^{7} = 3(7) - 3(2) = 15$

3. $\displaystyle\int_{-1}^{0} (x - 2) \, dx = \left[\dfrac{x^2}{2} - 2x\right]_{-1}^{0} = 0 - (\dfrac{1}{2} + 2) = -\dfrac{5}{2}$

4. $\displaystyle\int_{2}^{5} (-3v + 4) \, dv = \left[-\dfrac{3}{2}v^2 + 4v\right]_{2}^{5} = (-\dfrac{75}{2} + 20) - (-6 + 8) = -\dfrac{39}{2}$

5. $\displaystyle\int_{-1}^{1} (t^2 - 2) \, dt = \left[\dfrac{t^3}{3} - 2t\right]_{-1}^{1} = (\dfrac{1}{3} - 2) - (-\dfrac{1}{3} + 2) = -\dfrac{10}{3}$

6. $\displaystyle\int_{0}^{3} (3x^2 + x - 2) \, dx = \left[x^3 + \dfrac{x^2}{2} - 2x\right]_{0}^{3} = (27 + \dfrac{9}{2} - 6) - 0 = \dfrac{51}{2}$

7. Let $u = 2t - 1$, $\quad u' = 2$

$\displaystyle\int_{0}^{1} (2t - 1)^2 \, dt = \dfrac{1}{2}\int_{0}^{1} (2t - 1)^2(2) \, dt = \dfrac{1}{6}(2t - 1)^3 \Big]_{-1}^{1}$

$\qquad = \dfrac{1}{6}[1 - (-1)] = \dfrac{1}{3}$

8. $\displaystyle\int_{-1}^{1} (t^3 - 9t)\ dt = \left[\frac{1}{4}t^4 - \frac{9}{2}t^2\right]_{-1}^{1} = (\frac{1}{4} - \frac{9}{2}) - (\frac{1}{4} - \frac{9}{2}) = 0$

9. $\displaystyle\int_{1}^{2} (\frac{3}{x^2} - 1)\ dx = \left[-\frac{3}{x} - x\right]_{1}^{2} = (-\frac{3}{2} - 2) - (-3 - 1) = \frac{1}{2}$

10. $\displaystyle\int_{0}^{1} (3x^3 - 9x + 7)\ dx = \left[\frac{3x^4}{4} - \frac{9}{2}x^2 + 7x\right]_{0}^{1} = (\frac{3}{4} - \frac{9}{2} + 7) - 0 = \frac{13}{4}$

11. $\displaystyle\int_{1}^{2} (5x^4 + 5)\ dx = \left[x^5 + 5x\right]_{1}^{2} = (32 + 10) - (1 + 5) = 36$

12. $\displaystyle\int_{-3}^{3} v^{1/3}\ dv = \frac{3}{4}v^{4/3}\ \Big]_{-3}^{3} = \frac{3}{4}[(\sqrt[3]{3})^4 - (\sqrt[3]{-3})^4] = 0$

13. $\displaystyle\int_{-1}^{1} (\sqrt[3]{t} - 2)\ dt = \left[\frac{3}{4}t^{4/3} - 2t\right]_{-1}^{1} = (\frac{3}{4} - 2) - (\frac{3}{4} + 2) = -4$

14. $\displaystyle\int_{-2}^{-1} \sqrt{-\frac{2}{x}}\ dx = -\sqrt{2}\ \int_{-2}^{-1} (-x)^{-1/2}(-1)\ dx = -\sqrt{2}(2)\sqrt{-x}\ \Big]_{-2}^{-1}$

$= -2\sqrt{2}(1 - \sqrt{2}) = 4 - 2\sqrt{2}$

15. $\displaystyle\int_{1}^{4} \frac{u - 2}{\sqrt{2}}\ du = \int_{1}^{4} (u^{1/2} - 2u^{-1/2})\ du = \left[\frac{2}{3}u^{3/2} - 4u^{1/2}\right]_{1}^{4}$

$= \left[\frac{2}{3}(\sqrt{4})^3 - 4\sqrt{4}\right] - \left[\frac{2}{3} - 4\right] = \frac{2}{3}$

16. $\displaystyle\int_{-2}^{-1} (u - \frac{1}{u^2})\ du = \left[\frac{u^2}{2} + \frac{1}{u}\right]_{-2}^{-1} = (\frac{1}{2} - 1) - (2 - \frac{1}{2}) = -2$

17. $\displaystyle\int_{0}^{1} \frac{x - \sqrt{x}}{3}\ dx = \frac{1}{3}\int_{0}^{1} (x - x^{1/2})\ dx$

$= \frac{1}{3}(\frac{x^2}{2} - \frac{2}{3}x^{3/2})\ \Big]_{0}^{1} = \frac{1}{3}(\frac{1}{2} - \frac{2}{3}) = -\frac{1}{18}$

18. $\displaystyle\int_0^2 (2 - t)\sqrt{t}\ dt = \int_0^2 (2t^{1/2} - t^{3/2})\ dt$

$\displaystyle = \left[\frac{4}{3} t^{3/2} - \frac{2}{5} t^{5/2}\right]_0^2 = \left[\frac{t\sqrt{t}}{15}(20 - 6t)\right]_0^2$

$\displaystyle = \frac{2\sqrt{2}}{15}(20 - 12) = \frac{16\sqrt{2}}{15}$

19. $\displaystyle\int_{-1}^0 (t^{1/3} - t^{2/3})\ dt = \left[\frac{3}{4} t^{4/3} - \frac{3}{5} t^{5/3}\right]_{-1}^0 = 0 - (\frac{3}{4} + \frac{3}{5}) = -\frac{27}{20}$

20. $\displaystyle\int_{-8}^{-1} \frac{x - x^2}{2\sqrt[3]{x}}\ dx = \frac{1}{2}\int_{-8}^{-1} (x^{2/3} - x^{5/3})\ dx$

$\displaystyle = \frac{1}{2}(\frac{3}{5} x^{5/3} - \frac{3}{8} x^{8/3})\Big]_{-8}^{-1} = \left[\frac{x^{5/3}}{80}(24 - 15x)\right]_{-8}^{-1}$

$\displaystyle = -\frac{1}{80}(39) + \frac{32}{80}(144) = \frac{4569}{80}$

21. Let $u = 2x + 1$, $\quad u' = 2$

$\displaystyle\int_0^4 \frac{1}{\sqrt{2x + 1}}\ dx = \frac{1}{2}\int_0^4 (2x + 1)^{-1/2}(2)\ dx = \sqrt{2x + 1}\,\Big]_0^4 = \sqrt{9} - \sqrt{1} = 2$

22. Let $u = 1 - x^2$, $\quad u' = -2x$

$\displaystyle\int_0^1 x\sqrt{1 - x^2}\ dx = -\frac{1}{2}\int_0^1 (1 - x^2)^{1/2}(-2x)\ dx$

$\displaystyle = -\frac{1}{3}(1 - x^2)^{3/2}\Big]_0^1 = 0 + \frac{1}{3} = \frac{1}{3}$

23. Let $u = x^2 + 1$, $\quad u' = 2x$

$\displaystyle\int_{-1}^1 x(x^2 + 1)^3\ dx = \frac{1}{2}\int_{-1}^1 (x^2 + 1)^3(2x)\ dx = \frac{1}{8}(x^2 + 1)^4\Big]_{-1}^1 = 0$

24. Let $u = 1 + 2x^2$, $\quad u' = 4x$

$\displaystyle\int_0^2 \frac{x}{\sqrt{1 + 2x^2}}\ dx = \frac{1}{4}\int_0^2 (1 + 2x^2)^{-1/2}(4x)\ dx$

$\displaystyle = \frac{1}{2}\sqrt{1 + 2x^2}\,\Big]_0^2 = \frac{3}{2} - \frac{1}{2} = 1$

25. Let $u = 4 + x^2$, $u' = 2x$

$$\int_0^2 x\sqrt[3]{4 + x^2}\ dx = \frac{1}{2}\int_0^2 (4 + x^2)^{1/3}(2x)\ dx = \frac{3}{8}(4 + x^2)^{4/3}\bigg]_0^2$$

$$= \frac{3}{8}(8^{4/3} - 4^{4/3}) = 6 - \frac{3}{2}\sqrt[3]{4} \approx 3.619$$

26. Let $u = 1 + \sqrt{x}$, $u' = 1/(2\sqrt{x})$

$$\int_1^9 \frac{1}{\sqrt{x}\ (1 + \sqrt{x})^2}\ dx = 2\int_1^9 (1 + \sqrt{x})^{-2}(\frac{1}{2\sqrt{x}})\ dx = \left[-\frac{2}{1 + \sqrt{x}} \right]_1^9$$

$$= -\frac{1}{2} + 1 = \frac{1}{2}$$

27. $$\int_{-1}^1 |x|\ dx = \int_{-1}^0 (-x)\ dx + \int_0^1 x\ dx = \left[-\frac{x^2}{2} \right]_{-1}^0 + \left[\frac{x^2}{2} \right]_0^1 = \frac{1}{2} + \frac{1}{2} = 1$$

28. $$\int_0^3 |2x - 3|\ dx = \int_0^{3/2} (3 - 2x)\ dx + \int_{3/2}^3 (2x - 3)\ dx$$

$$= \left[3x - x^2 \right]_0^{3/2} + \left[x^2 - 3x \right]_{3/2}^3$$

$$= (\frac{9}{2} - \frac{9}{4}) - 0 + (9 - 9) - (\frac{9}{4} - \frac{9}{2}) = 2(\frac{9}{2} - \frac{9}{4}) = \frac{9}{2}$$

29. $$\int_0^4 |x^2 - 4x + 3|\ dx$$

$$= \int_0^1 (x^2 - 4x + 3)\ dx - \int_1^3 (x^2 - 4x + 3)\ dx + \int_3^4 (x^2 - 4x + 3)\ dx$$

$$= \left[\frac{x^3}{3} - 2x^2 + 3x \right]_0^1 - \left[\frac{x^3}{3} - 2x^2 + 3x \right]_1^3 + \left[\frac{x^3}{3} - 2x^2 + 3x \right]_3^4$$

$$= (\frac{1}{3} - 2 + 3) + (\frac{1}{3} - 2 + 3) + (\frac{64}{3} - 32 + 12) - (9 - 18 + 9)$$

$$= \frac{8}{3} + \frac{64}{3} - 20 = 4$$

30. $$\int_{-1}^1 |x^3|\ dx = -\int_{-1}^0 x^3\ dx + \int_0^1 x^3\ dx = \left[-\frac{x^4}{4} \right]_{-1}^0 + \left[\frac{x^4}{4} \right]_0^1 = \frac{1}{4} + \frac{1}{4} = \frac{1}{2}$$

31. Let $u = \sqrt{2 - x}$, $\quad x = 2 - u^2$, $\quad dx = -2u\,du$

$$\int_1^2 (x - 1)\sqrt{2 - x}\,dx = -2\int_1^0 (1 - u^2)u^2\,du = 2\int_0^1 (u^2 - u^4)\,du$$

$$= 2\left[\frac{u^3}{3} - \frac{u^5}{5}\right]_0^1 = \frac{4}{15}$$

32. Let $u = \sqrt{2x + 1}$, $\quad x = (u^2 - 1)/2$, $\quad dx = u\,du$

$$\int_0^4 \frac{x}{\sqrt{2x + 1}}\,dx = \frac{1}{2}\int_1^3 \frac{(u^2 - 1)}{u}u\,du = \frac{1}{2}\left[\frac{1}{3}u^3 - u\right]_1^3 = \frac{10}{3}$$

33. Let $u = \sqrt{x - 3}$, $\quad x = u^2 + 3$, $\quad dx = 2u\,du$

$$\int_3^7 x\sqrt{x - 3}\,dx = 2\int_0^2 (u^2 + 3)u^2\,du = 2\int_0^2 (u^4 + 3u^2)\,du$$

$$= 2\left[\frac{1}{5}u^5 + u^3\right]_0^2 = \frac{144}{5}$$

34. $$\int_0^1 \frac{1}{\sqrt{x} + \sqrt{x + 1}}\,dx = -\int_0^1 (\sqrt{x} - \sqrt{x + 1})\,dx \qquad \text{(Rationalize)}$$

$$= -\left[\frac{2}{3}x^{3/2} - \frac{2}{3}(x + 1)^{3/2}\right]_0^1 = \frac{4}{3}(\sqrt{2} - 1)$$

35. Let $u = \sqrt[3]{x + 1}$, $\quad x = u^3 - 1$, $\quad dx = 3u^2\,du$

$$\int_0^7 x\sqrt[3]{x + 1}\,dx = 3\int_1^2 (u^3 - 1)u^3\,du = 3\int_1^2 (u^6 - u^3)\,du$$

$$= 3\left[\frac{u^7}{7} - \frac{u^4}{4}\right]_1^2 = \frac{1209}{28}$$

36. Let $u = \sqrt[3]{x + 2}$, $\quad x = u^3 - 2$, $\quad dx = 3u^2\,du$

$$\int_{-2}^6 x^2\sqrt[3]{x + 2}\,dx = 3\int_0^2 (u^3 - 2)^2 u^3\,du = 3\int_0^2 (u^9 - 4u^6 + 4u^3)\,du$$

$$= 3\left[\frac{u^{10}}{10} - \frac{4u^7}{7} + u^4\right]_0^2 = \frac{4752}{35}$$

37. Let $u = \sqrt{x - 1}$, $\quad x = u^2 + 1$, $\quad dx = 2u\,du$

$$\int_1^5 x^2\sqrt{x - 1}\,dx = 2\int_0^2 (u^2 + 1)^2 u^2\,du = 2\int_0^2 (u^6 + 2u^4 + u^2)\,du$$

$$= 2\left[\frac{1}{7}u^7 + \frac{2}{5}u^5 + \frac{1}{3}u^3\right]_0^2 = \frac{7088}{105}$$

38. $\displaystyle\int_0^{\pi/2} \sin 2x \; dx = -\frac{1}{2}\cos 2x \Big]_0^{\pi/2} = -\frac{1}{2}(-1-1) = 1$

39. $\displaystyle\int_0^{\pi/2} \cos(\frac{2}{3}x) \; dx = \frac{3}{2}\sin(\frac{2}{3}x) \Big]_0^{\pi/2} = \frac{3}{2}(\frac{\sqrt{3}}{2}) = \frac{3\sqrt{3}}{4}$

40. $\displaystyle\int_{\pi/3}^{\pi/2} (x + \cos x) \; dx = \left[\frac{x^2}{2} + \sin x\right]_{\pi/3}^{\pi/2} = (\frac{\pi^2}{8} + 1) - (\frac{\pi^2}{18} + \frac{\sqrt{3}}{2})$

$\qquad\qquad = \dfrac{5\pi^2}{72} + \dfrac{2 - \sqrt{3}}{2}$

41. $\displaystyle\int_{\pi/2}^{2\pi/3} \sec^2(\frac{x}{2}) \; dx = 2\int_{\pi/2}^{2\pi/3} \sec^2(\frac{x}{2})(\frac{1}{2}) \; dx = 2 \tan(\frac{x}{2}) \Big]_{\pi/2}^{2\pi/3} = 2(\sqrt{3} - 1)$

42. Let $u = x/2$, $\quad u' = 1/2$

$\displaystyle\int_{\pi/3}^{\pi/2} \csc^2(\frac{x}{2}) \; dx = 2\int_{\pi/3}^{\pi/2} \csc^2(\frac{x}{2})(\frac{1}{2}) \; dx = -2 \cot(\frac{x}{2}) \Big]_{\pi/3}^{\pi/2} = 2(\sqrt{3} - 1)$

43. Let $u = 2x$, $\quad u' = 2$

$\displaystyle\int_{\pi/12}^{\pi/4} \csc 2x \cot 2x \; dx = \frac{1}{2}\int_{\pi/12}^{\pi/4} \csc 2x \cot 2x(2) \; dx = -\frac{1}{2}\csc 2x \Big]_{\pi/12}^{\pi/4} = \frac{1}{2}$

44. Let $u = \sin 2x$, $\quad u' = 2\cos 2x$

$\displaystyle\int_0^{\pi/8} \sin 2x \cos 2x \; dx = \frac{1}{2}\int_0^{\pi/8} \sin 2x(2\cos 2x) \; dx = \frac{1}{4}\sin^2 2x \Big]_0^{\pi/8} = \frac{1}{8}$

45. Let $u = 1 - x$, $\quad u' = -1$

$\displaystyle\int_0^1 \sec(1-x)\tan(1-x) \; dx = -\int_0^1 (-1)\sec(1-x)\tan(1-x) \; dx$

$\qquad\qquad = -\sec(1-x)\Big]_0^1 = -1 + \sec(1)$

46. $\displaystyle\int_0^{\pi/4} \frac{1 - \sin^2\theta}{\cos^2\theta} \; d\theta = \int_0^{\pi/4} d\theta = \theta \Big]_0^{\pi/4} = \pi/4$

47. $A = \displaystyle\int_0^1 (x - x^2) \; dx = \left[\frac{x^2}{2} - \frac{x^3}{3}\right]_0^1 = \frac{1}{6}$

48. $A = -\int_{-1}^{3} (x^2 - 2x - 3)\ dx = -\left[\dfrac{x^3}{3} - x^2 - 3x\right]_{-1}^{3}$

$= -(9 - 9 - 9) + (-\dfrac{1}{3} - 1 + 3) = \dfrac{32}{3}$

49. $A = \int_{-1}^{1} (1 - x^4)\ dx = \left[x - \dfrac{1}{5}x^5\right]_{-1}^{1} = \dfrac{8}{5}$

50. $A = \int_{1}^{2} \dfrac{1}{x^2}\ dx = -\dfrac{1}{x}\Big]_{1}^{2} = -\dfrac{1}{2} + 1 = \dfrac{1}{2}$

51. Let $u = 2x$, $\qquad u' = 2$

$A = \int_{0}^{4} \sqrt[3]{2x}\ dx = \dfrac{1}{2}\int_{0}^{4} (2x)^{1/3}2\ dx = \dfrac{3}{8}(2x)^{4/3}\Big]_{0}^{4} = 6$

52. $A = \int_{0}^{3} (3 - x)\sqrt{x}\ dx = \int_{0}^{3} (3x^{1/2} - x^{3/2})\ dx = \left[2x^{3/2} - \dfrac{2}{5}x^{5/2}\right]_{0}^{3}$

$= \dfrac{x\sqrt{x}}{5}(10 - 2x)\Big]_{0}^{3} = \dfrac{12\sqrt{3}}{5}$

53. $A = \int_{0}^{\pi} \cos\dfrac{x}{2}\ dx = 2\sin\dfrac{x}{2}\Big]_{0}^{\pi} = 2$

54. $A = \int_{0}^{\pi} (x + \sin x)\ dx = \left[\dfrac{x^2}{2} - \cos x\right]_{0}^{\pi} = \dfrac{\pi^2}{2} + 2 = \dfrac{\pi^2 + 4}{2}$

55. $A = \int_{0}^{\pi} (2\sin x + \sin 2x)\ dx = -\left[2\cos x + \dfrac{1}{2}\cos 2x\right]_{0}^{\pi} = 4$

56. $A = \int_{0}^{\pi} (\sin x + \cos 2x)\ dx = \left[-\cos x + \dfrac{1}{2}\sin 2x\right]_{0}^{\pi} = 2$

57. Since $y \geq 0$ on $[0,\ 2]$: $\quad A = \int_{0}^{2} (3x^2 + 1)\ dx = \left[x^3 + x\right]_{0}^{2} = 8 + 2 = 10$

58. Since $y \geq 0$ on $[0,\ 4]$: $\quad A = \int_{0}^{4} (1 + \sqrt{x})\ dx = \left[x + \dfrac{2}{3}x^{3/2}\right]_{0}^{4}$

$= 4 + \dfrac{2}{3}(8) = \dfrac{28}{3}$

59. Since $y \geq 0$ on $[0, 2]$: $A = \int_0^2 (x^3 + x) \, dx = \left[\frac{x^4}{4} + \frac{x^2}{2} \right]_0^2 = 4 + 2 = 6$

60. Since $y \geq 0$ on $[0, 3]$: $A = \int_0^3 (3x - x^2) \, dx = \left[\frac{3}{2} x^2 - \frac{x^3}{3} \right]_0^3 = \frac{9}{2}$

61. $\dfrac{1}{2 - (-2)} \displaystyle\int_{-2}^2 (4 - x^2) \, dx = \frac{1}{4} \left[4x - \frac{1}{3} x^3 \right]_{-2}^2$

$= \frac{1}{4} \left[(8 - \frac{8}{3}) - (-8 + \frac{8}{3}) \right] = \frac{8}{3}$

Average $= \dfrac{8}{3}$

$4 - x^2 = \dfrac{8}{3}$ when $x = \pm \dfrac{2\sqrt{3}}{3} \approx \pm 1.155$

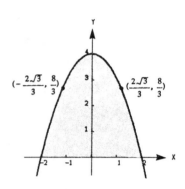

62. $\dfrac{1}{1 - 0} \displaystyle\int_0^1 (x^2 - 2x + 1) \, dx = \left[\frac{x^3}{3} - x^2 + x \right]_0^1$

$= \dfrac{1}{3} - 1 + 1 = \dfrac{1}{3}$

Average $= \dfrac{1}{3}$

$x^2 - 2x + 1 = \dfrac{1}{3}, \qquad 3x^2 - 6x + 2 = 0$

when $x = \dfrac{3 - \sqrt{3}}{3} \approx 0.423$

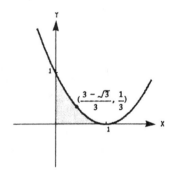

63. $\dfrac{1}{2 - 0} \displaystyle\int_0^2 x\sqrt{4 - x^2} \, dx = \frac{1}{2} \left[-\frac{1}{3}(4 - x^2)^{3/2} \right]_0^2 = \frac{4}{3}$

Average $= \dfrac{4}{3}$

$x\sqrt{4 - x^2} = \dfrac{4}{3}, \qquad x^2(4 - x^2) = \dfrac{16}{9}$

$9x^4 - 36x^2 + 16 = 0$

$x^2 = \dfrac{36 \pm \sqrt{720}}{18} = \dfrac{36 \pm 12\sqrt{5}}{18}$

$= 2 \pm \dfrac{2\sqrt{5}}{3}, \qquad x = \sqrt{2 \pm \dfrac{2\sqrt{5}}{3}}$

In the interval $[0, 2]$
$x \approx 0.714$ and $x \approx 1.868$

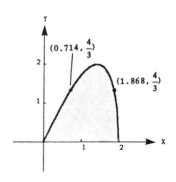

64. $\dfrac{1}{2 - (1/2)} \displaystyle\int_{1/2}^{2} \dfrac{x^2 + 1}{x^2}\, dx = \dfrac{2}{3} \displaystyle\int_{1/2}^{2} (1 + x^{-2})\, dx$

$\qquad = \dfrac{2}{3}\left[x - \dfrac{1}{x} \right]_{1/2}^{2} = \dfrac{2}{3}(\dfrac{3}{2} + \dfrac{3}{2}) = 2$

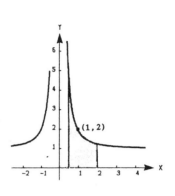

Average = 2

In the interval [1/2, 2]:

$\dfrac{x^2 + 1}{x^2} = 2$ when $x = 1$

65. $\dfrac{1}{4 - 0} \displaystyle\int_{0}^{4} (x - 2\sqrt{x})\, dx = \dfrac{1}{4}\left[\dfrac{x^2}{2} - \dfrac{4}{3}x^{3/2} \right]_{0}^{4} = \dfrac{1}{4}(8 - \dfrac{32}{3}) = -\dfrac{2}{3}$

Average = $-2/3$

$x - 2\sqrt{x} = -2/3, \qquad 3x - 6\sqrt{x} + 2 = 0$

when $\sqrt{x} = \dfrac{6 \pm \sqrt{36 - 24}}{6} = \dfrac{3 \pm \sqrt{3}}{3}$

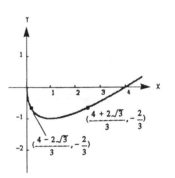

$x = (\dfrac{3 \pm \sqrt{3}}{3})^2 = \dfrac{4 \pm 2\sqrt{3}}{3}$

≈ 2.488 and 0.179

66. $\dfrac{1}{2 - 0} \displaystyle\int_{0}^{2} \dfrac{1}{(x - 3)^2}\, dx = -\dfrac{1}{2(x - 3)}\Big]_{0}^{2} = \dfrac{1}{3}$

Average = 1/3

$\dfrac{1}{(x - 3)^2} = \dfrac{1}{3}, \qquad (x - 3)^2 = 3$

$x - 3 = \pm\sqrt{3}, \qquad x = 3 \pm \sqrt{3}$

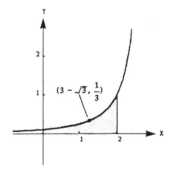

In the interval [0, 2]
$x = 3 - \sqrt{3} \approx 1.27$

67. $\dfrac{1}{\pi - 0} \displaystyle\int_{0}^{\pi} \sin x\, dx = -\dfrac{1}{\pi}\cos x \Big]_{0}^{\pi} = \dfrac{2}{\pi}$

Average = $\dfrac{2}{\pi}$

$\sin x = \dfrac{2}{\pi}$

$x \approx 0.690, 2.451$

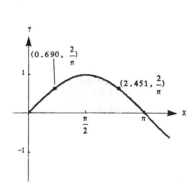

68. $\dfrac{1}{1/2 - 0} \displaystyle\int_0^{1/2} \cos \pi x \, dx = \dfrac{2}{\pi} \int_0^{1/2} (\cos \pi x)\,\pi \, dx$

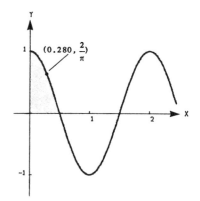

$\qquad\qquad = \dfrac{2}{\pi} \sin \pi x \Big]_0^{1/2}$

$\qquad\qquad = \dfrac{2}{\pi}(1 - 0) = \dfrac{2}{\pi}$

Average $= \dfrac{2}{\pi}$, $\qquad \cos \pi x = \dfrac{2}{\pi}$

$\pi x \approx 0.881$, $\qquad x \approx 0.280$

69. $\displaystyle\int_0^2 x^2 \, dx = \dfrac{8}{3}$

(a) $\displaystyle\int_{-2}^0 x^2 \, dx = \int_0^2 x^2 \, dx = \dfrac{8}{3}$

(b) $\displaystyle\int_{-2}^2 x^2 \, dx = 2\int_0^2 x^2 \, dx = \dfrac{16}{3}$

(c) $\displaystyle\int_0^2 (-x^2) \, dx = -\int_0^2 x^2 \, dx = -\dfrac{8}{3}$

(d) $\displaystyle\int_0^2 (x^2 + 1) \, dx = \int_0^2 x^2 \, dx + \int_0^2 dx = \dfrac{8}{3} + 2 = \dfrac{14}{3}$

(e) $\displaystyle\int_{-2}^0 3x^2 \, dx = 3\int_0^2 x^2 \, dx = 8$

70.(a) $\displaystyle\int_{-\pi/4}^{\pi/4} \sin x \, dx = 0$ since $\sin x$ is symmetric to the origin.

(b) $\displaystyle\int_{-\pi/4}^{\pi/4} \cos x \, dx = 2\int_0^{\pi/4} \cos x \, dx = 2 \sin x \Big]_0^{\pi/4} = \sqrt{2}$

since $\cos x$ is symmetric to the y-axis.

(c) $\displaystyle\int_{-\pi/2}^{\pi/2} \cos x \, dx = 2\int_0^{\pi/2} \cos x \, dx = 2 \sin x \Big]_0^{\pi/2} = 2$

(d) $\displaystyle\int_{-1.32}^{1.32} \sin 2x \, dx = 0$ since $\sin 2x$ is symmetric to the origin.

(e) $\displaystyle\int_{-\pi/2}^{\pi/2} \sin x \cos x \, dx = 0$ since $\sin(-x)\cos(-x) = -\sin x \cos x$ and

hence is symmetric to the origin.

71. $\dfrac{1}{5-0} \displaystyle\int_0^5 (0.1729t + 0.1522t^2 - 0.0374t^3)\, dt$

$= \dfrac{1}{5}(0.08645t^2 + 0.05073t^3 - 0.00935t^4) \Big]_0^5 = 0.5318$ liter

72. $\dfrac{1}{R-0} \displaystyle\int_0^R k(R^2 - r^2)\, dr = \dfrac{k}{R}(R^2 r - \dfrac{r^3}{3}) \Big]_0^R = \dfrac{2kR^2}{3}$

73.(a) $\dfrac{1}{6-0} \displaystyle\int_0^6 (53 + 5t - 0.3t^2)\, dt = \dfrac{1}{6}(53t + \dfrac{5}{2}t^2 - 0.1t^3) \Big]_0^6 = 64.4^\circ$

(b) $\dfrac{1}{12-0} \displaystyle\int_0^{12} (53 + 5t - 0.3t^2)\, dt = \dfrac{1}{12}(53t + \dfrac{5}{2}t^2 - 0.1t^3) \Big]_0^{12} = 68.6^\circ$

74. $\dfrac{1}{b-a} \displaystyle\int_a^b (217 + 13 \cos \dfrac{\pi(t-3)}{6})\, dt = \dfrac{1}{b-a} \left[217t + \dfrac{78}{\pi} \sin (\dfrac{\pi(t-3)}{6}) \right]_a^b$

(a) $\dfrac{1}{3} \left[217t + \dfrac{78}{\pi} \sin (\dfrac{\pi(t-3)}{6}) \right]_0^3 = \dfrac{1}{3}(651 + \dfrac{78}{\pi})$

≈ 225.28 million barrels

(b) $\dfrac{1}{3} \left[217t + \dfrac{78}{\pi} \sin (\dfrac{\pi(t-3)}{6}) \right]_3^6 = \dfrac{1}{3}(1302 + \dfrac{78}{\pi} - 651)$

≈ 225.28 million barrels

(c) $\dfrac{1}{12} \left[217t + \dfrac{78}{\pi} \sin (\dfrac{\pi(t-3)}{6}) \right]_0^{12} = \dfrac{1}{12}(2604 - \dfrac{78}{\pi} + \dfrac{78}{\pi})$

$= 217$ million barrels

75. $\dfrac{1}{b-a}\displaystyle\int_a^b\left[74.50+43.75\sin\dfrac{\pi t}{6}\right]dt=\dfrac{1}{b-a}\left[74.50t-\dfrac{262.5}{\pi}\cos\dfrac{\pi t}{6}\right]_a^b$

(a) $\dfrac{1}{3}\left[74.50t-\dfrac{262.5}{\pi}\cos\dfrac{\pi t}{6}\right]_0^3=\dfrac{1}{3}(223.5+\dfrac{262.5}{\pi})$

≈ 102.352 thousand units

(b) $\dfrac{1}{3}\left[74.50t-\dfrac{262.5}{\pi}\cos\dfrac{\pi t}{6}\right]_3^6=\dfrac{1}{3}(447+\dfrac{262.5}{\pi}-223.5)$

≈ 102.352 thousand units

(c) $\dfrac{1}{12}\left[74.50t-\dfrac{262.5}{\pi}\cos\dfrac{\pi t}{6}\right]_0^{12}=\dfrac{1}{12}(894-\dfrac{262.5}{\pi}+\dfrac{262.5}{\pi})$

$=74.5$ thousand units

76. $P=\dfrac{2}{\pi}\displaystyle\int_0^{\pi/2}\sin\theta\,d\theta=-\dfrac{2}{\pi}\cos\theta\Big]_0^{\pi/2}=-\dfrac{2}{\pi}(0-1)=\dfrac{2}{\pi}\approx 63.7\%$

77. $\dfrac{1}{b-a}\displaystyle\int_a^b\left[2\sin(60\pi t)+\cos(120\pi t)\right]dt$

$=\dfrac{1}{b-a}\left[-\dfrac{1}{30\pi}\cos(60\pi t)+\dfrac{1}{120\pi}\sin(120\pi t)\right]_a^b$

(a) $\dfrac{1}{1/60-0}\left[-\dfrac{1}{30\pi}\cos(60\pi t)+\dfrac{1}{120\pi}\sin(120\pi t)\right]_0^{1/60}$

$=60\left[(\dfrac{1}{30\pi}+0)-(-\dfrac{1}{30\pi})\right]=\dfrac{4}{\pi}\approx 1.273$ amps

(b) $\dfrac{1}{1/240-0}\left[-\dfrac{1}{30\pi}\cos(60\pi t)+\dfrac{1}{120\pi}\sin(120\pi t)\right]_0^{1/240}$

$=240\left[(-\dfrac{1}{30\sqrt{2}\pi}+\dfrac{1}{120\pi})-(-\dfrac{1}{30\pi})\right]=\dfrac{2}{\pi}(5-2\sqrt{2})\approx 1.382$ amps

(c) $\dfrac{1}{1/30-0}\left[-\dfrac{1}{30\pi}\cos(60\pi t)+\dfrac{1}{120\pi}\sin(120\pi t)\right]_0^{1/30}$

$=30\left[(-\dfrac{1}{30\pi})-(-\dfrac{1}{30\pi})\right]=0$ amps

5.7
Variable limits of integration and
the natural logarithmic function

1.(a) $\int_0^x (t + 2)\, dt = \left[\dfrac{t^2}{2} + 2t\right]_0^x = \dfrac{1}{2}x^2 + 2x$

 (b) $\dfrac{d}{dx}\left[\dfrac{1}{2}x^2 + 2x\right] = x + 2$

2.(a) $\int_4^x \sqrt{t}\, dt = \dfrac{2}{3}t^{3/2}\Big]_4^x = \dfrac{2}{3}x^{3/2} - \dfrac{16}{3} = \dfrac{2}{3}(x^{3/2} - 8)$

 (b) $\dfrac{d}{dx}\left[\dfrac{2}{3}x^{3/2} - \dfrac{16}{3}\right] = x^{1/2} = \sqrt{x}$

3.(a) $\int_7^x \sqrt[3]{t + 1}\, dt = \dfrac{3}{4}(t + 1)^{4/3}\Big]_7^x = \dfrac{3}{4}(x + 1)^{4/3} - 12$

 (b) $\dfrac{d}{dx}\left[\dfrac{3}{4}(x + 1)^{4/3} - 12\right] = (x + 1)^{1/3} = \sqrt[3]{x + 1}$

4.(a) $\int_0^x t(t^2 + 1)^3\, dt = \dfrac{1}{8}(t^2 + 1)^4\Big]_0^x = \dfrac{1}{8}(x^2 + 1)^4 - \dfrac{1}{8}$

$= \dfrac{1}{8}\left[(x^2 + 1)^4 - 1\right]$

 (b) $\dfrac{d}{dx}\left[\dfrac{1}{8}(x^2 + 1)^4 - \dfrac{1}{8}\right] = \dfrac{1}{8}(4)(2x)(x^2 + 1)^3 = x(x^2 + 1)^3$

5.(a) $\int_{\pi/4}^x \sec^2 t\, dt = \tan t\Big]_{\pi/4}^x = \tan x - 1$

 (b) $\dfrac{d}{dx}\left[\tan x - 1\right] = \sec^2 x$

6.(a) $\int_{\pi/6}^x \sec 2t \tan 2t\, dt = \dfrac{1}{2}\sec 2t\Big]_{\pi/6}^x = \dfrac{1}{2}\sec 2x - 1$

 (b) $\dfrac{d}{dx}\left[\dfrac{1}{2}\sec 2x - 1\right] = \dfrac{1}{2}(2)\sec 2x \tan 2x = \sec 2x \tan 2x$

7. $F(x) = \displaystyle\int_{-2}^{x} (t^2 - 2t + 5)\, dt$

   $F'(x) = x^2 - 2x + 5$

8. $F(x) = \displaystyle\int_{-1}^{x} \sqrt{t^4 + 1}\, dt$

   $F'(x) = \sqrt{x^4 + 1}$

9. $F(x) = \displaystyle\int_{1}^{x} \sqrt[4]{t}\, dt$

   $F'(x) = \sqrt[4]{x}$

10. $F(x) = \displaystyle\int_{0}^{x} \tan^4 t\, dt$

    $F'(x) = \tan^4 x$

11. $F(x) = \displaystyle\int_{0}^{x} t \cos t\, dt$

    $F'(x) = x \cos x$

12. $F(x) = \displaystyle\int_{1}^{x} \dfrac{t^2}{t^2 + 1}\, dt$

    $F'(x) = \dfrac{x^2}{x^2 + 1}$

13. $f(x) = \ln 2x$

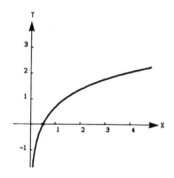

14. $f(x) = \ln |x|$

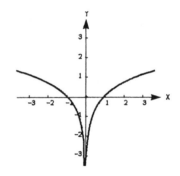

15. $f(x) = \ln (x - 1)$

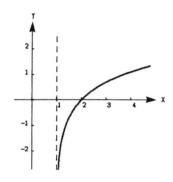

16. $f(x) = -2 \ln x$

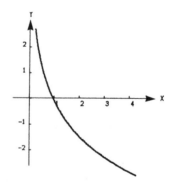

17.(a) $\ln 6 = \ln 2 + \ln 3 = 1.7917$

   (b) $\ln \dfrac{2}{3} = \ln 2 - \ln 3 = -0.4055$

   (c) $\ln 81 = 4 \ln 3 = 4.3944$

   (d) $\ln \sqrt{3} = \dfrac{1}{2} \ln 3 = 0.5493$

18.(a) $\ln 0.25 = \ln \frac{1}{4} = \ln 1 - 2 \ln 2 = -1.3862$

  (b) $\ln 24 = 3 \ln 2 + \ln 3 = 3.1779$

  (c) $\ln \sqrt[3]{12} = \frac{1}{3}[2 \ln 2 + \ln 3] = 0.8283$

  (d) $\ln \frac{1}{72} = \ln 1 - [3 \ln 2 + 2 \ln 3] = -4.2765$

19. $\ln \frac{2}{3} = \ln 2 - \ln 3$

20. $\ln xyz = \ln x + \ln y + \ln z$

21. $\ln \frac{xy}{z} = \ln x + \ln y - \ln z$

22. $\ln \sqrt{a-1} = \ln (a-1)^{1/2} = (1/2) \ln (a-1)$

23. $\ln \sqrt{2^3} = \ln 2^{3/2} = \frac{3}{2} \ln 2$

24. $\ln \frac{1}{5} = \ln 1 - \ln 5 = -\ln 5$

25. $\ln (\frac{x^2 - 1}{x^3})^3 = 3[\ln (x^2 - 1) - \ln x^3] = 3[\ln (x+1) + \ln(x-1) - 3 \ln x]$

26. $\ln 3e^2 = \ln 3 + 2 \ln e = 2 + \ln 3$

27. $\ln z(z-1)^2 = \ln z + \ln (z-1)^2 = \ln z + 2 \ln (z-1)$

28. $\ln \frac{1}{e} = \ln 1 - \ln e = -1$

29. $\ln (x-2) - \ln (x+2) = \ln \frac{x-2}{x+2}$

30. $3 \ln x + 2 \ln y - 4 \ln z = \ln x^3 + \ln y^2 - \ln z^4 = \ln \frac{x^3 y^2}{z^4}$

31. $\frac{1}{3}[2 \ln (x+3) + \ln x - \ln (x^2 - 1)] = \ln \sqrt[3]{\frac{x(x+3)^2}{x^2 - 1}}$

32. $2[\ln x - \ln (x+1) - \ln (x-1)] = \ln (\frac{x}{x^2 - 1})^2$

33. $2 \ln 3 - \frac{1}{2} \ln (x^2 + 1) = \ln \frac{9}{\sqrt{x^2 + 1}}$

34. $\frac{3}{2}[\ln (x^2 + 1) - \ln (x+1) - \ln (x-1)] = \frac{3}{2} \ln \frac{x^2 + 1}{(x+1)(x-1)}$

        $= \ln \sqrt{(\frac{x^2 + 1}{x^2 - 1})^3}$

5.8
The natural logarithmic function
and differentiation

1. $y = \ln x^3 = 3 \ln x$, $\qquad y' = \dfrac{3}{x}$, $\qquad$ At $(1, 0)$, $y' = 3$

2. $y = \ln x^{3/2} = \dfrac{3}{2} \ln x$, $\quad y' = \dfrac{3}{2x}$, $\qquad$ At $(1, 0)$, $y' = \dfrac{3}{2}$

3. $y = \ln x^2 = 2 \ln x$, $\qquad y' = \dfrac{2}{x}$, $\qquad$ At $(1, 0)$, $y' = 2$

4. $y = \ln x^{1/2} = \dfrac{1}{2} \ln x$, $\quad y' = \dfrac{1}{2x}$, $\qquad$ At $(1, 0)$, $y' = \dfrac{1}{2}$

5. $y = \ln x^2 = 2 \ln x$, $\qquad \dfrac{dy}{dx} = \dfrac{2}{x}$

6. $y = \ln (x^2 + 3)$, $\qquad \dfrac{dy}{dx} = \dfrac{2x}{x^2 + 3}$

7. $y = \ln \sqrt{x^4 - 4x} = \dfrac{1}{2} \ln (x^4 - 4x)$, $\qquad \dfrac{dy}{dx} = \dfrac{(4x^3 - 4)}{2(x^4 - 4x)} = \dfrac{2(x^3 - 1)}{x(x^3 - 4)}$

8. $y = \ln (1 - x)^{3/2} = \dfrac{3}{2} \ln (1 - x)$, $\qquad \dfrac{dy}{dx} = \dfrac{3}{2}(\dfrac{-1}{1 - x}) = \dfrac{3}{2(x - 1)}$

9. $y = (\ln x)^4$, $\qquad \dfrac{dy}{dx} = 4(\ln x)^3 (\dfrac{1}{x}) = \dfrac{4(\ln x)^3}{x}$

10. $y = x \ln x$, $\qquad \dfrac{dy}{dx} = x(\dfrac{1}{x}) + \ln x = 1 + \ln x$

11. $y = \ln x \sqrt{x^2 - 1} = \ln x + \dfrac{1}{2} \ln (x^2 - 1)$, $\qquad \dfrac{dy}{dx} = \dfrac{1}{x} + \dfrac{1}{2}(\dfrac{2x}{x^2 - 1}) = \dfrac{2x^2 - 1}{x(x^2 - 1)}$

12. $y = \ln \dfrac{x}{x + 1} = \ln x - \ln (x + 1)$, $\qquad \dfrac{dy}{dx} = \dfrac{1}{x} - \dfrac{1}{x + 1} = \dfrac{1}{x^2 + x}$

13. $y = \ln \dfrac{x}{x^2 + 1} = \ln x - \ln (x^2 + 1)$, $\qquad \dfrac{dy}{dx} = \dfrac{1}{x} - \dfrac{2x}{x^2 + 1} = \dfrac{1 - x^2}{x(x^2 + 1)}$

14. $y = \dfrac{\ln x}{x}$, $\qquad \dfrac{dy}{dx} = \dfrac{x(1/x) - \ln x}{x^2} = \dfrac{1 - \ln x}{x^2}$

15. $y = \dfrac{\ln x}{x^2}$, $\qquad \dfrac{dy}{dx} = \dfrac{x^2(1/x) - 2x \ln x}{x^4} = \dfrac{1 - 2 \ln x}{x^3}$

16. $y = \ln (\ln x)$, $\qquad \dfrac{dy}{dx} = \dfrac{1/x}{\ln x} = \dfrac{1}{x \ln x}$

17. $y = \ln(\ln x^2)$, $\quad \dfrac{dy}{dx} = \dfrac{(2x/x^2)}{\ln x^2} = \dfrac{2}{x \ln x^2} = \dfrac{1}{x \ln x}$

18. $y = \ln\sqrt{\dfrac{x-1}{x+1}} = \dfrac{1}{2}[\ln(x-1) - \ln(x+1)]$

$\dfrac{dy}{dx} = \dfrac{1}{2}\left[\dfrac{1}{x-1} - \dfrac{1}{x+1}\right] = \dfrac{1}{x^2-1}$

19. $y = \ln\sqrt{\dfrac{x+1}{x-1}} = \dfrac{1}{2}[\ln(x+1) - \ln(x-1)]$

$\dfrac{dy}{dx} = \dfrac{1}{2}\left[\dfrac{1}{x+1} - \dfrac{1}{x-1}\right] = \dfrac{1}{1-x^2}$

20. $y = \ln\sqrt{x^2-4} = \dfrac{1}{2}\ln(x^2-4)$, $\quad \dfrac{dy}{dx} = \dfrac{1}{2}(\dfrac{1}{x^2-4})(2x) = \dfrac{x}{x^2-4}$

21. $y = \ln\dfrac{\sqrt{4+x^2}}{x} = \dfrac{1}{2}\ln(4+x^2) - \ln x$

$\dfrac{dy}{dx} = \dfrac{x}{4+x^2} - \dfrac{1}{x} = \dfrac{-4}{x(x^2+4)}$

22. $y = \ln(x + \sqrt{4+x^2})$, $\quad \dfrac{dy}{dx} = \dfrac{1}{x+\sqrt{4+x^2}}(1 + \dfrac{x}{\sqrt{4+x^2}}) = \dfrac{1}{\sqrt{4+x^2}}$

23. $y = \dfrac{-\sqrt{x^2+1}}{x} + \ln(x+\sqrt{x^2+1})$

$\dfrac{dy}{dx} = \dfrac{-x(x/\sqrt{x^2+1}) + \sqrt{x^2+1}}{x^2} + (\dfrac{1}{x+\sqrt{x^2+1}})(1 + \dfrac{x}{\sqrt{x^2+1}})$

$= \dfrac{1}{x^2\sqrt{x^2+1}} + (\dfrac{1(x-\sqrt{x^2+1})}{-1})(\dfrac{\sqrt{x^2+1}+x}{\sqrt{x^2+1}}) = \dfrac{\sqrt{x^2+1}}{x^2}$

24. $y = \dfrac{-\sqrt{x^2+4}}{2x^2} - \dfrac{1}{4}\ln(\dfrac{2+\sqrt{x^2+4}}{x})$

$= \dfrac{-\sqrt{x^2+4}}{2x^2} - \dfrac{1}{4}\ln(2+\sqrt{x^2+4}) + \dfrac{1}{4}\ln x$

$\dfrac{dy}{dx} = \dfrac{-2x^2(x/\sqrt{x^2+4}) + 4x\sqrt{x^2+4}}{4x^4} - \dfrac{1}{4}(\dfrac{1}{2+\sqrt{x^2+4}})(\dfrac{x}{\sqrt{x^2+4}}) + \dfrac{1}{4x}$

$= \dfrac{\sqrt{x^2+4}}{x^3}$

25. $y = \ln|\sin x|$, $\quad \dfrac{dy}{dx} = \dfrac{\cos x}{\sin x} = \cot x$

26. $y = \ln|\sec x|$, $\quad \dfrac{dy}{dx} = \dfrac{\sec x \tan x}{\sec x} = \tan x$

27. $y = \ln \left| \dfrac{\cos x}{\cos x - 1} \right| = \ln |\cos x| - \ln |\cos x - 1|$

$\dfrac{dy}{dx} = \dfrac{-\sin x}{\cos x} - \dfrac{-\sin x}{\cos x - 1} = -\tan x + \dfrac{\sin x}{\cos x - 1}$

28. $y = \ln |\sec x + \tan x|$

$\dfrac{dy}{dx} = \dfrac{\sec x \tan x + \sec^2 x}{\sec x + \tan x} = \dfrac{\sec x(\sec x + \tan x)}{\sec x + \tan x} = \sec x$

29. $y = \ln \left| \dfrac{-1 + \sin x}{2 + \sin x} \right| = \ln |-1 + \sin x| - \ln |2 + \sin x|$

$\dfrac{dy}{dx} = \dfrac{\cos x}{-1 + \sin x} - \dfrac{\cos x}{2 + \sin x} = \dfrac{3 \cos x}{(\sin x - 1)(\sin x + 2)}$

30. $y = \ln \sqrt{1 + \sin^2 x} = \dfrac{1}{2} \ln (1 + \sin^2 x)$

$\dfrac{dy}{dx} = \left(\dfrac{1}{2}\right)\dfrac{2 \sin x \cos x}{1 + \sin^2 x} = \dfrac{\sin x \cos x}{1 + \sin^2 x}$

31. $y = x\sqrt{x^2 - 1}, \qquad \ln y = \ln x + \dfrac{1}{2} \ln (x^2 - 1)$

$\dfrac{1}{y}\left(\dfrac{dy}{dx}\right) = \dfrac{1}{x} + \dfrac{x}{x^2 - 1}, \qquad \dfrac{dy}{dx} = y\dfrac{2x^2 - 1}{x(x^2 - 1)} = \dfrac{2x^2 - 1}{\sqrt{x^2 - 1}}$

32. $y = \sqrt{(x - 1)(x - 2)(x - 3)}, \quad \ln y = \dfrac{1}{2}[\ln (x - 1) + \ln (x - 2) + \ln (x - 3)]$

$\dfrac{1}{y}\left(\dfrac{dy}{dx}\right) = \dfrac{1}{2}\left[\dfrac{1}{x - 1} + \dfrac{1}{x - 2} + \dfrac{1}{x - 3}\right] = \dfrac{1}{2}\left[\dfrac{3x^2 - 12x + 11}{(x - 1)(x - 2)(x - 3)}\right]$

$\dfrac{dy}{dx} = \dfrac{3x^2 - 12x + 11}{2y}$

33. $y = \dfrac{x^2 \sqrt{3x - 2}}{(x - 1)^2}, \qquad \ln y = \ln x^2 + \dfrac{1}{2} \ln (3x - 2) - 2 \ln (x - 1)$

$\dfrac{1}{y}\left(\dfrac{dy}{dx}\right) = \dfrac{2}{x} + \dfrac{3}{2(3x - 2)} - \dfrac{2}{x - 1}, \qquad \dfrac{dy}{dx} = y\left[\dfrac{3x^2 - 15x + 8}{2x(3x - 2)(x - 1)}\right]$

$= \dfrac{3x^3 - 15x^2 + 8x}{2(x - 1)^3 \sqrt{3x - 2}}$

34. $y = \sqrt[3]{\dfrac{x^2 + 1}{x^2 - 1}}$, $\qquad \ln y = \dfrac{1}{3}[\ln(x^2 + 1) - \ln(x + 1) - \ln(x - 1)]$

$\dfrac{1}{y}\left(\dfrac{dy}{dx}\right) = \dfrac{1}{3}\left[\dfrac{2x}{x^2 + 1} - \dfrac{1}{x + 1} - \dfrac{1}{x - 1}\right]$, $\qquad \dfrac{dy}{dx} = \dfrac{-4xy}{3(x^4 - 1)}$

35. $y = \dfrac{x(x - 1)^{3/2}}{\sqrt{x + 1}}$, $\qquad \ln y = \ln x + \dfrac{3}{2}\ln(x - 1) - \dfrac{1}{2}\ln(x + 1)$

$\dfrac{1}{y}\left(\dfrac{dy}{dx}\right) = \dfrac{1}{x} + \dfrac{3}{2}\left(\dfrac{1}{x - 1}\right) - \dfrac{1}{2}\left(\dfrac{1}{x + 1}\right)$

$\dfrac{dy}{dx} = \dfrac{y}{2}\left[\dfrac{2}{x} + \dfrac{3}{x - 1} - \dfrac{1}{x + 1}\right] = \dfrac{y}{2}\left[\dfrac{4x^2 + 4x - 2}{x(x^2 - 1)}\right] = \dfrac{(2x^2 + 2x - 1)\sqrt{x - 1}}{(x + 1)^{3/2}}$

36. $y = \dfrac{(x + 1)(x + 2)}{(x - 1)(x - 2)}$

$\ln y = \ln(x + 1) + \ln(x + 2) - \ln(x - 1) - \ln(x - 2)$

$\dfrac{1}{y}\left(\dfrac{dy}{dx}\right) = \dfrac{1}{x + 1} + \dfrac{1}{x + 2} - \dfrac{1}{x - 1} - \dfrac{1}{x - 2}$

$\dfrac{dy}{dx} = y\left[\dfrac{-2}{x^2 - 1} + \dfrac{-4}{x^2 - 4}\right] = y\left[\dfrac{-6x^2 + 12}{(x^2 - 1)(x^2 - 4)}\right] = \dfrac{6y(2 - x^2)}{(x^2 - 1)(x^2 - 4)}$

37. $y = 2(\ln x) + 3$, $\qquad y' = \dfrac{2}{x}$, $\qquad y'' = -\dfrac{2}{x^2}$, $\qquad xy'' + y' = x\left(-\dfrac{2}{x^2}\right) + \dfrac{2}{x} = 0$

38. $y = x(\ln x) - 4x$, $\qquad y' = x\left(\dfrac{1}{x}\right) + \ln x - 4 = -3 + \ln x$

$(x + y) - xy' = x + x\ln x - 4x - x(-3 + \ln x) = 0$

39. $y = \dfrac{x^2}{2} - \ln x$ $\qquad$ Domain: $(0, \infty)$

$y' = x - \dfrac{1}{x} = \dfrac{(x + 1)(x - 1)}{x} = 0$

when $x = -1$, $x = 1$

$y'' = 1 + \dfrac{1}{x^2} > 0$

$\left(1, \dfrac{1}{2}\right)$ is a relative minimum

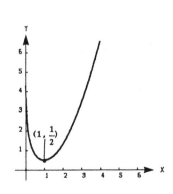

40. $y = x - \ln x$ Domain: $(0, \infty)$

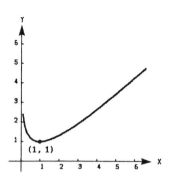

$y' = 1 - \dfrac{1}{x} = 0$ when $x = 1$

$y'' = \dfrac{1}{x^2} > 0$

$(1, 1)$ is a relative minimum

41. $y = x \ln x$ Domain: $(0, \infty)$

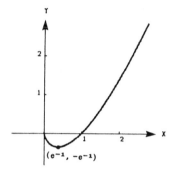

$y' = x(\dfrac{1}{x}) + \ln x = 1 + \ln x = 0$

   when $x = e^{-1}$

$y'' = \dfrac{1}{x} > 0$

$(e^{-1}, -e^{-1})$ is a relative minimum

42. $y = \dfrac{\ln x}{x}$ Domain: $(0, \infty)$

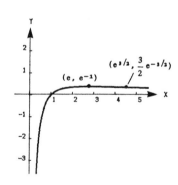

$y' = \dfrac{x(1/x) - \ln x}{x^2} = \dfrac{1 - \ln x}{x^2} = 0$

   when $x = e$

$y'' = \dfrac{x^2(-1/x) - (1 - \ln x)(2x)}{x^4}$

$= \dfrac{2(\ln x) - 3}{x^3} = 0$ when $x = e^{3/2}$

(e, e^{-1}) is a relative maximum

$(e^{3/2}, \dfrac{3}{2} e^{-3/2})$ is a point of inflection

43. $y = \dfrac{x}{\ln x}$ Domain: $(0, 1) \cup (1, \infty)$

$y' = \dfrac{(\ln x)(1) - (x)(1/x)}{(\ln x)^2} = \dfrac{\ln x - 1}{(\ln x)^2} = 0$

when $x = e$

(e, e) is a relative minimum

$y'' = \dfrac{2 - \ln x}{x(\ln x)^3} = 0$ when $x = e^2$

$(e^2, \dfrac{e^2}{2})$ is an inflection point

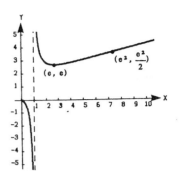

44. $y = x^2 \ln x$ Domain: $(0, \infty)$

$y' = x^2(\dfrac{1}{x}) + 2x(\ln x)$

$= x(1 + 2 \ln x) = 0$ when $x = e^{-1/2}$

$y'' = x(\dfrac{2}{x}) + (1 + 2 \ln x)$

$= 3 + 2(\ln x) = 0$ when $x = e^{-3/2}$

$(e^{-1/2}, -\dfrac{1}{2} e^{-1})$ is a relative minimum

$(e^{-3/2}, -\dfrac{3}{2} e^{-3})$ is a point of inflection

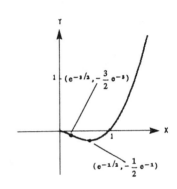

45. Find x such that $\ln x = -x$, $f(x) = (\ln x) + x = 0$, $f'(x) = \dfrac{1}{x} + 1$

$x_{n+1} = x_n - \dfrac{f(x_n)}{f'(x_n)}$

$= x_n \left[\dfrac{1 - \ln x_n}{1 + x_n} \right]$

n	1	2	3
x_n	0.5	0.5644	0.5671
$f(x_n)$	−0.1932	−0.0076	−0.0000

We approximate the root to be $x = 0.567$.

46. Find x such that $\ln x = 3 - x$

$f(x) = x + (\ln x) - 3 = 0$, $\quad f'(x) = 1 + \dfrac{1}{x}$

$x_{n+1} = x_n - \dfrac{f(x_n)}{f'(x_n)}$

$\quad = x_n \left[\dfrac{4 - \ln x_n}{1 + x_n} \right]$

n	1	2	3
x_n	2	2.2046	2.2079
$f(x_n)$	−0.3069	−0.0049	−0.0000

We approximate the root to be $x = 2.208$.

47. $f(x) = \dfrac{\ln x^n}{x} = \dfrac{n \ln x}{x}$

$f'(x) = n \left[\dfrac{x(1/x) - \ln x}{x^2} \right] = \dfrac{n(1 - \ln x)}{x^2} < 0$ if $1 - \ln x < 0$ and $n > 0$.

$\ln x > 1, \quad x > e$

48. $y = 10 \ln \left(\dfrac{10 + \sqrt{100 - x^2}}{x} \right) - \sqrt{100 - x^2}$

$\quad = 10 \left[\ln (10 + \sqrt{100 - x^2}) - \ln x \right] - \sqrt{100 - x^2}$

$\dfrac{dy}{dx} = 10 \left[\dfrac{-x}{\sqrt{100 - x^2}(10 + \sqrt{100 - x^2})} \right) - \dfrac{1}{x} \right] + \dfrac{x}{\sqrt{100 - x^2}} = -\dfrac{\sqrt{100 - x^2}}{x}$

(a) when $x = 10$, $\dfrac{dy}{dx} = 0$ $\qquad$ (b) when $x = 5$, $\dfrac{dy}{dx} = -\sqrt{3}$

49. $p(x) \approx \dfrac{x}{\ln x}$, $\qquad \dfrac{dp}{dx} \approx \dfrac{\ln x - 1}{(\ln x)^2}$

(a) $x = 1000$, $\quad$ 12.4 primes/100 numbers

(b) $x = 1,000,000$, $\quad$ 6.7 primes/100 numbers

(c) $x = 1,000,000,000$, $\quad$ 4.6 primes/100 numbers

The natural logarithmic function and integration

1. $\int \dfrac{1}{x+1}\,dx = \ln|x+1| + C \qquad (u = x+1,\ u' = 1)$

2. $\int \dfrac{1}{x-5}\,dx = \ln|x-5| + C \qquad (u = x-5,\ u' = 1)$

3. $\int \dfrac{1}{3-2x}\,dx = -\dfrac{1}{2}\int \dfrac{1}{3-2x}(-2)\,dx \qquad (u = 3-2x,\ u' = -2)$

$\qquad = -\dfrac{1}{2}\ln|3-2x| + C$

4. $\int \dfrac{1}{6x+1}\,dx = \dfrac{1}{6}\int \dfrac{1}{6x+1}(6)\,dx \qquad (u = 6x+1,\ u' = 6)$

$\qquad = \dfrac{1}{6}\ln|6x+1| + C$

5. $\int \dfrac{x}{x^2+1}\,dx = \dfrac{1}{2}\int \dfrac{1}{x^2+1}(2x)\,dx \qquad (u = x^2+1,\ u' = 2x)$

$\qquad = \dfrac{1}{2}\ln(x^2+1) + C = \ln\sqrt{x^2+1} + C$

6. $\int \dfrac{x^2}{3-x^3}\,dx = -\dfrac{1}{3}\int \dfrac{1}{3-x^3}(-3x^2)\,dx \qquad (u = 3-x^3,\ u' = -3x^2)$

$\qquad = -\dfrac{1}{3}\ln|3-x^3| + C$

7. $\int \dfrac{x^2-4}{x}\,dx = \int \left(x - \dfrac{4}{x}\right)\,dx = \dfrac{x^2}{2} - 4\ln|x| + C$

8. $\int \dfrac{x+5}{x}\,dx = \int \left(1 + \dfrac{5}{x}\right)\,dx = x + 5\ln|x| + C$

9. $\int_1^e \dfrac{\ln x}{2x}\,dx = \dfrac{1}{2}\int_1^e \dfrac{\ln x}{x}\,dx \qquad (u = \ln x,\ u' = 1/x)$

$\qquad = \left[\dfrac{1}{4}\ln^2|x|\right]_1^e = \dfrac{1}{4}$

10. $\displaystyle\int_e^{e^2} \frac{1}{x \ln x}\, dx = \int_e^{e^2} \left(\frac{1}{\ln x}\right)\frac{1}{x}\, dx$ $(u = \ln x, \quad u' = 1/x)$

$\qquad = \Big[\ln |\ln |x||\Big]_e^{e^2} = \ln 2$

11. $\displaystyle\int_1^e \frac{(1 + \ln x)^2}{x}\, dx = \left[\frac{1}{3}(1 + \ln |x|)^3\right]_1^e = \frac{7}{3}$ $(u = 1 + \ln x, \quad u' = 1/x)$

12. $\displaystyle\int_0^1 \frac{x - 1}{x + 1}\, dx = \int_0^1 1\, dx + \int_0^1 \frac{-2}{x + 1}\, dx = \Big[x - 2\ln |x + 1|\Big]_0^1 = 1 - 2\ln 2$

13. $\displaystyle\int_0^2 \frac{x^2 - 2}{x + 1}\, dx = \int_0^2 \left(x - 1 - \frac{1}{x + 1}\right) dx = \left[\frac{1}{2}x^2 - x - \ln |x + 1|\right]_0^2 = -\ln 3$

14. $\displaystyle\int \frac{1}{(x + 1)^2}\, dx = \int (x + 1)^{-2}\, dx$ $(u = x + 1, \quad u' = 1)$

$\qquad = -(x + 1)^{-1} + C = \frac{-1}{x + 1} + C$

15. $\displaystyle\int \frac{1}{\sqrt{x + 1}}\, dx = \int (x + 1)^{-1/2}\, dx$ $(u = x + 1, \quad u' = 1)$

$\qquad = 2(x + 1)^{1/2} + C = 2\sqrt{x + 1} + C$

16. $\displaystyle\int \frac{x + 3}{x^2 + 6x + 7}\, dx = \frac{1}{2}\int \frac{2x + 6}{x^2 + 6x + 7}\, dx$ $(u = x^2 + 6x + 7, \ u' = 2(x + 3))$

$\qquad = \frac{1}{2}\ln |x^2 + 6x + 7| + C$

17. $\displaystyle\int \frac{x^2 + 2x + 3}{x^3 + 3x^2 + 9x + 1}\, dx$ $(u = x^3 + 3x^2 + 9x + 1, \ u' = 3(x^2 + 2x + 3))$

$\qquad = \frac{1}{3}\int \frac{3(x^2 + 2x + 3)}{x^3 + 3x^2 + 9x + 1}\, dx = \frac{1}{3}\ln |x^3 + 3x^2 + 9x + 1| + C$

18. $\displaystyle\int \frac{(\ln x)^2}{x}\, dx = \frac{1}{3}(\ln x)^3 + C$ $(u = \ln x, \quad u' = 1/x)$

19. $\int \dfrac{1}{x^{2/3}(1 + x^{1/3})}\, dx \qquad (u = 1 + x^{1/3}, \quad u' = 1/(3x^{2/3}))$

$= 3 \int \dfrac{1}{1 + x^{1/3}} (\dfrac{1}{3x^{2/3}})\, dx = 3 \ln |1 + x^{1/3}| + C$

20. $\int \dfrac{1}{x \ln x^2}\, dx = \dfrac{1}{2} \int \dfrac{1}{x \ln x}\, dx = \dfrac{1}{2} \ln |\ln |x|| + C \qquad (u = \ln x, \quad u' = 1/x)$

21. $\int \dfrac{1}{1 + \sqrt{x}}\, dx \qquad (u = 1 + \sqrt{x}, \quad du = \dfrac{dx}{2\sqrt{x}} \implies 2(u - 1)\, du = dx)$

$= 2 \int \dfrac{u - 1}{u}\, du = 2 \int 1\, du - 2 \int \dfrac{1}{u}\, du = 2u - 2 \ln u + C$

$= 2(1 + \sqrt{x}) - 2 \ln (1 + \sqrt{x}) + C$

22. $\int \dfrac{1 - \sqrt{x}}{1 + \sqrt{x}}\, dx \qquad (u = 1 + \sqrt{x}, \quad du = \dfrac{1}{2\sqrt{x}}\, dx \implies 2(u - 1)\, du = dx)$

$= 2 \int \dfrac{(2 - u)(u - 1)}{u}\, du = 2 \int \dfrac{-u^2 + 3u - 2}{u}\, du = 2 \int (-u + 3 - \dfrac{2}{u})\, du$

$= 2 \left[-\dfrac{u^2}{2} + 3u - 2 \ln |u| \right] + C = -u^2 + 6u - 4 \ln |u| + C$

$= -(1 + \sqrt{x})^2 + 6(1 + \sqrt{x}) - 4 \ln (1 + \sqrt{x}) + C$

23. $\int \dfrac{\sqrt{x}}{\sqrt{x} - 3}\, dx \qquad (u = \sqrt{x} - 3, \quad du = \dfrac{1}{2\sqrt{x}}\, dx \implies 2(u + 3)\, du = dx)$

$= 2 \int \dfrac{(u + 3)^2}{u}\, du = 2 \int \dfrac{u^2 + 6u + 9}{u}\, du = 2 \int (u + 6 + \dfrac{9}{u})\, du$

$= 2 \left[\dfrac{u^2}{2} + 6u + 9 \ln |u| \right] + C = u^2 + 12u + 18 \ln |u| + C$

$= (\sqrt{x} - 3)^2 + 12(\sqrt{x} - 3) + 18 \ln |\sqrt{x} - 3| + C$

24. $\displaystyle\int_0^2 \frac{1}{1+\sqrt{2x}}\ dx \qquad (u = 1 + \sqrt{2x}, \quad du = \frac{1}{\sqrt{2x}}\ dx \quad \Longrightarrow \quad (u-1)\ du = dx)$

$$= \int_1^3 \frac{(u-1)}{u}\ du = \int_1^3 (1 - \frac{1}{u})\ du = \left[u - \ln u \right]_1^3$$

$$= (3 - \ln 3) - (1 - \ln 1) = 2 - \ln 3 \approx 0.901$$

25. $\displaystyle\int \frac{\sqrt{x}}{1 - x\sqrt{x}}\ dx \qquad (u = 1 - x\sqrt{x} = 1 - x^{3/2}, \quad u' = -\frac{3}{2}x^{1/2})$

$$= -\frac{2}{3} \int \frac{1}{1 - x\sqrt{x}}(-\frac{3}{2}\sqrt{x})\ dx = -\frac{2}{3}\ln |1 - x\sqrt{x}| + C$$

26. $\displaystyle\int \frac{2x}{(x-1)^2}\ dx = \int \frac{2x - 2 + 2}{(x-1)^2}\ dx = \int \frac{2(x-1)}{(x-1)^2}\ dx + 2 \int \frac{1}{(x-1)^2}\ dx$

$$= 2 \int \frac{1}{x-1}\ dx + 2 \int \frac{1}{(x-1)^2}\ dx = 2\ln |x-1| - \frac{2}{(x-1)} + C$$

27. $\displaystyle\int \frac{x(x-2)}{(x-1)^3}\ dx = \int \frac{x^2 - 2x + 1 - 1}{(x-1)^3}\ dx = \int \frac{(x-1)^2}{(x-1)^3}\ dx - \int \frac{1}{(x-1)^3}\ dx$

$$= \int \frac{1}{x-1}\ dx - \int \frac{1}{(x-1)^3}\ dx = \ln |x-1| + \frac{1}{2(x-1)^2} + C$$

28. $\displaystyle\int \tan 5x\ dx = \frac{1}{5} \int \frac{5\sin 5x}{\cos 5x}\ dx = -\frac{1}{5}\ln |\cos 5x| + C$

29. $\displaystyle\int \csc 2x\ dx = \frac{1}{2} \int (\csc 2x)(2)\ dx = -\frac{1}{2}\ln |\csc 2x + \cot 2x| + C$

30. $\displaystyle\int \sec\frac{x}{2}\ dx = 2 \int \sec\frac{x}{2}(\frac{1}{2})\ dx = 2\ln \left| \sec\frac{x}{2} + \tan\frac{x}{2} \right| + C$

31. $\displaystyle\int \cos (1-x)\ dx = -\sin (1-x) + C$

32. $\displaystyle\int \frac{\tan^2 2x}{\sec 2x}\, dx = \int \frac{\sec^2 2x - 1}{\sec 2x}\, dx = \int (\sec 2x - \cos 2x)\, dx$

$\displaystyle\qquad = \frac{1}{2}\int (\sec 2x)(2)\, dx - \frac{1}{2}\int (\cos 2x)(2)\, dx$

$\displaystyle\qquad = \frac{1}{2}[\ln|\sec 2x + \tan 2x| - \sin 2x] + C$

33. $\displaystyle\int \frac{\sec x\, \tan x}{\sec x - 1}\, dx = \ln|\sec x - 1| + C$

34. $\displaystyle\int \frac{\sin x}{1 + \cos x}\, dx = -\int \frac{-\sin x}{1 + \cos x}\, dx = -\ln|1 + \cos x| + C$

35. $\displaystyle\int \frac{\cos t}{1 + \sin t}\, dt = \ln|1 + \sin t| + C$

36. $\displaystyle\int (\sec t + \tan t)\, dt = \ln|\sec t + \tan t| - \ln|\cos t| + C$

$\displaystyle\qquad = \ln\left| \frac{\sec t + \tan t}{\cos t} \right| + C = \ln|\sec t(\sec t + \tan t)| + C$

37. $\displaystyle\int (\csc x - \sin x)\, dx = -\ln|\csc x + \cot x| + \cos x + C$

38. $\displaystyle\int \frac{\sin^2 x - \cos^2 x}{\cos x}\, dx = \int \frac{1 - 2\cos^2 x}{\cos x}\, dx = \int (\sec x - 2\cos x)\, dx$

$\displaystyle\qquad = \ln|\sec x + \tan x| - 2\sin x + C$

39. $\displaystyle\int \frac{1 - \cos\theta}{\theta - \sin\theta}\, d\theta = \ln|\theta - \sin\theta| + C$

40. $\displaystyle\int (\csc 2\theta - \cot 2\theta)^2\, d\theta = \int (\csc^2 2\theta - 2\csc 2\theta \cot 2\theta + \cot^2 2\theta)\, d\theta$

$\displaystyle\qquad = \int (2\csc^2 2\theta - 2\csc 2\theta \cot 2\theta - 1)\, d\theta$

$\displaystyle\qquad = -\cot 2\theta + \csc 2\theta - \theta + C$

41. $-\ln |\cos x| + C = \ln \left| \dfrac{1}{\cos x} \right| + C = \ln |\sec x| + C$

42. $\ln |\sin x| + C = \ln \left| \dfrac{1}{\csc x} \right| + C = -\ln |\csc x| + C$

43. $\ln |\sec x + \tan x| + C = \ln \left| \dfrac{(\sec x + \tan x)(\sec x - \tan x)}{(\sec x - \tan x)} \right| + C$

$\qquad = \ln \left| \dfrac{\sec^2 x - \tan^2 x}{\sec x - \tan x} \right| + C = \ln \left| \dfrac{1}{\sec x - \tan x} \right| + C$

$\qquad = -\ln |\sec x - \tan x| + C$

44. $-\ln |\csc x + \cot x| + C = -\ln \left| \dfrac{(\csc x + \cot x)(\csc x - \cot x)}{(\csc x - \cot x)} \right| + C$

$\qquad = -\ln \left| \dfrac{\csc^2 x - \cot^2 x}{\csc x - \cot x} \right| + C = -\ln \left| \dfrac{1}{\csc x - \cot x} \right| + C$

$\qquad = \ln |\csc x - \cot x| + C$

45. $A = \displaystyle\int_1^4 \dfrac{x^2 + 4}{x} \, dx = \int_1^4 \left(x + \dfrac{4}{x} \right) dx = \left[\dfrac{x^2}{2} + 4 \ln x \right]_1^4 = (8 + 4 \ln 4) - \dfrac{1}{2}$

$\qquad = \dfrac{15}{2} + 8 \ln 2 \approx 13.045$ square units

46. $A = \displaystyle\int_1^5 \dfrac{x + 5}{x} \, dx = \int_1^5 \left(1 + \dfrac{5}{x} \right) dx = \left[x + 5 \ln x \right]_1^5 = 4 + 5 \ln 5$

$\qquad \approx 12.047$ square units

47. $P = \displaystyle\int \dfrac{3000}{1 + 0.25t} \, dt = (3000)(4) \int \dfrac{0.25}{1 + 0.25t} \, dt = 12{,}000 \ln |1 + 0.25t| + C$

$\quad P(0) = 12{,}000 \ln |1 + 0.25(0)| + C = 1000, \qquad C = 1000$

$\quad P = 12{,}000 \ln |1 + 0.25t| + 1000 = 1000[12 \ln |1 + 0.25t| + 1]$

$\quad P(3) = 1000[12(\ln 1.75) + 1] \approx 7{,}715$

48. $\dfrac{1}{50 - 40} \displaystyle\int_{40}^{50} \dfrac{90{,}000}{400 + 3x} \, dx = \left[3000 \ln |400 + 3x| \right]_{40}^{50} \approx \$168.27$

Review Exercises for Chapter 5

1. $\displaystyle\int \frac{2}{3\sqrt[3]{x}}\,dx = \frac{2}{3}\int x^{-1/3}\,dx = x^{2/3} + C$

2. $u = 3x, \qquad u' = 3$

$\displaystyle\int \frac{2}{\sqrt[3]{3x}}\,dx = \frac{2}{3}\int (3x)^{-1/3}(3)\,dx = (3x)^{2/3} + C$

3. $\displaystyle\int (2x^2 + x - 1)\,dx = \frac{2}{3}x^3 + \frac{1}{2}x^2 - x + C$

4. $\displaystyle\int \frac{x^3 - 2x^2 + 1}{x^2}\,dx = \int (x - 2 + x^{-2})\,dx = \frac{1}{2}x^2 - 2x - \frac{1}{x} + C$

5. $\displaystyle\int \frac{(1 + x)^2}{\sqrt{x}}\,dx \;=\; \int (x^{-1/2} + 2x^{1/2} + x^{3/2})\,dx$

$\displaystyle\quad = 2x^{1/2} + \frac{4}{3}x^{3/2} + \frac{2}{5}x^{5/2} + C = \frac{2\sqrt{x}}{15}(15 + 10x + 3x^2) + C$

6. $u = x^3 + 3, \qquad u' = 3x^2$

$\displaystyle\int x^2\sqrt{x^3 + 3}\,dx \;=\; \frac{1}{3}\int (x^3 + 3)^{1/2}(3x^2)\,dx = \frac{2}{9}(x^3 + 3)^{3/2} + C$

7. $u = x^3 + 3, \qquad u' = 3x^2$

$\displaystyle\int \frac{x^2}{\sqrt{x^3 + 3}}\,dx \;=\; \frac{1}{3}\int (x^3 + 3)^{-1/2}(3x^2)\,dx = \frac{2}{3}\sqrt{x^3 + 3} + C$

8. $\displaystyle\int \frac{x^2 + 2x}{(x + 1)^2}\,dx = \int \frac{(x^2 + 2x + 1) - 1}{(x + 1)^2}\,dx = \int \left[1 - \frac{1}{(x + 1)^2}\right]dx$

$\displaystyle\quad = \int 1\,dx - \int (x + 1)^{-2}\,dx = x + (x + 1)^{-1} + C = x + \frac{1}{x + 1} + C$

9. $\displaystyle\int (x^2 + 1)^3\,dx = \int (x^6 + 3x^4 + 3x^2 + 1)\,dx$

$\displaystyle\quad = \frac{1}{7}x^7 + \frac{3}{5}x^5 + x^3 + x + C$

10. $u = 2 - 5x, \qquad u' = -5$

$$\int \sqrt{2 - 5x}\ dx = -\frac{1}{5} \int (2 - 5x)^{1/2}(-5)\,dx = -\frac{2}{15}(2 - 5x)^{3/2} + C$$

11. $$\int \frac{\sin x}{1 + \cos x}\ dx = -\int \frac{-\sin x}{1 + \cos x}\ dx = -\ln |1 + \cos x| + C$$

12. $$\int \sin^3 x \cos x\ dx = \frac{1}{4}\sin^4 x + C \qquad (u = \sin x, \quad u' = \cos x)$$

13. $$\int \frac{2x}{1 + 4x^2}\ dx = \frac{1}{4} \int \frac{8x}{1 + 4x^2}\ dx = \frac{1}{4}\ln (1 + 4x^2) + C = \ln \sqrt[4]{1 + 4x^2} + C$$

14. $$\int \frac{\cos x}{\sqrt{\sin x}}\ dx = \int (\sin x)^{-1/2} \cos x\ dx = 2(\sin x)^{1/2} + C = 2\sqrt{\sin x} + C$$

15. $$\int \tan^n x \sec^2 x\ dx = \frac{\tan^{n+1} x}{n + 1} + C, \qquad n \neq -1$$

16. $$\int \frac{\sin \theta}{\sqrt{1 - \cos \theta}}\ d\theta = \int (1 - \cos \theta)^{-1/2} \sin \theta\ d\theta = 2(1 - \cos \theta)^{1/2} + C$$

$$= 2\sqrt{1 - \cos \theta} + C$$

17. $$\int \sec 2x \tan 2x\ dx = \frac{1}{2} \int (\sec 2x \tan 2x)(2)\ dx = \frac{1}{2} \sec 2x + C$$

18. $$\int x \sin 3x^2\ dx = \frac{1}{6} \int (\sin 3x^2)(6x)\ dx = -\frac{1}{6} \cos 3x^2 + C$$

19. $u = 7x - 2, \qquad u' = 7$

$$\int \frac{1}{7x - 2}\ dx = \frac{1}{7} \int \frac{1}{7x - 2}(7)\ dx = \frac{1}{7}\ln |7x - 2| + C$$

20. $u = x^2 - 1, \qquad u' = 2x$

$$\int \frac{x}{x^2 - 1}\ dx = \frac{1}{2} \int \frac{2x}{x^2 - 1}\ dx = \frac{1}{2}\ln |x^2 - 1| + C$$

21. $\int \dfrac{1}{x \ln 3x} \, dx = \ln |\ln 3x| + C \qquad (u = \ln 3x, \quad u' = \dfrac{1}{x})$

22. $u = \ln x, \qquad u' = 1/x$

$\int \dfrac{\ln \sqrt{x}}{x} \, dx = \dfrac{1}{2} \int (\ln x)(\dfrac{1}{x}) \, dx = \dfrac{1}{4}(\ln x)^2 + C$

23. $\int \dfrac{x^2 + 3}{x} \, dx = \int (x + \dfrac{3}{x}) \, dx = \dfrac{x^2}{2} + 3 \ln |x| + C$

24. $\int \dfrac{x^3 + 1}{x^2} \, dx = \int (x + \dfrac{1}{x^2}) \, dx = \dfrac{1}{2} x^2 - \dfrac{1}{x} + C$

25. $u = x^3 - 1, \qquad u' = 3x^2$

$\int \dfrac{x^2}{x^3 - 1} \, dx = \dfrac{1}{3} \int \dfrac{3x^2}{x^3 - 1} \, dx = \dfrac{1}{3} \ln |x^3 - 1| + C$

26. $\int (x + \dfrac{1}{x})^2 \, dx = \int (x^2 + 2 + x^{-2}) dx = \dfrac{1}{3} x^3 + 2x - \dfrac{1}{x} + C$

27. $u = \ln x, \qquad u' = 1/x$

$\int \dfrac{1}{x\sqrt{\ln x}} \, dx = \int (\ln x)^{-1/2}(\dfrac{1}{x}) \, dx = 2\sqrt{\ln x} + C$

28. $\int \dfrac{x + 2}{2x + 3} \, dx = \int \left[\dfrac{1}{2} + \dfrac{1/2}{2x + 3}\right] dx = \dfrac{1}{2}\left[\int dx + \int \dfrac{1}{2x + 3} \, dx\right]$

$= \dfrac{1}{2}\left[x + \dfrac{1}{2} \ln |2x + 3|\right] + C = \dfrac{1}{4}(2x + \ln |2x + 3|) + C$

29. (a) $\sum\limits_{i=1}^{10} (2i - 1)$ \qquad (b) $\sum\limits_{i=1}^{n} i^3$ \qquad (c) $\sum\limits_{i=1}^{10} (4i + 2)$

30. $x_1 = 2, \ x_2 = -1, \ x_3 = 5, \ x_4 = 3, \text{ and } x_5 = 7$

(a) $\dfrac{1}{5} \sum\limits_{i=1}^{5} x_i = \dfrac{1}{5}(2 - 1 + 5 + 3 + 7) = \dfrac{16}{5}$

(b) $\sum\limits_{i=1}^{5} \dfrac{1}{x_i} = \dfrac{1}{2} - 1 + \dfrac{1}{5} + \dfrac{1}{3} + \dfrac{1}{7} = \dfrac{37}{210}$

30.(c) $\displaystyle\sum_{i=1}^{5} (2x_i - x_i{}^2) = [2(2) - (2)^2] + [2(-1) - (-1)^2] + [2(5) - (5)^2]$

$+ [2(3) - (3)^2] + [2(7) - (7)^2] = -56$

(d) $\displaystyle\sum_{i=2}^{5} (x_i - x_{i-1}) = (-1 - 2) + (5 - (-1)) + (3 - 5) + (7 - 3) = 5$

31. $\displaystyle\int_0^4 (2 + x)\,dx = \left[2x + \frac{x^2}{2}\right]_0^4 = 8 + \frac{16}{2} = 16$

32. $\displaystyle\int_{-1}^1 (t^2 + 2)\,dt = \left[\frac{t^3}{3} + 2t\right]_{-1}^1 = \frac{14}{3}$

33. $\displaystyle\int_{-1}^1 (4t^3 - 2t)\,dt = \left[t^4 - t^2\right]_{-1}^1 = 0$

34. $u = x^2 - 8, \quad u' = 2x$

$\displaystyle\int_3^6 \frac{x}{3\sqrt{x^2 - 8}}\,dx = \frac{1}{6}\int_3^6 (x^2 - 8)^{-1/2}(2x)\,dx$

$= \frac{1}{3}(x^2 - 8)^{1/2}\Big]_3^6 = \frac{1}{3}(2\sqrt{7} - 1)$

35. $u = 1 + x, \quad u' = 1$

$\displaystyle\int_0^3 \frac{1}{\sqrt{1 + x}}\,dx = \int_0^3 (1 + x)^{-1/2}\,dx = 2(1 + x)^{1/2}\Big]_0^3 = 4 - 2 = 2$

36. $u = x^3 + 1, \quad u' = 3x^2$

$\displaystyle\int_0^1 x^2(x^3 + 1)^3\,dx = \frac{1}{3}\int_0^1 (x^3 + 1)^3(3x^2)\,dx = \frac{1}{12}(x^3 + 1)^4\Big]_0^1$

$= \frac{1}{12}(16 - 1) = \frac{5}{4}$

37. $\displaystyle\int_4^9 x\sqrt{x}\,dx = \int_4^9 x^{3/2}\,dx = \frac{2}{5}x^{5/2}\Big]_4^9 = \frac{2}{5}[(\sqrt{9})^5 - (\sqrt{4})^5]$

$= \frac{2}{5}(243 - 32) = \frac{422}{5}$

38. $\displaystyle\int_1^2 \left(\frac{1}{x^2} - \frac{1}{x^3}\right)dx = \int_1^2 (x^{-2} - x^{-3})\,dx = \left[-\frac{1}{x} + \frac{1}{2x^2}\right]_1^2$

$= \left(-\frac{1}{2} + \frac{1}{8}\right) - \left(-1 + \frac{1}{2}\right) = \frac{1}{8}$

39. $\displaystyle\int_1^4 \frac{x+1}{x}\,dx = \int_1^4 \left(1 + \frac{1}{x}\right)dx = \left[x + \ln|x|\right]_1^4 = 3 + \ln 4$

40. $\displaystyle\int_1^e \frac{\ln x}{x}\,dx = \int_1^e (\ln x)^1\left(\frac{1}{x}\right)dx = \frac{1}{2}(\ln x)^2\Big]_1^e = \frac{1}{2}$

41. $\displaystyle\int_0^{\pi/3} \sec\theta\,d\theta = \ln|\sec\theta + \tan\theta|\Big]_0^{\pi/3} = \ln(2 + \sqrt{3})$

42. $\displaystyle\int_0^{\pi/4} \tan\left(\frac{\pi}{4} - x\right)dx = \ln\left|\cos\left(\frac{\pi}{4} - x\right)\right|\Big]_0^{\pi/4} = 0 - \ln\left(\frac{1}{\sqrt{2}}\right) = \frac{1}{2}\ln 2$

43. $\displaystyle\int_{-\sqrt{\pi/4}}^{\sqrt{\pi/4}} x\tan x^2\,dx = -\frac{1}{2}\ln|\cos x^2|\Big]_{-\sqrt{\pi/4}}^{\sqrt{\pi/4}} = 0$

44. $\displaystyle 2\pi\int_0^1 (y+1)\sqrt{1-y}\,dy \qquad (u = \sqrt{1-y},\quad y = 1 - u^2,\quad dy = -2u\,du)$

$$= 2\pi\int_1^0 (2 - u^2)(u)(-2u)\,du = -4\pi\int_1^0 (2u^2 - u^4)\,du$$

$$= -4\pi\left(\frac{2}{3}u^3 - \frac{1}{5}u^5\right)\Big]_1^0 = \frac{28\pi}{15}$$

45. $\displaystyle\int_0^1 \frac{1 - \sqrt{y}}{1 + \sqrt{y}}\,dy \qquad (u = \sqrt{y},\quad y = u^2,\quad dy = 2u\,du)$

$$= \int_0^1 \frac{1 - u}{1 + u}2u\,du = 2\int_0^1 \frac{u - u^2}{1 + u}\,du = 2\int_0^1 \left(-u + 2 - \frac{2}{1 + u}\right)du$$

$$= 2\left(-\frac{1}{2}u^2 + 2u - 2\ln|1 + u|\right)\Big]_0^1 = 3 - 4\ln 2$$

46. $\displaystyle\int_0^1 \frac{1}{1 + \sqrt{y}}\,dy \qquad (u = 1 + \sqrt{y},\quad y = (u - 1)^2,\quad dy = 2(u - 1)\,du)$

$$= \int_1^2 \frac{2(u - 1)}{u}\,du = 2\int_1^2 \left(1 - \frac{1}{u}\right)du = 2(u - \ln|u|)\Big]_1^2 = 2(1 - \ln 2)$$

47. $2\pi \displaystyle\int_{-1}^{0} x^2 \sqrt{x+1}\, dx \qquad (u = \sqrt{x+1}, \quad x = u^2 - 1, \quad dx = 2u\, du)$

$$= 2\pi \int_{0}^{1} (u^2 - 1)^2 u(2u)\, du = 4\pi \int_{0}^{1} (u^6 - 2u^4 + u^2)\, du$$

$$= 4\pi(\frac{1}{7} u^7 - \frac{2}{5} u^5 + \frac{1}{3} u^3)\Big]_{0}^{1} = \frac{32\pi}{105}$$

48. $\displaystyle\int_{3}^{4} \frac{1}{x-2}\, dx - \int_{3}^{4} \frac{1}{x+2}\, dx = \ln|x-2|\;\Big]_{3}^{4} - \ln|x+2|\;\Big]_{3}^{4}$

$$= \ln 2 - \ln 6 + \ln 5 = \ln\frac{5}{3}$$

49. $y = \ln\sqrt{x} = \dfrac{1}{2}\ln x, \qquad \dfrac{dy}{dx} = \dfrac{1}{2x}$

50. $y = \ln\dfrac{x(x-1)}{x-2} = \ln x + \ln(x-1) - \ln(x-2)$

$$\frac{dy}{dx} = \frac{1}{x} + \frac{1}{x-1} - \frac{1}{x-2} = \frac{x^2 - 4x + 2}{x^3 - 3x^2 + 2x}$$

51. $y = x\sqrt{\ln x}$

$$\frac{dy}{dx} = (\frac{x}{2})(\ln x)^{-1/2}(\frac{1}{x}) + \sqrt{\ln x} = \frac{1}{2\sqrt{\ln x}} + \sqrt{\ln x} = \frac{1 + 2\ln x}{2\sqrt{\ln x}}$$

52. $y = \ln[x(x^2 - 2)^{2/3}] = \ln x + \dfrac{2}{3}\ln(x^2 - 2)$

$$\frac{dy}{dx} = \frac{1}{x} + \frac{2}{3}(\frac{2x}{x^2 - 2}) = \frac{7x^2 - 6}{3x^3 - 6x}$$

53. $y(\ln x) + y^2 = 0, \qquad y(\dfrac{1}{x}) + (\ln x)(\dfrac{dy}{dx}) + 2y(\dfrac{dy}{dx}) = 0$

$$(2y + \ln x)\,\frac{dy}{dx} = \frac{-y}{x}, \qquad \frac{dy}{dx} = \frac{-y}{x(2y + \ln x)}$$

54. $\ln(x+y) = x, \qquad \dfrac{1}{x+y}(1 + \dfrac{dy}{dx}) = 1, \qquad (\dfrac{1}{x+y})\dfrac{dy}{dx} = 1 - \dfrac{1}{x+y}$

$$\frac{dy}{dx} = x + y - 1$$

55. $\ln y = x\ln x, \qquad \dfrac{1}{y}(\dfrac{dy}{dx}) = x(\dfrac{1}{x}) + (\ln x)(1), \qquad \dfrac{dy}{dx} = y(1 + \ln x)$

56. $\ln y = x \ln (\sin x)$, $\quad \dfrac{1}{y}(\dfrac{dy}{dx}) = x(\dfrac{\cos x}{\sin x}) + (1)[\ln (\sin x)]$

$\dfrac{dy}{dx} = y[x \cot x + \ln (\sin x)]$

57. $y = \dfrac{1}{b^2}\left[\ln (a + bx) + \dfrac{a}{a + bx}\right]$

$\dfrac{dy}{dx} = \dfrac{1}{b^2}\left[\dfrac{b}{a + bx} - \dfrac{ab}{(a + bx)^2}\right] = \dfrac{x}{(a + bx)^2}$

58. $y = \dfrac{1}{b^2}[a + bx - a \ln (a + bx)]$, $\quad \dfrac{dy}{dx} = \dfrac{1}{b^2}\left[b - \dfrac{ab}{a + bx}\right] = \dfrac{x}{a + bx}$

59. $y = -\dfrac{1}{a} \ln (\dfrac{a + bx}{x}) = -\dfrac{1}{a}[\ln (a + bx) - \ln x]$

$\dfrac{dy}{dx} = -\dfrac{1}{a}\left[\dfrac{b}{a + bx} - \dfrac{1}{x}\right] = \dfrac{1}{x(a + bx)}$

60. $y = -\dfrac{1}{ax} + \dfrac{b}{a^2} \ln \dfrac{a + bx}{x} = -\dfrac{1}{ax} + \dfrac{b}{a^2}[\ln (a + bx) - \ln x]$

$\dfrac{dy}{dx} = \dfrac{1}{ax^2} + (\dfrac{b}{a^2})\dfrac{-a}{x(a + bx)} = \dfrac{1}{x^2(a + bx)}$

61. $f'(x) = -2x$, $(-1, 1)$, $\quad f(x) = \displaystyle\int -2x \, dx = -x^2 + C$

when $x = -1$: $y = -1 + C = 1$, $\quad C = 2$, $\quad y = 2 - x^2$

62. $f''(x) = 6(x - 1)$, $\quad$ tangent to $3x - y - 5 = 0$ at $(2, 1)$

$f'(x) = \displaystyle\int 6(x - 1) \, dx = 3(x - 1)^2 + C_1$

$f'(2) = 3 + C_1 = 3$ when $C_1 = 0$, $\quad f'(x) = 3(x - 1)^2$

$f(x) = \displaystyle\int 3(x - 1)^2 \, dx = (x - 1)^3 + C_2$, $\quad f(2) = 1 + C_2 = 1$ when $C_2 = 0$

$f(x) = (x - 1)^3$

63. $a(t) = a$, $\quad v(t) = \displaystyle\int a\ dt = at + C_1$, $\quad v(0) = 0 + C_1 = 0$ when $C_1 = 0$

$v(t) = at$, $\quad s(t) = \displaystyle\int at\ dt = \dfrac{a}{2}t^2 + C_2$,

$s(0) = 0 + C_2 = 0$ when $C_2 = 0$, $\quad s(30) = \dfrac{a}{2}(30)^2 = 3600$ or

$a = \dfrac{2(3600)}{(30)^2} = 8$ ft/sec^2, $\quad v(30) = 8(30) = 240$ ft/sec

64. 45 mph = 66 ft/sec $\quad$ 30 mph = 44 ft/sec, $\quad a(t) = -a$

$v(t) = -at + 66$ since $v(0) = 66$ ft/sec

$s(t) = -\dfrac{a}{2}t^2 + 66t$ since $s(0) = 0$

Solving the system $v(t) = -at + 66 = 44$, $\quad s(t) = -\dfrac{a}{2}t^2 + 66t = 264$

we obtain $t = 24/5$ and $a = 55/12$. We now solve $-\dfrac{55}{12}t + 66 = 0$

and get $t = \dfrac{792}{55}$. $\quad s(\dfrac{792}{55}) = -\dfrac{1}{2}(\dfrac{55}{12})(\dfrac{792}{55})^2 + 66(\dfrac{792}{55}) \approx 475.2$ ft

Stopping distance from 30 mph to rest = $475.2 - 264 = 211.2$ ft.

65. $a(t) = -32$ $\qquad$ (a) $\quad v(t) = -32t + 96 = 0$ when $t = 3$ sec

$v(t) = -32t + 96$ $\qquad$ (b) $\quad s(3) = -144 + 288 = 144$ ft

$s(t) = -16t^2 + 96t$ $\qquad$ (c) $\quad v(t) = -32t + 96 = 96/2$ when $t = 3/2$ sec

$\qquad$ (d) $\quad s(3/2) = -16(9/4) + 96(3/2) = 108$ ft

66. $a(t) = -32$ $\qquad$ (a) $\quad v(t) = -32t + 128 = 0$ when $t = 4$ sec

$v(t) = -32t + 128$ $\qquad$ (b) $\quad s(4) = 256$ ft

$s(t) = -16t^2 + 128t$ $\qquad$ (c) $\quad v(t) = -32t + 128 = \dfrac{128}{2}$ when $t = 2$ sec

$\qquad$ (d) $\quad s(2) = 192$ ft

67.(a) $S = m(\frac{b}{4})(\frac{b}{4}) + m(\frac{2b}{4})(\frac{b}{4}) + m(\frac{3b}{4})(\frac{b}{4}) + m(\frac{4b}{4})(\frac{b}{4})$

$= \frac{mb^2}{16}(1 + 2 + 3 + 4) = \frac{5mb^2}{8}$

$s = m(0)(\frac{b}{4}) + m(\frac{b}{4})(\frac{b}{4}) + m(\frac{2b}{4})(\frac{b}{4}) + m(\frac{3b}{4})(\frac{b}{4})$

$= \frac{mb^2}{16}(1 + 2 + 3) = \frac{3mb^2}{8}$

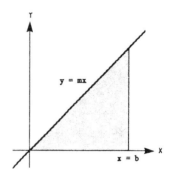

(b) $S(n) = \sum_{i=1}^{n} f(\frac{bi}{n})(\frac{b}{n}) = \sum_{i=1}^{n} (\frac{mbi}{n})(\frac{b}{n}) = m(\frac{b}{n})^2 \sum_{i=1}^{n} i$

$= \frac{mb^2}{n^2}(\frac{n(n + 1)}{2}) = \frac{mb^2(n + 1)}{2n}$

$s(n) = \sum_{i=0}^{n-1} f(\frac{bi}{n})(\frac{b}{n}) = \sum_{i=0}^{n-1} m(\frac{bi}{n})(\frac{b}{n}) = m(\frac{b}{n})^2 \sum_{i=0}^{n-1} i$

$= \frac{mb^2}{n^2}(\frac{(n - 1)n}{2}) = \frac{mb^2(n - 1)}{2n}$

(c) $\text{Area} = \lim_{n \to \infty} \frac{mb^2(n + 1)}{2n} = \lim_{n \to \infty} \frac{mb^2(n - 1)}{2n} = \frac{1}{2}mb^2$

(d) $\int_{0}^{b} mx\ dx = \frac{1}{2}mx^2 \Big]_{0}^{b} = \frac{1}{2}mb^2$

68.(a) $S(n) = \sum_{i=1}^{n} f(1 + \frac{2i}{n})(\frac{2}{n}) = \sum_{i=1}^{n} (1 + \frac{2i}{n})^3(\frac{2}{n})$

$= \sum_{i=1}^{n} (1 + \frac{6i}{n} + \frac{12i^2}{n^2} + \frac{8i^3}{n^3})(\frac{2}{n})$

$= 2 + \frac{12}{n^2}\sum_{i=1}^{n} i + \frac{24}{n^3}\sum_{i=1}^{n} i^2 + \frac{16}{n^4}\sum_{i=1}^{n} i^3$

$= 2 + 6(1 + \frac{1}{n}) + 4(2 + \frac{3}{n} + \frac{1}{n^2}) + 4(1 + \frac{2}{n} + \frac{1}{n^2})$

Area = $\lim_{n \to \infty}$ $S(n) = 2 + 6 + 8 + 4 = 20$

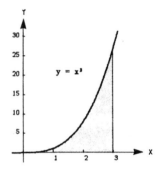

(b) $\int_{1}^{3} x^3 dx = \frac{x^4}{4}\Big]_{1}^{3} = \frac{81}{4} - \frac{1}{4} = 20$

69. $\int_{1}^{3} (2x - 1) dx = \left[x^2 - x\right]_{1}^{3} = 6$

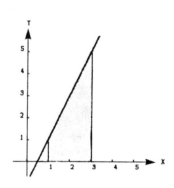

70. $\int_{0}^{2} (x + 4) dx = \left[\frac{x^2}{2} + 4x\right]_{0}^{2} = 10$

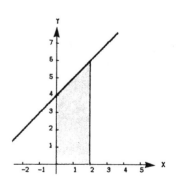

71. $\displaystyle\int_{3}^{4} (x^2 - 9)\ dx = \left[\frac{x^3}{3} - 9x\right]_{3}^{4}$

$$= (\frac{64}{3} - 36) - (9 - 27)$$

$$= \frac{64}{3} - \frac{54}{3} = \frac{10}{3}$$

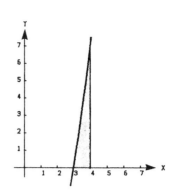

72. $\displaystyle\int_{-1}^{2} (-x^2 + x + 2)\ dx = \left[-\frac{x^3}{3} + \frac{x^2}{2} + 2x\right]_{-1}^{2}$

$$= (-\frac{8}{3} + 2 + 4) - (\frac{1}{3} + \frac{1}{2} - 2)$$

$$= \frac{10}{3} + \frac{7}{6} = \frac{9}{2}$$

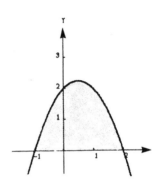

73. $\displaystyle\int_{0}^{1} (x - x^3)\ dx = \left[\frac{x^2}{2} - \frac{x^4}{4}\right]_{0}^{1}$

$$= \frac{1}{2} - \frac{1}{4} = \frac{1}{4}$$

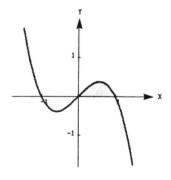

74. $\displaystyle\int_{0}^{1} \sqrt{x}(1 - x)\ dx = \int_{0}^{1} (x^{1/2} - x^{3/2})\ dx$

$$= \left[\frac{2}{3}x^{3/2} - \frac{2}{5}x^{5/2}\right]_{0}^{1}$$

$$= \frac{2}{3} - \frac{2}{5} = \frac{4}{15}$$

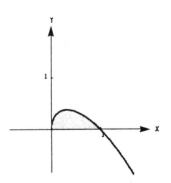

75. $\int_0^{\pi/3} \tan x \, dx = -\ln|\cos x| \Big]_0^{\pi/3}$

$= -\ln\frac{1}{2} = \ln 2$

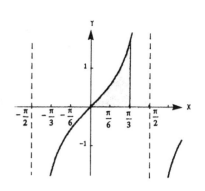

76. $\int_0^1 \frac{1}{x+1} \, dx = \ln|x+1| \Big]_0^1 = \ln 2$

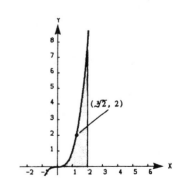

77. $\frac{1}{10-5} \int_5^{10} \frac{1}{\sqrt{x-1}} \, dx = \frac{2}{5}\sqrt{x-1} \Big]_5^{10} = \frac{2}{5}$

$\frac{1}{\sqrt{x-1}} = \frac{2}{5}, \qquad \sqrt{x-1} = \frac{5}{2}$

$x - 1 = \frac{25}{4}, \qquad x = \frac{29}{4}$

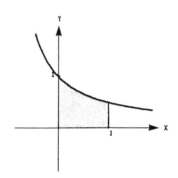

78. $\frac{1}{2-0} \int_0^2 x^3 \, dx = \frac{x^4}{8} \Big]_0^2 = 2$

$x^3 = 2, \qquad x = \sqrt[3]{2}$

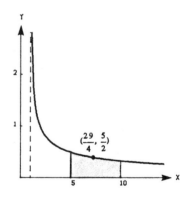

79. $\dfrac{1}{4 - 0} \displaystyle\int_0^4 x \, dx = \dfrac{x^2}{8} \Big]_0^4 = 2$

$x = 2$

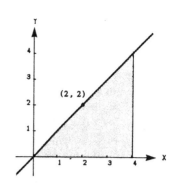

80. $\dfrac{1}{2 - 1} \displaystyle\int_1^2 \left(x^2 - \dfrac{1}{x^2}\right) dx = \left(\dfrac{x^3}{3} + \dfrac{1}{x}\right) \Big]_1^2$

$= \left(\dfrac{8}{3} + \dfrac{1}{2}\right) - \left(\dfrac{1}{3} + 1\right) = \dfrac{11}{6}$

$x^2 - \dfrac{1}{x^2} = \dfrac{11}{6}, \qquad 6x^4 - 6 = 11x^2$

$6x^4 - 11x^2 - 6 = 0$ when $x^2 = \dfrac{11 \pm \sqrt{265}}{12}$

In the interval $[1, 2]$: $x = \sqrt{\dfrac{11 + \sqrt{265}}{12}} \approx 1.508$

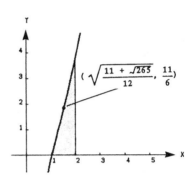

81. $V = \displaystyle\int_0^3 0.85 \sin \dfrac{\pi t}{3} \, dt = -\dfrac{3}{\pi}(0.85) \cos \dfrac{\pi t}{3} \Big]_0^3$

$= -\dfrac{2.55}{\pi}(-1 - 1) = \dfrac{5.1}{\pi} \approx 1.6234 \text{ liters}$

82. $\displaystyle\int_0^3 1.75 \sin \dfrac{\pi t}{2} \, dt = -\dfrac{2}{\pi}(1.75) \cos \dfrac{\pi t}{2} \Big]_0^2$

$= -\dfrac{2}{\pi}(1.75)(-1 - 1) = \dfrac{7}{\pi} \approx 2.2282 \text{ liters}$

Increase is $\dfrac{1.9}{\pi}$ liters

83. (a) $C = 0.1 \int_8^{20} \left[12 \sin \dfrac{\pi(t-8)}{12} \right] dt = -\dfrac{14.4}{\pi} \cos \dfrac{\pi(t-8)}{12} \Big]_8^{20}$

$= \dfrac{-14.4}{\pi}(-1-1) \approx \$9.17$

(b) $C = 0.1 \int_{10}^{18} \left[12 \sin \dfrac{\pi(t-8)}{12} - 0.6 \right] dt = \left[\dfrac{-14.4}{\pi} \cos \dfrac{\pi(t-8)}{12} - 0.6t \right]_{10}^{18}$

$= \left[\dfrac{-14.4}{\pi}(\dfrac{-\sqrt{3}}{2}) - 10.8 \right] - \left[\dfrac{-14.4}{\pi}(\dfrac{\sqrt{3}}{2}) - 6 \right] \approx \$3.14$

Savings $\approx 9.17 - 3.14 = \$6.03$

84. $\dfrac{1}{365} \int_0^{365} 100{,}000 \left[1 + \sin \dfrac{2\pi(t-60)}{365} \right] dt$

$\dfrac{100{,}000}{365} \left[t - \dfrac{365}{2\pi} \cos \dfrac{2\pi(t-60)}{365} \right]_0^{365} = 100{,}000$

85. $p = 1 + 0.1t + 0.02t^2$

$C = \dfrac{15{,}000}{M} \int_t^{t+1} (1 + 0.1t + 0.02t^2) \, dt = \dfrac{15{,}000}{M} \left[t + 0.05t^2 + \dfrac{0.02}{3} t^3 \right]_t^{t+1}$

(a) Since 1983 is represented by $t = 0$, 1985 is represented by $t = 2$

$C = \dfrac{15{,}000}{M} \left[t + 0.05t^2 + \dfrac{0.02}{3} t^3 \right]_2^3 = \dfrac{20{,}650}{M}$

(b) 1990 is represented by $t = 7$

$C = \dfrac{15{,}000}{M} \left[t + 0.05t^2 + \dfrac{0.02}{3} t^3 \right]_7^8 = \dfrac{43{,}150}{M}$

86. $u = \sqrt{1-x}, \qquad x = 1 - u^2, \qquad dx = -2u \, du$

$Pa, b = \int_a^b \dfrac{15}{4} x \sqrt{1-x} \, dx \; = \dfrac{15}{4} \int_{\sqrt{1-a}}^{\sqrt{1-b}} (1 - u^2) u (-2u) \, du$

$= -\dfrac{15}{2} \int_{\sqrt{1-a}}^{\sqrt{1-b}} (u^2 - u^4) \, du = -\dfrac{15}{2}(\dfrac{u^3}{3} - \dfrac{u^5}{5}) \Big]_{\sqrt{1-a}}^{\sqrt{1-b}}$

$= -\dfrac{u^3}{2}(5 - 3u^2) \Big]_{\sqrt{1-a}}^{\sqrt{1-b}} = -\dfrac{(1-x)^{3/2}}{2}(3x + 2) \Big]_a^b$

86.(a) $P_{0.50, 0.75} = -\dfrac{(1 - x)^{3/2}}{2}(3x + 2) \Big]_{0.50}^{0.75} = 35.3\%$

(b) $P_{0, b} = -\dfrac{(1 - x)^{3/2}}{2}(3x + 2) \Big]_{0}^{b} = -\dfrac{(1 - b)^{3/2}}{2}(3b + 2) + 1 = 0.5$

$(1 - b)^{3/2}(3b + 2) = 1, \quad b \approx 58.6\%$

87. $u = \sqrt{1 - x}, \qquad x = 1 - u^2, \qquad dx = -2u \, du$

$P_{a, b} = \displaystyle\int_{a}^{b} \dfrac{1155}{32} x^3 (1 - x)^{3/2} \, dx = \dfrac{1155}{32} \int_{\sqrt{1-a}}^{\sqrt{1-b}} (1 - u^2)^3 u^3 (-2u) \, du$

$\quad = \dfrac{1155}{16} \displaystyle\int_{\sqrt{1-a}}^{\sqrt{1-b}} (u^{10} - 3u^8 + 3u^6 - u^4) \, du$

$\quad = \dfrac{1155}{16} \left(\dfrac{u^{11}}{11} - \dfrac{u^9}{3} + \dfrac{3u^7}{7} - \dfrac{u^5}{5} \right) \Bigg]_{\sqrt{1-a}}^{\sqrt{1-b}}$

(a) $P_{0, 0.25} = \dfrac{1155}{16} \left(\dfrac{u^{11}}{11} - \dfrac{u^9}{3} + \dfrac{3u^7}{7} - \dfrac{u^5}{5} \right) \Bigg]_{1}^{\sqrt{0.75}} \approx 0.025 = 2.5\%$

(b) $P_{0.50, 1} = \dfrac{1155}{16} \left(\dfrac{u^{11}}{11} - \dfrac{u^9}{3} + \dfrac{3u^7}{7} - \dfrac{u^5}{5} \right) \Bigg]_{\sqrt{0.50}}^{0} \approx 0.736 = 73.6\%$

6 *Inverse functions*

6.1
Inverse functions

1. $f(x) = x^3$, $\qquad g(x) = \sqrt[3]{x}$

$\quad f(g(x)) = f(\sqrt[3]{x}) = (\sqrt[3]{x})^3 = x$

$\quad g(f(x)) = g(x^3) = \sqrt[3]{x^3} = x$

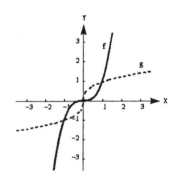

2. $f(x) = \dfrac{1}{x}$, $\qquad g(x) = \dfrac{1}{x}$

$\quad f(g(x)) = g(f(x)) = \dfrac{1}{1/x} = x$

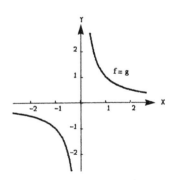

3. $f(x) = 5x + 1$, $\qquad g(x) = \dfrac{x - 1}{5}$

$\quad f(g(x)) = f(\dfrac{x - 1}{5}) = 5(\dfrac{x - 1}{5}) + 1 = x$

$\quad g(f(x)) = g(5x + 1) = \dfrac{(5x + 1) - 1}{5} = x$

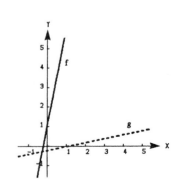

4. $f(x) = 3 - 4x, \qquad g(x) = \dfrac{3 - x}{4}$

    $f(g(x)) = f(\dfrac{3 - x}{4}) = 3 - 4(\dfrac{3 - x}{4}) = x$

    $g(f(x)) = g(3 - 4x) = \dfrac{3 - (3 - 4x)}{4} = x$

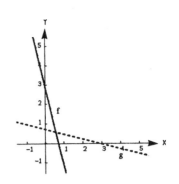

5. $f(x) = \sqrt{x - 4}, \qquad g(x) = x^2 + 4, \quad x \geq 0$

    $f(g(x)) = f(x^2 + 4) = \sqrt{(x^2 + 4) - 4}$

             $= \sqrt{x^2} = x$

    $g(f(x)) = g(\sqrt{x - 4}) = (\sqrt{x - 4})^2 + 4$

             $= x - 4 + 4 = x$

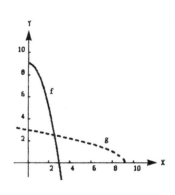

6. $f(x) = 9 - x^2, \qquad x \geq 0$

    $g(x) = \sqrt{9 - x}, \qquad x \leq 9$

    $f(g(x)) = f(\sqrt{9 - x}) = 9 - (\sqrt{9 - x})^2$

             $= 9 - (9 - x) = x$

    $g(f(x)) = g(9 - x^2) = \sqrt{9 - (9 - x^2)}$

             $= \sqrt{x^2} = x$

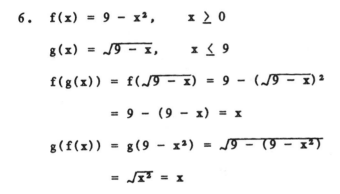
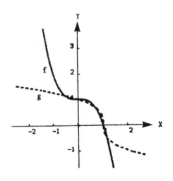

7. $f(x) = 1 - x^3, \qquad g(x) = \sqrt[3]{1 - x}$

    $f(g(x)) = f(\sqrt[3]{1 - x}) = 1 - (\sqrt[3]{1 - x})^3$

             $= 1 - (1 - x) = x$

    $g(f(x)) = g(1 - x^3) = \sqrt[3]{1 - (1 - x^3)}$

             $= \sqrt[3]{x^3} = x$

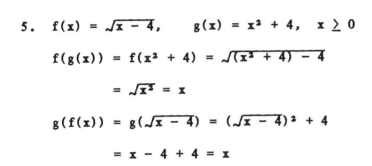
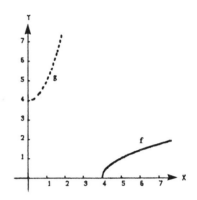
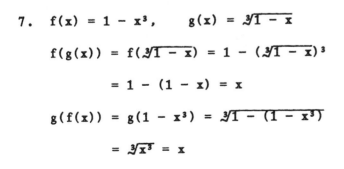

8. $f(x) = \dfrac{1}{1 + x^2}, \quad x \geq 0, \qquad g(x) = \sqrt{\dfrac{1 - x}{x}}, \quad 0 < x \leq 1$

$f(g(x)) = f\left(\sqrt{\dfrac{1 - x}{x}}\right)$

$= \dfrac{1}{1 + (\sqrt{(1 - x)/x})^2} = \dfrac{1}{1 + (1 - x)/x}$

$= \dfrac{x}{x + (1 - x)} = \dfrac{x}{1} = x$

$g(f(x)) = g\left(\dfrac{1}{1 + x^2}\right) = \sqrt{\dfrac{1 - (1 + x^2)^{-1}}{(1 + x^2)^{-1}}}$

$= \sqrt{\dfrac{(1 + x^2) - 1}{1}} = \sqrt{x^2} = x$

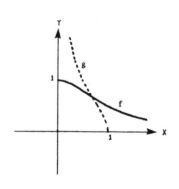

9. $f(x) = 2x - 3 = y$

$x = (y + 3)/2$

$f^{-1}(x) = \dfrac{x + 3}{2}$

10. $f(x) = 3x = y, \quad x = y/3$

$f^{-1}(x) = x/3$

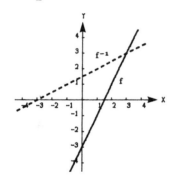

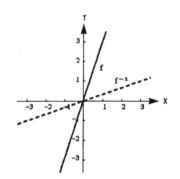

11. $f(x) = x^5 = y, \quad x = \sqrt[5]{y}$

$f^{-1}(x) = \sqrt[5]{x}$

12. $f(x) = x^3 + 1 = y, \quad x = \sqrt[3]{y - 1}$

$f^{-1}(x) = \sqrt[3]{x - 1}$

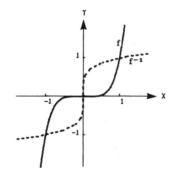

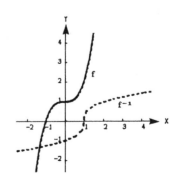

13. $f(x) = \sqrt{x} = y$, $x = y^2$

$f^{-1}(x) = x^2$

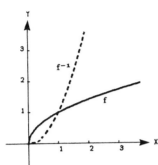

14. $f(x) = x^2 = y, (0 \leq x)$, $x = \sqrt{y}$

$f^{-1}(x) = \sqrt{x}$

15. $f(x) = \sqrt{4 - x^2} = y$, $0 \leq x \leq 2$

$x = \sqrt{4 - y^2}$

$f^{-1}(x) = \sqrt{4 - x^2}$, $0 \leq x \leq 2$

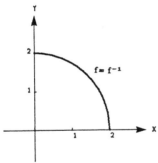

16. $f(x) = \sqrt{x^2 - 4} = y$, $x \geq 2$

$x = \sqrt{y^2 + 4}$

$f^{-1}(x) = \sqrt{x^2 + 4}$, $x \geq 0$

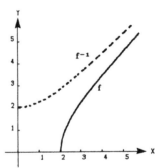

17. $f(x) = \sqrt[3]{x - 1} = y$

$x = y^3 + 1$

$f^{-1}(x) = x^3 + 1$

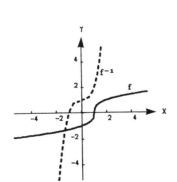

18. $f(x) = 3\sqrt[5]{2x - 1} = y$

$x = \dfrac{y^5 + 243}{486}$

$f^{-1}(x) = \dfrac{x^5 + 243}{486}$

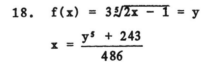
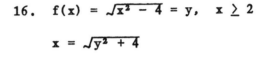
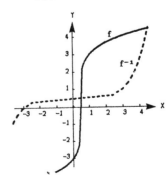

19. $f(x) = x^{2/3} = y, \quad x \geq 0$
    $x = y^{3/2}$
    $f^{-1}(x) = x^{3/2}, \quad x \geq 0$

20. $f(x) = x^{3/5}, \quad x = y^{5/3}$

    $f^{-1}(x) = x^{5/3}$

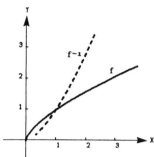

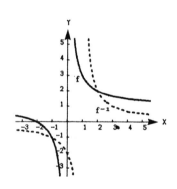

21. $f(x) = \dfrac{x}{\sqrt{x^2 + 7}} = y$

    $x = \dfrac{\sqrt{7}y}{\sqrt{1 - y^2}}$

    $f^{-1}(x) = \dfrac{\sqrt{7}\,x}{\sqrt{1 - x^2}}, \quad -1 < x < 1$

22. $f(x) = \dfrac{x + 2}{x} = y, \quad x = \dfrac{2}{y - 1}$

    $f^{-1}(x) = \dfrac{2}{x - 1}$

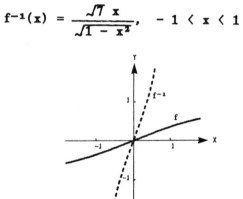

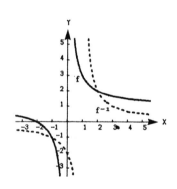

23. $f(x) = \dfrac{3}{4}x + 6$

    One-to-one, has an inverse

24. $f(x) = 5x - 3$

    One-to-one, has an inverse

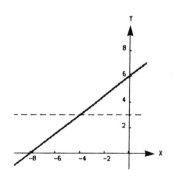

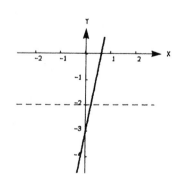

25. $f(\theta) = \sin \theta$

Not one-to-one, does not have an inverse

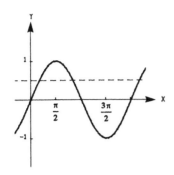

26. $F(x) = \dfrac{x^2}{x^2 + 4}$

Not one-to-one, does not have an inverse

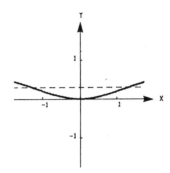

27. $h(s) = \dfrac{1}{s - 2} - 3$

One-to-one, has an inverse

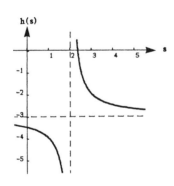

28. $g(t) = \dfrac{t}{\sqrt{t^2 + 1}}$

One-to-one, has an inverse

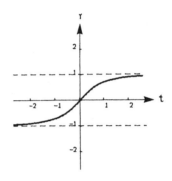

29. $f(x) = \ln x$

One-to-one, has an inverse

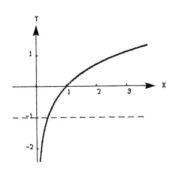

30. $f(x) = 3x\sqrt{x + 1}$

Not one-to-one, does not have an inverse

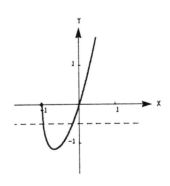

31. $f(x) = (x + a)^3 + b,$ $f'(x) = 3(x + a)^2 \geq 0$ for all x

    f is increasing on $(-\infty, \infty)$, therefore f is strictly monotonic and has an inverse.

32. $f(x) = \cos\dfrac{3x}{2},$ $f'(x) = -\dfrac{3}{2} \sin\dfrac{3x}{2} = 0$ when $x = 0, \dfrac{2\pi}{3}, \dfrac{4\pi}{3}, \ldots$

    f is not strictly monotonic on $(-\infty, \infty)$, therefore f does not have an inverse.

33. $f(x) = \dfrac{x^4}{4} - 2x^2,$ $f'(x) = x^3 - 4x = 0$ when $x = 0, 2, -2$

    f is not strictly monotonic on $(-\infty, \infty)$, therefore f does not have an inverse.

34. $f(x) = x^3 - 6x^2 + 12x - 8$
    $f'(x) = 3x^2 - 12x + 12 = 3(x - 2)^2 \geq 0$ for all x

    f is increasing on $(-\infty, \infty)$, therefore f is strictly monotonic and has an inverse.

35. $f(x) = 2 - x - x^3,$ $f'(x) = -1 - 3x^2 < 0$ for all x

    f is decreasing on $(-\infty, \infty)$, therefore f is strictly monotonic and has an inverse.

36. $f(x) = \ln(x - 3), x > 3,$ $f'(x) = \dfrac{1}{x - 3} > 0$ for $x > 3$

    f is increasing on $(3, \infty)$, therefore f is strictly monotonic and has an inverse.

37. $f(x) = (x - 4)^2$ on $[4, \infty),$ $f'(x) = 2(x - 4) > 0$ on $(4, \infty)$

    f is increasing on $[4, \infty)$, therefore f is strictly monotonic and has an inverse.

38. $f(x) = |x + 2|$ on $[-2, \infty),$ $f'(x) = \dfrac{|x + 2|}{x + 2}(1) = 1 > 0$ on $(-2, \infty)$

    f is increasing on $[-2, \infty)$, therefore f is strictly monotonic and has an inverse.

39. $f(x) = \dfrac{4}{x^2}$ on $(0, \infty),$ $f'(x) = -\dfrac{8}{x^3} < 0$ on $(0, \infty)$

    f is decreasing on $(0, \infty)$, therefore f is strictly monotonic and has an inverse.

40. $f(x) = \tan x$ on $(-\frac{\pi}{2}, \frac{\pi}{2})$, $\qquad f'(x) = \sec^2 x > 0$ on $(-\frac{\pi}{2}, \frac{\pi}{2})$

   f is increasing on $(-\frac{\pi}{2}, \frac{\pi}{2})$, therefore f is strictly monotonic and has an inverse.

41. $f(x) = \cos x$ on $[0, \pi]$, $\qquad f'(x) = -\sin x < 0$ on $(0, \pi)$

   f is decreasing on $[0, \pi]$, therefore f is strictly monotonic and has an inverse.

42. $f(x) = \sec x$ on $[0, \frac{\pi}{2})$, $(\frac{\pi}{2}, \pi]$, $\quad f'(x) = \sec x \tan x > 0$ on $(0, \frac{\pi}{2})$, $(\frac{\pi}{2}, \pi)$

   f is increasing on $[0, \frac{\pi}{2})$, $(\frac{\pi}{2}, \pi]$ therefore f is strictly monotonic and has an inverse.

43. $f(x) = x^3$, $(\frac{1}{2}, \frac{1}{8})$, $\qquad f'(x) = 3x^2$, $\qquad f'(\frac{1}{2}) = \frac{3}{4}$

   $f^{-1}(x) = \sqrt[3]{x}$, $(\frac{1}{8}, \frac{1}{2})$, $\quad (f^{-1})'(x) = \dfrac{1}{3\sqrt[3]{x^2}}$, $\quad (f^{-1})'(\frac{1}{8}) = \frac{4}{3}$

44. $f(x) = 3 - 4x$, $(1, -1)$, $\qquad f'(x) = -4$, $\qquad f'(1) = -4$

   $f^{-1}(x) = \dfrac{3 - x}{4}$, $(-1, 1)$, $\quad (f^{-1})'(x) = -\frac{1}{4}$, $\quad (f^{-1})'(-1) = -\frac{1}{4}$

45. $f(x) = \sqrt{x - 4}$, $(5, 1)$, $\qquad f'(x) = \dfrac{1}{2\sqrt{x - 4}}$, $\qquad f'(5) = \frac{1}{2}$

   $f^{-1}(x) = x^2 + 4$, $(1, 5)$, $\quad (f^{-1})'(x) = 2x$, $\quad (f^{-1})'(1) = 2$

46. $f(x) = \dfrac{1}{1 + x^2}$, $(1, \frac{1}{2})$, $\qquad f'(x) = \dfrac{-2x}{(1 + x^2)^2}$, $\qquad f'(1) = -\frac{1}{2}$

   $f^{-1}(x) = \sqrt{\dfrac{1 - x}{x}}$, $(\frac{1}{2}, 1)$, $\quad (f^{-1})'(x) = -\dfrac{1}{2x^2}\sqrt{\dfrac{x}{1 - x}}$

   $(f^{-1})'(\frac{1}{2}) = -2$

47. $f(x) = \tan x$, $\qquad f'(x) = \sec^2 x > 0$

   $f(x)$ is not one-to-one since f is not continuous at $x = \dfrac{(2n - 1)\pi}{2}$.

   Example: $f(0) = f(\pi) = 0$

48. $f(x) = \dfrac{x}{x^2 - 4}$, $\qquad f'(x) = -\dfrac{x^2 + 4}{(x^2 - 4)^2} < 0$

f is not one-to-one since f is not continuous at $x = \pm\, 2$.

Example: $f\left(\dfrac{1 + \sqrt{17}}{2}\right) = f\left(\dfrac{1 - \sqrt{17}}{2}\right) = 1$

6.2
Exponential functions and differentiation

1.(a) $2^3 = 8$, $\quad \log_2 8 = 3$ $\qquad$ **(b)** $3^{-1} = \dfrac{1}{3}$, $\quad \log_3 \dfrac{1}{3} = -1$

2.(a) $27^{2/3} = 9$, $\quad \log_{27} 9 = \dfrac{2}{3}$ $\qquad$ **(b)** $16^{3/4} = 8$, $\quad \log_{16} 8 = \dfrac{3}{4}$

3.(a) $\log_{10} 0.01 = -2$, $10^{-2} = 0.01$ $\quad$ **(b)** $\log_{0.5} 8 = -3$, $\quad \left(\dfrac{1}{2}\right)^{-3} = 8$

4.(a) $e^0 = 1$, $\quad \ln 1 = 0$ $\qquad$ **(b)** $e^2 = 7.389\ldots$, $\quad \ln 7.389\ldots = 2$

5.(a) $\ln 2 = 0.6931\ldots$, $e^{0.6931\ldots} = 2$
$\quad$ **(b)** $\ln 8.4 = 2.128\ldots$, $e^{2.128\ldots} = 8.4$

6.(a) $\ln 0.5 = -0.6931\ldots$, $\qquad e^{-0.6931\ldots} = \dfrac{1}{2}$

$\quad$ **(b)** $49^{1/2} = 7$, $\qquad \log_{49} 7 = \dfrac{1}{2}$

7.(a) $\log_{10} 1000 = x$, $\quad 10^x = 1000 = 10^3$, $\quad x = 3$

$\quad$ **(b)** $\log_{10} 0.1 = x$, $\quad 10^x = \dfrac{1}{10} = 10^{-1}$, $\quad x = -1$

8.(a) $\log_4 \dfrac{1}{64} = x$, $\quad 4^x = \dfrac{1}{64} = 4^{-3}$, $\quad x = -3$

$\quad$ **(b)** $\log_5 25 = x$, $\quad 5^x = 25 = 5^2$, $\quad x = 2$

9.(a) $\log_3 x = -1$, $\quad 3^{-1} = x$, $\quad x = \dfrac{1}{3}$

$\quad$ **(b)** $\log_2 x = -4$, $\quad 2^{-4} = x$, $\quad x = \dfrac{1}{16}$

10.(a) $\log_b 27 = 3$, $\quad b^3 = 27 = 3^3$, $\quad b = 3$
$\quad\;\,$ **(b)** $\log_b 125 = 3$, $\quad b^3 = 125 = 5^3$, $\quad b = 5$

11.(a) $\log_{27} x = -\dfrac{2}{3}$, $\qquad 27^{-2/3} = \dfrac{1}{(\sqrt[3]{27})^2} = x$, $\qquad x = \dfrac{1}{9}$

   (b) $\ln e^x = 3$, $\qquad e^3 = e^x$, $\qquad x = 3$

12.(a) $e^{\ln x} = 4$, $\qquad \ln x = \ln 4$, $\qquad x = 4$

   (b) $\ln x = 2$, $\qquad e^2 = x$, $\qquad x = 7.389\ldots$

13.(a) $x^2 - x = \log_5 25$, $\quad 5^{(x^2-x)} = 5^2$, $\quad x^2 - x = 2$, $\quad x^2 - x - 2 = 0$

       $(x - 2)(x + 1) = 0$ when $x = 2, -1$

   (b) $3x + 5 = \log_2 64$, $\qquad 2^{(3x+5)} = 64 = 2^6$, $\qquad 3x + 5 = 6$, $\qquad x = \dfrac{1}{3}$

14.(a) $\log_3 x + \log_3 (x - 2) = 1$, $\qquad \log_3 x(x - 2) = 1$, $\qquad x(x - 2) = 3$

       $x^2 - 2x - 3 = 0$, $\qquad (x - 3)(x + 1) = 0$ when $x = 3, -1$

       Since $\log_3 (-1)$ is undefined, $x = 3$

   (b) $\log_{10} (x + 3) - \log_{10} x = 1$, $\qquad \log_{10} \dfrac{x + 3}{x} = 1$, $\qquad \dfrac{x + 3}{x} = 10$

       $x + 3 = 10x$, $\qquad 9x = 3$, $\qquad x = 1/3$

15. $y = 3^x$

16. $y = 3^{x-1}$

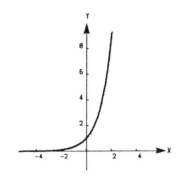

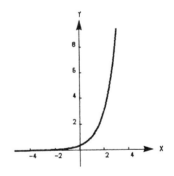

17. $y = \left(\dfrac{1}{3}\right)^x$

18. $y = 2^{x^2}$

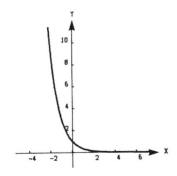

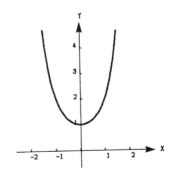

19. $y = e^{-x^2}$

Symmetric with respect to the
y-axis.
$y = 0$ horizontal asymptote

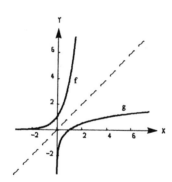

20. $y = e^{-x/2}$

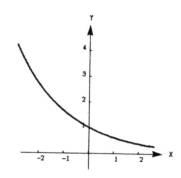

21. $f(x) = 4^x$

$g(x) = \log_4 x, \qquad x = 4^y$

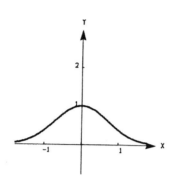

22. $f(x) = 3^x,$

$g(x) = \log_3 x, \qquad x = 3^y$

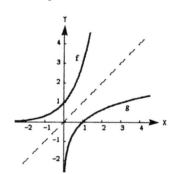
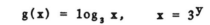

23. $f(x) = e^{2x}$

$g(x) = \ln \sqrt{x} = \frac{1}{2} \ln x$

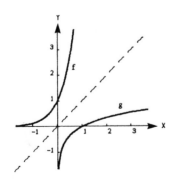

24. $f(x) = e^{x/3}$

$g(x) = \ln x^3$

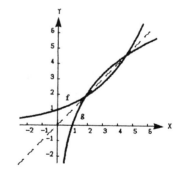
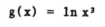

25. $f(x) = e^x - 1$
    $g(x) = \ln(x + 1)$

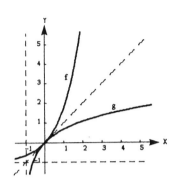

26. $f(x) = e^{x-1}$
    $g(x) = 1 + \ln x$

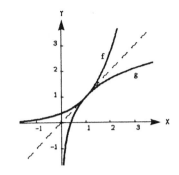

27.

x	1	10^{-1}	10^{-2}	10^{-4}	10^{-6}
$(1 + x)^{1/x}$	2	2.5937	2.7048	2.7181	2.71828

28. $\log_{10} 2 = 0.30103$, $\quad \dfrac{\ln 2}{\ln 10} = \dfrac{0.69315}{2.30259} = 0.30103$

29. (a) $\dfrac{271,801}{99,990} = 2.7\overline{1828}$, e $- \dfrac{271,801}{99,990} \approx 0.0000000003$

   (b) $\dfrac{299}{110} = 2.7\overline{18}$, e $- \dfrac{299}{110} \approx 0.0001$

30. (a) $1 + 1 + \dfrac{1}{2} + \dfrac{1}{6} + \dfrac{1}{24} = 2.7083\overline{3}$

   (b) $1 + 1 + \dfrac{1}{2} + \dfrac{1}{6} + \dfrac{1}{24} + \dfrac{1}{120} + \dfrac{1}{720} + \dfrac{1}{5040} = 2.71\overline{825396}$

31. $2P = Pe^{0.075t}$, $\quad \ln 2 = 0.075t$, $\quad t = \dfrac{\ln 2}{0.075} = 9.24$ years

    $3P = Pe^{0.075t}$, $\quad \ln 3 = 0.075t$, $\quad t = \dfrac{\ln 3}{0.075} = 14.65$ years

32. From 31 we see that $t = \dfrac{\ln 2}{r}$

r	2%	4%	6%	8%	10%	12%
t(years)	34.66	17.33	11.55	8.66	6.93	5.78

33.(a) $1000(1 + 0.075)^{10} = \$2061.03$ (b) $1000(1 + \dfrac{0.075}{2})^{20} = \$2088.15$

(c) $1000(1 + \dfrac{0.075}{4})^{40} \approx \$2102.35$ (d) $1000(1 + \dfrac{0.075}{12})^{120} = \$2112.06$

(e) $1000(1 + \dfrac{0.075}{365})^{3650} = \$2116.83$ (f) $1000e^{0.75} = \$2117.00$

34.(a) $2500(1 + 0.12)^{20} = \$24,115.73$ (b) $2500(1 + \dfrac{0.12}{2})^{40} = \$25,714.29$

(c) $2500(1 + \dfrac{0.12}{4})^{80} = \$26,602.23$ (d) $2500(1 + \dfrac{0.12}{12})^{240} = \$27,231.38$

(e) $2500(1 + \dfrac{0.12}{365})^{7300} = \$27,547.07$ (f) $2500e^{2.4} = \$27,557.94$

35.(a) $y = e^{3x}$
$y' = 3e^{3x}$
At $(0, 1)$, $y' = 3$

(b) $y = e^{-3x}$
$y' = -3e^{-3x}$
At $(0, 1)$, $y' = -3$

36.(a) $y = e^{2x}$
$y' = 2e^{2x}$
At $(0, 1)$, $y' = 2$

(b) $y = e^{-2x}$
$y' = -2e^{-2x}$
At $(0, 1)$, $y' = -2$

37. $y = e^{2x}$, $\dfrac{dy}{dx} = 2e^{2x}$

38. $y = e^{1-x}$, $\dfrac{dy}{dx} = -e^{1-x}$

39. $y = e^{-2x+x^2}$, $\dfrac{dy}{dx} = 2(x - 1)e^{-2x+x^2}$

40. $y = e^{-x^2}$, $\dfrac{dy}{dx} = -2xe^{-x^2}$

41. $y = e^{\sqrt{x}}$, $\dfrac{dy}{dx} = \dfrac{e^{\sqrt{x}}}{2\sqrt{x}}$

42. $y = x^2e^{-x}$, $\dfrac{dy}{dx} = -x^2e^{-x} + 2xe^{-x} = xe^{-x}(2 - x)$

43. $y = (e^{-x} + e^x)^3$, $\dfrac{dy}{dx} = 3(e^{-x} + e^x)^2(e^x - e^{-x})$

44. $y = e^{-1/x^2}$, $\dfrac{dy}{dx} = \dfrac{2e^{-1/x^2}}{x^3}$

45. $y = \ln e^{x^2} = x^2$, $\dfrac{dy}{dx} = 2x$

46. $y = \ln \left(\dfrac{1 + e^x}{1 - e^x}\right) = \ln (1 + e^x) - \ln (1 - e^x)$

$\dfrac{dy}{dx} = \dfrac{e^x}{1 + e^x} - \dfrac{-e^x}{1 - e^x} = \dfrac{2e^x}{1 - e^{2x}}$

47. $y = \ln (1 + e^{2x})$, $\qquad \dfrac{dy}{dx} = \dfrac{2e^{2x}}{1 + e^{2x}}$

48. $y = \dfrac{2}{e^x + e^{-x}} = 2(e^x + e^{-x})^{-1}$

$\dfrac{dy}{dx} = -2(e^x + e^{-x})^{-2}(e^x - e^{-x}) = \dfrac{-2(e^x - e^{-x})}{(e^x + e^{-x})^2}$

49. $y = \ln \left(\dfrac{e^x + e^{-x}}{2}\right) = \ln (e^x + e^{-x}) - \ln 2$, $\qquad \dfrac{dy}{dx} = \dfrac{(e^x - e^{-x})}{e^x + e^{-x}} = \dfrac{e^{2x} - 1}{e^{2x} + 1}$

50. $y = xe^x - e^x = e^x(x - 1)$, $\qquad \dfrac{dy}{dx} = e^x + e^x(x - 1) = xe^x$

51. $y = x^2 e^x - 2xe^x + 2e^x = e^x(x^2 - 2x + 2)$

$\dfrac{dy}{dx} = e^x(2x - 2) + e^x(x^2 - 2x + 2) = x^2 e^x$

52. $y = \dfrac{e^x - e^{-x}}{2}$, $\qquad \dfrac{dy}{dx} = \dfrac{e^x + e^{-x}}{2}$

53. $y = 5^{x-2}$, $\qquad \dfrac{dy}{dx} = (\ln 5)5^{x-2}$

54. $y = x7^{-3x}$, $\qquad \dfrac{dy}{dx} = x(\ln 7)(7^{-3x})(-3) + 7^{-3x} = 7^{-3x}(1 - 3x \ln 7)$

55. $y = e^{-x} \ln x$, $\qquad \dfrac{dy}{dx} = e^{-x}\left(\dfrac{1}{x}\right) - e^{-x} \ln x = e^{-x}\left(\dfrac{1}{x} - \ln x\right)$

56. $y = 2^{x^2}3^{-x}$, $\qquad \dfrac{dy}{dx} = 2^{x^2}(\ln 3)(3^{-x})(-1) + 3^{-x}(\ln 2)(2^{x^2})(2x)$

$= 2^{x^2}3^{-x}(2x \ln 2 - \ln 3)$

57. $y = e^x(\sin x + \cos x)$, $\qquad \dfrac{dy}{dx} = e^x(\cos x - \sin x) + (\sin x + \cos x)(e^x)$

$= e^x(2 \cos x) = 2e^x \cos x$

58. $y = e^{\tan x}$, $\qquad \dfrac{dy}{dx} = e^{\tan x} \sec^2 x$

59. $y = \tan^2(e^x)$, $\quad \dfrac{dy}{dx} = 2[\tan(e^x)]^1[e^x \sec^2(e^x)] = 2e^x \tan(e^x)\sec^2(e^x)$

60. $y = \ln e^x = x$, $\quad \dfrac{dy}{dx} = 1$

61. $y = \log_3 x$, $\quad \dfrac{dy}{dx} = \dfrac{1}{x \ln 3}$

62. $y = \log_3 \dfrac{x\sqrt{x-1}}{2} = \log_3 x + \dfrac{1}{2}\log_3(x-1) - \log_3 2$

$\dfrac{dy}{dx} = \dfrac{1}{x \ln 3} + \dfrac{1}{2(x-1)\ln 3} - 0 = \dfrac{1}{\ln 3}\left[\dfrac{1}{x} + \dfrac{1}{2(x-1)}\right]$

$\qquad = \dfrac{1}{\ln 3}\left[\dfrac{3x-2}{2x(x-1)}\right]$

63. $y = \log_5 \sqrt{x^2 - 1} = \dfrac{1}{2}\log_5(x^2 - 1)$

$\dfrac{dy}{dx} = \dfrac{1}{2(\ln 5)}\left(\dfrac{2x}{x^2-1}\right) = \dfrac{x}{(\ln 5)(x^2-1)}$

64. $y = \log_{10}\dfrac{x^2-1}{x} = \log_{10}(x^2-1) - \log_{10}x$

$\dfrac{dy}{dx} = \dfrac{1}{\ln 10}\left(\dfrac{2x}{x^2-1}\right) - \dfrac{1}{\ln 10}\left(\dfrac{1}{x}\right) = \dfrac{1}{\ln 10}\left[\dfrac{2x}{x^2-1} - \dfrac{1}{x}\right]$

$\qquad = \dfrac{1}{\ln 10}\left[\dfrac{x^2+1}{x(x^2-1)}\right]$

65. $y = x^{2/x}$, $\quad \ln y = \dfrac{2}{x}\ln x$, $\quad \dfrac{1}{y}\left(\dfrac{dy}{dx}\right) = \dfrac{2}{x}\left(\dfrac{1}{x}\right) + \ln x\left(-\dfrac{2}{x^2}\right)$

$\dfrac{dy}{dx} = \dfrac{2y}{x^2}(1 - \ln x) = 2x^{(2/x)-2}(1 - \ln x)$

66. $y = x^{x-1}$, $\quad \ln y = (x-1)(\ln x)$, $\quad \dfrac{1}{y}\left(\dfrac{dy}{dx}\right) = (x-1)\left(\dfrac{1}{x}\right) + \ln x$

$\dfrac{dy}{dx} = y\left[\dfrac{x-1}{x} + \ln x\right] = x^{x-2}(x - 1 + x\ln x)$

67. $y = (x-2)^{x+1}$, $\quad \ln y = (x+1)\ln(x-2)$

$\dfrac{1}{y}\left(\dfrac{dy}{dx}\right) = (x+1)\left(\dfrac{1}{x-2}\right) + \ln(x-2)$

$\dfrac{dy}{dx} = y\left[\dfrac{x+1}{x-2} + \ln(x-2)\right] = (x-2)^{x+1}\left[\dfrac{x+1}{x-2} + \ln(x-2)\right]$

68. $y = (1 + x)^{1/x}$, $\quad \ln y = \frac{1}{x} \ln (1 + x)$

$$\frac{1}{y}\left(\frac{dy}{dx}\right) = \frac{1}{x}\left(\frac{1}{1 + x}\right) + \ln (1 + x)\left(-\frac{1}{x^2}\right)$$

$$\frac{dy}{dx} = \frac{y}{x}\left[\frac{1}{x + 1} - \frac{\ln (x + 1)}{x}\right]$$

69. $y = e^x(\cos \sqrt{2}\, x + \sin \sqrt{2}\, x)$

$y' = e^x(-\sqrt{2} \sin \sqrt{2}\, x + \sqrt{2} \cos \sqrt{2}\, x) + e^x(\cos \sqrt{2}\, x + \sin \sqrt{2}\, x)$

$\quad = e^x[(1 + \sqrt{2}) \cos \sqrt{2}\, x + (1 - \sqrt{2}) \sin \sqrt{2}\, x]$

$y'' = e^x[-(\sqrt{2} + 2) \sin \sqrt{2}\, x + (\sqrt{2} - 2) \cos \sqrt{2}\, x]$

$\qquad + e^x[(1 + \sqrt{2}) \cos \sqrt{2}\, x + (1 - \sqrt{2}) \sin \sqrt{2}\, x]$

$\quad = e^x[(-1 - 2\sqrt{2}) \sin \sqrt{2}x + (-1 + 2\sqrt{2}) \cos \sqrt{2}\, x]$

$-2y' + 3y = -2e^x[(1 + \sqrt{2}) \cos \sqrt{2}\, x + (1 - \sqrt{2}) \sin \sqrt{2}\, x\,]$

$\qquad\qquad + 3e^x[\cos \sqrt{2}\, x + \sin \sqrt{2}\, x]$

$\qquad = e^x[(1 - 2\sqrt{2}) \cos \sqrt{2}\, x + (1 + 2\sqrt{2}) \sin \sqrt{2}\, x] = -y''$

Therefore, $-2y' + 3y = -y''$ ➡ $y'' - 2y' + 3y = 0$

70. $y = e^x(3 \cos 2x - 4 \sin 2x)$

$y' = e^x(-6 \sin 2x - 8 \cos 2x) + e^x(3 \cos 2x - 4 \sin 2x)$

$\quad = e^x(-10 \sin 2x - 5 \cos 2x) = -5e^x(2 \sin 2x + \cos 2x)$

$y'' = -5e^x(4 \cos 2x - 2 \sin 2x) - 5e^x(2 \sin 2x + \cos 2x)$

$\quad = -5e^x(5 \cos 2x) = -25e^x \cos 2x$

$y'' - 2y' = -5e^x(5 \cos 2x) - 2(-5e^x)(2 \sin 2x + \cos 2x)$

$\qquad = -5e^x[3 \cos 2x - 4 \sin 2x] = -5y$

Therefore, $y'' - 2y' = -5y$ ➡ $y'' - 2y' + 5y = 0$

71. $f(x) = \dfrac{1}{\sqrt{2\pi}} e^{-x^2/2}$, $f'(x) = \dfrac{-x}{\sqrt{2\pi}} e^{-x^2/2} = 0$ when $x = 0$

$f''(x) = \dfrac{1}{\sqrt{2\pi}}(x^2 e^{-x^2/2} - e^{-x^2/2})$

$ = \dfrac{e^{-x^2/2}}{\sqrt{2\pi}}(x^2 - 1) = 0$ when $x = -1, 1$

$\left(0, \dfrac{1}{\sqrt{2\pi}}\right)$ is a relative maximum

$\left(\pm 1, \dfrac{1}{\sqrt{2e\pi}}\right)$ are points of inflection

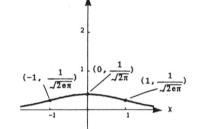

72. $f(x) = \dfrac{e^x - e^{-x}}{2}$

$f'(x) = \dfrac{e^x + e^{-x}}{2} > 0$

$f''(x) = \dfrac{e^x - e^{-x}}{2} = 0$ when $x = 0$

$(0, 0)$ is a point of inflection

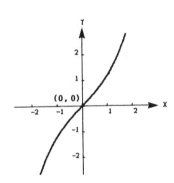

73. $f(x) = \dfrac{e^x + e^{-x}}{2}$

$f'(x) = \dfrac{e^x - e^{-x}}{2} = 0$ when $x = 0$

$f''(x) = \dfrac{e^x + e^{-x}}{2} > 0$

$(0, 1)$ is a relative minimum

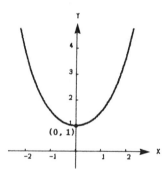

74. $f(x) = xe^{-x}$

$f'(x) = -xe^{-x} + e^{-x} = e^{-x}(1 - x) = 0$
when $x = 1$

$f''(x) = -e^{-x} + (-e^{-x})(1 - x)$

$ = e^{-x}(x - 2) = 0$ when $x = 2$

$(1, e^{-1})$ is a relative maximum

$(2, 2e^{-2})$ is a point of inflection

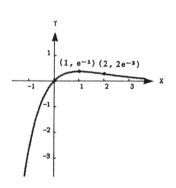

75. $f(x) = x^2 e^{-x}$

$f'(x) = -x^2 e^{-x} + 2xe^{-x} = xe^{-x}(2 - x) = 0$ when $x = 0, 2$

$f''(x) = -e^{-x}(2x - x^2) + e^{-x}(2 - 2x) = e^{-x}(x^2 - 4x + 2) = 0$

when $x = 2 \pm \sqrt{2}$

(0, 0) is a relative minimum

$(2, 4e^{-2})$ is a relative maximum

$x = 2 \pm \sqrt{2}, \quad y = (2 \pm \sqrt{2})^2 e^{-(2 \pm \sqrt{2})}$

(3.414, 0.384), (0.586, 0.191) are points of inflection

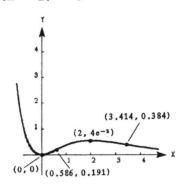

76. $f(x) = -2 + e^{3x}(4 - 2x)$

$f'(x) = e^{3x}(-2) + 3e^{3x}(4 - 2x)$

$\qquad = e^{3x}(10 - 6x) = 0$ when $x = \dfrac{5}{3}$

$f''(x) = e^{3x}(-6) + 3e^{3x}(10 - 6x)$

$\qquad = e^{3x}(24 - 18x) = 0$ when $x = \dfrac{4}{3}$

$(\dfrac{5}{3}, 96.942)$ is a relative maximum

$(\dfrac{4}{3}, 70.798)$ is a point of inflection

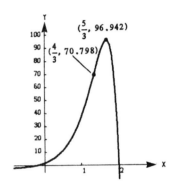

77. $y = e^{-x}$,   point: (0, 1)

$y' = -e^{-x}$ (slope of tangent line)

$-1/y' = e^x$ (slope of normal line)

At (0, 1), $e^x = 1$

$y = x + 1$

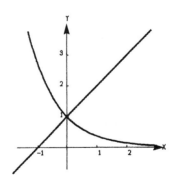

78. $y = e^{-x}$

$y' = -e^{-x}$ (slope of tangent line)

$-1/y' = e^{x}$ (slope of normal line)

$y - e^{-x_0} = e^{x_0} (x - x_0)$

We want $(0, 0)$ to satisfy the equation:

$-e^{-x_0} = -x_0 e^{x_0}, \qquad 1 = x_0 e^{2x_0}$

$x_0 e^{2x_0} - 1 = 0$ when $x_0 \approx 0.4263$

$(0.4263, e^{-0.4263})$

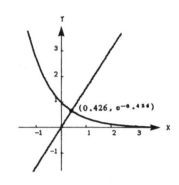

79. $A = 2xe^{-x^2}$

$\dfrac{dA}{dx} = -4x^2 e^{-x^2} + 2e^{-x^2}$

$\qquad = 2e^{-x^2}(1 - 2x^2) = 0$

when $x = \dfrac{\sqrt{2}}{2}$

$A = \sqrt{2}\, e^{-1/2}$

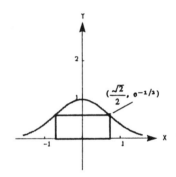

80. $e^{-x} = x \implies f(x) = x - e^{-x}, \qquad f'(x) = 1 + e^{-x}$

$x_{n+1} = x_n - \dfrac{f(x_n)}{f'(x_n)} = x_n - \dfrac{x_n - e^{-x_n}}{1 + e^{-x_n}}, \qquad x_1 = 1$

$x_2 = x_1 - \dfrac{f(x_1)}{f'(x_1)} = 0.5379$

$x_3 = x_2 - \dfrac{f(x_2)}{f'(x_2)} = 0.5670$

$x_4 = x_3 - \dfrac{f(x_3)}{f'(x_3)} = 0.5671$

We approximate the root of f
to be $x = 0.567$

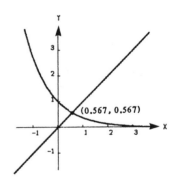

81.(a) $\lim_{t \to \infty} 6.7e^{(-48.1)/t} = 6.7e^0 = 6.7$ million ft^3

(b) $V' = \dfrac{322.27}{t^2} e^{-(48.1)/t}$

$V'(20) = 0.073$ million ft^3/yr

$V'(60) = 0.040$ million ft^3/yr

82.(a) $\lim_{t \to \infty} \dfrac{157}{1 + 5.4e^{-0.12t}} = 157$ words per minute

(b) $N' = \dfrac{101.736e^{-0.12t}}{[1 + 5.4e^{-0.12t}]^2}$

$N'(5) = 3.554$ words per week

$N'(25) = 3.146$ words per week

83. (a) $\lim_{n \to \infty} \dfrac{0.83}{1 + e^{-0.2n}} = 0.83 = 83\%$

(b) $P' = \dfrac{0.166e^{-0.2n}}{[1 + e^{-0.2n}]^2}$

$P'(3) = 0.038, \qquad P'(10) = 0.017$

84. $p(t) = \dfrac{10,000}{1 + 19e^{-t/5}}, \quad p'(t) = \dfrac{e^{-t/5}}{(1 + 19e^{t/5})^2} \left(\dfrac{19}{5}\right)(10,000)$

$= \dfrac{38000\, e^{-t/5}}{(1 + 19e^{-t/5})^2}$

$p'(1) = 113.5, \qquad p'(10) = 403.2$

$p''(t) = -\dfrac{38000}{5}(e^{-t/5})\left[\dfrac{1 - 19e^{-t/5}}{(1 + 19e^{-t/5})^3}\right] = 0$

$19e^{-t/5} = 1, \qquad \dfrac{t}{5} = \ln 19, \qquad t = 5\ln 19 \approx 14.72$

6.3

Integration of exponential functions: Growth and decay

1. Let $u = -2x$, $\quad u' = -2$

$$\int_0^1 e^{-2x}\,dx = -\frac{1}{2}\int_0^1 e^{-2x}(-2)\,dx = -\frac{1}{2}e^{-2x}\Big]_0^1 = \frac{1}{2}(1 - e^{-2}) = \frac{e^2 - 1}{2e^2}$$

2. Let $u = 1 - x$, $\quad u' = -1$

$$\int_1^2 e^{1-x}\,dx = -\int_1^2 e^{1-x}(-1)\,dx = -e^{1-x}\Big]_1^2 = 1 - e^{-1} = \frac{e - 1}{e}$$

3. $\displaystyle\int_0^2 (x^2 - 1)e^{x^3-3x+1}\,dx = \frac{1}{3}\int_0^2 e^{x^3-3x+1}\,3(x^2 - 1)\,dx$

$$= \frac{1}{3}e^{x^3-3x+1}\Big]_0^2 = \frac{e}{3}(e^2 - 1)$$

4. Let $u = x^3$, $\quad u' = 3x^2$

$$\int x^2 e^{x^3}\,dx = \frac{1}{3}\int e^{x^3}(3x^2)\,dx = \frac{1}{3}e^{x^3} + C$$

5. Let $u = 1 + e^{-x}$, $\quad u' = -e^{-x}$

$$\int \frac{e^{-x}}{1 + e^{-x}}\,dx = -\int \frac{-e^{-x}}{1 + e^{-x}}\,dx = -\ln(1 + e^{-x}) + C = x - \ln(e^x + 1) + C$$

6. Let $u = 1 + e^{2x}$, $\quad u' = 2e^{2x}$

$$\int \frac{e^{2x}}{1 + e^{2x}}\,dx = \frac{1}{2}\int \frac{2e^{2x}}{1 + e^{2x}}\,dx = \frac{1}{2}\ln(1 + e^{2x}) + C$$

7. Let $u = ax^2$, $\quad u' = 2ax$

$$\int xe^{ax^2}\,dx = \frac{1}{2a}\int e^{ax^2}(2ax)\,dx = \frac{1}{2a}e^{ax^2} + C$$

8. Let $u = \dfrac{-x^2}{2}$, $\quad u' = -x$

$$\int_0^{\sqrt{2}} xe^{-x^2/2}\,dx = -\int_0^{\sqrt{2}} e^{-x^2/2}\,(-x)\,dx = -e^{-x^2/2}\Big]_0^{\sqrt{2}} = 1 - e^{-1} = \frac{e - 1}{e}$$

9. Let $u = 3/x$, $\quad u' = -3/x^2$

$$\int_1^3 \frac{e^{3/x}}{x^2}\,dx = -\frac{1}{3}\int_1^3 e^{3/x}\left(-\frac{3}{x^2}\right)\,dx = -\frac{1}{3}e^{3/x}\Big]_1^3 = \frac{e}{3}(e^2 - 1)$$

10. $\displaystyle\int (e^x - e^{-x})^2 \, dx = \int (e^{2x} - 2 + e^{-2x}) \, dx = \frac{1}{2}e^{2x} - 2x + (-\frac{1}{2})e^{-2x} + C$

$$= \frac{1}{2}(e^{2x} - 4x - e^{-2x}) + C$$

11. $\displaystyle\int_{-1}^{2} 2^x \, dx = \frac{1}{\ln 2}(2^x) \Big]_{-1}^{2} = \frac{7}{2 \ln 2} = \frac{7}{\ln 4}$

12. Let $u = -x^2, \qquad u' = -2x$

$$\int x5^{-x^2} \, dx = -\frac{1}{2} \int 5^{-x^2}(-2x) \, dx = \frac{-1}{2 \ln 5}(5^{-x^2}) + C$$

13. Let $u = 1 - e^x, \qquad u' = -e^x$

$$\int e^x \sqrt{1 - e^x} \, dx \quad = -\int (1 - e^x)^{1/2}(-e^x) \, dx = -\frac{2}{3}(1 - e^x)^{3/2} + C$$

14. Let $u = e^x + e^{-x}, \qquad u' = e^x - e^{-x}$

$$\int \frac{e^x - e^{-x}}{e^x + e^{-x}} \, dx = \ln (e^x + e^{-x}) + C$$

15. Let $u = e^x - e^{-x}, \qquad u' = e^x + e^{-x}$

$$\int \frac{e^x + e^{-x}}{e^x - e^{-x}} \, dx = \ln |e^x - e^{-x}| + C$$

16. Let $u = e^x + e^{-x}, \qquad u' = e^x - e^{-x}$

$$\int \frac{2e^x - 2e^{-x}}{(e^x + e^{-x})^2} = 2 \int (e^x + e^{-x})^{-2}(e^x - e^{-x}) \, dx = \frac{-2}{e^x + e^{-x}} + C$$

17. $\displaystyle\int \frac{5 - e^x}{e^{2x}} \, dx = \int 5e^{-2x} \, dx - \int e^{-x} \, dx = -\frac{5}{2}e^{-2x} + e^{-x} + C$

18. Let $u = (3 - x)^2, \qquad u' = -2(3 - x)$

$$\int (3 - x)7^{(3-x)^2} \, dx = -\frac{1}{2} \int 7^{(3-x)^2}(-2)(3 - x) \, dx = \frac{-1}{2 \ln 7}(7^{(3-x)^2}) + C$$

19. $\displaystyle\int_{-2}^{0} (3^3 - 5^2) \, dx = (3^3 - 5^2)x \Big]_{-2}^{0} = 4$

20. $\displaystyle\int \frac{e^{2x} + 2e^x + 1}{e^x}\ dx = \int (e^x + 2 + e^{-x})\ dx = e^x + 2x - e^{-x} + C$

21. $\displaystyle\int e^{\sin \pi x} \cos \pi x\ dx = \frac{1}{\pi} \int e^{\sin \pi x}\ (\pi \cos \pi x)\ dx = \frac{1}{\pi} e^{\sin \pi x} + C$

22. $\displaystyle\int e^{\tan 2x} \sec^2 2x\ dx = \frac{1}{2} \int e^{\tan 2x}(2 \sec^2 2x)\ dx = \frac{1}{2} e^{\tan 2x} + C$

23. $\displaystyle\int e^{-x} \tan(e^{-x})\ dx = - \int [\tan(e^{-x})](-e^{-x})\ dx = \ln |\cos(e^{-x})| + C$

24. $\displaystyle\int \ln (e^{2x-1})\ dx = \int (2x - 1)\ dx = x^2 - x + C$

25. $\displaystyle\int_0^5 e^x\ dx = e^x \Big]_0^5$
$$= e^5 - 1 \approx 147.413$$

26. $\displaystyle\int_a^b e^{-x}\ dx = -e^{-x} \Big]_a^b = e^{-a} - e^{-b}$

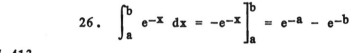

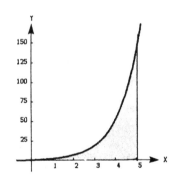

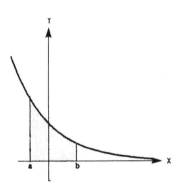

27. $\displaystyle\int_0^{\sqrt{2}} xe^{-(x^2/2)}\ dx = -e^{-(x^2/2)} \Big]_0^{\sqrt{2}}$
$$= -e^{-1} + 1 \approx 0.632$$

28. $\displaystyle\int_0^3 3^x\ dx = \frac{3^x}{\ln 3} \Big]_0^3 = \frac{26}{\ln 3}$

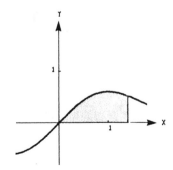

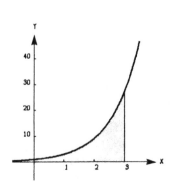

29. $\int_0^x e^t \, dt \geq \int_0^x 1 \, dt, \quad e^t \Big]_0^x \geq t \Big]_0^x, \quad e^x - 1 \geq x \implies e^x \geq 1 + x, \quad x \geq 0$

30.(a) $e^x \geq 1 + x, \qquad \int_0^x e^t \, dt \geq \int_0^x (1 + t) \, dt$

$e^t \Big]_0^x \geq (t + \dfrac{t^2}{2}) \Big]_0^x$

$e^x - 1 \geq x + \dfrac{x^2}{2} \implies e^x \geq 1 + x + \dfrac{x^2}{2}, \quad x \geq 0$

When $x = 1$, $2.718281828 > 2.5$

(b) $e^x \geq 1 + x + \dfrac{x^2}{2}, \qquad \int_0^x e^t \, dt \geq \int_0^x (1 + t + \dfrac{t^2}{2}) \, dt$

$e^t \Big]_0^x \geq t + \dfrac{t^2}{2} + \dfrac{t^3}{6} \Big]_0^x$

$e^x - 1 \geq x + \dfrac{x^2}{2} + \dfrac{x^3}{6} \implies e^x \geq 1 + x + \dfrac{x^2}{2} + \dfrac{x^3}{6}, \quad x \geq 0$

When $x = 1$, $2.718281828 > 2.666\overline{6}$

(c) $e^x \geq 1 + x + \dfrac{x^2}{2} + \dfrac{x^3}{6}, \qquad \int_0^x e^t \, dt > \int_0^x (1 + t + \dfrac{t^2}{2} + \dfrac{t^3}{6}) \, dt$

$e^x - 1 \geq x + \dfrac{x^2}{2} + \dfrac{x^3}{6} + \dfrac{x^4}{24} \implies e^x \geq 1 + x + \dfrac{x^2}{2} + \dfrac{x^3}{6} + \dfrac{x^4}{24}, \quad x \geq 0$

When $x = 1$, $2.718281828 > 2.70833\overline{3}$

(d) $e^x \geq 1 + x + \dfrac{x^2}{2} + \dfrac{x^3}{6} + \dfrac{x^4}{24}$

$\int_0^x e^t \, dt \geq \int_0^x (1 + t + \dfrac{t^2}{2} + \dfrac{t^3}{6} + \dfrac{t^4}{24}) \, dt$

$e^x - 1 \geq x + \dfrac{x^2}{2} + \dfrac{x^3}{6} + \dfrac{x^4}{24} + \dfrac{x^5}{120}$

$e^x \geq 1 + x + \dfrac{x^2}{2} + \dfrac{x^3}{6} + \dfrac{x^4}{24} + \dfrac{x^5}{120}, \quad x \geq 0$

When $x = 1$, $2.718281828 > 2.71666\overline{6}$

31.(a) $y = Ce^{kt}$, $(0, \frac{1}{2})$ and $(5, 5)$, $C = \frac{1}{2}$, $y = \frac{1}{2}e^{kt}$

$5 = \frac{1}{2}e^{5k}$, $k = \frac{\ln 10}{5} \approx 0.4605$, $y = \frac{1}{2}e^{0.4605t}$

(b) $y = Ce^{kt}$, $(0, 4)$ and $(5, \frac{1}{2})$, $C = 4$, $y = 4e^{kt}$, $\frac{1}{2} = 4e^{5k}$

$k = \frac{\ln(1/8)}{5} \approx -0.4159$, $y = 4e^{-0.4159t}$

32.(a) $y = Ce^{kt}$, $(1, 1)$ and $(5, 5)$, $1 = Ce^{k}$ and $5 = Ce^{5k}$, $5Ce^{k} = Ce^{5k}$

$5e^{k} = e^{5k}$, $5 = e^{4k}$, $k = \frac{\ln 5}{4} \approx 0.4024$, $y = Ce^{0.4024t}$

$1 = Ce^{0.4024t}$, $C \approx 0.6687$, $y = 0.6687e^{0.4024t}$

(b) $y = Ce^{kt}$, $(3, \frac{1}{2})$ and $(4, 5)$, $\frac{1}{2} = Ce^{3k}$ and $5 = Ce^{4k}$

$2Ce^{3k} = \frac{1}{5}Ce^{4k}$, $10e^{3k} = e^{4k}$, $10 = e^{k}$, $k = \ln 10 \approx 2.3026$

$y = Ce^{2.3026t}$, $5 = Ce^{2.3026(4)}$, $C = 0.0005$, $y = 0.0005e^{2.3026t}$

33. $A = Pe^{rt}$, $t = 7.75$, $P = \$750$

(a) $2P = Pe^{7.75r}$, $r = \frac{\ln 2}{7.75} \approx 8.94\%$

(b) $A = 750e^{[(\ln 2)/7.75](10)} \approx \$1834.37$

34. $A = Pe^{rt}$, $t = 5$, $P = \$10,000$

(a) $2P = Pe^{5r}$, $r = \frac{\ln 2}{5} \approx 13.86\%$

(b) $A = 10,000e^{[(\ln 2)/5](1)} \approx \$11,486.98$

35. $y = Ce^{kt}$, $(0, 742000)$ and $(2, 632000)$, $C = 742,000$

$632,000 = 742,000e^{2k}$, $k = \frac{\ln(632/742)}{2} \approx -0.0802$

$y = 742,000e^{-0.0802t}$, $y(3) \approx \$583,327.77$

36. $y = Ce^{kt}$, $(0, 760)$ and $(1000, 672.71)$, $C = 760$

$672.71 = 760e^{1000k}$, $k = \frac{\ln(672.71/760)}{1000} \approx -0.000122$

$y = 760e^{-0.000122t}$, $y(3000) \approx 527.06$ mm Hg

37.(a) $19 = 30(1 - e^{20k})$, $30e^{20k} = 11$, $k = \dfrac{\ln (11/30)}{20} \approx -0.0502$

$N = 30(1 - e^{-0.0502t})$

(b) $25 = 30(1 - e^{-0.0502t})$, $e^{-0.0502t} = \dfrac{1}{6}$, $t = \dfrac{-\ln 6}{-0.0502} \approx 36$ days

38.(a) $20 = 30(1 - e^{30k})$, $30e^{30k} = 10$, $k = \dfrac{\ln (1/3)}{30} = \dfrac{-\ln 3}{30} \approx -0.0366$

$N = 30(1 - e^{-0.0366t})$

(b) $25 = 30(1 - e^{-0.0366t})$, $e^{-0.0366t} = \dfrac{1}{6}$, $t = \dfrac{-\ln 6}{-0.0366} \approx 49$ days

39. $S = Ce^{k/t}$

(a) $S = 5$ when $t = 1$ (c)

$5 = Ce^{k}$ and $\lim\limits_{t \to \infty} Ce^{k/t} = C = 30$

$5 = 30e^{k}$

$k = \ln\dfrac{1}{6} \approx -1.7918$

$S = 30e^{-1.7918/t}$

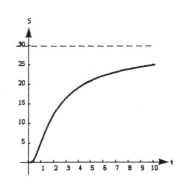

(b) $S(5) = 30e^{-1.7918/5} \approx 20.9646$
(rounded to 20,965 units)

40. $S = 30(1 - e^{kt})$

(a) $5 = 30(1 - e^{k(1)}) \Longrightarrow k = \ln\dfrac{5}{6}$ (d)

$S = 30(1 - e^{\ln(5/6)t})$

$\approx 30(1 - e^{-0.1823t})$

(b) 30,000 units

(c) 17,944 units

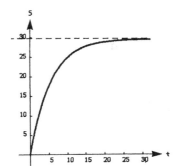

41. $y = Ce^{kt}$, $(0, 100)$ and $(5, 300)$, $C = 100$, $300 = 100e^{5k}$

$k = \dfrac{\ln 3}{5} \approx 0.2197$, $y = 100e^{0.2197t}$, $y(10) \approx 900$

42. $200 = 100e^{0.2197t}$, $t = \dfrac{\ln 2}{0.2197} \approx 3.15$ hours

43. Since $\frac{dy}{dt} = ky$, $y = Ce^{kt}$

    When $t = 0$ (1960), $y = 2500$ thus $C = 2500$

    When $t = 10$(1970), $y = 3350$ therefore $3350 = 2500e^{10k}$

    $k = \dfrac{\ln(3350/2500)}{10} = \dfrac{\ln 67 - \ln 50}{10}$ and $y = 2500e^{t(\ln 67 - \ln 50)/10}$

    When $t = 30$ (1990), $y = 2500e^{(\ln 67 - \ln 50)(30)/10} \approx 6015$

44. $5000 = 2500e^{t(\ln 67 - \ln 50)/10}$, $\qquad \ln 2 = \dfrac{\ln 67 - \ln 50}{10} t$

    $t = \dfrac{10 \ln 2}{\ln 67 - \ln 50} \approx 23.68$ years

45. Since $\frac{dy}{dx} = ky$, $y = Ce^{kt}$ or $y = y_0 e^{ky}$,

    $\frac{1}{2} y_0 = y_0 e^{1600k}$, $\qquad k = \dfrac{-\ln 2}{1600}$, $\qquad y = y_0 e^{-(\ln 2)t/1600}$

    When $t = 100$, $y = y_0 e^{-(\ln 2)/16} = y_0(0.9576)$ therefore 95.76% of the present amount still exists.

46. Since $\frac{dy}{dt} = ky$, $y = Ce^{kt}$ or $y = y_0 e^{kt}$

    When $t = 1$, $0.9957 y_0 = y_0 e^k$, $k = \ln(0.9957)$ and $y = y_0 e^{[\ln(0.9957)]t}$

    Solving $\frac{1}{2} y_0 = y_0 e^{[\ln(0.9957)]t}$, $\qquad -\ln 2 = [\ln(0.9957)]t$

    therefore $t = \dfrac{-\ln 2}{\ln(0.9957)} \approx 160.85$ years

47. Since $\frac{dy}{dt} = k(y - 20)$, $\qquad \displaystyle\int \frac{1}{y - 20}\left(\frac{dy}{dt}\right) dt = \int k \, dt$

    $\ln(y - 20) = kt + C$, $\qquad y = Ce^{kt} + 20$

    When $t = 0$, $y = 72$ therefore $C = 52$

    When $t = 1$, $y = 48$, therefore $48 = 52e^k + 20$, $e^k = \dfrac{28}{52} = \dfrac{7}{13}$, $\quad k = \ln \dfrac{7}{13}$

    thus $y = 52e^{[(\ln 7)/13]t} + 20$

    When $t = 5$, $y = 52e^{5(\ln 7)/13} + 20 = 22.35°$

48. Since $\frac{dy}{dt} = k(y - 70)$, $\int \frac{1}{y - 70}(\frac{dy}{dt}) \, dt = \int k \, dt$, $\quad \ln(y - 70) = kt + C$

When $t = 0$, $y = 350$ thus $C = \ln 280$

$kt = \ln(y - 70) - \ln 280$, $\quad kt = \ln(\frac{y - 70}{280})$

When $t = 45$, $y = 150$ thus $k = \frac{1}{45} \ln(\frac{150 - 70}{280}) = \frac{1}{45} \ln \frac{2}{7}$

When $y = 80^\circ$, $t = \frac{\ln[(80 - 70)/280]}{(1/45) \ln(2/7)} = \frac{-45 \ln 28}{(\ln 2 - \ln 7)} \approx 119.7$ min

49. Since $\frac{dy}{dt} = k(y - T)$, $\int \frac{1}{y - T}(\frac{dy}{dt}) \, dt = \int k \, dt$, $\quad \ln(y - T) = kt + C$

When $t = 0$ min, $y = 68^\circ$, thus $\ln(68 - T) = k(0) + C$, $\quad C = \ln(68 - T)$

When $t = 0.5$ min, $y = 53^\circ$, thus $\ln(53 - T) = k(0.5) + \ln(68 - T)$

$k = \ln \left[\frac{53 - T}{68 - T}\right]^2$ $\quad$ When $t = 1$ min, $y = 42^\circ$ thus

$\ln(42 - T) = k(1) + \ln(68 - T)$, $\quad \ln(42 - T) = \ln \left[\frac{53 - T}{68 - T}\right]^2 + \ln(68 - T)$

$\ln(42 - T) = \ln \frac{(53 - T)^2}{68 - T}$, $\quad 42 - T = \frac{(53 - T)^2}{68 - T}$

$2856 - 110T + T^2 = 2809 - 106T + T^2$, $\quad -4T = -47$, $\quad T = \frac{47}{4} = 11.75^\circ$

6.4
Inverse trigonometric functions and differentiation

1. $\arcsin \frac{1}{2} = \frac{\pi}{6}$

2. $\arcsin 0 = 0$

3. $\arccos \frac{1}{2} = \frac{\pi}{3}$

4. $\arccos 0 = \frac{\pi}{2}$

5. $\arctan \frac{\sqrt{3}}{3} = \frac{\pi}{6}$

6. $\text{arccot}(-1) = \frac{3\pi}{4}$

7. $\text{arccsc} \sqrt{2} = \frac{\pi}{4}$

8. $\arcsin(-0.39) \approx -0.401$

9. $\text{arcsec}(1.269) = \arccos(\frac{1}{1.269})$ $\quad$ 10. $\arctan(-3) \approx -1.249$

≈ 0.663

11.(a) $\sin(\arcsin \frac{1}{2}) = \sin(\frac{\pi}{6}) = \frac{1}{2}$

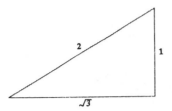

(b) $\cos(2\arcsin \frac{1}{2}) = \cos 2(\frac{\pi}{6}) = \cos \frac{\pi}{3} = \frac{1}{2}$

By constructing a triangle for the $\arcsin \frac{1}{2}$ both above and below we see that 2θ is the angle of one corner of an equilateral triangle. Therefore

$$2\arcsin \frac{1}{2} = \frac{\pi}{3}$$

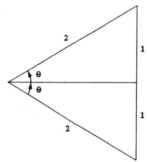

12.(a) $\tan(\arccos \frac{\sqrt{2}}{2}) = \tan(\frac{\pi}{4}) = 1$ (b) $\cot(\arcsin(-\frac{1}{2}) = \cot(-\frac{\pi}{6}) = -\sqrt{3}$

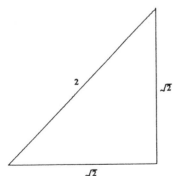

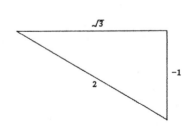

13.(a) $\sin[\arctan(\frac{3}{4})] = \frac{3}{5}$ **(b)** $\sec[\arcsin(\frac{4}{5})] = \frac{5}{3}$

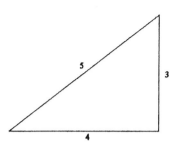   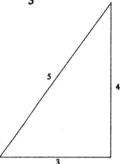

14.(a) $\tan[\text{arccot } 2] = \frac{1}{2}$ **(b)** $\cos[\text{arcsec}(\sqrt{5})] = \frac{\sqrt{5}}{5}$

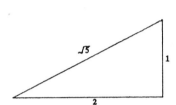   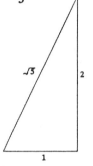

15.(a) $\cos[\arcsin\frac{5}{13}] = \frac{12}{13}$ **(b)** $\csc[\arctan(-\frac{5}{12})] = -\frac{13}{5}$

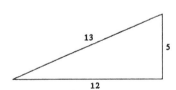   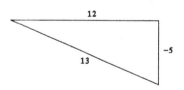

16.(a) $\sec[\arctan(-\frac{3}{5})] = \frac{\sqrt{34}}{5}$ **(b)** $\tan[\arcsin(-\frac{5}{6})] = -\frac{5\sqrt{11}}{11}$

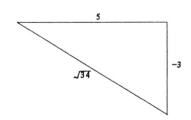   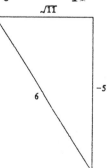

17. $y = \tan(\arctan x)$
    $\theta = \arctan x$
    $y = \tan \theta = x$

18. $y = \sin(\arccos x)$
    $\theta = \arccos x$
    $y = \sin \theta = \sqrt{1 - x^2}$

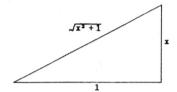

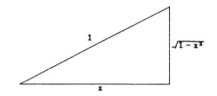

19. $y = \cos(\arcsin 2x)$
    $\theta = \arcsin 2x$
    $y = \cos \theta = \sqrt{1 - 4x^2}$

20. $y = \sec(\arctan 3x)$
    $\theta = \arctan 3x$
    $y = \sec \theta = \sqrt{9x^2 + 1}$

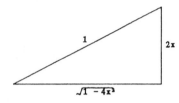

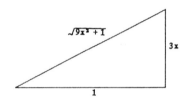

21. $y = \sin(\text{arcsec } x)$
    $\theta = \text{arcsec } x$

    $y = \sin \theta = \dfrac{\sqrt{x^2 - 1}}{x}$

22. $y = \cos(\text{arccot } x)$
    $\theta = \text{arccot } x$

    $y = \cos \theta = \dfrac{x}{\sqrt{x^2 + 1}}$

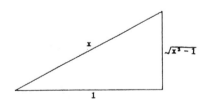

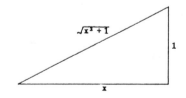

23. $y = \tan \left[\text{arcsec} \dfrac{x}{3}\right]$

$\theta = \text{arcsec} \dfrac{x}{3}$

$y = \tan \theta = \dfrac{\sqrt{x^2 - 9}}{3}$

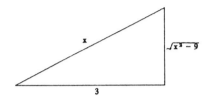

24. $y = \sec \left[\arcsin (x - 1)\right]$

$\theta = \arcsin (x - 1)$

$y = \sec \theta = \dfrac{1}{\sqrt{2x - x^2}}$

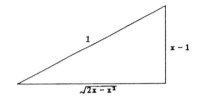

25. $y = \csc \left[\arctan \dfrac{x}{\sqrt{2}}\right]$

$\theta = \arctan \dfrac{x}{\sqrt{2}}$

$y = \csc \theta = \dfrac{\sqrt{x^2 + 2}}{x}$

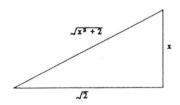

26. $y = \cos \left[\arcsin \dfrac{x - h}{r}\right]$

$\theta = \arcsin \dfrac{x - h}{r}$

$y = \cos \theta = \dfrac{\sqrt{r^2 - (x - h)^2}}{r}$

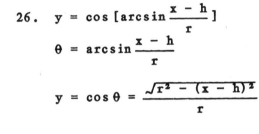

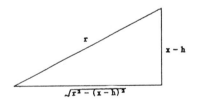

27. $\arctan \dfrac{9}{x} = \arcsin \dfrac{9}{\sqrt{x^2 + 81}}$

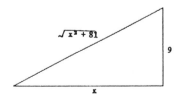

28. $\arcsin \dfrac{\sqrt{36 - x^2}}{6} = \arccos \dfrac{x}{6}$

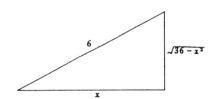

29.(a) $\arccsc x = \arcsin \dfrac{1}{x}, \quad |x| \geq 1 \qquad$ Let $y = \arccsc x$

Then for $-\dfrac{\pi}{2} \leq y < 0$ and $0 < y \leq \dfrac{\pi}{2}, \quad \csc y = x \Longrightarrow \sin y = \dfrac{1}{x}$

thus $y = \arcsin \dfrac{1}{x}$

(b) $\text{arccot } x = \arctan \dfrac{1}{x}, \quad x > 0 \qquad$ Let $y = \text{arccot } x$

Then for $0 < y < \dfrac{\pi}{2}, \quad \cot y = x \Longrightarrow \tan y = \dfrac{1}{x}$

thus $y = \arctan \dfrac{1}{x}$

30.(a) $\arcsin (-x) = -\arcsin x, \quad |x| \leq 1 \qquad$ Let $y = \arcsin (-x)$

Then $-x = \sin y \Longrightarrow x = -\sin y \Longrightarrow x = \sin (-y)$

thus $-y = \arcsin x \Longrightarrow y = -\arcsin x$

therefore $\arcsin (-x) = -\arcsin x$

(b) $\arccos (-x) = \pi - \arccos x, \quad |x| \leq 1 \qquad$ Let $y = \arccos (-x)$

Then $-x = \cos y \Longrightarrow x = -\cos y \Longrightarrow x = \cos (\pi - y)$

thus $\pi - y = \arccos x \Longrightarrow y = \pi - \arccos x$

therefore $\arccos (-x) = \pi - \arccos x$

31. $f(x) = \arcsin (x - 1)$

$x - 1 = \sin y, \quad x = 1 + \sin y$

Domain: $[0, 2]$

Range: $[-\dfrac{\pi}{2}, \dfrac{\pi}{2}]$

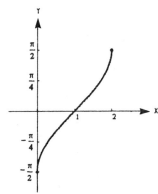

32. $f(x) = \arctan x + \dfrac{\pi}{2}$

$x = \tan (y - \dfrac{\pi}{2})$

Domain: $(-\infty, \infty)$

Range: $(0, \pi)$

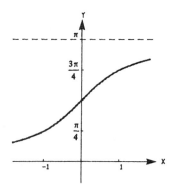

33. $f(x) = \text{arcsec } 2x$

$2x = \sec y$

$x = \frac{1}{2} \sec y$

Domain: $(-\infty, -\frac{1}{2}], [\frac{1}{2}, \infty)$

Range: $[0, \frac{\pi}{2}), (\frac{\pi}{2}, \pi]$

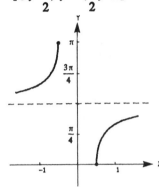

34. $f(x) = \text{arccos } (\frac{x}{4})$

$\frac{x}{4} = \cos y$

$x = 4 \cos y$

Domain: $[-4, 4]$

Range: $[0, \pi]$

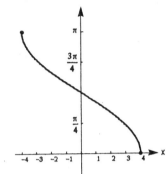

35. $\arcsin (3x - \pi) = \frac{1}{2}$, $3x - \pi = \sin (\frac{1}{2})$, $x = \frac{1}{3}[\sin (\frac{1}{2}) + \pi] \approx 1.207$

36. $\arctan (2x) = -1$, $2x = \tan (-1)$, $x = \frac{1}{2} \tan (-1) \approx -0.779$

37. $\arcsin \sqrt{2x} = \arccos \sqrt{x}$

$\sqrt{2x} = \sin (\arccos \sqrt{x})$

$\sqrt{2x} = \sqrt{1 - x}$, $0 \le x \le 1$

$2x = 1 - x$, $3x = 1$, $x = \frac{1}{3}$

38. $\arccos x = \text{arcsec } x$

$x = \cos (\text{arcsec } x)$

$x = \frac{1}{x}$, $x^2 = 1$

$x = \pm 1$

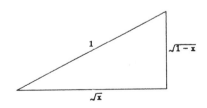

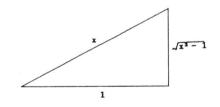

39. $f(x) = \arcsin 2x$, $\qquad f'(x) = \dfrac{2}{\sqrt{1 - 4x^2}}$

40. $f(x) = \arcsin x^2$, $\qquad f'(x) = \dfrac{2x}{\sqrt{1 - x^4}}$

41. $f(x) = 2 \arcsin (x - 1)$, $\qquad f'(x) = \dfrac{2}{\sqrt{1 - (x - 1)^2}} = \dfrac{2}{\sqrt{2x - x^2}}$

42. $f(x) = \arccos \sqrt{x}$, $\qquad f'(x) = (\dfrac{-1}{\sqrt{1 - x}})(\dfrac{1}{2\sqrt{x}}) = \dfrac{-1}{2\sqrt{x}\,\sqrt{1 - x}}$

43. $f(x) = 3 \arccos \dfrac{x}{2}$, $\qquad f'(x) = \dfrac{-3(1/2)}{\sqrt{1 - (x^2/4)}} = \dfrac{-3}{\sqrt{4 - x^2}}$

44. $f(x) = \arctan \sqrt{x}$, $\qquad f'(x) = (\dfrac{1}{1 + x})(\dfrac{1}{2\sqrt{x}}) = \dfrac{1}{2\sqrt{x}(1 + x)}$

45. $f(x) = \arctan 5x$, $\qquad f'(x) = \dfrac{5}{1 + 25x^2}$

46. $f(x) = x \arctan x$, $\qquad f'(x) = \dfrac{x}{1 + x^2} + \arctan x$

47. $f(x) = \arccos \dfrac{1}{x} = \operatorname{arcsec} x$

 $\qquad f'(x) = (\dfrac{-1}{\sqrt{1 - (1/x^2)}})(-\dfrac{1}{x^2}) = \dfrac{1}{x^2\sqrt{x^2 - 1}/|x|} = \dfrac{1}{|x|\sqrt{x^2 - 1}}$

48. $f(x) = \operatorname{arcsec} 2x$, $\qquad f'(x) = \dfrac{2}{|2x|\sqrt{4x^2 - 1}} = \dfrac{1}{|x|\sqrt{4x^2 - 1}}$

49. $f(x) = \arcsin x + \arccos x = \dfrac{\pi}{2}$, $\qquad f'(x) = 0$

50. $f(x) = \operatorname{arcsec} x + \operatorname{arccsc} x = \dfrac{\pi}{2}$, $\qquad f'(x) = 0$

51. $h(t) = \sin (\arccos t) = \sqrt{1 - t^2}$, $\qquad h'(t) = \dfrac{1}{2}(1 - t^2)^{-1/2}(-2t) = \dfrac{-t}{\sqrt{1 - t^2}}$

52. $g(t) = \tan (\arcsin t) = \dfrac{t}{\sqrt{1 - t^2}}$

 $\qquad g'(t) = \dfrac{\sqrt{1 - t^2} - t(-t)/\sqrt{1 - t^2}}{1 - t^2} = \dfrac{1}{(1 - t^2)^{3/2}}$

53. $f(t) = \dfrac{1}{\sqrt{6}} \arctan \dfrac{\sqrt{6}t}{2}$, $\qquad f'(t) = \dfrac{1}{\sqrt{6}}(\dfrac{1}{1 + (3/2)t^2})\dfrac{\sqrt{6}}{2} = \dfrac{1}{2 + 3t^2}$

54. $f(x) = \dfrac{1}{2}\left[\dfrac{1}{2}\ln\dfrac{x+1}{x-1} - \arctan x\right] = \dfrac{1}{2}\left[\dfrac{1}{2}[\ln(x+1) - \ln(x-1)] - \arctan x\right]$

$f'(x) = \dfrac{1}{2}\left[\dfrac{1}{2(x+1)} - \dfrac{\cdot 1}{2(x-1)} - \dfrac{1}{1+x^2}\right] = \dfrac{1}{4}\left[\dfrac{-2}{x^2-1} - \dfrac{2}{1+x^2}\right] = \dfrac{x^2}{1-x^4}$

55. $f(x) = \dfrac{1}{2}\left[\dfrac{1}{2}\ln\dfrac{x+1}{x-1} + \arctan x\right]$

$\qquad = \dfrac{1}{4}[\ln(x+1) - \ln(x-1)] + \dfrac{1}{2}\arctan x$

$f'(x) = \dfrac{1}{4}\left[\dfrac{1}{x+1} - \dfrac{1}{x-1}\right] + \dfrac{1/2}{1+x^2} = \dfrac{1}{1-x^4}$

56. $f(x) = \dfrac{1}{2}\,[x\sqrt{1-x^2} + \arcsin x]$

$f'(x) = \dfrac{1}{2}\left[x\left(\dfrac{-x}{\sqrt{1-x^2}}\right) + \sqrt{1-x^2} + \dfrac{1}{\sqrt{1-x^2}}\right] = \sqrt{1-x^2}$

57. $f(x) = x\arcsin x + \sqrt{1-x^2}$

$f'(x) = x\left(\dfrac{1}{\sqrt{1-x^2}}\right) + \arcsin x - \dfrac{x}{\sqrt{1-x^2}} = \arcsin x$

58. $f(x) = x\arctan(2x) - \dfrac{1}{4}\ln(1+4x^2)$

$f'(x) = \dfrac{2x}{1+4x^2} + \arctan(2x) - \dfrac{1}{4}\left(\dfrac{8x}{1+4x^2}\right) = \arctan(2x)$

59. $f(x) = \arcsin x,\qquad f'(x) = \dfrac{1}{\sqrt{1-x^2}},\qquad f''(x) = \dfrac{-x}{(1-x^2)^{3/2}} = 0$

when $x = 0$. Point of inflection $(0, 0)$

60. $f(x) = \operatorname{arccot} 2x,\qquad f'(x) = \dfrac{-2}{1+4x^2},\qquad f''(x) = \dfrac{16x}{(1+4x^2)^2} = 0$

when $x = 0$. Point of inflection $\left(0, \dfrac{\pi}{2}\right)$

61. $f(x) = \operatorname{arcsec} x - x$, $\quad f'(x) = \dfrac{1}{|x|\sqrt{x^2 - 1}} - 1 = 0$

when $|x|\sqrt{x^2 - 1} = 1$, $\quad x^2(x^2 - 1) = 1$, $\quad x^4 - x^2 - 1 = 0$

when $x^2 = \dfrac{1 \pm \sqrt{5}}{2}$ or $x = \pm\sqrt{\dfrac{1 + \sqrt{5}}{2}} = \pm 1.272$

$(1.272, -0.606)$ is a relative maximum
$(-1.272, 3.747)$ is a relative minimum

62. $f(x) = \arcsin x - 2x$, $\quad f'(x) = \dfrac{1}{\sqrt{1 - x^2}} - 2 = 0$

when $\sqrt{1 - x^2} = \dfrac{1}{2}$ or $x = \pm\dfrac{\sqrt{3}}{2}$, $\quad f''(x) = \dfrac{x}{(1 - x^2)^{3/2}}$

$f''(\sqrt{3}/2) > 0$ therefore $(\sqrt{3}/2, -0.68)$ is a relative minimum
$f''(-\sqrt{3}/2) < 0$ therefore $(-\sqrt{3}/2, 0.68)$ is a relative maximum

63. $y = \arccos x$, $y = \arctan x$

The point of intersection is given by $f(x) = \arccos x - \arctan x = 0$

$\cos(\arccos x) = \cos(\arctan x)$

$x = \dfrac{1}{\sqrt{1 + x^2}}$

$x^2(1 + x^2) = 1$

$x^4 + x^2 - 1 = 0$ when $x^2 = \dfrac{-1 \pm \sqrt{5}}{2}$

Therefore, $x = \pm\sqrt{\dfrac{-1 + \sqrt{5}}{2}} \approx \pm 0.7862$

$(0.7862, 0.6662)$ is the point of intersection

(Since $f(-0.7862) = \pi \neq 0$)

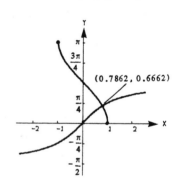

64. $y = \arcsin x$, $y = \arccos x$

The point of intersection is given by $f(x) = \arcsin x - \arccos x = 0$

$\sin(\arcsin x) = \sin(\arccos x)$

$x = \sqrt{1 - x^2}$

$x^2 = 1 - x^2$, $\quad x = \pm\dfrac{1}{\sqrt{2}} = \pm\dfrac{\sqrt{2}}{2}$

$\left(\dfrac{\sqrt{2}}{2}, \dfrac{\pi}{4}\right)$ is the point of intersection

(Since $f(-\dfrac{\sqrt{2}}{2}) = -\pi$)

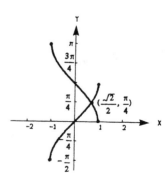

65. $\sin \theta = \dfrac{10}{L}$, $\qquad \dfrac{dL}{dt} = 1.5$ ft/sec

$\theta = \arcsin \dfrac{10}{L}$

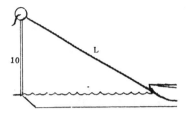

$\dfrac{d\theta}{dt} = \dfrac{1}{\sqrt{1 - (100/L^2)}}(-\dfrac{10}{L^2})\dfrac{dL}{dt}$

$\qquad = \dfrac{-10}{L\sqrt{L^2 - 100}}(\dfrac{dL}{dt})$

$\qquad = \dfrac{-10}{20\sqrt{300}}(-\dfrac{3}{2}) = \dfrac{\sqrt{3}}{40}$ rad/sec

66. $\tan \theta = \dfrac{h}{300}$, $\qquad \dfrac{dh}{dt} = 5$ ft/sec

$\theta = \arctan(\dfrac{h}{300})$

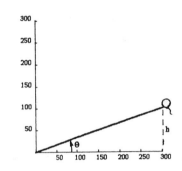

$\dfrac{d\theta}{dt} = \dfrac{1/300}{1 + (h^2/300^2)}(\dfrac{dh}{dt})$

$\qquad = \dfrac{300}{300^2 + h^2}(\dfrac{dh}{dt})$

$\qquad = \dfrac{1500}{300^2 + h^2} = \dfrac{3}{200}$ rad/sec

when $h = 100$

6.5

Inverse trigonometric functions:

Integration and completing the square

1. Let $u = 3x$, $\qquad u' = 3$

$\displaystyle\int_0^{1/6} \dfrac{1}{\sqrt{1 - 9x^2}}\, dx = \dfrac{1}{3}\int_0^{1/6} \dfrac{1}{\sqrt{1 - (3x)^2}}(3)\, dx = \dfrac{1}{3}\arcsin(3x)\Big]_0^{1/6} = \dfrac{\pi}{18}$

2. $\displaystyle\int_0^1 \dfrac{1}{\sqrt{4 - x^2}}\, dx = \arcsin\dfrac{x}{2}\Big]_0^1 = \dfrac{\pi}{6}$

3. Let $u = 2x$, $\qquad u' = 2$

$\displaystyle\int_0^{\sqrt{3}/2} \dfrac{1}{1 + 4x^2}\, dx = \dfrac{1}{2}\int_0^{\sqrt{3}/2} \dfrac{2}{1 + (2x)^2}\, dx = \dfrac{1}{2}\arctan(2x)\Big]_0^{\sqrt{3}/2} = \dfrac{\pi}{6}$

4. $\displaystyle\int_{\sqrt{3}}^3 \dfrac{1}{9 + x^2}\, dx = \dfrac{1}{3}\arctan\dfrac{x}{3}\Big]_{\sqrt{3}}^3 = \dfrac{\pi}{36}$

5. $\displaystyle\int \frac{1}{x\sqrt{4x^2-1}}\,dx = \int \frac{2}{|2x|\sqrt{(2x)^2-1}}\,dx = \operatorname{arcsec}(2x) + C$

6. $\displaystyle\int \frac{1}{4+(x-1)^2}\,dx = \frac{1}{2}\arctan\frac{x-1}{2} + C$

7. $\displaystyle\int \frac{x^3}{x^2+1}\,dx = \int\left[x - \frac{x}{x^2+1}\right]dx = \int x\,dx - \frac{1}{2}\int \frac{2x}{x^2+1}\,dx$

$\displaystyle\qquad = \frac{1}{2}x^2 - \frac{1}{2}\ln(x^2+1) + C$

8. $\displaystyle\int \frac{x^4-1}{x^2+1}\,dx = \int (x^2-1)\,dx = \frac{1}{3}x^3 - x + C$

9. $\displaystyle\int \frac{1}{\sqrt{1-(x+1)^2}}\,dx = \arcsin(x+1) + C$

10. Let $u = t^2$, $\quad u' = 2t$

$\displaystyle\int \frac{t}{t^4+16}\,dt = \frac{1}{2}\int \frac{1}{(4)^2+(t^2)^2}(2t)\,dt = \frac{1}{8}\arctan\frac{t^2}{4} + C$

11. Let $u = t^2$, $\quad u' = 2t$

$\displaystyle\int \frac{t}{\sqrt{1-t^4}}\,dt = \frac{1}{2}\int \frac{1}{\sqrt{1-(t^2)^2}}(2t)\,dt = \frac{1}{2}\arcsin(t^2) + C$

12. Let $u = x^2$, $\quad u' = 2x$

$\displaystyle\int \frac{1}{x\sqrt{x^4-4}}\,dx = \frac{1}{2}\int \frac{1}{x^2\sqrt{(x^2)^2-2^2}}(2x)\,dx = \frac{1}{2}\operatorname{arcsec}\frac{x^2}{2} + C$

13. Let $u = \arctan x$, $\quad u' = 1/(1+x^2)$

$\displaystyle\int \frac{\arctan x}{1+x^2}\,dx = \frac{1}{2}\arctan^2 x + C$

14. $\displaystyle\int \frac{1}{(x-1)\sqrt{(x-1)^2-4}}\,dx = \frac{1}{2}\operatorname{arcsec}\frac{x-1}{2} + C$

15. Let $u = \arcsin x$, $\quad u' = 1/\sqrt{1-x^2}$

$\displaystyle\int_0^{1/\sqrt{2}} \frac{\arcsin x}{\sqrt{1-x^2}}\,dx = \frac{1}{2}\arcsin^2 x\,\Big]_0^{1/\sqrt{2}} = \frac{\pi^2}{32} \approx 0.308$

16. Let $u = \arccos x$, $\quad u' = -1/\sqrt{1 - x^2}$

$$\int_0^{1/\sqrt{2}} \frac{\arccos x}{\sqrt{1 - x^2}}\, dx = -\int \frac{-\arccos x}{\sqrt{1 - x^2}}\, dx = -\frac{1}{2}\arccos^2 x \Big]_0^{1/\sqrt{2}} = \frac{3\pi^2}{32}$$

17. Let $u = 1 - x^2$, $\quad u' = -2x$

$$\int_{-1/2}^0 \frac{x}{\sqrt{1 - x^2}}\, dx = -\frac{1}{2}\int_{-1/2}^0 (1 - x^2)^{-1/2}(-2x)\, dx = -\sqrt{1 - x^2}\Big]_{-1/2}^0$$

$$= \frac{(\sqrt{3} - 2)}{2} \approx -0.134$$

18. Let $u = 1 + x^2$, $\quad u' = 2x$

$$\int_{-\sqrt{3}}^0 \frac{x}{1 + x^2}\, dx = \frac{1}{2}\int_{-\sqrt{3}}^0 \frac{1}{1 + x^2}(2x)\, dx = \ln\sqrt{1 + x^2}\Big]_{-\sqrt{3}}^0 = -\ln 2$$

19. Let $u = e^x$, $\quad u' = e^x$

$$\int \frac{e^x}{\sqrt{1 - e^{2x}}}\, dx = \arcsin e^x + C$$

20. Let $u = \sin x$, $\quad u' = \cos x$

$$\int \frac{\cos x}{\sqrt{4 - \sin^2 x}}\, dx = \arcsin \frac{\sin x}{2} + C$$

21. $$\int \frac{1}{9 + (x - 3)^2}\, dx = \frac{1}{3}\arctan\frac{x - 3}{3} + C$$

22. $$\int \frac{x + 1}{x^2 + 1}\, dx = \frac{1}{2}\int \frac{2x}{x^2 + 1}\, dx + \int \frac{1}{1 + x^2}\, dx$$

$$= \frac{1}{2}\ln(x^2 + 1) + \arctan x + C$$

23. Let $u = \sqrt{x}$, $\quad u' = 1/(2\sqrt{x})$

$$\int \frac{1}{\sqrt{x}(1 + x)}\, dx = 2\int \left(\frac{1}{1 + (\sqrt{x})^2}\right)\left(\frac{1}{2\sqrt{x}}\right)\, dx = 2\arctan\sqrt{x} + C$$

24. $$\int_1^2 \frac{1}{3 + (x - 2)^2}\, dx = \int_1^2 \frac{1}{(\sqrt{3})^2 + (x - 2)^2}\, dx$$

$$= \frac{1}{\sqrt{3}}\arctan\left(\frac{x - 2}{\sqrt{3}}\right)\Big]_1^2 = \frac{\sqrt{3}\,\pi}{18}$$

25. Let $u = \cos x$, $\quad u' = -\sin x$

$$\int_{\pi/2}^{\pi} \frac{\sin x}{1 + \cos^2 x}\, dx = -\int_{\pi/2}^{\pi} \frac{-\sin x}{1 + \cos^2 x}\, dx$$

$$= -\arctan(\cos x)\Big]_{\pi/2}^{\pi} = \frac{\pi}{4}$$

26. Let $u = e^{2x}$, $\quad u' = 2e^{2x}$

$$\int \frac{e^{2x}}{4 + e^{4x}}\, dx = \frac{1}{2}\int \frac{2e^{2x}}{4 + (e^{2x})^2}\, dx = \frac{1}{4}\arctan\frac{e^{2x}}{2} + C$$

27. $$\int_0^2 \frac{1}{x^2 - 2x + 2}\, dx = \int_0^2 \frac{1}{1 + (x-1)^2}\, dx = \arctan(x-1)\Big]_0^2 = \frac{\pi}{2}$$

28. $$\int_{-3}^{-1} \frac{1}{x^2 + 6x + 13}\, dx = \int_{-3}^{-1} \frac{1}{(x+3)^2 + 4}\, dx = \frac{1}{2}\arctan\frac{x+3}{2}\Big]_{-3}^{-1} = \frac{\pi}{8}$$

29. $$\int \frac{2x}{x^2 + 6x + 13}\, dx = \int \frac{2x + 6}{x^2 + 6x + 13}\, dx - 6\int \frac{1}{x^2 + 6x + 13}\, dx$$

$$= \int \frac{2x + 6}{x^2 + 6x + 13}\, dx - 6\int \frac{1}{4 + (x+3)^2}\, dx$$

$$= \ln|x^2 + 6x + 13| - 3\arctan\frac{x+3}{2} + C$$

30. $$\int \frac{2x - 5}{x^2 + 2x + 2}\, dx = \int \frac{2x + 2}{x^2 + 2x + 2}\, dx - 7\int \frac{1}{1 + (x+1)^2}\, dx$$

$$= \ln|x^2 + 2x + 2| - 7\arctan(x+1) + C$$

31. $$\int \frac{1}{\sqrt{-x^2 - 4x}}\, dx = \int \frac{1}{\sqrt{4 - (x+2)^2}}\, dx = \arcsin\frac{x+2}{2} + C$$

32. Let $u = -x^2 - 4x$, $\quad u' = -2x - 4$

$$\int \frac{x + 2}{\sqrt{-x^2 - 4x}}\, dx = -\frac{1}{2}\int (-x^2 - 4x)^{-1/2}(-2x - 4)\, dx = -\sqrt{-x^2 - 4x} + C$$

33. $$\int \frac{1}{\sqrt{-x^2 + 2x}}\, dx = \int \frac{1}{\sqrt{1 - (x-1)^2}}\, dx = \arcsin(x - 1) + C$$

34. Let $u = x^2 - 2x$, $u' = 2x - 2$

$$\int \frac{x - 1}{\sqrt{x^2 - 2x}}\, dx = \frac{1}{2} \int (x^2 - 2x)^{-1/2}(2x - 2)\, dx = \sqrt{x^2 - 2x} + C$$

35. $$\int_2^3 \frac{2x - 3}{\sqrt{4x - x^2}}\, dx = \int_2^3 \frac{2x - 4}{\sqrt{4x - x^2}}\, dx + \int_2^3 \frac{1}{\sqrt{4x - x^2}}\, dx$$

$$= -\int_2^3 (4x - x^2)^{-1/2}(4 - 2x)\, dx + \int_2^3 \frac{1}{\sqrt{4 - (x - 2)^2}}\, dx$$

$$= \left[-2\sqrt{4x - x^2} + \arcsin \frac{x - 2}{2} \right]_2^3 = 4 - 2\sqrt{3} + \frac{\pi}{6}$$

36. $$\int \frac{1}{(x - 1)\sqrt{x^2 - 2x}}\, dx = \int \frac{1}{(x - 1)\sqrt{(x - 1)^2 - 1}}\, dx = \operatorname{arcsec}|x - 1| + C$$

37. Let $u = x^2 + 1$, $u' = 2x$

$$\int \frac{x}{x^4 + 2x^2 + 2}\, dx = \frac{1}{2} \int \frac{2x}{(x^2 + 1)^2 + 1}\, dx = \frac{1}{2} \arctan(x^2 + 1) + C$$

38. Let $u = x^2 - 4$, $u' = 2x$

$$\int \frac{x}{\sqrt{9 + 8x^2 - x^4}}\, dx = \frac{1}{2} \int \frac{2x}{\sqrt{25 - (x^2 - 4)^2}}\, dx = \frac{1}{2} \arcsin \frac{x^2 - 4}{5} + C$$

39. Let $u = 4(x - \frac{1}{2})$, $u' = 4$

$$\int \frac{1}{\sqrt{-16x^2 + 16x - 3}}\, dx = \frac{1}{4} \int \frac{4}{\sqrt{1 - [4(x - \frac{1}{2})]^2}}\, dx = \frac{1}{4} \arcsin (4x - 2) + C$$

40. Let $u = 3(x - 1)$, $u' = 3$

$$\int \frac{1}{(x - 1)\sqrt{9x^2 - 18x + 5}}\, dx = \int \frac{3}{3(x - 1)\sqrt{[3(x - 1)]^2 - 4}}\, dx$$

$$= \frac{1}{2} \operatorname{arcsec} \frac{|3(x - 1)|}{2} + C$$

41. Let $u = \sqrt{x - 1}$, $u^2 + 1 = x$, $2u\, du = dx$

$$\int \frac{\sqrt{x - 1}}{x}\, dx = \int \frac{2u^2}{u^2 + 1}\, du = \int \frac{2u^2 + 2 - 2}{u^2 + 1}\, du = 2 \int du - 2 \int \frac{1}{u^2 + 1}\, du$$

$$= 2u - 2 \arctan u + C = 2\sqrt{x - 1} - 2 \arctan \sqrt{x - 1} + C$$

42. Let $u = \sqrt{x-2}$, $u^2 + 2 = x$, $2u\,du = dx$

$$\int \frac{\sqrt{x-2}}{x+1}\,dx = \int \frac{2u^2}{u^2+3}\,du = \int \frac{2u^2+6-6}{u^2+3}\,du$$

$$= 2\int du - 6\int \frac{1}{u^2+3}\,du = 2u - \frac{6}{\sqrt{3}}\arctan\frac{u}{\sqrt{3}} + C$$

$$= 2\sqrt{x-2} - 2\sqrt{3}\arctan\sqrt{\frac{x-2}{3}} + C$$

43. Let $u = \sqrt{e^t - 3}$, $u^2 + 3 = e^t$, $2u\,du = e^t\,dt$, $\dfrac{2u\,du}{u^2+3} = dt$

$$\int \sqrt{e^t - 3}\,dt = \int \frac{2u^2}{u^2+3}\,du \qquad \text{(Same as in Exercise 42.)}$$

$$= 2u - 2\sqrt{3}\arctan\frac{u}{\sqrt{3}} + C$$

$$= 2\sqrt{e^t - 3} - 2\sqrt{3}\arctan\sqrt{\frac{e^t - 3}{3}} + C$$

44. Let $u = \sqrt{t}$, $du = \dfrac{1}{2\sqrt{t}}\,dt$

$$\int \frac{1}{t\sqrt{t} + \sqrt{t}}\,dt = 2\int \frac{1}{2\sqrt{t}(t+1)}\,dt = 2\arctan\sqrt{t} + C$$

45. $A = \displaystyle\int_0^1 \frac{1}{1+x^2}\,dx$

$$= \arctan x \Big]_0^1 = \frac{\pi}{4}$$

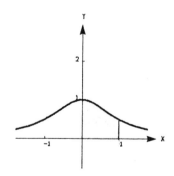

46. $A = \displaystyle\int_0^1 \frac{1}{\sqrt{4-x^2}}\,dx$

$$= \arcsin\frac{x}{2}\Big]_0^1 = \frac{\pi}{6}$$

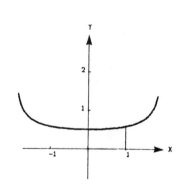

47. $A = \int_1^3 \dfrac{1}{x^2 - 2x + 5}\, dx$

$= \int_1^3 \dfrac{1}{(x - 1)^2 + 4}\, dx$

$= \dfrac{1}{2} \arctan \dfrac{x - 1}{2}\Big]_1^3 = \dfrac{\pi}{8}$

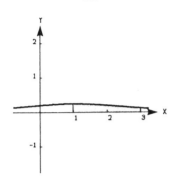

48. $A = \int_0^2 \dfrac{1}{\sqrt{3 + 2x - x^2}}\, dx$

$= \int_0^2 \dfrac{1}{\sqrt{4 - (x - 1)^2}}\, dx$

$= \arcsin \dfrac{x - 1}{2}\Big]_0^2 = \dfrac{\pi}{3}$

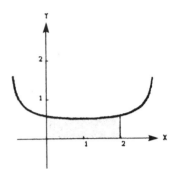

49. $\displaystyle\int \dfrac{1}{32 + kv^2}\, dv = -\int dt, \qquad \dfrac{1}{\sqrt{32k}} \arctan\left(\sqrt{\dfrac{k}{32}}\, v\right) = -t + C_1$

$\arctan\left(\sqrt{\dfrac{k}{32}}\, v\right) = -\sqrt{32k}\, t + C, \qquad \sqrt{\dfrac{k}{32}}\, v = \tan(C - \sqrt{32k}\, t)$

$v = \sqrt{\dfrac{32}{k}} \tan(C - \sqrt{32k}\, t)$

When $t = 0$, $v = 500$, $C = \arctan\left(\sqrt{\dfrac{k}{32}}\, 500\right)$ and we have

$v = \sqrt{\dfrac{32}{k}} \tan\left[\arctan\left(500 \sqrt{\dfrac{k}{32}}\right) - \sqrt{32k}\, t\right]$

50. $\displaystyle\int \dfrac{1}{\sqrt{A^2 - y^2}}\, dy = \int \sqrt{\dfrac{k}{m}}\, dt, \quad \arcsin \dfrac{y}{A} = \sqrt{\dfrac{k}{m}}\, t + C$

$y = A \sin\left(\sqrt{\dfrac{k}{m}}\, t + C\right)$

Since $y = 0$ when $t = 0$, $C = 0$ and we have $y = A \sin\left(\sqrt{\dfrac{k}{m}}\, t\right)$

51.(a) $\displaystyle\int \dfrac{1}{\sqrt{1 - x^2}}\, dx = \arcsin x + C$ $\hspace{2cm}$ $(u = x)$

(b) $\displaystyle\int \dfrac{x}{\sqrt{1 - x^2}}\, dx = -\sqrt{1 - x^2} + C$ $\hspace{1.5cm}$ $(u = 1 - x^2)$

(c) $\displaystyle\int \dfrac{1}{x\sqrt{1 - x^2}}\, dx$ cannot be evaluated using the Basic Integration Formulas

52.(a) $\int e^{x^2}\,dx$ cannot be evaluated using the Basic Integration Formulas

(b) $\int xe^{x^2}\,dx = \frac{1}{2}e^{x^2} + C$ $\qquad\qquad (u = x^2)$

(c) $\int \frac{1}{x^2}e^{1/x}\,dx = -e^{1/x} + C$ $\qquad\qquad (u = \frac{1}{x})$

53.(a) $\int \sqrt{x-1}\,dx = \frac{2}{3}(x-1)^{3/2} + C$ $\qquad (u = x - 1)$

(b) Let $u = \sqrt{x-1}$ then $x = u^2 + 1$ and $dx = 2u\,du$

$\int x\sqrt{x-1}\,dx = \int (u^2+1)(u)(2u)\,du = 2\int (u^4 + u^2)\,du$

$= 2(\frac{u^5}{5} + \frac{u^3}{3}) + C = \frac{2}{15}u^3(3u^2 + 5) + C$

$= \frac{2}{15}(x-1)^{3/2}[3(x-1)+5] + C = \frac{2}{15}(x-1)^{3/2}(3x+2) + C$

(c) Let $u = \sqrt{x-1}$ then $x = u^2 + 1$ and $dx = 2u\,du$

$\int \frac{x}{\sqrt{x-1}}\,dx = \int \frac{u^2+1}{u}(2u)\,du = 2\int (u^2+1)\,du$

$= 2(\frac{u^3}{3} + u) + C = \frac{2}{3}u(u^2+3) + C = \frac{2}{3}\sqrt{x-1}(x+2) + C$

Note: In (b) and (c) substitution was necessary before the Basic Formulas could be used.

54.(a) $\int \frac{1}{1+x^4}\,dx$ cannot be evaluated using the Basic Integration Formulas

(b) $\int \frac{x}{1+x^4}\,dx = \frac{1}{2}\int \frac{2x}{1+(x^2)^2}\,dx = \frac{1}{2}\arctan(x^2) + C$

(c) $\int \frac{x^3}{1+x^4}\,dx = \frac{1}{4}\int \frac{4x^3}{1+x^4}\,dx = \frac{1}{4}\ln(1+x^4) + C$

55.(a) $\frac{d}{dx}\left[\frac{1}{a}\arctan\frac{u}{a} + C\right]$

$= \frac{1}{a}\left[\frac{u'/a}{1+(u/a)^2}\right] = \frac{1}{a^2}\left[\frac{u'}{(a^2+u^2)/a^2}\right] = \frac{u'}{a^2+u^2}$

Thus $\int \frac{du}{a^2+u^2} = \int \frac{u'}{a^2+u^2}\,dx = \frac{1}{a}\arctan\frac{u}{a} + C$

55.(b)   Assume $u > 0$

$$\frac{d}{dx}\left[\frac{1}{a}\text{arcsec}\frac{u}{a} + C\right] = \frac{1}{a}\left[\frac{u'/a}{(u/a)\sqrt{(u/a)^2 - 1}}\right]$$

$$= \frac{1}{a}\left[\frac{u'}{u\sqrt{(u^2 - a^2)/a^2}}\right] = \frac{u'}{u\sqrt{u^2 - a^2}}$$

Thus $\displaystyle\int \frac{du}{u\sqrt{u^2 - a^2}} = \int \frac{u'}{u\sqrt{u^2 - a^2}}\,dx = \frac{1}{a}\text{arcsec}\frac{|u|}{a} + C$

6.6
Hyperbolic functions

1. (a) $\sinh 3 = \dfrac{e^3 - e^{-3}}{2} \approx 10.018$

    (b) $\tanh(-2) = \dfrac{\sinh(-2)}{\cosh(-2)} = \dfrac{e^{-2} - e^2}{e^{-2} + e^2} \approx -0.964$

2. (a) $\cosh(0) = \dfrac{e^0 + e^0}{2} = 1$

    (b) $\text{sech}(1) = \dfrac{2}{e^2 + e^{-2}} \approx 0.648$

3. (a) $\text{csch}(\ln 2) = \dfrac{2}{e^{\ln 2} - e^{-\ln 2}} = \dfrac{2}{2 - (1/2)} = \dfrac{4}{3}$

    (b) $\coth(\ln 5) = \dfrac{\cosh(\ln 5)}{\sinh(\ln 5)} = \dfrac{e^{\ln 5} + e^{-\ln 5}}{e^{\ln 5} - e^{-\ln 5}} = \dfrac{5 + (1/5)}{5 - (1/5)} = \dfrac{13}{12}$

4. (a) $\sinh^{-1}(0) = 0$
    (b) $\tanh^{-1}(0) = 0$

5. (a) $\cosh^{-1}(2) = \ln(2 + \sqrt{3}) \approx 1.317$

    (b) $\text{sech}^{-1}\left(\dfrac{2}{3}\right) = \ln\left(\dfrac{1 + \sqrt{1 - (4/9)}}{2/3}\right) \approx 0.962$

6. (a) $\text{csch}^{-1}(2) = \ln\left(\dfrac{1 + \sqrt{5}}{2}\right) \approx 0.481$

    (b) $\coth^{-1}(3) = \dfrac{1}{2}\ln\left(\dfrac{4}{2}\right) \approx 0.347$

7. $\tanh^2 x + \text{sech}^2 x = \left(\dfrac{e^x - e^{-x}}{e^x + e^{-x}}\right)^2 + \left(\dfrac{2}{e^x + e^{-x}}\right)^2$

$\qquad = \dfrac{e^{2x} - 2 + e^{-2x} + 4}{(e^x + e^{-x})^2} = \dfrac{e^{2x} + 2 + e^{-2x}}{e^{2x} + 2 + e^{-2x}} = 1$

8. $\dfrac{1 + \cosh 2x}{2} = \dfrac{1 + (e^{2x} + e^{-2x})/2}{2} = \dfrac{e^{2x} + 2 + e^{-2x}}{4}$

$\qquad = \left(\dfrac{e^x + e^{-x}}{2}\right)^2 = \cosh^2 x$

9. $\sinh x \cosh y + \cosh x \sinh y = \left(\dfrac{e^x - e^{-x}}{2}\right)\left(\dfrac{e^y + e^{-y}}{2}\right) + \left(\dfrac{e^x + e^{-x}}{2}\right)\left(\dfrac{e^y - e^{-y}}{2}\right)$

$\qquad = \dfrac{1}{4}[e^{x+y} - e^{-x+y} + e^{x-y} - e^{-(x+y)} + e^{x+y} + e^{-x+y} - e^{x-y} - e^{-(x+y)}]$

$\qquad = \dfrac{1}{4}[2(e^{x+y} - e^{-(x+y)})] = \dfrac{e^{(x+y)} - e^{-(x+y)}}{2} = \sinh(x + y)$

10. $2 \sinh x \cosh x = 2\left(\dfrac{e^x - e^{-x}}{2}\right)\left(\dfrac{e^x + e^{-x}}{2}\right) = \dfrac{e^{2x} - e^{-2x}}{2} = \sinh 2x$

11. $3 \sinh x + 4 \sinh^3 x = \sinh x (3 + 4 \sinh^2 x)$

$\qquad = \left(\dfrac{e^x - e^{-x}}{2}\right)\left[3 + 4\left(\dfrac{e^x - e^{-x}}{2}\right)^2\right]$

$\qquad = \left(\dfrac{e^x - e^{-x}}{2}\right)[3 + e^{2x} - 2 + e^{-2x}] = \dfrac{1}{2}(e^x - e^{-x})(e^{2x} + e^{-2x} + 1)$

$\qquad = \dfrac{1}{2}[e^{3x} + e^{-x} + e^x - e^x - e^{-3x} - e^{-x}] = \dfrac{e^{3x} - e^{-3x}}{2} = \sinh(3x)$

12. $2 \cosh\left(\dfrac{x + y}{2}\right) \cosh\left(\dfrac{x - y}{2}\right) = 2\left[\dfrac{e^{(x+y)/2} + e^{-(x+y)/2}}{2}\right]\left[\dfrac{e^{(x-y)/2} + e^{-(x-y)/2}}{2}\right]$

$\qquad = 2\left[\dfrac{e^x + e^y + e^{-y} + e^{-x}}{4}\right] = \dfrac{e^x + e^{-x}}{2} + \dfrac{e^y + e^{-y}}{2}$

$\qquad = \cosh x + \cosh y$

13. $y = \sinh(1 - x^2), \qquad y' = -2x \cosh(1 - x^2)$

14. $y = \coth(3x), \qquad y' = -3 \,\text{csch}^2(3x)$

15. $y = \ln(\sinh x), \qquad y' = \dfrac{1}{\sinh x}(\cosh x) = \coth x$

16. $y = \ln(\cosh x)$, $\quad y' = \dfrac{1}{\cosh x}(\sinh x) = \tanh x$

17. $y = \ln(\tanh \dfrac{x}{2})$

$y' = \dfrac{(1/2)}{\tanh(x/2)} \operatorname{sech}^2(\dfrac{x}{2}) = \dfrac{1}{2\sinh(x/2)\cosh(x/2)} = \dfrac{1}{\sinh x} = \operatorname{csch} x$

18. $y = x\sinh x - \cosh x$, $\quad y' = x\cosh x + \sinh x - \sinh x = x\cosh x$

19. $y = \dfrac{1}{4}\sinh(2x) - \dfrac{x}{2}$, $\quad y' = \dfrac{1}{2}\cosh(2x) - \dfrac{1}{2} = \dfrac{\cosh(2x) - 1}{2} = \sinh^2 x$

20. $y = x - \coth x$, $\quad y' = 1 + \operatorname{csch}^2 x = \coth^2 x$

21. $y = \arctan(\sinh x)$, $\quad y' = \dfrac{1}{1 + \sinh^2 x}(\cosh x) = \dfrac{\cosh x}{\cosh^2 x} = \operatorname{sech} x$

22. $y = e^{\sinh x}$, $\quad y' = (\cosh x)(e^{\sinh x})$

23. $y = x^{\cosh x}$, $\quad \ln y = \cosh x \ln x$, $\quad \dfrac{1}{y}(\dfrac{dy}{dx}) = \dfrac{\cosh x}{x} + \sinh x \ln x$

$\dfrac{dy}{dx} = \dfrac{y}{x}[\cosh x + x(\sinh x)\ln x] = x^{-1+\cosh x}[\cosh x + x\sinh x(\ln x)]$

24. $y = \operatorname{sech}^2 3x$

$y' = -2\operatorname{sech}(3x)\operatorname{sech}(3x)\tan(3x)(3) = -6\operatorname{sech}^2 3x \tanh 3x$

25. $y = (\cosh x - \sinh x)^2$

$y' = 2(\cosh x - \sinh x)(\sinh x - \cosh x) = -2(\cosh x - \sinh x)^2 = -2e^{-2x}$

26. $y = \operatorname{sech}(x+1)$, $\quad y' = -\operatorname{sech}(x+1)\tanh(x+1)$

27. $y = \cosh^{-1}(3x)$, $\quad y' = \dfrac{3}{\sqrt{9x^2 - 1}}$

28. $y = \tanh^{-1}(\dfrac{x}{2})$, $\quad y' = \dfrac{1}{1 - (x/2)^2}(\dfrac{1}{2}) = \dfrac{2}{4 - x^2}$

29. $y = \sinh^{-1}(\tan x)$, $\quad y' = \dfrac{1}{\sqrt{\tan^2 x + 1}}(\sec^2 x) = |\sec x|$

30. $y = \text{sech}^{-1}(\cos 2x), \quad 0 < x < \dfrac{\pi}{4}$

$y' = \dfrac{-1}{\cos 2x\sqrt{1 - \cos^2 2x}}(-2\sin 2x) = \dfrac{2\sin 2x}{\cos 2x\,|\sin 2x|} = \dfrac{2}{\cos 2x}$

(since $\sin 2x \geq 0$ for $0 < x < \dfrac{\pi}{4}$) $= 2\sec 2x$

31. $y = \coth^{-1}(\sin 2x), \quad y' = \dfrac{1}{1 - \sin^2 2x}(2\cos 2x) = 2\sec 2x$

32. $y = (\text{csch}^{-1} x)^2, \quad y' = 2\,\text{csch}^{-1} x\left[\dfrac{-1}{|x|\sqrt{1 + x^2}}\right] = \dfrac{-2\,\text{csch}^{-1} x}{|x|\sqrt{1 + x^2}}$

33. $y = 2x\sinh^{-1}(2x) - \sqrt{1 + 4x^2}$

$y' = 2x\left(\dfrac{2}{\sqrt{1 + 4x^2}}\right) + 2\sinh^{-1}(2x) - \dfrac{4x}{\sqrt{1 + 4x^2}} = 2\sinh^{-1}(2x)$

34. $y = x\tanh^{-1} x + \ln\sqrt{1 - x^2} = x\tanh^{-1} x + \dfrac{1}{2}\ln(1 - x^2)$

$y' = x\left(\dfrac{1}{1 - x^2}\right) + \tanh^{-1} x + \dfrac{-x}{1 - x^2} = \tanh^{-1} x$

35. Let $u = 1 - 2x, \quad u' = -2$

$\displaystyle\int \sinh(1 - 2x)\,dx = -\dfrac{1}{2}\int \sinh(1 - 2x)(-2)\,dx = -\dfrac{1}{2}\cosh(1 - 2x) + C$

36. Let $u = \sqrt{x}, \quad u' = 1/(2\sqrt{x})$

$\displaystyle\int \dfrac{\cosh\sqrt{x}}{\sqrt{x}}\,dx = 2\int \cosh\sqrt{x}\left(\dfrac{1}{2\sqrt{x}}\right)dx = 2\sinh\sqrt{x} + C$

37. Let $u = \cosh(x - 1), \quad u' = \sinh(x - 1)$

$\displaystyle\int \cosh^2(x - 1)\sinh(x - 1)\,dx = \dfrac{1}{3}\cosh^3(x - 1) + C$

38. Let $u = \cosh x, \quad u' = \sinh x$

$\displaystyle\int \dfrac{\sinh x}{1 + \sinh^2 x}\,dx = \int \dfrac{\sinh x}{\cosh^2 x}\,dx = \dfrac{-1}{\cosh x} + C = -\text{sech}\,x + C$

39. Let $u = \sinh x, \quad u' = \cosh x$

$\displaystyle\int \dfrac{\cosh x}{\sinh x}\,dx = \ln|\sinh x| + C$

40. Let $u = 2x - 1, \quad u' = 2$

$\displaystyle\int \text{sech}^2(2x - 1)\,dx = \dfrac{1}{2}\int \text{sech}^2(2x - 1)(2)\,dx = \dfrac{1}{2}\tanh(2x - 1) + C$

41. Let $u = x^2/2$, $\quad u' = x$

$$\int x \operatorname{csch}^2 \frac{x^2}{2}\, dx = \int \operatorname{csch}^2 \frac{x^2}{2}\, x\, dx = -\coth \frac{x^2}{2} + C$$

42. Let $u = \operatorname{sech} x$, $\quad u' = -\operatorname{sech} x \tanh x$

$$\int \operatorname{sech}^3 x \tanh x\, dx = -\int \operatorname{sech}^2 x(-\operatorname{sech} x \tanh x)\, dx = -\frac{1}{3}\operatorname{sech}^3 x + C$$

43. Let $u = 1/x$, $\quad u' = -1/x^2$

$$\int \frac{\operatorname{csch}(1/x)\coth(1/x)}{x^2}\, dx = -\int \operatorname{csch}\frac{1}{x}\coth\frac{1}{x}\left(-\frac{1}{x^2}\right)\, dx = \operatorname{csch}\frac{1}{x} + C$$

44. $$\int \sinh^2 x\, dx = \frac{1}{2}\int (\cosh 2x - 1)\, dx = \frac{1}{2}\left[\frac{1}{2}\sinh 2x - x\right] + C$$

$$= \frac{1}{4}(-2x + \sinh 2x) + C$$

45. $$\int_0^4 \frac{1}{25 - x^2}\, dx = \frac{1}{10}\ln\left|\frac{5 + x}{5 - x}\right|\Big]_0^4 = \frac{1}{10}\ln 9 = \frac{1}{5}\ln 3$$

46. $$\int_0^4 \frac{1}{\sqrt{25 - x^2}}\, dx = \arcsin\frac{x}{5}\Big]_0^4 = \arcsin\frac{4}{5}$$

47. Let $u = 2x$, $\quad u' = 2$

$$\int_0^{\sqrt{2}/4} \frac{1}{\sqrt{1 - (2x)^2}}\, (2)\, dx = \arcsin(2x)\Big]_0^{\sqrt{2}/4} = \frac{\pi}{4}$$

48. $$\int \frac{2}{x\sqrt{1 + 4x^2}}\, dx = 2\int \frac{1}{(2x)\sqrt{1 + (2x)^2}}(2)\, dx = -2\ln\left[\frac{1 + \sqrt{1 + 4x^2}}{|2x|}\right] + C$$

49. Let $u = x^2$, $\quad u' = 2x$

$$\int \frac{x}{x^4 + 1}\, dx = \frac{1}{2}\int \frac{2x}{(x^2)^2 + 1}\, dx = \frac{1}{2}\arctan(x^2) + C$$

50. Let $u = \sinh x$, $\quad u' = \cosh x$

$$\int \frac{\cosh x}{\sqrt{9 - \sinh^2 x}}\, dx = \arcsin\left(\frac{\sinh x}{3}\right) + C = \arcsin\left(\frac{e^x - e^{-x}}{6}\right) + C$$

51. $\displaystyle \int \frac{1}{\sqrt{1 + e^{2x}}}\, dx \;=\; \int \frac{e^x}{e^x\sqrt{1 + (e^x)^2}}\, dx = -\text{csch}^{-1}(e^x) + C$

$\displaystyle \qquad\qquad = -\ln\left(\frac{1 + \sqrt{1 + e^{2x}}}{e^x}\right) + C$

52. Let $u = e^x$, $\qquad u' = e^x$

$\displaystyle \int \frac{e^x}{1 - e^{2x}}\, dx = \int \frac{1}{1 - (e^x)^2}(e^x)\, dx = \tanh^{-1}(e^x) + C$

$\displaystyle \qquad\qquad = \frac{1}{2}\ln\left|\frac{1 + e^x}{1 - e^x}\right| + C$

53. Let $u = \sqrt{x}$, $\qquad u' = 1/(2\sqrt{x})$

$\displaystyle \int \frac{1}{\sqrt{x}\,\sqrt{1 + x}}\, dx \;=\; 2 \int \frac{1}{\sqrt{1 + (\sqrt{x})^2}}\left(\frac{1}{2\sqrt{x}}\right)\, dx = 2\sinh^{-1}\sqrt{x} + C$

$\displaystyle \qquad\qquad = 2\ln(\sqrt{x} + \sqrt{1 + x}) + C$

54. Let $u = x^{3/2}$, $\qquad u' = \dfrac{3}{2}\sqrt{x}$

$\displaystyle \int \frac{\sqrt{x}}{\sqrt{1 + x^3}}\, dx \;=\; \frac{2}{3} \int \frac{1}{\sqrt{1 + (x^{3/2})^2}}\left(\frac{3}{2}\sqrt{x}\right)\, dx = \frac{2}{3}\sinh^{-1}(x^{3/2}) + C$

$\displaystyle \qquad\qquad = \frac{2}{3}\ln(x^{3/2} + \sqrt{1 + x^3}) + C$

55. $\displaystyle \int \frac{1}{(x - 1)\sqrt{x^2 - 2x + 2}}\, dx \;=\; \int \frac{1}{(x - 1)\sqrt{(x - 1)^2 + 1}}\, dx$

$\displaystyle \qquad\qquad = -\ln\left[\frac{1 + \sqrt{(x - 1)^2 + 1}}{|x - 1|}\right] + C$

56. $\displaystyle \int \frac{-1}{4x - x^2}\, dx = \int \frac{1}{(x - 2)^2 - 4}\, dx = \frac{1}{4}\ln\left|\frac{(x - 2) - 2}{(x - 2) + 2}\right| = \frac{1}{4}\ln\left|\frac{x - 4}{x}\right| + C$

57. $\displaystyle \int \frac{1}{1 - 4x - 2x^2}\, dx = \int \frac{1}{3 - 2(x + 1)^2}\, dx$

$\displaystyle \qquad\qquad = \frac{-1}{\sqrt{2}} \int \frac{\sqrt{2}}{[\sqrt{2}(x + 1)]^2 - (\sqrt{3})^2}\, dx$

$\displaystyle \qquad\qquad = \frac{-1}{2\sqrt{6}}\ln\left|\frac{\sqrt{2}(x + 1) - \sqrt{3}}{\sqrt{2}(x + 1) + \sqrt{3}}\right| + C = \frac{1}{2\sqrt{6}}\ln\left|\frac{\sqrt{2}(x + 1) + \sqrt{3}}{\sqrt{2}(x + 1) - \sqrt{3}}\right| + C$

58. $\displaystyle\int \frac{1}{(x+1)\sqrt{2x^2+4x+8}}\,dx = \int \frac{1}{(x+1)\sqrt{2(x+1)^2+6}}\,dx$

$\displaystyle\qquad = \frac{1}{\sqrt{2}}\int \frac{1}{(x+1)\sqrt{(x+1)^2+(\sqrt{3})^2}}\,dx$

$\displaystyle\qquad = -\frac{1}{\sqrt{6}}\ln\left(\frac{\sqrt{3}+\sqrt{(x+1)^2+3}}{x+1}\right)+C$

59. Let $u = 4x - 1$, $\quad u' = 4$

$\displaystyle\int \frac{1}{\sqrt{80+8x-16x^2}}\,dx = \frac{1}{4}\int \frac{4}{\sqrt{81-(4x-1)^2}}\,dx$

$\displaystyle\qquad = \frac{1}{4}\arcsin\frac{4x-1}{9}+C$

60. Let $u = 2(x-1)$, $\quad u' = 2$

$\displaystyle\int \frac{1}{(x-1)\sqrt{-4x^2+8x-1}}\,dx = \int \frac{2}{2(x-1)\sqrt{(\sqrt{3})^2-[2(x-1)]^2}}\,dx$

$\displaystyle\qquad = -\frac{1}{\sqrt{3}}\ln\left|\frac{\sqrt{3}+\sqrt{-4x^2+8x-1}}{2(x-1)}\right|+C$

61. $\displaystyle\int \frac{x^3-21x}{5+4x-x^2}\,dx = \int \left[-x-4+\frac{20}{5+4x-x^2}\right]dx$

$\displaystyle\qquad = \int (-x-4)\,dx - 20\int \frac{1}{(x-2)^2-3^2}\,dx$

$\displaystyle\qquad = -\frac{x^2}{2}-4x-\frac{20}{6}\ln\left|\frac{(x-2)-3}{(x-2)+3}\right|+C$

$\displaystyle\qquad = -\frac{x^2}{2}-4x-\frac{10}{3}\ln\left|\frac{x-5}{x+1}\right|+C$

62. $\displaystyle\int \frac{1-2x}{4x-x^2}\,dx = \int \frac{4-2x}{4x-x^2}\,dx + 3\int \frac{1}{(x-2)^2-4}\,dx$

$\displaystyle\qquad = \ln|4x-x^2| + \frac{3}{4}\ln\left|\frac{(x-2)-2}{(x-2)+2}\right|+C$

$\displaystyle\qquad = \ln|4x-x^2| + \frac{3}{4}\ln\left|\frac{x-4}{x}\right|+C$

63. $y = x \cosh x - \sinh x$

$y' = x \sinh x = 0$ when $x = 0$

$y'' = x \cosh x + \sinh x = 0$ when $x = 0$

Point of inflection $(0, 0)$

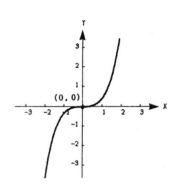

64. $y = x - \tanh x$

$y' = 1 - \text{sech}^2 x = 0$ when $x = 0$

$y'' = 2 \text{sech}^2 x \tanh x = 0$ when $x = 0$

Point of inflection $(0, 0)$

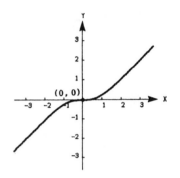

65. $y = a \sinh x, \quad y' = a \cosh x, \quad y'' = a \sinh x, \quad y''' = a \cosh x$

Therefore $y''' - y' = 0$

66. $y = a \cosh x, \quad y' = a \sinh x, \quad y'' = a \cosh x$

Therefore $y'' - y = 0$

67. $y = a \, \text{sech}^{-1} \left(\dfrac{x}{a}\right) - \sqrt{a^2 - x^2}$

$$\frac{dy}{dx} = \frac{-1}{(x/a)\sqrt{1 - (x^2/a^2)}} + \frac{x}{\sqrt{a^2 - x^2}}$$

$$= \frac{-a^2}{x\sqrt{a^2 - x^2}} + \frac{x}{\sqrt{a^2 - x^2}} = \frac{-\sqrt{a^2 - x^2}}{x}$$

68. Equation of tangent line:

$$y - a\,\text{sech}^{-1}\frac{x_0}{a} + \sqrt{a^2 - x_0^2} = \frac{-\sqrt{a^2 - x_0^2}}{x_0}(x - x_0)$$

When $x = 0$, $y = a\,\text{sech}^{-1}\frac{x_0}{a} - \sqrt{a^2 - x_0^2} + \sqrt{a^2 - x_0^2} = a\,\text{sech}^{-1}\frac{x_0}{a}$

Hence Q is the point $(0, a\,\text{sech}^{-1}\frac{x_0}{a})$

Distance from P to Q: $d = \sqrt{x_0^2 + (-\sqrt{a^2 - x_0^2})^2} = a$

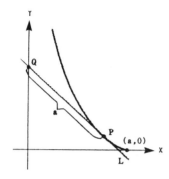

69. $\displaystyle\int \frac{3k}{16}\,dt = \int \frac{1}{x^2 - 12x + 32}\,dx$

$$\frac{3kt}{16} = \int \frac{1}{(x-6)^2 - 4}\,dx = \frac{1}{4}\ln\left|\frac{(x-6) - 2}{(x-6) + 2}\right| + C = \frac{1}{4}\ln\left|\frac{x - 8}{x - 4}\right| + C$$

When $x = 0$, $t = 0$, $C = -\dfrac{1}{4}\ln(2)$

When $x = 1$, $t = 10$, $\dfrac{30k}{16} = \dfrac{1}{4}\ln\left|\dfrac{-7}{-3}\right| - \dfrac{1}{4}\ln(2) = \dfrac{1}{4}\ln(\dfrac{7}{6})$, $k = \dfrac{4}{300}\ln(\dfrac{7}{6})$

When $t = 20$, we have $(\dfrac{3}{16})(\dfrac{4}{30})\ln(\dfrac{7}{6})(20) = \dfrac{1}{4}\ln\dfrac{x-8}{2x-8}$

$\ln(\dfrac{7}{6})^2 = \ln\dfrac{x-8}{2x-8}$, $\dfrac{49}{36} = \dfrac{x-8}{2x-8}$, $62x = 104$, $x = \dfrac{104}{62} \approx 1.677$

70. Area $= 2\displaystyle\int_0^{20}\left[31 - 20\cosh\frac{x}{20}\right]dx = 2\left[31x - 400\sinh\frac{x}{20}\right]_0^{20}$

$= 1240 - 800\sinh 1 \approx 299.839 \text{ ft}^2$

Volume $= 100(1240 - 800\sinh 1) = 12{,}400 - 8{,}000\sinh 1 \approx 29{,}983.9 \text{ ft}^3$

71. $y = \tanh x = \dfrac{e^x - e^{-x}}{e^x + e^{-x}}$

$y' = \dfrac{(e^x + e^{-x})(e^x + e^{-x}) - (e^x - e^{-x})(e^x - e^{-x})}{(e^x + e^{-x})^2}$

$= \dfrac{e^{2x} + 2 + e^{-2x} - e^{2x} + 2 - e^{-2x}}{(e^x + e^{-x})^2} = \left[\dfrac{2}{e^x + e^{-x}}\right]^2 = \operatorname{sech}^2 x$

72. $y = \operatorname{sech} x = \dfrac{2}{e^x + e^{-x}}$

$y' = -2(e^x + e^{-x})^{-2}(e^x - e^{-x}) = \left(\dfrac{-2}{e^x + e^{-x}}\right)\left(\dfrac{e^x - e^{-x}}{e^x + e^{-x}}\right) = -\operatorname{sech} x \tanh x$

73. $y = \cosh^{-1} x$, $\qquad \cosh y = x$, $\qquad (\sinh y)(y') = 1$

$y' = \dfrac{1}{\sinh y} = \dfrac{1}{\sqrt{\cosh^2 y - 1}} = \dfrac{1}{\sqrt{x^2 - 1}}$

74. $y = \tanh^{-1} x$, $\qquad \tanh y = x$, $\qquad (\operatorname{sech}^2 y)(y') = 1$

$y' = \dfrac{1}{\operatorname{sech}^2 y} = \dfrac{1}{1 - \tanh^2 y} = \dfrac{1}{1 - x^2}$

75. $y = \operatorname{sech}^{-1} x$, $\qquad \operatorname{sech} y = x$, $\qquad -(\operatorname{sech} y)(\tanh y)y' = 1$

$y' = \dfrac{-1}{(\operatorname{sech} y)(\tanh y)} = \dfrac{-1}{(\operatorname{sech} y)\sqrt{1 - \operatorname{sech}^2 y}} = \dfrac{-1}{x\sqrt{1 - x^2}}$

76. $y = \sinh^{-1} x$, $\qquad \sinh y = x$, $\qquad (\cosh y)y' = 1$

$y' = \dfrac{1}{\cosh y} = \dfrac{1}{\sqrt{\sinh^2 y + 1}} = \dfrac{1}{\sqrt{x^2 + 1}}$

Review Exercises for Chapter 6

1.(a) $f(x) = \dfrac{1}{2}x - 3$ **(b)**

$y = \dfrac{1}{2}x - 3$, $\qquad 2(y + 3) = x$

$f^{-1}(x) = 2x + 6$

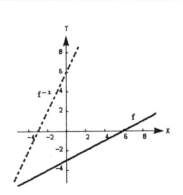

2.(a) $f(x) = 5x - 7$ (b)

$y = 5x - 7, \quad \dfrac{y + 7}{5} = x$

$f^{-1}(x) = \dfrac{x + 7}{5}$

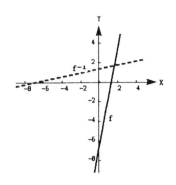

3.(a) $f(x) = \sqrt{x + 1}$ (b)

$y = \sqrt{x + 1}, \quad y^2 - 1 = x$

$f^{-1}(x) = x^2 - 1, \quad x \geq 0$

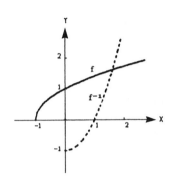

4.(a) $f(x) = x^3 + 2$ (b)

$y = x^3 + 2, \quad \sqrt[3]{y - 2} = x$

$f^{-1}(x) = \sqrt[3]{x - 2}$

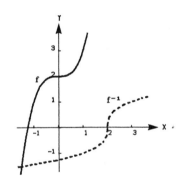

5.(a) $f(x) = x^2 - 5, \quad x \geq 0$ (b)

$y = x^2 - 5, \quad \sqrt{y + 5} = x$

$f^{-1}(x) = \sqrt{x + 5}$

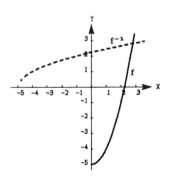

6.(a) $f(x) = \sqrt[3]{x + 1}$ (b)

$y = \sqrt[3]{x + 1}, \quad y^3 - 1 = x$

$f^{-1}(x) = x^3 - 1$

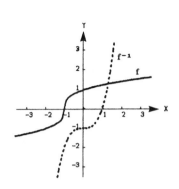

7. $f(x) = 2(x - 4)^2, \quad x \geq 4, \quad y = 2(x - 4)^2, \quad \sqrt{y/2} + 4 = x$

$f^{-1}(x) = \sqrt{x/2} + 4$

8. $f(x) = |x - 2| = x - 2, \quad x \geq 2, \quad y = x - 2, \quad y + 2 = x$

$f^{-1}(x) = x + 2, \quad x \geq 0$

9. $e^{\ln x} = 3, \quad x = 3$

10. $\ln x + \ln (x - 3) = 0, \quad \ln [x(x - 3)] = 0, \quad x^2 - 3x = e^0$

$x^2 - 3x - 1 = 0, \quad x = \dfrac{3 \pm \sqrt{13}}{2}, \quad x = \dfrac{3 + \sqrt{13}}{2}$

11. $\log_3 x + \log_3 (x - 1) - \log_3 (x - 2) = 2, \quad \log_3 \dfrac{x(x - 1)}{x - 2} = 2$

$\dfrac{x(x - 1)}{x - 2} = 3^2, \quad x^2 - x = 9(x - 2), \quad x^2 - 10x + 18 = 0$

$x = \dfrac{10 \pm \sqrt{28}}{2} = 5 \pm \sqrt{7}$

12. $\log_x 125 = 3, \quad 125 = x^3, \quad x = 5$

13. $y = \ln (e^{-x^2}) = -x^2, \quad y' = -2x$

14. $y = \ln \left(\dfrac{e^x}{1 + e^x}\right) = \ln e^x - \ln (1 + e^x) = x - \ln (1 + e^x)$

$y' = 1 - \dfrac{e^x}{1 + e^x} = \dfrac{1}{1 + e^x}$

15. $y = x^2 e^x, \quad y' = x^2 e^x + 2x e^x = x e^x (x + 2)$

16. $y = e^{-x^2/2}$, $y' = -xe^{-x^2/2}$

17. $y = \sqrt{e^{2x} + e^{-2x}}$

$y' = \dfrac{1}{2}(e^{2x} + e^{-2x})^{-1/2}(2e^{2x} - 2e^{-2x}) = \dfrac{e^{2x} - e^{-2x}}{\sqrt{e^{2x} + e^{-2x}}}$

18. $y = x^{2x+1}$, $\ln y = (2x + 1)\ln x$, $\dfrac{y'}{y} = \dfrac{2x + 1}{x} + 2\ln x$

$y' = y\left[\dfrac{2x + 1}{x} + 2\ln x\right] = x^{2x+1}\left[\dfrac{2x + 1}{x} + 2\ln x\right]$

19. $y = 3^{x-1}$, $y' = 3^{x-1}\ln 3$

20. $y = 4^x e^x$, $y' = 4^x e^x + (\ln 4)4^x e^x = 4^x e^x(1 + \ln 4)$

21. $ye^x + xe^y = xy$, $ye^x + y'e^x + xy'e^y + e^y = y + xy'$

$y'(e^x + xe^y - x) = y - ye^x - e^y$, $y' = \dfrac{y - ye^x - e^y}{e^x + xe^y - x}$

22. $y = \dfrac{x^2}{e^x}$, $y' = \dfrac{e^x(2x) - x^2 e^x}{e^{2x}} = \dfrac{x(2 - x)}{e^x}$

23. $\cos x^2 = xe^y$, $-2x\sin x^2 = xe^y y' + e^y$

$y' = -\dfrac{2x\sin x^2 + e^y}{xe^y} = -2x\tan x^2 - \dfrac{1}{x}$

24. $y = \dfrac{1}{2}e^{\sin 2x}$, $y' = \cos 2x\, e^{\sin 2x}$

25. $y = \tan(\arcsin x) = \dfrac{x}{\sqrt{1 - x^2}}$

$y' = \dfrac{(1 - x^2)^{1/2} + x^2(1 - x^2)^{-1/2}}{1 - x^2} = (1 - x^2)^{-3/2}$

26. $y = \arctan(x^2 - 1)$, $y' = \dfrac{2x}{1 + (x^2 - 1)^2} = \dfrac{2x}{x^4 - 2x^2 + 2}$

27. $y = x\operatorname{arcsec} x$, $y' = \dfrac{x}{|x|\sqrt{x^2 - 1}} + \operatorname{arcsec} x$

28. $y = \dfrac{1}{2}\arctan e^{2x}$, $y' = \dfrac{1}{2}\left(\dfrac{1}{1 + e^{4x}}\right)(2e^{2x}) = \dfrac{e^{2x}}{1 + e^{4x}}$

29. $y = x(\arcsin x)^2 - 2x + 2\sqrt{1 - x^2}\,\arcsin x$

$$y' = \frac{2x\arcsin x}{\sqrt{1 - x^2}} + (\arcsin x)^2 - 2 + \frac{2\sqrt{1 - x^2}}{\sqrt{1 - x^2}} + \frac{-2x\arcsin x}{\sqrt{1 - x^2}}$$

$$= (\arcsin x)^2$$

30. $y = \sqrt{x^2 - 4} - 2\,\text{arcsec}\,\dfrac{x}{2}, \quad 0 < x < \dfrac{\pi}{4}$

$$y' = \frac{x}{\sqrt{x^2 - 4}} - \frac{1}{|x|/2\,\sqrt{(x/2)^2 - 1}} = \frac{x}{\sqrt{x^2 - 4}} - \frac{4}{|x|\sqrt{x^2 - 4}}$$

$$= \frac{x^2 - 4}{|x|\sqrt{x^2 - 4}} = \frac{\sqrt{x^2 - 4}}{|x|}$$

31. (a) $y = x^a, \quad y' = ax^{a-1}$ (b) $y = a^x, \quad y' = (\ln a)a^x$

   (c) $y = x^x, \quad y' = x^x(1 + \ln x)$ (d) $y = a^a, \quad y' = 0$

32. $y = e^x(a\cos 3x + b\sin 3x)$

$$y' = e^x(-3a\sin 3x + 3b\cos 3x) + e^x(a\cos 3x + b\sin 3x)$$

$$= e^x[(-3a + b)\sin 3x + (a + 3b)\cos 3x]$$

$$y'' = e^x[3(-3a + b)\cos 3x - 3(a + 3b)\sin 3x]$$

$$+ e^x[(-3a + b)\sin 3x + (a + 3b)\cos 3x]$$

$$= e^x[(-6a - 8b)\sin 3x + (-8a + 6b)\cos 3x]$$

$$y'' - 2y' + 10y = e^x\{[(-6a - 8b) - 2(-3a + b) + 10b]\sin 3x$$

$$+ [(-8a + 6b) - 2(a + 3b) + 10a]\cos 3x\} = 0$$

33. $\displaystyle\int xe^{-3x^2}\,dx = -\frac{1}{6}\int e^{-3x^2}(-6x)\,dx = -\frac{1}{6}e^{-3x^2} + C$

34. $\displaystyle\int \frac{e^{1/x}}{x^2}\,dx = -\int e^{1/x}\left(-\frac{1}{x^2}\right)\,dx = -e^{1/x} + C$

35. $\displaystyle\int \frac{e^{4x} - e^{2x} + 1}{e^x}\,dx = \int (e^{3x} - e^x + e^{-x})\,dx$

$$= \frac{1}{3}e^{3x} - e^x - e^{-x} + C = \frac{e^{4x} - 3e^{2x} - 3}{3e^x} + C$$

36. Let $u = e^{2x} + e^{-2x}$, $\quad u' = 2e^{2x} - 2e^{-2x}$

$$\int \frac{e^{2x} - e^{-2x}}{e^{2x} + e^{-2x}} \, dx = \frac{1}{2} \int \frac{2e^{2x} - 2e^{-2x}}{e^{2x} + e^{-2x}} \, dx = \frac{1}{2} \ln (e^{2x} + e^{-2x}) + C$$

37. $\displaystyle \int \frac{e^x}{e^x - 1} \, dx = \ln |e^x - 1| + C$

38. $\displaystyle \int x^2 e^{x^3+1} \, dx = \frac{1}{3} \int e^{x^3+1}(3x^2) \, dx = \frac{1}{3} e^{x^3+1} + C$

39. $\displaystyle \int xe^{-x^2/2} \, dx = - \int e^{-x^2/2}(-x) \, dx = -e^{-x^2/2} + C$

40. $\displaystyle \int \frac{x - 1}{3x^2 - 6x - 1} \, dx = \frac{1}{6} \int \frac{6x - 6}{3x^2 - 6x - 1} \, dx = \frac{1}{6} \ln |3x^2 - 6x - 1| + C$

41. $\displaystyle \int \frac{e^{-2x}}{1 + e^{-2x}} \, dx = -\frac{1}{2} \int \frac{-2e^{-2x}}{1 + e^{-2x}} \, dx = -\frac{1}{2} \ln (1 + e^{-2x}) + C$

42. Let $u = \cos(1/x)$, $\quad u' = -\sin(1/x)(-1/x^2) = (1/x^2)\sin(1/x)$

$$\int \frac{\tan(1/x)}{x^2} \, dx = \int \frac{(1/x^2)\sin(1/x)}{\cos(1/x)} \, dx = \ln \left| \cos \left(\frac{1}{x}\right) \right| + C$$

43. Let $u = e^{2x}$, $\quad u' = 2e^{2x}$

$$\int \frac{1}{e^{2x} + e^{-2x}} \, dx = \int \frac{e^{2x}}{1 + e^{4x}} \, dx = \frac{1}{2} \int \frac{1}{1 + (e^{2x})^2}(2e^{2x}) \, dx$$

$$= \frac{1}{2} \arctan(e^{2x}) + C$$

44. Let $u = 5x$, $\quad u' = 5$

$$\int \frac{1}{3 + 25x^2} \, dx = \frac{1}{5} \int \frac{1}{(\sqrt{3})^2 + (5x)^2}(5) \, dx = \frac{1}{5\sqrt{3}} \arctan \frac{5x}{\sqrt{3}} + C$$

45. Let $u = x^2$, $\quad u' = 2x$

$$\int \frac{x}{\sqrt{1 - x^4}} \, dx = \frac{1}{2} \int \frac{1}{\sqrt{1 - (x^2)^2}}(2x) \, dx = \frac{1}{2} \arcsin x^2 + C$$

46. $\displaystyle \int \frac{1}{16 + x^2} \, dx = \frac{1}{4} \arctan \frac{x}{4} + C$

47. Let $u = 16 + x^2$, $\quad u' = 2x$

$$\int \frac{x}{16 + x^2}\ dx = \frac{1}{2} \int \frac{1}{16 + x^2}(2x)\ dx = \frac{1}{2}\ln(16 + x^2) + C$$

48. Let $u = \arcsin x$, $\quad u' = \dfrac{1}{\sqrt{1 - x^2}}$

$$\int \frac{\arcsin x}{\sqrt{1 - x^2}}\ dx = \frac{1}{2}(\arcsin x)^2 + C$$

49. Let $u = \arctan\left(\dfrac{x}{2}\right)$, $\quad u' = \dfrac{2}{4 + x^2}$

$$\int \frac{\arctan(x/2)}{4 + x^2}\ dx = \frac{1}{2}\int \left(\arctan \frac{x}{2}\right)\left(\frac{2}{4 + x^2}\right)\ dx = \frac{1}{4}\left[\arctan \frac{x}{2}\right]^2 + C$$

50. $\displaystyle \int \frac{4 - x}{\sqrt{4 - x^2}}\ dx = 4 \int \frac{1}{\sqrt{4 - x^2}}\ dx + \frac{1}{2}\int (4 - x^2)^{-1/2}(-2x)\ dx$

$$= 4\arcsin \frac{x}{2} + \sqrt{4 - x^2} + C$$

51. $\text{Area} = \displaystyle\int_0^4 xe^{-x^2}\ dx = -\frac{1}{2}e^{-x^2}\bigg]_0^4 = -\frac{1}{2}[e^{-16} - 1] \approx 0.500$

52. $\text{Area} = \displaystyle\int_0^4 3e^{-x/2}\ dx = -6e^{-x/2}\bigg]_0^4 = -6[e^{-2} - 1] \approx 5.188$

53.(a)　$A = 500e^{(0.05)(1)} \approx \$525.64$

(b)　$A = 500e^{(0.05)(10)} \approx \$824.36$

(c)　$A = 500e^{(0.05)(100)} \approx \$74,206.58$

54. $2P = Pe^{10r}$, $\quad 2 = e^{10r}$, $\quad \ln 2 = 10r$, $\quad r = \dfrac{\ln 2}{10} \approx 6.93\%$

55. $10,000 = Pe^{(0.07)(15)}$, $\quad P = \dfrac{10,000}{e^{1.05}} \approx \$3499.38$

56. $\dfrac{1}{5}\displaystyle\int 2500e^{0.12t}\ dt = \dfrac{12500}{3}e^{0.12t}\bigg]_0^5 \approx \$3425.50$

57.(a)　$2P_0 = P_0 e^{0.025t}$, $\quad 2 = e^{0.025t}$, $\quad t = \dfrac{\ln 2}{0.025} \approx 27.7 \text{ yrs}$

(b)　$3P_0 = P_0 e^{0.025t}$, $\quad 3 = e^{0.025t}$, $\quad t = \dfrac{\ln 3}{0.025} \approx 43.94 \text{ yrs}$

58. $P(h) = 30e^{kh}$, $\quad P(18,000) = 30e^{18,000k} = 15$, $\quad k = \dfrac{\ln(1/2)}{18,000} \approx -0.00004$

$P(h) = 30e^{-(h \ln 2)/18,000}$

$P(35,000) = 30e^{-(35,000 \ln 2)/18,000} \approx 7.79$

59. $p(t) = 80e^{-0.5t} + 20$

$p'(t) = -40e^{-0.5t}$, $\quad p'(1) = -40e^{-0.5(1)} \approx -24.26\%$

$p'(2) = -40e^{-0.5(2)} \approx -14.72\%$

60. (a) $\dfrac{1}{2} \displaystyle\int_0^2 (80e^{-0.5t} + 20)\, dt = \dfrac{1}{2}(-160e^{-0.5t} + 20t) \Big]_0^2$

$= 10(-8e^{-0.5t} + t) \Big]_0^2 \approx 70.57\%$

(b) $\dfrac{1}{2} \displaystyle\int_2^4 (80e^{-0.5t} + 20)\, dt = 10(-8e^{-0.5t} + t) \Big]_2^4 \approx 38.60\%$

61. $\displaystyle\int_{t_1}^{t_2} \dfrac{1}{20} e^{-t/20}\, dt = -e^{-t/20} \Big]_{t_1}^{t_2}$

(a) $\displaystyle\int_0^{10} \dfrac{1}{20} e^{-t/20}\, dt = -e^{-t/20} \Big]_0^{10} \approx 0.3935$

(b) $\displaystyle\int_0^{30} \dfrac{1}{20} e^{-t/20}\, dt = -e^{-t/20} \Big]_0^{30} \approx 0.7769$

(c) $\displaystyle\int_{15}^{30} \dfrac{1}{20} e^{-t/20}\, dt = -e^{-t/20} \Big]_{15}^{30} \approx 0.2492$

(d) $\displaystyle\int_0^{60} \dfrac{1}{20} e^{-t/20}\, dt = -e^{-t/20} \Big]_0^{60} \approx 0.9502$

62. $\displaystyle\int_{t_1}^{t_2} \dfrac{1}{3} e^{-t/3}\, dt = -e^{-t/3} \Big]_{t_1}^{t_2}$

0 – 2	2 – 4	4 – 6	6 – 8	8 – 10
0.4866	0.2498	0.1283	0.0658	0.0338

63.(a) $P = \dfrac{1}{100}\left[25 + \displaystyle\int_{2.5}^{10} \dfrac{25}{x}\, dx\right] = \dfrac{1}{100}\left[25 + \Big[25 \ln x\Big]_{2.5}^{10}\right]$

$= \dfrac{1}{4}(1 + \ln 4) \approx 0.60$

(b) $P = \dfrac{1}{100}\left[50 + \displaystyle\int_{5}^{10} \dfrac{50}{x}\, dx\right] = \dfrac{1}{100}\left[50 + \Big[50 \ln x\Big]_{5}^{10}\right]$

$= \dfrac{1}{2}(1 + \ln 2) \approx 0.85$

64. $A = A_0 e^{kt}$, $A_0 = 500$, When $t = 40$ $A = 300$, $300 = 500e^{40k}$

$\ln \dfrac{3}{5} = 40k$, $k \approx -0.0128$, $A = 500e^{-0.0128t}$

65. $\dfrac{dv}{dt} = -kv$, $v = v_0 e^{-kt}$, $s = \displaystyle\int v_0 e^{-kt}\, dt = \dfrac{-v_0}{k}(e^{-kt}) + C$

$s(0) = 0 = \dfrac{-v_0}{k} + C$, $C = \dfrac{v_0}{k}$, $s = \dfrac{-v_0}{k}(e^{-kt}) + \dfrac{v_0}{k} = \dfrac{v_0}{k}(1 - e^{-kt})$

66. $\displaystyle\int \dfrac{1}{kv - 32}\, dv = \int dt$, $\dfrac{1}{k}\ln|kv - 32| = t + C_1$, $kv - 32 = Ce^{kt}$

$kv = 32 + Ce^{kt}$, $v(t) = \dfrac{1}{k}(32 + Ce^{kt})$

$v(0) = \dfrac{1}{k}(32 + C) = 0 \implies C = -32$, $v(t) = \dfrac{32}{k}(1 - e^{kt})$

(Note that $k < 0$ since the object is moving downward.

67. From Exercise 66, we have $v(t) = \dfrac{1}{k}(32 + Ce^{kt})$

$v(0) = \dfrac{1}{k}(32 + C) = v_0 \implies C = v_0 k - 32$

$v(t) = \dfrac{1}{k}[32 + (v_0 k - 32)e^{kt}]$

68. $v(t) = \dfrac{32}{k}(1 - e^{kt})$ $(k < 0)$

$\displaystyle\lim_{t \to \infty} \dfrac{32}{k}(1 - e^{kt}) = \dfrac{32}{k}(1 - 0) = \dfrac{32}{k}$

The velocity does not increase indefinitely but approaches a limiting velocity $32/k$ ft/sec.

69. $s(t) = \dfrac{32}{k} \displaystyle\int (1 - e^{kt})\, dt = \dfrac{32}{k}(t - \dfrac{1}{k}e^{kt}) + C$

$s(0) = \dfrac{32}{k}(-\dfrac{1}{k}) + C = s_0 \implies C = s_0 + \dfrac{32}{k^2}$

$s(t) = \dfrac{32}{k}(t - \dfrac{1}{k}e^{kt}) + s_0 + \dfrac{32}{k^2}$

$\qquad = \dfrac{32t}{k} + \dfrac{32}{k^2}(1 - e^{kt}) + s_0$

70. $y = Ce^{-0.012(s)}, \qquad 28 = Ce^{-0.6}, \qquad C = 28e^{0.6} \approx 51.02$

(a) $y = 28e^{0.6-0.012s}$

(b)

speed(s)	50	55	60	65	70
mileage(y)	28	26.4	24.8	23.4	22.0

7 Applications of integration

7.1
Area of a region between two curves

1. $A = \int_0^6 [0 - (x^2 - 6x)] \, dx = -\left[\frac{x^3}{3} - 3x^2\right]_0^6 = 36$

2. $A = \int_{-2}^2 [(2x + 5) - (x^2 + 2x + 1)] \, dx = \int_{-2}^2 (-x^2 + 4) \, dx$

   $= \left[-\frac{x^3}{3} + 4x\right]_{-2}^2 = \frac{32}{3}$

3. $A = \int_0^3 [(-x^2 + 2x + 3) - (x^2 - 4x + 3)] \, dx = \int_0^3 (-2x^2 + 6x) \, dx$

   $= \left[-\frac{2x^3}{3} + 3x^2\right]_0^3 = 9$

4. $A = \int_0^1 (x^2 - x^3) \, dx = \left[\frac{x^3}{3} - \frac{x^4}{4}\right]_0^1 = \frac{1}{12}$

5. $A = 2\int_{-1}^0 3(x^3 - x) \, dx = 6\left[\frac{x^4}{4} - \frac{x^2}{2}\right]_{-1}^0 = 6\left(\frac{1}{4}\right) = \frac{3}{2}$

6. $A = 2\int_0^1 [(x - 1)^3 - (x - 1)] \, dx = 2\left[\frac{(x - 1)^4}{4} - \left(\frac{x^2}{2} - x\right)\right]_0^1 = \frac{1}{2}$

7. The points of intersection are given by:

$x^2 - 4x = 0$

$x(x - 4) = 0$ when $x = 0, 4$

$A = \int_0^4 [g(x) - f(x)]\, dx = -\int_0^4 (x^2 - 4x)\, dx$

$= -\left[\dfrac{x^3}{3} - 2x^2\right]_0^4 = \dfrac{32}{3}$

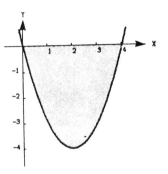

8. The points of intersection are given by:

$3 - 2x - x^2 = 0$

$(3 + x)(1 - x) = 0$ when $x = -3, 1$

$A = \int_{-3}^1 [f(x) - g(x)]\, dx$

$= \int_{-3}^1 (3 - 2x - x^2)\, dx$

$= \left[3x - x^2 - \dfrac{x^3}{3}\right]_{-3}^1 = \dfrac{32}{3}$

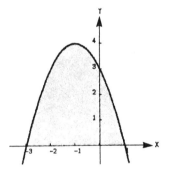

9. The points of intersection are given by:

$x^2 + 2x + 1 = 3x + 3$

$(x - 2)(x + 1) = 0$ when $x = -1, 2$

$A = \int_{-1}^2 [g(x) - f(x)]\, dx$

$= \int_{-1}^2 [(3x + 3) - (x^2 + 2x + 1)]\, dx$

$= \int_{-1}^2 (2 + x - x^2)\, dx$

$= \left[2x + \dfrac{x^2}{2} - \dfrac{x^3}{3}\right]_{-1}^2 = \dfrac{9}{2}$

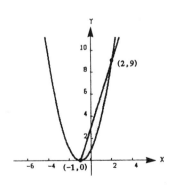

10. The points of intersection are given by:

$-x^2 + 4x + 2 = x + 2$

$x(3 - x) = 0$ when $x = 0, 3$

$A = \int_0^3 [f(x) - g(x)]\ dx$

$= \int_0^3 [(-x^2 + 4x + 2) - (x + 2)]\ dx$

$= \int_0^3 (-x^2 + 3x)\ dx$

$= \left[\dfrac{-x^3}{3} + \dfrac{3}{2} x^2 \right]_0^3 = \dfrac{9}{2}$

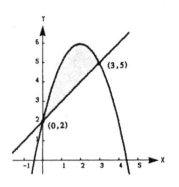

11. The points of intersection are given by:

$x = 2 - x$ and $x = 0$ and $2 - x = 0$

$x = 1 \qquad\qquad x = 0 \qquad x = 2$

$A = \int_0^1 [(2 - y) - (y)]\ dy = \left[2y - y^2 \right]_0^1 = 1$

(Note that if we integrate with respect to x we need two integrals.)

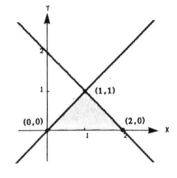

12. $A = \int_1^5 \left(\dfrac{1}{x^2} - 0 \right)\ dx$

$= -\dfrac{1}{x} \Big]_1^5 = \dfrac{4}{5}$

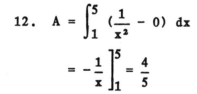

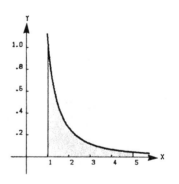

13. The points of intersection are given by: $3x^2 + 2x = 8$

$(3x - 4)(x + 2) = 0$ when $x = -2, \dfrac{4}{3}$

$$A = \int_{-2}^{4/3} [g(x) - f(x)]\, dx$$

$$= \int_{-2}^{4/3} (8 - 2x - 3x^2)\, dx$$

$$= \left[8x - x^2 - x^3 \right]_{-2}^{4/3} = \frac{500}{27}$$

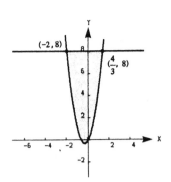

14. The points of intersection are given by: $x^3 - 3x^2 + 3x = x^2$

$x(x - 1)(x - 3) = 0$ when $x = 0, 1, 3$

$$A = \int_0^1 [f(x) - g(x)]\, dx + \int_1^3 [g(x) - f(x)]\, dx$$

$$= \int_0^1 [(x^3 - 3x^2 + 3x) - x^2]\, dx + \int_1^3 [x^2 - (x^3 - 3x^2 + 3x)]\, dx$$

$$= \int_0^1 (x^3 - 4x^2 + 3x)\, dx$$

$$+ \int_1^3 (-x^3 + 4x^2 - 3x)\, dx$$

$$= \left[\frac{x^4}{4} - \frac{4}{3}x^3 + \frac{3}{2}x^2 \right]_0^1$$

$$+ \left[\frac{-x^4}{4} + \frac{4}{3}x^3 - \frac{3}{2}x^2 \right]_1^3 = \frac{37}{12}$$

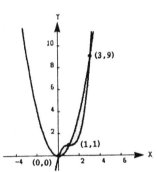

15. The point of intersection is given by: $x^3 - 2x + 1 = -2x$

$x^3 + 1 = 0$ when $x = -1$

$$A = \int_{-1}^1 [f(x) - g(x)]\, dx$$

$$= \int_{-1}^1 [(x^3 - 2x + 1) - (-2x)]\, dx$$

$$= \int_{-1}^1 (x^3 + 1)\, dx$$

$$= \left[\frac{x^4}{4} + x \right]_{-1}^1 = 2$$

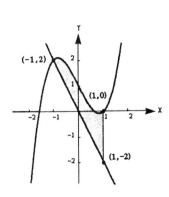

16. The points of intersection are given by:

$\sqrt[3]{x} = x, \qquad x = -1, 0, 1$

$A = 2 \int_0^1 [f(x) - g(x)] \, dx$

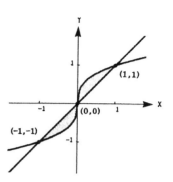

$= 2 \int_0^1 (\sqrt[3]{x} - x) \, dx = 2 \int_0^1 (x^{1/3} - x) \, dx$

$= 2 \left[\frac{3}{4} x^{4/3} - \frac{1}{2} x^2 \right]_0^1 = \frac{1}{2}$

17. The points of intersection are given by:

$\sqrt{3x} + 1 = x + 1$

$\sqrt{3x} = x$ when $x = 0, 3$

$A = \int_0^3 [f(x) - g(x)] \, dx$

$= \int_0^3 [(\sqrt{3x} + 1) - (x + 1)] \, dx$

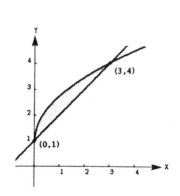

$= \int_0^3 [(3x)^{1/2} - x] \, dx$

$= \left[\frac{2}{9}(3x)^{3/2} - \frac{x^2}{2} \right]_0^3 = \frac{3}{2}$

18. The points of intersection are given by:

$x^2 + 5x - 6 = 6x - 6$

$x(x - 1) = 0$ when $x = 0, 1$

$A = \int_0^1 [g(x) - f(x)] \, dx$

$= \int_0^1 [(6x - 6) - (x^2 + 5x - 6)] \, dx$

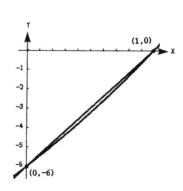

$= \int_0^1 (x - x^2) \, dx$

$= \left[\frac{x^2}{2} - \frac{x^3}{3} \right]_0^1 = \frac{1}{6}$

19. The points of intersection are given by:

$$x^2 - 4x + 3 = 3 + 4x - x^2$$

$$8x = 2x^2 \text{ when } x = 0, 4$$

$$A = \int_0^4 [(3 + 4x - x^2) - (x^2 - 4x + 3)] \, dx$$

$$= \int_0^4 (-2x^2 + 8x) \, dx = \left[-\frac{2x^3}{3} + 4x^2 \right]_0^4$$

$$= \frac{64}{3}$$

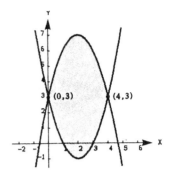

20. The points of intersection are given by:

$$x^4 - 2x^2 = 2x^2$$

$$x^2(x^2 - 4) = 0 \text{ when } x = -2, 0, 2$$

$$A = 2 \int_0^2 [2x^2 - (x^4 - 2x^2)] \, dx$$

$$= 2 \int_0^2 (4x^2 - x^4) \, dx$$

$$= 2 \left[\frac{4x^3}{3} - \frac{x^5}{5} \right]_0^2 = \frac{128}{15}$$

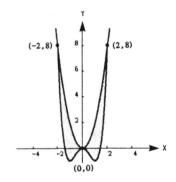

21. The points of intersection are given by:

$$y^2 = y + 2$$

$$(y - 2)(y + 1) = 0 \text{ when } y = -1, 2$$

$$A = \int_{-1}^2 [g(y) - f(y)] \, dy$$

$$= \int_{-1}^2 [(y + 2) - y^2] \, dy$$

$$= \left[2y + \frac{y^2}{2} - \frac{y^3}{3} \right]_{-1}^2 = \frac{9}{2}$$

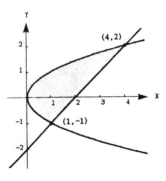

22. The points of intersection are given by:

$2y - y^2 = -y$

$y(y - 3) = 0$ when $y = 0, 3$

$A = \int_0^3 [f(y) - g(y)]\, dy$

$= \int_0^3 [(2y - y^2) - (-y)]\, dy$

$= \int_0^3 (3y - y^2)\, dy$

$= \left[\frac{3}{2} y^2 - \frac{1}{3} y^3 \right]_0^3 = \frac{9}{2}$

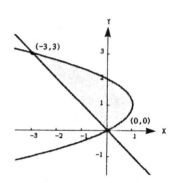

23. $A = \int_{-1}^2 [f(y) - g(y)]\, dy$

$= \int_{-1}^2 [(y^2 + 1) - 0]\, dy$

$= \left[\frac{y^3}{3} + y \right]_{-1}^2 = 6$

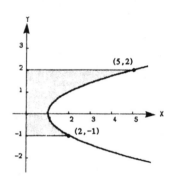

24. $A = \int_0^3 [f(y) - g(y)]\, dy$

$= \int_0^3 \left[\frac{y}{\sqrt{16 - y^2}} - 0 \right] dy$

$= -\frac{1}{2} \int_0^3 (16 - y^2)^{-1/2}(-2y)\, dy$

$= -\sqrt{16 - y^2}\, \Big]_0^3 = 4 - \sqrt{7}$

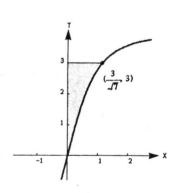

25. The points of intersection are given by:

$$\frac{1}{x} = -x^2 + 4x - 2 \quad \text{or} \quad x^3 - 4x^2 + 2x + 1 = 0$$

when $x = 1$, $(3 \pm \sqrt{13})/2$

$$A = \int_1^{(3+\sqrt{13})/2} [g(x) - f(x)] \, dx$$

$$= \int_1^{(3+\sqrt{13})/2} (-x^2 + 4x - 2 - \frac{1}{x}) \, dx$$

$$= \left[-\frac{1}{3}x^3 + 2x^2 - 2x - \ln x \right]_1^{(3+\sqrt{13})/2}$$

$$= \frac{7}{3} + \frac{\sqrt{13}}{3} - \ln \frac{3 + \sqrt{13}}{2} \approx 2.340$$

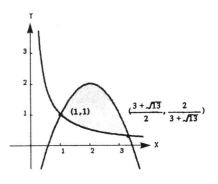

26. From the graph we see that f and g intersect twice at $x = 0$ and $x = 1$.

$$A = \int_0^1 [g(x) - f(x)] \, dx$$

$$= \int_0^1 [(2x + 1) - 3^x] \, dx$$

$$= \left[x^2 + x - \frac{1}{\ln 3}(3^x) \right]_0^1$$

$$= 2(1 - \frac{1}{\ln 3})$$

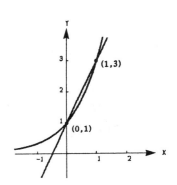

27. The points of intersection are given by:

$$\frac{1}{1 + x^2} = \frac{x^2}{2}, \quad x^4 + x^2 - 2 = 0, \quad x^2 = \frac{-1 \pm \sqrt{1 + 8}}{2}$$

$$x = \pm \sqrt{\frac{-1 + 3}{2}} = \pm 1$$

$$A = 2 \int_0^1 [f(x) - g(x)] \, dx$$

$$= 2 \int_0^1 \left[\frac{1}{1 + x^2} - \frac{x^2}{2} \right] dx$$

$$= 2 \left[\arctan x - \frac{x^3}{6} \right]_0^1$$

$$= 2(\frac{\pi}{4} - \frac{1}{6}) = \frac{\pi}{2} - \frac{1}{3}$$

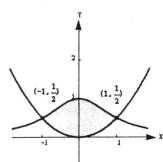

28. $A = 2 \displaystyle\int_0^{\pi/3} [f(x) - g(x)]\, dx$

$= 2 \displaystyle\int_0^{\pi/3} [2 - \sec x]\, dx$

$= 2 \left[2x - \ln |\sec x + \tan x| \right]_0^{\pi/3}$

$= 2(\dfrac{2\pi}{3} - \ln (2 + \sqrt{3})) \approx 6.823$

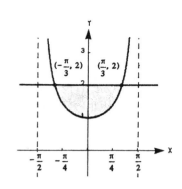

29. $A = 2 \displaystyle\int_0^{\pi/3} [f(x) - g(x)]\, dx$

$= 2 \displaystyle\int_0^{\pi/3} (2 \sin x - \tan x)\, dx$

$= 2 \left[-2 \cos x + \ln |\cos x| \right]_0^{\pi/3}$

$= 2(1 - \ln 2)$

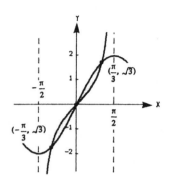

30. $A = 2 \displaystyle\int_{\pi/6}^{\pi/2} [f(x) - g(x)]\, dx$

$= 2 \displaystyle\int_{\pi/6}^{\pi/2} [\sin 2x - \cos x]\, dx$

$= 2 \left[-\dfrac{1}{2} \cos 2x - \sin x \right]_{\pi/6}^{\pi/2}$

$= \dfrac{1}{2}$

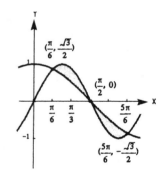

31. $A = \displaystyle\int_0^{\pi} [(2 \sin x + \sin 2x) - 0]\, dx$

$= \left[-2 \cos x - \dfrac{1}{2} \cos 2x \right]_0^{\pi}$

$= 4$

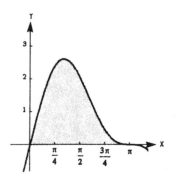

32. $A = \displaystyle\int_0^\pi [(2\sin x + \cos 2x) - 0]\, dx$

$= \left[-2\cos x + \dfrac{1}{2}\sin 2x \right]_0^\pi$

$= 4$

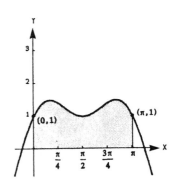

33. $A = \displaystyle\int_0^1 [xe^{-x^2} - 0]\, dx$

$= -\dfrac{1}{2} e^{-x^2}\, \Big]_0^1 = \dfrac{1}{2}\left(1 - \dfrac{1}{e}\right)$

≈ 0.316

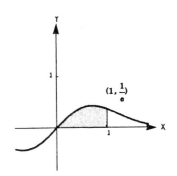

34. $A = \displaystyle\int_1^3 \left[\dfrac{1}{x^2} e^{1/x} - 0 \right] dx$

$= -e^{1/x}\, \Big]_1^3$

$= e - e^{1/3}$

≈ 1.3227

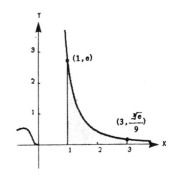

35. $A = \displaystyle\int_0^3 \left[\dfrac{6x}{x^2 + 1} - 0 \right] dx$

$= 3\ln(x^2 + 1)\, \Big]_0^3$

$= 3\ln 10 \approx 6.908$

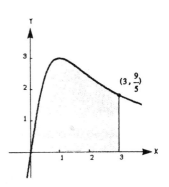

36. $A = \displaystyle\int_1^4 \left[\dfrac{4}{y} - 0 \right] dy$

$= 4 \ln |y| \Big]_1^4$

$= 4 \ln 4 = 8 \ln 2 \approx 5.545$

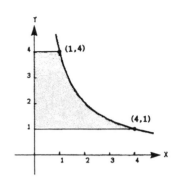

37. $A = \displaystyle\int_0^4 x \; dx = \dfrac{x^2}{2} \Big]_0^4 = 8$

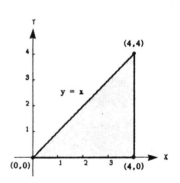

38. $A = \displaystyle\int_0^4 \left[\left(\dfrac{y}{2} + 4 \right) - \dfrac{3}{2} y \right] dy$

$= \displaystyle\int_0^4 (4 - y) \; dy$

$= \left[4y - \dfrac{y^2}{2} \right]_0^4 = 8$

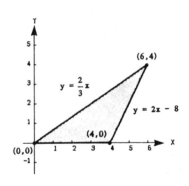

39. $A = \displaystyle\int_0^c \left[\left(\dfrac{b - a}{c} y + a \right) - \dfrac{b}{c} y \right] dy$

$= \displaystyle\int_0^c \left(-\dfrac{a}{c} y + a \right) dy$

$= \left[-\dfrac{a}{2c} y^2 + ac \right]_0^c$

$= -\dfrac{ac}{2} + ac = \dfrac{ac}{2}$

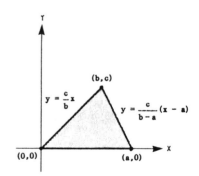

40. $A = \int_2^4 \left[(\frac{9}{2}x - 12) - (x - 5) \right] dx$

$\quad + \int_4^6 \left[(-\frac{5}{2}x + 16) - (x - 5) \right] dx$

$\quad = \int_2^4 (\frac{7}{2}x - 7) \ dx + \int_4^6 (-\frac{7}{2}x + 21) \ dx$

$\quad = \left[\frac{7}{4}x^2 - 7x \right]_2^4 + \left[-\frac{7}{4}x^2 + 21x \right]_4^6 = 14$

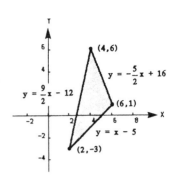

41. $A = \int_{-3}^3 (9 - x^2) \ dx = 36$

$\int_{-\sqrt{9-b}}^{\sqrt{9-b}} [(9 - x^2) - b] \ dx = 18$

$\int_0^{\sqrt{9-b}} [(9 - b) - x^2] \ dx = 9$

$\left[(9 - b)x - \frac{x^3}{3} \right]_0^{\sqrt{9-b}} = 9$

$\frac{2}{3}(9 - b)^{3/2} = 9, \quad (9 - b)^{3/2} = \frac{27}{2}$

$9 - b = \frac{9}{\sqrt[3]{4}}, \quad b = 9 - \frac{9}{\sqrt[3]{4}} \approx 3.330$

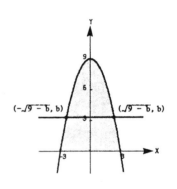

42. $A = 2 \int_0^9 (9 - x) \ dx = 2 \left[9x - \frac{x^2}{2} \right]_0^9 = 81$

$2 \int_0^{9-b} [(9 - x) - b] \ dx = \frac{81}{2}$

$2 \int_0^{9-b} [(9 - b) - x] \ dx = \frac{81}{2}$

$2 \left[(9 - b)x - \frac{x^2}{2} \right]_0^{9-b} = \frac{81}{2}$

$(9 - b)(9 - b) = \frac{81}{2}$

$9 - b = \frac{9}{\sqrt{2}}, \quad b = 9 - \frac{9}{\sqrt{2}} \approx 2.636$

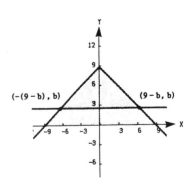

429

43. $x^4 - 2x^2 + 1 \leq 1 - x^2$ on $[-1, 1]$

$$A = \int_{-1}^{1} [(1 - x^2) - (x^4 - 2x^2 + 1)] \, dx$$

$$= \int_{-1}^{1} (x^2 - x^4) \, dx$$

$$= \left[\frac{x^3}{3} - \frac{x^5}{5} \right]_{-1}^{1} = \frac{4}{15}$$

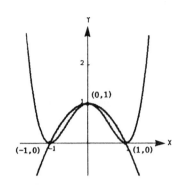

44. $x^3 \geq x$ on $[-1, 0]$ $x^3 \leq x$ on $[0, 1]$

Both functions are symmetric to the origin.

$$\int_{-1}^{0} (x^3 - x) \, dx = -\int_{0}^{1} (x^3 - x) \, dx$$

Thus $\int_{-1}^{1} (x^3 - x) \, dx = 0$

$$A = 2 \int_{0}^{1} (x - x^3) \, dx$$

$$= 2 \left[\frac{x^2}{2} - \frac{x^4}{4} \right]_{0}^{1} = \frac{1}{2}$$

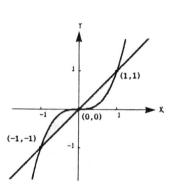

45. $\int_{0}^{5} [(7.21 + 0.58t) - (7.21 + 0.45t)] \, dt = \int_{0}^{5} 0.13t \, dt$

$$= \frac{0.13t^2}{2} \Big]_{0}^{5} = \$1.625 \text{ billion}$$

46. $\int_{0}^{5} [(7.21 + 0.26t + 0.02t^2) - (7.21 + 0.1t + 0.01t^2)] \, dt$

$$= \int_{0}^{5} (0.01t^2 + 0.16t) \, dt = \left[\frac{0.01t^3}{3} + \frac{0.16t^2}{2} \right]_{0}^{5}$$

$$= \frac{29}{12} \text{ billion} \approx \$2.4167 \text{ billion}$$

47. $50 - 0.5x = 0.125x$, $x = 80$, $p(x) = 10$

Point of equilibrium $(80, 10)$

$$CS = \int_0^{80} [(50 - 0.5x) - 10]\, dx = \left[-\frac{0.5x^2}{2} + 40x \right]_0^{80} = 1600$$

$$PS = \int_0^{80} [10 - 0.125x]\, dx = \left[10x - \frac{0.125x^2}{2} \right]_0^{80} = 400$$

48. $1000 - 0.4x^2 = 42x$, $x = 20$, $p(x) = 840$

Point of equilibrium $(20, 840)$

$$CS = \int_0^{20} [(1000 - 0.4x^2) - 840]\, dx = 160x - \frac{0.4x^3}{3} \bigg]_0^{20} \approx 2133.33$$

$$PS = \int_0^{20} [840 - 42x]\, dx = 840x - 21x^2 \bigg]_0^{20} = 8400$$

49. $\dfrac{10{,}000}{\sqrt{x + 100}} = 100\sqrt{0.05x + 10}$, $100 = \sqrt{(x + 100)(0.05x + 10)}$

$10{,}000 = 0.05x^2 + 15x + 1000$

$0 = x^2 + 300x - 18{,}000$, $0 = (x + 600)(x - 300)$

$x = 300$, $p(x) = 500$,   Point of equilibrium: $(300, 500)$

$$CS = \int_0^{300} \left(\frac{10{,}000}{\sqrt{x + 100}} - 500 \right) dx = \left[20{,}000\sqrt{x + 100} - 500x \right]_0^{300}$$

$$= 250{,}000 - 200{,}000 = 50{,}000$$

$$PS = \int_0^{300} (500 - 100\sqrt{0.05x + 10})\, dx = \left[500x - \frac{4000}{3}(0.05x + 10)^{3/2} \right]_0^{300}$$

$$= \frac{-50{,}000}{3} + \frac{40{,}000\sqrt{10}}{3} = \frac{10{,}000}{3}(4\sqrt{10} - 5) \approx 25{,}497$$

50. $\sqrt{25 - 0.1x} = \sqrt{9 + 0.1x} - 2,$ $x^2 - 160x = 0,\ 0 < x < 250$

$x(x - 160) = 0,\ x = 160,\ p(x) = 3,$ Point of equilibrium: (160, 3)

Consumer's Surplus $= \displaystyle\int_0^{160} (\sqrt{25 - 0.1x} - 3)\ dx$

$= \left[-\dfrac{20}{3}(25 - 0.1x)^{3/2} - 3x \right]_0^{160} = \dfrac{520}{3}$

Producer's Surplus $= \displaystyle\int_0^{160} [3 - (\sqrt{0.1x + 9} - 2)]\ dx$

$= \left[5x - \dfrac{20}{3}(0.1x + 9)^{3/2} \right]_0^{160} = \dfrac{440}{3}$

7.2
Volume: The disc method

1. $V = \pi \displaystyle\int_0^1 (-x + 1)^2\ dx = \pi \int_0^1 (x^2 - 2x + 1)\ dx = \pi\left[\dfrac{x^3}{3} - x^2 + x\right]_0^1 = \dfrac{\pi}{3}$

2. $V = \pi \displaystyle\int_0^2 (4 - x^2)^2\ dx = \pi \int_0^2 (x^4 - 8x^2 + 16)\ dx$

$= \pi\left[\dfrac{x^5}{5} - \dfrac{8x^3}{3} + 16x\right]_0^2 = \dfrac{256\pi}{15}$

3. $V = \pi \displaystyle\int_0^2 (\sqrt{4 - x^2})^2\ dx = \pi \int_0^2 (4 - x^2)\ dx = \pi\left[4x - \dfrac{x^3}{3}\right]_0^2 = \dfrac{16\pi}{3}$

4. $V = \pi \displaystyle\int_0^1 (x^2)^2\ dx = \pi \int_0^1 x^4\ dx = \pi\dfrac{x^5}{5}\bigg]_0^1 = \dfrac{\pi}{5}$

5. $V = \pi \displaystyle\int_1^4 (\sqrt{x})^2\ dx = \pi \int_1^4 x\ dx = \pi\dfrac{x^2}{2}\bigg]_1^4 = \dfrac{15\pi}{2}$

6. $V = \pi \displaystyle\int_{-2}^2 (\sqrt{4 - x^2})^2\ dx = 2\pi \int_0^2 (4 - x^2)\ dx = 2\pi\left[4x - \dfrac{x^3}{3}\right]_0^2 = \dfrac{32\pi}{3}$

7. $V = \pi \displaystyle\int_0^1 [(x^2)^2 - (x^3)^2]\ dx = \pi \int_0^1 (x^4 - x^6)\ dx = \pi\left[\dfrac{x^5}{5} - \dfrac{x^7}{7}\right]_0^1 = \dfrac{2\pi}{35}$

8. $V = \pi \int_{-2}^{2} \left[(4 - \frac{x^2}{2})^2 - (2)^2 \right] dx = 2\pi \int_{0}^{2} (\frac{x^4}{4} - 4x^2 + 12) \ dx$

$= 2\pi \left[\frac{x^5}{20} - \frac{4x^3}{3} + 12x \right]_{0}^{2} = \frac{448\pi}{15}$

9. $y = x^2 \implies x = \sqrt{y}, \quad V = \pi \int_{0}^{4} (\sqrt{y})^2 \ dy = \pi \int_{0}^{4} y \ dy = \pi \frac{y^2}{2} \Big]_{0}^{4} = 8\pi$

10. $y = \sqrt{16 - x^2} \implies x = \sqrt{16 - y^2}$

$V = \pi \int_{0}^{4} (\sqrt{16 - y^2})^2 \ dy = \pi \int_{0}^{4} (16 - y^2) \ dy = \pi \left[16y - \frac{y^3}{3} \right]_{0}^{4} = \frac{128\pi}{3}$

11. $y = x^{2/3} \implies x = y^{3/2}$

$V = \pi \int_{0}^{1} (y^{3/2})^2 \ dy = \pi \int_{0}^{1} y^3 \ dy = \pi \frac{y^4}{4} \Big]_{0}^{1} = \frac{\pi}{4}$

12. $V = \pi \int_{1}^{4} (-y^2 + 4y)^2 \ dy = \pi \int_{1}^{4} (y^4 - 8y^3 + 16y^2) \ dy$

$= \pi \left[\frac{y^5}{5} - 2y^4 + \frac{16y^3}{3} \right]_{1}^{4} = \frac{459\pi}{15}$

13. $y = \sqrt{x}, \ y = 0, \ x = 4$

   (a) $R(x) = \sqrt{x}, \qquad r(x) = 0$ (b) $R(y) = 4, \qquad r(y) = y^2$

$\qquad V = \pi \int_{0}^{4} (\sqrt{x})^2 \ dx = \pi \int_{0}^{4} x \ dx \qquad V = \pi \int_{0}^{2} (16 - y^4) \ dy$

$\qquad = \frac{\pi}{2} x^2 \Big]_{0}^{4} = 8\pi \qquad\qquad\qquad = \pi \left[16y - \frac{1}{5} y^5 \right]_{0}^{2} = \frac{128\pi}{5}$

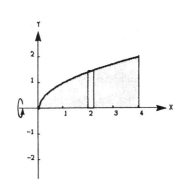
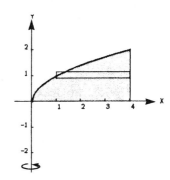

13.(c) $R(y) = 4 - y^2$, $r(y) = 0$

$$V = \pi \int_0^2 (4 - y^2)^2 \, dy$$

$$= \pi \int_0^2 (16 - 8y^2 + y^4) \, dy$$

$$= \pi \left[16y - \frac{8}{3}y^3 + \frac{1}{5}y^5 \right]_0^2$$

$$= \frac{256\pi}{15}$$

(d) $R(y) = 6 - y^2$, $r(y) = 2$

$$V = \pi \int_0^2 [(6 - y^2)^2 - 4] \, dy$$

$$= \pi \int_0^2 (32 - 12y^2 + y^4) \, dy$$

$$= \pi \left[32y - 4y^3 + \frac{1}{5}y^5 \right]_0^2$$

$$= \frac{192\pi}{5}$$

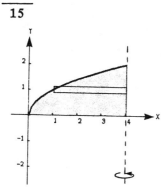

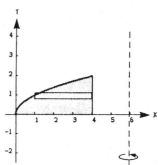

14. $y = 2x^2$, $y = 0$, $x = 2$

(a) $R(y) = 2$, $r(y) = \sqrt{y/2}$

$$V = \pi \int_0^8 \left(4 - \frac{y}{2}\right) dy$$

$$= \pi \left[4y - \frac{y^2}{4} \right]_0^8 = 16\pi$$

(b) $R(x) = 2x^2$, $r(x) = 0$

$$V = \pi \int_0^2 4x^4 \, dx = \pi \left[\frac{4x^5}{5} \right]_0^2$$

$$= \frac{128\pi}{5}$$

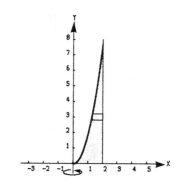

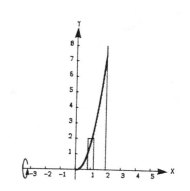

14.(c) $R(x) = 8, \qquad r(x) = 8 - 2x^2$

$$V = \pi \int_0^2 [64 - (64 - 32x^2 + 4x^4)] \, dx$$

$$= \pi \int_0^2 (32x^2 - 4x^4) \, dx$$

$$= 4\pi \int_0^2 (8x^2 - x^4) \, dx$$

$$= 4\pi \left[\frac{8}{3}x^3 - \frac{1}{5}x^5 \right]_0^2 = \frac{896\pi}{15}$$

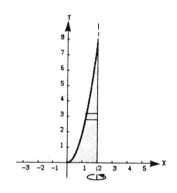

(d) $R(y) = 2 - \sqrt{y/2}, \qquad r(y) = 0$

$$V = \pi \int_0^8 \left(2 - \sqrt{\frac{y}{2}}\right)^2 \, dy$$

$$= \pi \int_0^8 \left(4 - 4\sqrt{\frac{y}{2}} + \frac{y}{2}\right) \, dy$$

$$= \pi \left[4y - \frac{4\sqrt{2}}{3}y^{3/2} + \frac{y^2}{4} \right]_0^8 = \frac{16\pi}{3}$$

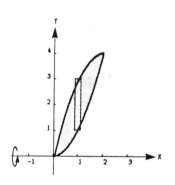

15. $y = x^2, \ y = 4x - x^2$

(a) $R(x) = 4x - x^2, \ r(x) = x^2$

$$V = \pi \int_0^2 [(4x - x^2)^2 - x^4] \, dx$$

$$= \pi \int_0^2 (16x^2 - 8x^3) \, dx$$

$$= \pi \left[\frac{16}{3}x^3 - 2x^4 \right]_0^2 = \frac{32\pi}{3}$$

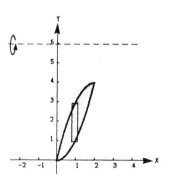

(b) $R(x) = 6 - x^2, \ r(x) = 6 - (4x - x^2)$

$$V = \pi \int_0^2 [(6 - x^2)^2 - (6 - 4x + x^2)^2] \, dx$$

$$= 8\pi \int_0^2 (x^3 - 5x^2 + 6x) \, dx$$

$$= 8\pi \left[\frac{x^4}{4} - \frac{5}{3}x^3 + 3x^2 \right]_0^2 = \frac{64\pi}{3}$$

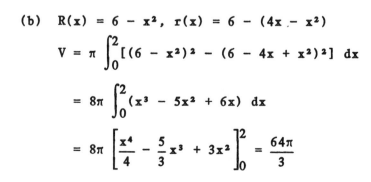

16.(a) $y = 6 - 2x - x^2$, $y = x + 6$

$R(x) = 6 - 2x - x^2$, $r(x) = x + 6$

$V = \pi \displaystyle\int_{-3}^{0} [(6 - 2x - x^2)^2 - (x + 6)^2] \, dx$

$= \pi \displaystyle\int_{-3}^{0} (x^4 + 4x^3 - 9x^2 - 36x) \, dx$

$= \pi \left[\dfrac{1}{5}x^5 + x^4 - 3x^3 - 18x^2 \right]_{-3}^{0} = \dfrac{243\pi}{5}$

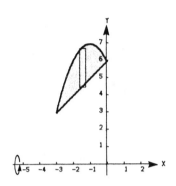

(b) $R(x) = (6 - 2x - x^2) - 3$

$r(x) = (x + 6) - 3$

$V = \pi \displaystyle\int_{-3}^{0} [(3 - 2x - x^2)^2 - (x + 3)^2] \, dx$

$= \pi \displaystyle\int_{-3}^{0} (x^4 + 4x^3 - 3x^2 - 18x) \, dx$

$= \pi \left[\dfrac{1}{5}x^5 + x^4 - x^3 - 9x^2 \right]_{-3}^{0} = \dfrac{108\pi}{5}$

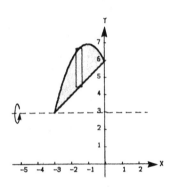

17. $R(x) = 4 - x$, $r(x) = 1$

$V = \pi \displaystyle\int_{0}^{3} [(4 - x)^2 - (1)^2] \, dx$

$= \pi \displaystyle\int_{0}^{3} (x^2 - 8x + 15) \, dx$

$= \pi \left[\dfrac{x^3}{3} - 4x^2 + 15x \right]_{0}^{3} = 18\pi$

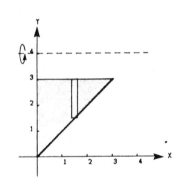

18. $R(x) = 4 - x^2$, $r(x) = 0$

$V = \pi \displaystyle\int_{0}^{2} (4 - x^2)^2 \, dx$

$= \pi \displaystyle\int_{0}^{2} (x^4 - 8x^2 + 16) \, dx$

$= \pi \left[\dfrac{x^5}{5} - \dfrac{8x^3}{3} + 16x \right]_{0}^{2} = \dfrac{256\pi}{15}$

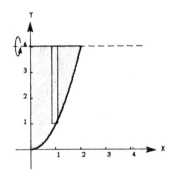

19. $R(x) = 4, \qquad r(x) = 4 - \dfrac{1}{x}$

$$V = \pi \int_1^4 [(4)^2 - (4 - \frac{1}{x})^2]\ dx$$

$$= \pi \int_1^4 (\frac{8}{x} - \frac{1}{x^2})\ dx$$

$$= \pi \left[8 \ln |x| + \frac{1}{x} \right]_1^4$$

$$= \pi\ (8 \ln 4 - \frac{3}{4})$$

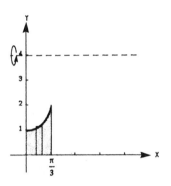

20. $R(x) = 4, \qquad r(x) = 4 - \sec x$

$$V = \pi \int_0^{\pi/3} [(4)^2 - (4 - \sec x)^2]\ dx$$

$$= \pi \int_0^{\pi/3} (8 \sec x - \sec^2 x)\ dx$$

$$= \pi \left[8 \ln |\sec x + \tan x| - \tan x \right]_0^{\pi/3}$$

$$= \pi\ [(8 \ln |2 + \sqrt{3}| - \sqrt{3}) - (8 \ln |1 + 0| - 0)]$$

$$= \pi\ (8 \ln (2 + \sqrt{3}) - \sqrt{3})$$

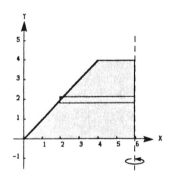

21. $R(y) = 6 - y, \qquad r(y) = 0$

$$V = \pi \int_0^4 (6 - y)^2\ dy$$

$$= \pi \int_0^4 (y^2 - 12y + 36)\ dy$$

$$= \pi \left[\frac{y^3}{3} - 6y^2 + 36y \right]_0^4 = \frac{208\pi}{3}$$

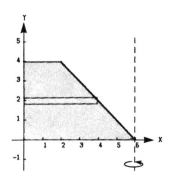

22. $R(y) = 6, \qquad r(y) = 6 - (6 - y) = y$

$$V = \pi \int_0^4 [(6)^2 - (y)^2]\ dy$$

$$= \pi \left[36y - \frac{y^3}{3} \right]_0^4 = \frac{368\pi}{3}$$

23. $R(y) = 6 - y^2$, $\quad r(y) = 2$

$$V = \pi \int_{-2}^{2} [(6 - y^2)^2 - (2)^2] \, dy$$

$$= 2\pi \int_{0}^{2} (y^4 - 12y^2 + 32) \, dy$$

$$= 2\pi \left[\frac{y^5}{5} - 4y^3 + 32y \right]_{0}^{2} = \frac{384\pi}{5}$$

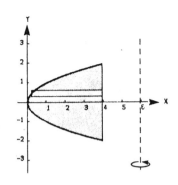

24. $R(y) = 6 - \dfrac{6}{y}$, $\quad r(y) = 0$

$$V = \pi \int_{2}^{6} \left(6 - \frac{6}{y}\right)^2 \, dy$$

$$= 36\pi \int_{2}^{6} \left(1 - \frac{2}{y} + \frac{1}{y^2}\right) \, dy$$

$$= 36\pi \left[y - 2 \ln |y| - \frac{1}{y} \right]_{2}^{6}$$

$$= 36\pi \left[\left(\frac{35}{6} - 2 \ln 6\right) - \left(\frac{3}{2} - 2 \ln 2\right) \right]$$

$$= 36\pi\left(\frac{13}{3} + 2 \ln \frac{1}{3}\right) = 12\pi(13 - 6 \ln 3)$$

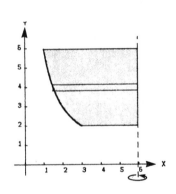

25. $R(x) = \dfrac{1}{\sqrt{x + 1}}$, $\quad r(x) = 0$

$$V = \pi \int_{0}^{3} \left(\frac{1}{\sqrt{x + 1}}\right)^2 \, dx$$

$$= \pi \int_{0}^{3} \frac{1}{x + 1} \, dx$$

$$= \pi \ln |x + 1| \Big]_{0}^{3} = \pi \ln 4$$

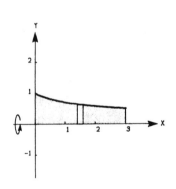

26. $R(x) = x \sqrt{4 - x^2}$, $\quad r(x) = 0$

$$V = 2\pi \int_{0}^{2} [x\sqrt{4 - x^2}]^2 \, dx$$

$$= 2\pi \int_{0}^{2} (4x^2 - x^4) \, dx$$

$$= 2\pi \left[\frac{4x^3}{3} - \frac{x^5}{5} \right]_{0}^{2} = \frac{128\pi}{15}$$

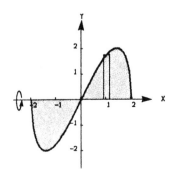

438

27. $R(x) = \dfrac{1}{x}, \qquad r(x) = 0$

$V = \pi \displaystyle\int_1^4 \left(\dfrac{1}{x}\right)^2 \, dx$

$\quad = \pi \left[-\dfrac{1}{x}\right]_1^4 = \dfrac{3\pi}{4}$

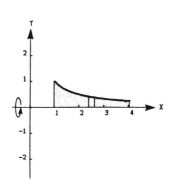

28. $R(x) = \dfrac{3}{x+1}, \qquad r(x) = 0$

$V = \pi \displaystyle\int_0^8 \left(\dfrac{3}{x+1}\right)^2 \, dx$

$\quad = 9\pi \displaystyle\int_0^8 (x+1)^{-2} \, dx$

$\quad = 9\pi \left[-\dfrac{1}{x+1}\right]_0^8 = 8\pi$

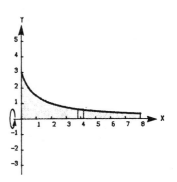

29. $R(x) = e^{-x}, \qquad r(x) = 0$

$V = \pi \displaystyle\int_0^1 (e^{-x})^2 \, dx$

$\quad = \pi \displaystyle\int_0^1 e^{-2x} \, dx$

$\quad = -\dfrac{\pi}{2} e^{-2x} \Big]_0^1 = \dfrac{\pi}{2}(1 - e^{-2})$

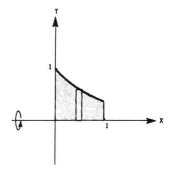

30. $R(x) = e^{x/2}, \qquad r(x) = 0$

$V = \pi \displaystyle\int_0^4 (e^{x/2})^2 \, dx$

$\quad = \pi \displaystyle\int_0^4 e^x \, dx$

$\quad = \pi e^x \Big]_0^4 = \pi(e^4 - 1)$

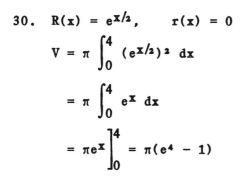

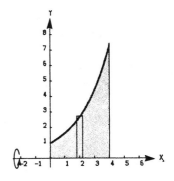

31. $R(x) = \sqrt{\sin x}$, $\qquad r(x) = 0$

$$V = \pi \int_0^{\pi/2} (\sqrt{\sin x})^2 \, dx$$

$$= \pi \int_0^{\pi/2} \sin x \, dx$$

$$= -\pi \cos x \Big]_0^{\pi/2} = \pi$$

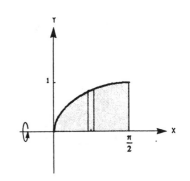

32. $R(x) = \sqrt{\cos x}$, $\qquad r(x) = 0$

$$V = \pi \int_0^{\pi/2} (\sqrt{\cos x})^2 \, dx$$

$$= \pi \int_0^{\pi/2} \cos x \, dx$$

$$= \pi \sin x \Big]_0^{\pi/2} = \pi$$

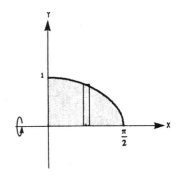

33. $R(x) = 4x - x^2$

$$V = \pi \int_0^4 (4x - x^2)^2 \, dx$$

$$= \pi \int_0^4 (16x^2 - 8x^3 + x^4) \, dx$$

$$= \pi \left[\frac{16}{3} x^3 - 2x^4 + \frac{x^5}{5} \right]_0^4 = \frac{512\pi}{15}$$

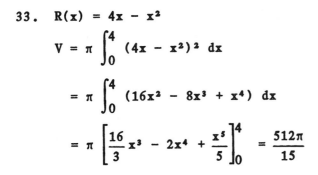
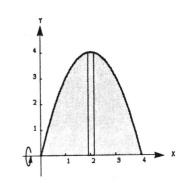

34. Completing the square, we have:

$$4x - x^2 = 4 - (x^2 - 4x + 4) = 4 - (x - 2)^2$$

Thus, $y = 4 - x^2$ has the same volume as in Exercise 11 since the solid has been translated horizontally only.

35. $R(x) = \dfrac{3}{5} \sqrt{25 - x^2}, \quad r(x) = 0$

$V = \dfrac{9\pi}{25} \displaystyle\int_{-5}^{5} (25 - x^2)\, dx$

$= \dfrac{18\pi}{25} \displaystyle\int_{0}^{5} (25 - x^2)\, dx$

$= \dfrac{18\pi}{25} \left[25x - \dfrac{x^3}{3} \right]_{0}^{5} = 60\pi$

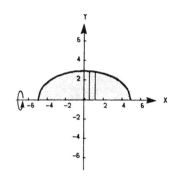

36. $R(y) = \dfrac{5}{3} \sqrt{9 - y^2}, \quad r(y) = 0$

$V = \dfrac{25\pi}{9} \displaystyle\int_{-3}^{3} (9 - y^2)\, dy$

$= \dfrac{50\pi}{9} \displaystyle\int_{0}^{3} (9 - y^2)\, dy$

$= \dfrac{50\pi}{9} \left[9y - \dfrac{y^3}{3} \right]_{0}^{3} = 100\pi$

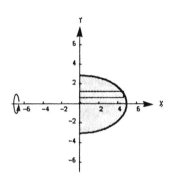

37. $R(x) = \dfrac{1}{2} x, \quad r(x) = 0$

$V = \pi \displaystyle\int_{0}^{6} \dfrac{1}{4} x^2\, dx$

$= \dfrac{\pi}{12} x^3 \Big]_{0}^{6} = 18\pi$

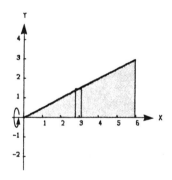

38. $R(x) = \dfrac{r}{h} x, \quad r(x) = 0$

$V = \pi \displaystyle\int_{0}^{h} \dfrac{r^2}{h^2} x^2\, dx$

$= \dfrac{r^2 \pi}{3h^2} x^3 \Big]_{0}^{h} = \dfrac{r^2 \pi}{3h^2} h^3 = \dfrac{1}{3} \pi r^2 h$

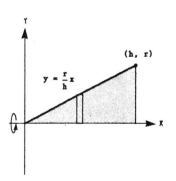

39. $R(x) = \sqrt{r^2 - x^2}, \qquad r(x) = 0$

$$V = \pi \int_{-r}^{r} (r^2 - x^2) \, dx$$

$$= 2\pi \int_{0}^{r} (r^2 - x^2) \, dx$$

$$= 2\pi \left[r^2 x - \frac{1}{3} x^3 \right]_{0}^{r}$$

$$= 2\pi (r^3 - \frac{1}{3} r^3) = \frac{4}{3} \pi r^3$$

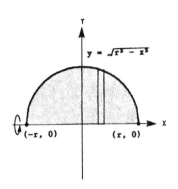

40. $x = \sqrt{r^2 - y^2}, \qquad R(y) = \sqrt{r^2 - y^2}, \qquad r(y) = 0$

$$V = \pi \int_{h}^{r} (\sqrt{r^2 - y^2})^2 \, dy = \pi \int_{h}^{r} (r^2 - y^2) \, dy$$

$$= \pi \left[r^2 y - \frac{y^3}{3} \right]_{h}^{r}$$

$$= \pi \left[(r^3 - \frac{r^3}{3}) - (r^2 h - \frac{h^3}{3}) \right]$$

$$= \pi (\frac{2r^3}{3} - r^2 h + \frac{h^3}{3})$$

$$= \frac{\pi}{3} (2r^3 - 3r^2 h + h^3)$$

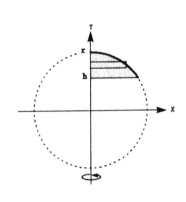

41. $x = r - \frac{r}{H} y = r(1 - \frac{y}{H}), \qquad R(y) = r(1 - \frac{y}{H}), \qquad r(y) = 0$

$$V = \pi \int_{0}^{h} [r(1 - \frac{y}{H})]^2 \, dy$$

$$= \pi r^2 \int_{0}^{h} (1 - \frac{2}{H} y + \frac{1}{H^2} y^2) \, dy$$

$$= \pi r^2 \left[y - \frac{1}{H} y^2 + \frac{1}{3H^2} y^3 \right]_{0}^{h}$$

$$= \pi r^2 (h - \frac{h^2}{H} + \frac{h^3}{3H^2})$$

$$= \frac{\pi r^2 h}{3H^2} (3H^2 - 3Hh + h^2)$$

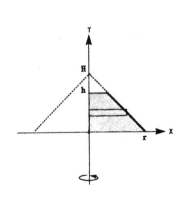

42. $V = \pi \int_0^4 (\sqrt{x})^2 \, dx$

    $= \pi \int_0^4 x \, dx = \pi \frac{x^2}{2} \Big]_0^4 = 8\pi$

    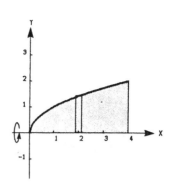

    Let $c \varepsilon (0, 4)$, $0 < c < 4$ and set

    $\pi \int_0^c x \, dx = 4\pi = \pi \frac{x^2}{2} \Big]_0^c = \frac{\pi c^2}{2} = 4\pi$

    $c^2 = 8$, $\quad c = \sqrt{8} = 2\sqrt{2}$

    Thus when $x = 2\sqrt{2}$ the solid is divided into two parts of equal volume.

43. Set $\pi \int_0^c x \, dx = \frac{8\pi}{3}$ (one third of the volume)

    then $\frac{\pi c^2}{2} = \frac{8\pi}{3}$, $\quad c^2 = \frac{16}{3}$, $\quad c = \frac{4}{\sqrt{3}} = \frac{4\sqrt{3}}{3}$

    To find the other value set

    $\pi \int_0^d x \, dx = \frac{16\pi}{3}$ (two thirds of the volume)

    then $\frac{\pi d^2}{2} = \frac{16\pi}{3}$, $\quad d^2 = \frac{32}{3}$, $\quad d = \frac{\sqrt{32}}{\sqrt{3}} = \frac{4\sqrt{6}}{3}$

    The x values that divide the solid into three parts of equal volume

    are $x = \frac{4\sqrt{3}}{3}$ and $x = \frac{4\sqrt{6}}{3}$.

44. Total volume: $V = \frac{4\pi(50)^3}{3} = \frac{500,000\pi}{3} \text{ ft}^3$

    Volume of water: $0.216V = 36,000\pi = \pi \int_{-50}^{y_0} (\sqrt{2500 - y^2})^2 \, dy$

    $= \pi \int_{-50}^{y_0} (2500 - y^2) \, dy = \pi \left[2500y - \frac{y^3}{3} \right]_{-50}^{y_0} = \pi(2500y_0 - \frac{y_0^3}{3} + \frac{250,000}{3})$

    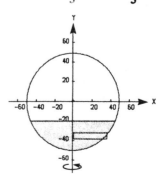

    $108,000 = 7500y_0 - y_0^3 + 250,000$

    $y_0^3 - 7500y_0 - 142,000 = 0$

    $(y_0 + 20)(y_0^2 - 20y_0 - 7100) = 0$

    $y_0 = -20$, $\quad 10 \pm 60\sqrt{2}$

    Since $-50 < y_0 < 50$, $\quad y_0 = -20$

    depth $= [-20 - (-50)] = 30 \text{ ft}$

45. $V = \pi \int_0^2 (\frac{1}{8} x^2 \sqrt{2 - x})^2 \, dx = \frac{\pi}{64} \int_0^2 x^4 (2 - x) \, dx$

$= \frac{\pi}{64} \left[\frac{2x^5}{5} - \frac{x^6}{6} \right]_0^2 = \frac{\pi}{30}$

46.(a) $\pi \int_0^h r^2 \, dx$ is the volume of **(b)** $\pi \int_{-b}^b (a \sqrt{1 - \frac{x^2}{b^2}})^2 \, dx$ is the

a right circular cylinder with radius r and height h.

volume of an ellipsoid with axes 2a and 2b.

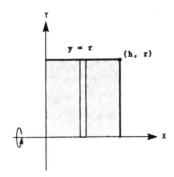
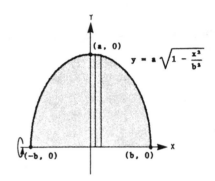

(c) $\pi \int_{-r}^r (\sqrt{r^2 - x^2})^2 \, dx$ is **(d)** $\pi \int_0^h (\frac{rx}{h})^2 \, dx$ is the volume of a

the volume of a sphere with radius r.

right circular cone with the radius of the base as r and height h.

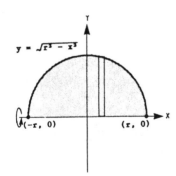
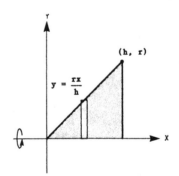

(e) $\pi \int_{-r}^r [(R + \sqrt{r^2 - x^2})^2 - (R - \sqrt{r^2 - x^2})^2] \, dx$

is the volume of a torus with the radius of its circular cross section as r and the distance from the axis of the torus to the center of its cross section as R.

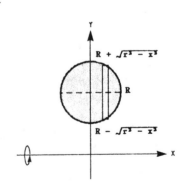

47.

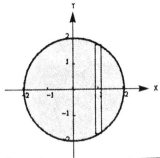

Base of cross section = $2\sqrt{4 - x^2}$

(a) $A(x) = b^2 = (2\sqrt{4 - x^2})^2$

$$V = \int_{-2}^{2} 4(4 - x^2) \; dx$$

$$= 4 \left[4x - \frac{x^3}{3} \right]_{-2}^{2} = \frac{128}{3}$$

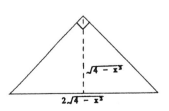

(b) $A(x) = \frac{1}{2} bh = \frac{1}{2}(2\sqrt{4 - x^2})(\sqrt{3}\sqrt{4 - x^2})$

$$= \sqrt{3}(4 - x^2)$$

$$V = \sqrt{3} \int_{-2}^{2} (4 - x^2) \; dx$$

$$= \sqrt{3} \left[4x - \frac{x^3}{3} \right]_{-2}^{2} = \frac{32\sqrt{3}}{3}$$

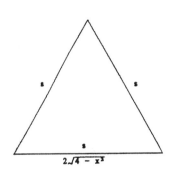

(c) $A(x) = \frac{1}{2}\pi r^2 = \frac{\pi}{2}(\sqrt{4 - x^2})^2$

$$= \frac{\pi}{2}(4 - x^2)$$

$$V = \frac{\pi}{2} \int_{-2}^{2} (4 - x^2) \; dx = \frac{\pi}{2} \left[4x - \frac{x^3}{3} \right]_{-2}^{2}$$

$$= \frac{16\pi}{3}$$

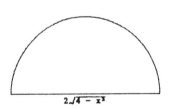

(d) $A(x) = \frac{1}{2} bh = \frac{1}{2}(2\sqrt{4 - x^2})(\sqrt{4 - x^2})$

$$= 4 - x^2$$

$$V = \int_{-2}^{2} (4 - x^2) \; dx = \left[4x - \frac{x^3}{3} \right]_{-2}^{2} = \frac{32}{3}$$

445

48.

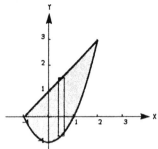

Base of cross section
$(x + 1) - (x^2 - 1) = 2 + x - x^2$

(a) $A(x) = b^2 = (2 + x - x^2)^2$

$$= 4 + 4x - 3x^2 - 2x^3 + x^4$$

$$V = \int_{-1}^{2} (4 + 4x - 3x^2 - 2x^3 + x^4)\, dx$$

$$= \left[4x + 2x^2 - x^3 - \frac{1}{2}x^4 + \frac{1}{5}x^5 \right]_{-1}^{2} = \frac{81}{10}$$

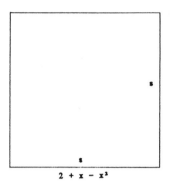

$2 + x - x^2$

(b) $A(x) = bh = 2 + x - x^2$

$$V = \int_{-1}^{2} (2 + x - x^2)\, dx$$

$$= \left[2x + \frac{x^2}{2} - \frac{x^3}{3} \right]_{-1}^{2} = \frac{9}{2}$$

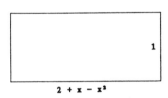

$2 + x - x^2$

(c) $A(x) = \frac{1}{2}\pi ab = (\frac{1}{2})\pi(2)(\frac{2 + x - x^2}{2})$

$$= \frac{\pi}{2}(2 + x - x^2)$$

$$V = \frac{\pi}{2} \int_{-1}^{2} (2 + x - x^2)\, dx$$

$$= \frac{\pi}{2} \left[2x + \frac{x^2}{2} - \frac{x^3}{3} \right]_{-1}^{2}$$

$$= \frac{\pi}{2}(\frac{9}{2}) = \frac{9\pi}{4}$$

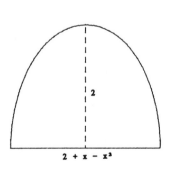

$2 + x - x^2$

48.(d) $A(x) = \dfrac{1}{2}bh = \dfrac{1}{2}(2 + x - x^2)\dfrac{\sqrt{3}(2 + x - x^2)}{2}$

$$= \dfrac{\sqrt{3}}{4}(2 + x - x^2)^2$$

$$V = \dfrac{\sqrt{3}}{4}\int_{-1}^{2}(2 + x - x^2)^2\ dx$$

$$= \dfrac{\sqrt{3}}{4}\left[4x + 2x^2 - x^3 - \dfrac{1}{2}x^4 + \dfrac{1}{5}x^5\right]_{-1}^{2}$$

$$= \dfrac{\sqrt{3}}{4}\left(\dfrac{81}{10}\right) = \dfrac{81\sqrt{3}}{40}$$

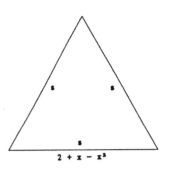

$2 + x - x^2$

49.

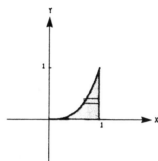

Base of cross section $= 1 - \sqrt[3]{y}$

(a) $A(y) = b^2 = (1 - \sqrt[3]{y})^2$

$$V = \int_{0}^{1}(1 - \sqrt[3]{y})^2\ dy$$

$$= \int_{0}^{1}(1 - 2y^{1/3} + y^{2/3})\ dy$$

$$= \left[y - \dfrac{3}{2}y^{4/3} + \dfrac{3}{5}y^{5/3}\right]_{0}^{1} = \dfrac{1}{10}$$

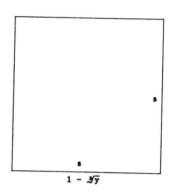

$1 - \sqrt[3]{y}$

(b) $A(y) = \dfrac{1}{2}\pi r^2 = \dfrac{1}{2}\pi\left(\dfrac{1 - \sqrt[3]{y}}{2}\right)^2$

$$= \dfrac{1}{8}\pi(1 - \sqrt[3]{y})^2$$

$$V = \dfrac{1}{8}\pi\int_{0}^{1}(1 - \sqrt[3]{y})^2\ dy$$

$$= \dfrac{\pi}{8}\left(\dfrac{1}{10}\right) = \dfrac{\pi}{80}$$

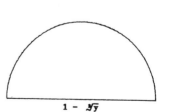

$1 - \sqrt[3]{y}$

49. (continued)

(c) $A(y) = \frac{1}{2}bh = \frac{1}{2}(1 - \sqrt[3]{y})(\frac{\sqrt{3}}{2})(1 - \sqrt[3]{y})$

$\qquad = \frac{\sqrt{3}}{4}(1 - \sqrt[3]{y})^2$

$\qquad V = \frac{\sqrt{3}}{4}\int_0^1 (1 - \sqrt[3]{y})^2\, dy$

$\qquad = \frac{\sqrt{3}}{4}(\frac{1}{10}) = \frac{\sqrt{3}}{40}$

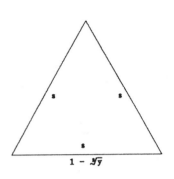

(d) $A(y) = \frac{h}{2}(b_1 + b_2)$

$\qquad = \frac{1}{2}\left[\frac{1 - \sqrt[3]{y}}{2} + (1 - \sqrt[3]{y})\right](\frac{1 - \sqrt[3]{y}}{2})$

$\qquad = \frac{3}{8}(1 - \sqrt[3]{y})^2$

$\qquad V = \frac{3}{8}\int_0^1 (1 - \sqrt[3]{y})^2\, dy = \frac{3}{8}(\frac{1}{10}) = \frac{3}{80}$

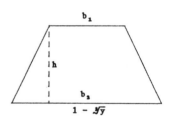

(e) $A(y) = \frac{1}{2}\pi ab = \frac{\pi}{2}(2)(1 - \sqrt[3]{y})\frac{1 - \sqrt[3]{y}}{2}$

$\qquad = \frac{\pi}{2}(1 - \sqrt[3]{y})^2$

$\qquad V = \frac{\pi}{2}\int_0^1 (1 - \sqrt[3]{y})^2\, dy$

$\qquad = \frac{\pi}{2}(\frac{1}{10}) = \frac{\pi}{20}$

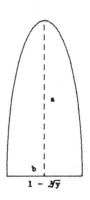

50. Since the cross sections are square

$A(y) = b^2 = (\sqrt{r^2 - y^2})^2$

$V = 8\int_0^r (r^2 - y^2)\, dy$

$\quad = 8\left[r^2 y - \frac{1}{3}y^3\right]_0^r$

$\quad = \frac{16}{3}r^3$

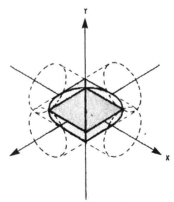

51. Since the cross sections are isosceles right triangles

$$A(x) = \frac{1}{2}bh = \frac{1}{2}(\sqrt{r^2 - y^2})(\sqrt{r^2 - y^2})$$

$$= \frac{1}{2}(r^2 - y^2)$$

$$V = \frac{1}{2}\int_{-r}^{r}(r^2 - y^2)\,dy = \int_{-r}^{r}(r^2 - y^2)\,dy$$

$$= \left[r^2 y - \frac{y^3}{3}\right]_0^r = \frac{2}{3}r^3$$

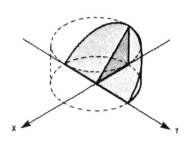

7.3

Volume: The shell method

1. $p(x) = x, \qquad h(x) = x$

$$V = 2\pi\int_0^2 x(x)\,dx = \frac{2\pi x^3}{3}\bigg]_0^2 = \frac{16\pi}{3}$$

2. $p(x) = x, \qquad h(x) = 1 - x$

$$V = 2\pi\int_0^1 x(1 - x)\,dx = 2\pi\int_0^1 (x - x^2)\,dx = 2\pi\left[\frac{x^2}{2} - \frac{x^3}{3}\right]_0^1 = \frac{\pi}{3}$$

3. $p(y) = y, \qquad h(y) = 2 - y$

$$V = 2\pi\int_0^2 y(2 - y)\,dy = 2\pi\int_0^2 (2y - y^2)\,dy = 2\pi\left[y^2 - \frac{y^3}{3}\right]_0^2 = \frac{8\pi}{3}$$

4. $p(y) = -y, \qquad h(y) = 4 - (2 - y) = 2 + y$

$$V = 2\pi\int_{-2}^0 (-y)(2 + y)\,dy = 2\pi\int_{-2}^0 (-2y - y^2)\,dy = 2\pi\left[-y^2 - \frac{y^3}{3}\right]_{-2}^0 = \frac{8\pi}{3}$$

5. $p(x) = x, \qquad h(x) = \sqrt{x}$

$$V = 2\pi\int_0^4 x\sqrt{x}\,dx = 2\pi\int_0^4 x^{3/2}\,dx = \frac{4\pi}{5}x^{5/2}\bigg]_0^4 = \frac{128\pi}{5}$$

6. $p(x) = x, \qquad h(x) = 8 - (x^2 + 4) = 4 - x^2$

$$V = 2\pi\int_0^2 x(4 - x^2)\,dx = 2\pi\int_0^2 (4x - x^3)\,dx = 2\pi\left[2x^2 - \frac{x^4}{4}\right]_0^2 = 8\pi$$

7. $p(x) = x$

$h(x) = x^2$

$V = 2\pi \displaystyle\int_0^2 x^3 \, dx$

$\quad = \dfrac{\pi}{2} x^4 \, \Big]_0^2 = 8\pi$

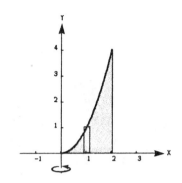

8. $p(x) = x$

$h(x) = x^2$

$V = 2\pi \displaystyle\int_0^4 x^3 \, dx$

$\quad = \dfrac{\pi}{2} x^4 \, \Big]_0^4 = 128\pi$

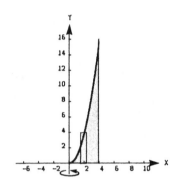

9. $p(x) = x$

$h(x) = (4x - x^2) - x^2 = 4x - 2x^2$

$V = 2\pi \displaystyle\int_0^2 x(4x - 2x^2) \, dx$

$\quad = 4\pi \displaystyle\int_0^2 (2x^2 - x^3) \, dx$

$\quad = 4\pi \left[\dfrac{2}{3} x^3 - \dfrac{1}{4} x^4 \right]_0^2 = \dfrac{16\pi}{3}$

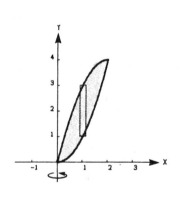

10. $p(x) = 2 - x$

$h(x) = 4x - 2x^2$

$V = 2\pi \displaystyle\int_0^2 (2 - x)(4x - 2x^2) \, dx$

$\quad = 2\pi \displaystyle\int_0^2 (8x - 8x^2 + 2x^3) \, dx$

$\quad = 2\pi \left[4x^2 - \dfrac{8}{3} x^3 + \dfrac{1}{2} x^4 \right]_0^2 = \dfrac{16\pi}{3}$

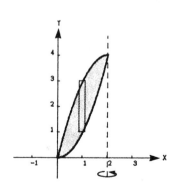

450

11. $p(x) = 4 - x, \qquad h(x) = 4x - 2x^2$

$V = 2\pi \int_0^2 (4 - x)(4x - 2x^2) \, dx$

$= 2\pi (2) \int_0^2 (x^3 - 6x^2 + 8x) \, dx$

$= 4\pi \left[\dfrac{x^4}{4} - 2x^3 + 4x^2 \right]_0^2 = 16\pi$

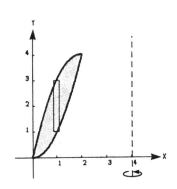

12. $p(y) = y$ and $h(y) = 1$ if $0 \le y < \dfrac{1}{2}$

$p(y) = y$ and $h(y) = \dfrac{1}{y} - 1$ if $\dfrac{1}{2} \le y \le 1$

$V = 2\pi \int_0^{1/2} y \, dy + 2\pi \int_{1/2}^1 (1 - y) \, dy$

$= 2\pi \left[\dfrac{y^2}{2} \right]_0^{1/2} + 2\pi \left[y - \dfrac{y^2}{2} \right]_{1/2}^1 = \dfrac{\pi}{2}$

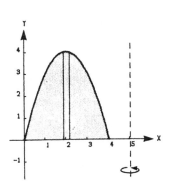

13. $p(x) = 5 - x, \qquad h(x) = 4x - x^2$

$V = 2\pi \int_0^4 (5 - x)(4x - x^2) \, dx$

$= 2\pi \int_0^4 (x^3 - 9x^2 + 20x) \, dx$

$= 2\pi \left[\dfrac{x^4}{4} - 3x^3 + 10x^2 \right]_0^4 = 64\pi$

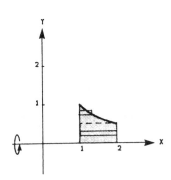

14. $p(y) = y, \qquad h(y) = 9 - y^2$

$V = 2\pi \int_0^3 y(9 - y^2) \, dy$

$= 2\pi \int_0^3 (9y - y^3) \, dy$

$= 2\pi \left[\dfrac{9}{2} y^2 - \dfrac{1}{4} y^4 \right]_0^3 = \dfrac{81\pi}{2}$

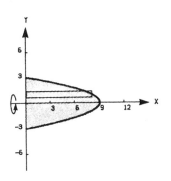

15. $p(x) = x$

$h(x) = 4 - (4x - x^2) = x^2 - 4x + 4$

$V = 2\pi \int_0^2 (x^3 - 4x^2 + 4x)\ dx$

$= 2\pi \left[\dfrac{x^4}{4} - \dfrac{4}{3}x^3 + 2x^2 \right]_0^2 = \dfrac{8\pi}{3}$

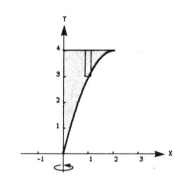

16. $p(x) = x$

$h(x) = 4 - x^2$

$V = 2\pi \int_0^2 (4x - x^3)\ dx$

$= 2\pi \left[2x^2 - \dfrac{1}{4}x^4 \right]_0^2 = 8\pi$

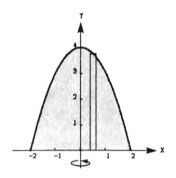

17. $p(x) = 6 - x, \qquad h(x) = \sqrt{x}$

$V = 2\pi \int_0^4 (6 - x)\sqrt{x}\ dx$

$= 2\pi \int_0^4 (6x^{1/2} - x^{3/2})\ dx$

$= 2\pi \left[4x^{3/2} - \dfrac{2}{5}x^{5/2} \right]_0^4 = \dfrac{192\pi}{5}$

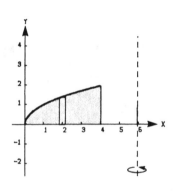

18. $p(x) = x, \qquad h(x) = 4 - 2x$

$V = 2\pi \int_0^2 x(4 - 2x)\ dx$

$= 2\pi \int_0^2 (4x - 2x^2)\ dx$

$= 2\pi \left[2x^2 - \dfrac{2}{3}x^3 \right]_0^2 = \dfrac{16\pi}{3}$

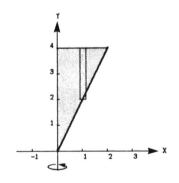

19. $p(x) = x$, $\qquad h(x) = \dfrac{1}{\sqrt{2\pi}}\, e^{-x^2/2}$

$$V = 2\pi \int_0^1 x\left(\frac{1}{\sqrt{2\pi}}\, e^{-x^2/2}\right)\, dx$$

$$= \sqrt{2\pi} \int_0^1 e^{-x^2/2}\, x\, dx$$

$$= -\sqrt{2\pi}\, e^{-x^2/2}\, \Big]_0^1$$

$$= \sqrt{2\pi}\left(1 - \frac{1}{\sqrt{e}}\right) \approx 0.986$$

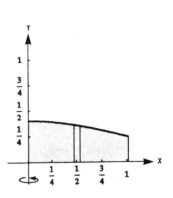

20. $p(x) = x$, $\qquad h(x) = \dfrac{\sin x}{x}$

$$V = 2\pi \int_0^\pi x \left(\frac{\sin x}{x}\right)\, dx$$

$$= 2\pi \int_0^\pi \sin x\, dx$$

$$= -2\pi \cos x\, \Big]_0^\pi = 4\pi$$

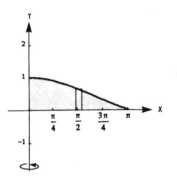

21.(a) <u>Disc</u>

$R(x) = x^3$, $\qquad r(x) = 0$

$$V = \pi \int_0^2 x^6\, dx = \pi \frac{x^7}{7}\, \Big]_0^2$$

$$= \frac{128\pi}{7}$$

(b) <u>Shell</u>

$p(x) = x$, $\qquad h(x) = x^3$

$$V = 2\pi \int_0^2 x^4\, dx = 2\pi \frac{x^5}{5}\, \Big]_0^2$$

$$= \frac{64\pi}{5}$$

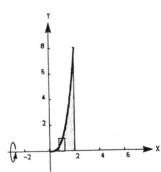

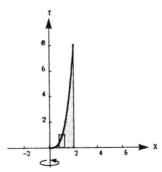

21.(c) <u>Shell</u>

$$p(x) = 4 - x, \qquad h(x) = x^3$$

$$V = 2\pi \int_0^2 (4 - x)x^3 \, dx$$

$$= 2\pi \int_0^2 (4x^3 - x^4) \, dx$$

$$= 2\pi \left[x^4 - \frac{1}{5} x^5 \right]_0^2 = \frac{96\pi}{5}$$

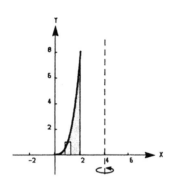

(d) <u>Disc</u>

$$R(x) = 8, \qquad r(x) = 8 - x^3$$

$$V = \pi \int_0^2 [64 - (8 - x^3)^2] \, dx$$

$$= \pi \int_0^2 (16x^3 - x^6) \, dx$$

$$= \pi \left[4x^4 - \frac{x^7}{7} \right]_0^2 = \frac{320\pi}{7}$$

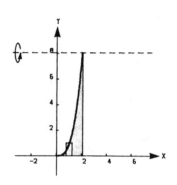

22.(a) <u>Disc</u>

$$R(x) = \frac{1}{x^2}, \qquad r(x) = 0$$

$$V = \pi \int_1^4 \left(\frac{1}{x^2}\right)^2 \, dx$$

$$= \pi \left(-\frac{1}{3x^3}\right) \Bigg]_1^4$$

$$= \frac{21\pi}{64}$$

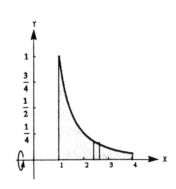

(b) <u>Shell</u>

$$p(x) = x, \qquad h(x) = \frac{1}{x^2}$$

$$V = 2\pi \int_1^4 x\left(\frac{1}{x^2}\right) \, dx$$

$$= 2\pi \int_1^4 \frac{1}{x} \, dx$$

$$= 2\pi \ln |x| \Bigg]_1^4 = 2\pi \ln 4$$

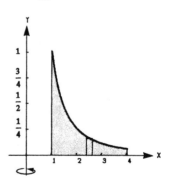

22.(c) <u>Shell</u>

$$p(x) = 4 - x, \qquad h(x) = \frac{1}{x^2}$$

$$V = 2\pi \int_1^4 (4 - x)(\frac{1}{x^2}) \, dx$$

$$= 2\pi \int_1^4 (\frac{4}{x^2} - \frac{1}{x}) \, dx$$

$$= 2\pi \left[-\frac{4}{x} - \ln |x| \right]_1^4 = 2\pi(3 - \ln 4)$$

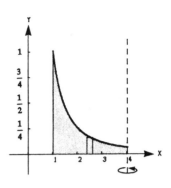

23.(a) <u>Shell</u>

$$p(y) = y, \qquad h(y) = (a^{1/2} - y^{1/2})^2$$

$$V = 2\pi \int_0^a y(a - 2a^{1/2}y^{1/2} + y) \, dy$$

$$= 2\pi \int_0^a (ay - 2a^{1/2}y^{3/2} + y^2) \, dy$$

$$= 2\pi \left[\frac{a}{2}y^2 - \frac{4a^{1/2}}{5} y^{5/2} + \frac{y^3}{3} \right]_0^a$$

$$= 2\pi \left[\frac{a^3}{2} - \frac{4a^3}{5} + \frac{a^3}{3} \right] = \frac{\pi a^3}{15}$$

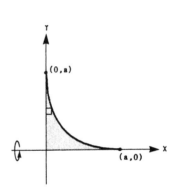

(b) Same as part (a) by symmetry

(c) $p(x) = a - x$

$$h(x) = (a^{1/2} - x^{1/2})^2$$

$$V = 2\pi \int_0^a (a - x)(a^{1/2} - x^{1/2})^2 \, dx$$

$$= 2\pi \int_0^a (a^2 - 2a^{3/2}x^{1/2} + 2a^{1/2} x^{3/2} - x^2) \, dx$$

$$= 2\pi \left[a^2x - \frac{4}{3} a^{3/2}x^{3/2} + \frac{4}{5} a^{1/2}x^{5/2} - \frac{1}{3} x^3 \right]_0^a$$

$$= \frac{4\pi a^3}{15}$$

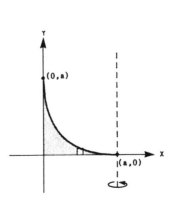

24. $R(x) = (a^{2/3} - x^{2/3})^{3/2}, \qquad r(x) = 0$

$$V = \pi \int_{-a}^{a} (a^{2/3} - x^{2/3})^3 \, dx$$

$$= 2\pi \int_0^a (a^2 - 3a^{4/3}x^{2/3} + 3a^{2/3}x^{4/3} - x^2) \, dx$$

$$= 2\pi \left[a^2 x - \frac{9}{5}a^{4/3}x^{5/3} + \frac{9}{7}a^{2/3}x^{7/3} - \frac{1}{3}x^3 \right]_0^a$$

$$= 2\pi(a^3 - \frac{9}{5}a^3 + \frac{9}{7}a^3 - \frac{1}{3}a^3) = \frac{32\pi a^3}{105}$$

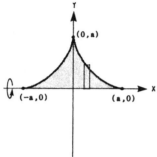

25. $p(x) = x, \qquad h(x) = 2 - \frac{1}{2}x^2$

$$V = 2\pi \int_0^2 x(2 - \frac{1}{2}x^2) \, dx = 2\pi \int_0^2 (2x - \frac{1}{2}x^3) \, dx$$

$$= 2\pi \left[x^2 - \frac{1}{8}x^4 \right]_0^2 = 4\pi \ \text{(total volume)}$$

Now find x_0 such that

$$\pi = 2\pi \int_0^{x_0} (2x - \frac{1}{2}x^3) \, dx$$

$$1 = 2 \left[x^2 - \frac{1}{8}x^4 \right]_0^{x_0}, \qquad 1 = 2x_0^2 - \frac{1}{4}x_0^4$$

$$x_0^4 - 8x_0^2 + 4 = 0, \qquad x_0^2 = 4 \pm 2\sqrt{3}$$

$$\text{diameter} = 2\sqrt{4 - 2\sqrt{3}} \approx 1.464$$

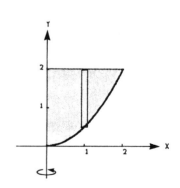

26. Total volume of the hemisphere is $\frac{1}{2}(\frac{4}{3})\pi r^3 = \frac{2}{3}\pi(3)^3 = 18\pi$

By the shell method $[p(x) = x, \; h(x) = \sqrt{9 - x^2}]$. find x_0 such that:

$6\pi = 2\pi \displaystyle\int_0^{x_0} x\sqrt{9 - x^2} \; dx$

$6 = -\displaystyle\int_0^{x_0} (9 - x^2)^{1/2}(-2x) \; dx$

$\quad = -\dfrac{2}{3}(9 - x^2)^{3/2} \Big]_0^{x_0}$

$\quad = 18 - \dfrac{2}{3}(9 - x_0{}^2)^{3/2}$

$(9 - x_0{}^2)^{3/2} = 18, \quad x_0 = \sqrt{9 - 18^{2/3}}$

diameter $= 2\sqrt{9 - 18^{2/3}} \approx 2.920$

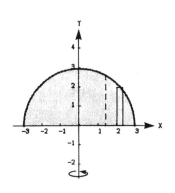

27. <u>Disc</u>

$R(y) = \sqrt{r^2 - y^2}, \qquad r(y) = 0$

$V = \pi \displaystyle\int_{r-h}^{r} (r^2 - y^2) \; dy$

$\quad = \pi \left[r^2 y - \dfrac{y^3}{3} \right]_{r-h}^{r}$

$\quad = \dfrac{1}{3}\pi h^2(3r - h)$

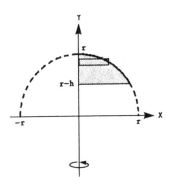

28.(a) $2\pi \displaystyle\int_0^{r} hx\left(1 - \dfrac{x}{r}\right) \; dx$ is the

volume of a right circular cone with the radius of the base as r and height h.

(b) $2\pi \displaystyle\int_{-r}^{r} (R - x)(2\sqrt{r^2 - x^2}) \; dx$

is the volume of a torus with the radius of its circular cross section as r and the distance from the axis of the torus to the center of its cross section as R.

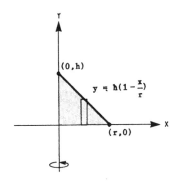

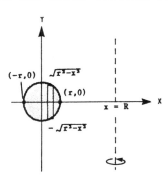

28.(c) $\quad 2\pi \displaystyle\int_0^r 2x\sqrt{r^2 - x^2}\ dx$

is the volume of a sphere with radius r.

(d) $\quad 2\pi \displaystyle\int_0^r hx\ dx$ is the volume of a right circular cylinder with a radius of r and a height of h.

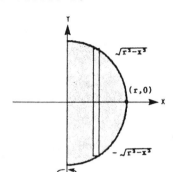

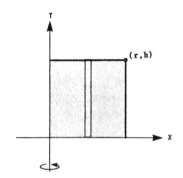

(e) $\quad 2\pi \displaystyle\int_0^b 2ax\sqrt{1 - (x^2/b^2)}\ dx$

is the volume of an ellipsoid with axes 2a and 2b.

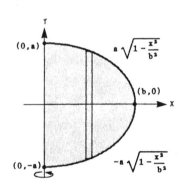

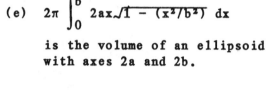

7.4 Work

1. $W = Fd = (100)(10) = 1000$ ft·lb

2. $W = Fd = (2400)(6) = 14400$ ft·lb

3. $W = Fd = (25)(12) = 300$ ft·lb

4. $W = Fd = [9(2000)]\left[\dfrac{1}{2}(5280)\right] = 47,520,000$ ft·lb

5. $F(x) = kx, \qquad 5 = k(4), \qquad k = \dfrac{5}{4}$

$W = \displaystyle\int_0^7 \dfrac{5}{4}x\ dx = \dfrac{5}{8}x^2\Big]_0^7 = \dfrac{245}{8}$ in·lb

6. $W = \int_5^9 \frac{5}{4}x\,dx = \frac{5}{8}x^2\Big]_5^9 = 35$ in·lb

7. $F(x) = kx,\quad 60 = 12k,\quad k = 5$

$W = \int_9^{15} 5x\,dx = \frac{5}{2}x^2\Big]_9^{15} = 360$ in·lb

8. $F(x) = kx,\quad 200 = 2k,\quad k = 100$

$W = \int_0^2 100x\,dx = 50x^2\Big]_0^2 = 200$ ft·lb

9. $F(x) = kx,\quad 15 = 6k,\quad k = \frac{5}{2}$

$W = \int_0^{12} \frac{5}{2}x\,dx = \frac{5}{4}x^2\Big]_0^{12} = 180$ in·lb

10. $F(x) = kx,\quad 15 = k$

$W = 2\int_0^4 15x\,dx = 15x^2\Big]_0^4 = 240$ ft·lb

11. Weight of each layer: $62.4(20)\,\Delta y$,   Distance: $4 - y$

(a) $W = \int_2^4 62.4(20)(4 - y)\,dy$

$= 4992y - 624y^2\Big]_2^4 = 2496$ ft·lb

(b) $W = \int_0^4 62.4(20)(4 - y)\,dy$

$= 4992y - 624y^2\Big]_0^4 = 9984$ ft·lb

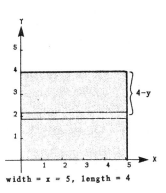

width = x = 5, length = 4

12. (a) $W = \int_2^4 42(20)(4 - y)\,dy$ (b) $W = \int_0^4 42(20)(4 - y)\,dy$

$= 3360y - 420y^2\Big]_2^4$ $= 3360y - 420y^2\Big]_0^4$

$= 1680$ ft·lb $= 6720$ ft·lb

13. Volume of disc of water: $\pi(64)\ \Delta y$
    Weight of disc of water: $62.4(64\pi\ \Delta y)$
    Distance the disc of water is moved: $15 - y$

$$W = \int_0^{12} (15 - y)(62.4)(64\pi)\ dy = 3993.6\pi \int_0^{12} (15 - y)\ dy$$

$$= 3993.6\pi \left[15y - \frac{y^2}{2} \right]_0^{12} = 431{,}308.8\pi\ \text{ft}\cdot\text{lb}$$

14. $$W = \int_{20}^{26} y(62.4)\pi(64)\ dy = 3993.6\pi \int_{20}^{26} y\ dy$$

$$= 1996.8\pi y^2 \Big]_{20}^{26} = 551{,}116.8\pi\ \text{ft}\cdot\text{lbs}$$

15. Volume of disc: $\pi(\sqrt{36 - y^2})^2\ \Delta y$
    Weight of disc: $62.4\pi(36 - y^2)\ \Delta y$
    Distance: y

$$W = 62.4\pi \int_0^6 y(36 - y^2)\ dy$$

$$= 62.4\pi \int_0^6 (36y - y^3)\ dy$$

$$= 62.4\pi \left[18y^2 - \frac{1}{4}y^4 \right]_0^6$$

$$= 20{,}217.6\pi\ \text{ft}\cdot\text{lb}$$

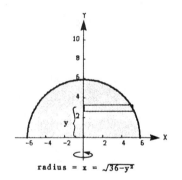

radius = $x = \sqrt{36 - y^2}$

16. Volume of disc: $\pi(36 - y^2)\ \Delta y$
    Weight of disc: $62.4\pi(36 - y^2)\ \Delta y$
    Distance: $-y$

$$W = \int_{-2}^{0} 62.4\pi\ (36 - y^2)(-y)\ dy$$

$$= -62.4\pi \left[18y^2 - \frac{1}{4}y^4 \right]_{-2}^{0}$$

$$= 4243.2\pi\ \text{ft}\cdot\text{lb}$$

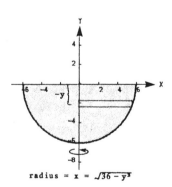

radius = $x = \sqrt{36 - y^2}$

17. Volume of disc: $\pi(\frac{2}{3}y)^2 \, \Delta y$

   Weight of disc: $62.4\pi(\frac{2}{3}y)^2 \, \Delta y$

   Distance: $6 - y$

   $$W = \frac{4(62.4)\pi}{9} \int_0^6 (6 - y)y^2 \, dy$$

   $$= \frac{4}{9}(62.4)\pi \left[2y^3 - \frac{1}{4}y^4 \right]_0^6$$

   $$= 2995.2\pi \ \ \text{ft·lb}$$

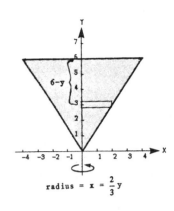

radius = $x = \frac{2}{3}y$

18. Volume of disc: $\pi(\frac{2}{3}y)^2 \, \Delta y$

   Weight of disc: $62.4\pi \, (\frac{2}{3}y)^2 \, \Delta y$

   Distance: y

   (a) $$W = \frac{4}{9}(62.4)\pi \int_0^2 y^3 \, dy$$

   $$= \frac{4}{9}(62.4)\pi(\frac{1}{4}y^4) \Big]_0^2$$

   $$= 110.9\pi \ \ \text{ft·lb}$$

   (b) $$W = \frac{4}{9}(62.4)\pi \int_4^6 y^3 \, dy$$

   $$= \frac{4}{9}(62.4)\pi(\frac{1}{4}y^4) \Big]_4^6$$

   $$= 7{,}210.7\pi \ \ \text{ft·lb}$$

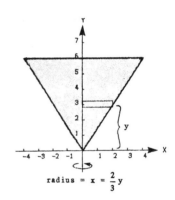

radius = $x = \frac{2}{3}y$

19. Volume of layer: $V = lwh = 4(2)\sqrt{(9/4) - y^2} \, \Delta y$

   Weight of layer: $W = 42(8)\sqrt{(9/4) - y^2} \, \Delta y$

   Distance: $\frac{13}{2} - y$

   Work $$W = \int_{-1.5}^{1.5} 42(8)\sqrt{(9/4) - y^2}(\frac{13}{2} - y) \, dy$$

   $$= 336 \left[\frac{13}{2} \int_{-1.5}^{1.5} \sqrt{(9/4) - y^2} \, dy \right.$$

   $$\left. - \int_{-1.5}^{1.5} \sqrt{(9/4) - y^2} \, y \, dy \right]$$

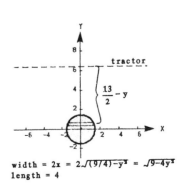

width = $2x = 2\sqrt{(9/4) - y^2} = \sqrt{9 - 4y^2}$
length = 4

19. (continued)
     The second integral is zero since the integrand is odd and the limits
     of integration are symmetric to the origin. The first integral
     represents the area of a semicircle of radius 3/2. Thus the work is:

$$W = 336(\frac{13}{2})\pi(\frac{3}{2})^2(\frac{1}{2}) = 2457\pi \text{ ft}\cdot\text{lb}$$

$$\approx 7718.89 \text{ ft}\cdot\text{lb}$$

20. Volume of layer: $V = 12(2)\sqrt{(25/4) - y^2}\ \Delta y$

     Weight of layer: $W = 42(24)\sqrt{(25/4) - y^2}\ \Delta y$

     Distance: $\frac{19}{2} - y$

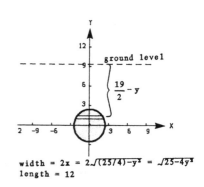

width = $2x = 2\sqrt{(25/4)-y^2} = \sqrt{25-4y^2}$
length = 12

$$\text{Work } W = \int_{-2.5}^{2.5} 42(24)\sqrt{(25/4) - y^2}(\frac{19}{2} - y)\ dy$$

$$= 1008\left[\frac{19}{2}\int_{-2.5}^{2.5}\sqrt{(25/4) - y^2}\ dy\right.$$

$$\left. + \int_{-2.5}^{2.5}\sqrt{(25/4) - y^2}\ (-y)\ dy\right]$$

The second integral is zero since the integrand is odd and the limits
of integration are symmetric to the origin. The first integral
represents the area of a semicircle of radius 5/2. Thus the work is:

$$W = 1008(\frac{19}{2})\pi(\frac{5}{2})^2(\frac{1}{2}) = 29,925\pi \text{ ft}\cdot\text{lb} \approx 94,012.16 \text{ ft}\cdot\text{lb}$$

21. Assume that the earth has a radius of 4000 miles.

$$F(x) = \frac{k}{x^2}, \qquad 4 = \frac{k}{(4000)^2}, \qquad k = 64,000,000, \qquad F(x) = \frac{64,000,000}{x^2}$$

(a) $$W = \int_{4000}^{4200}\frac{64,000,000}{x^2}\ dx = -\frac{64,000,000}{x}\Big]_{4000}^{4200} \approx -15238.095 + 16000$$

$$= 761.905 \text{ mi}\cdot\text{tons} \approx 8.046 \times 10^9 \text{ ft}\cdot\text{lb}$$

(b) $$W = \int_{4000}^{4400}\frac{64,000,000}{x^2}\ dx = -\frac{64,000,000}{x}\Big]_{4000}^{4400} \approx -14545.455 + 16000$$

$$= 1454.545 \text{ mi}\cdot\text{tons} \approx 1.536 \times 10^{10} \text{ ft}\cdot\text{lb}$$

22. Assume that the earth has a radius of 4000 miles.

$$F(x) = \frac{k}{x^2}, \qquad 10 = \frac{k}{(4000)^2}, \qquad k = 160,000,000, \qquad F(x) = \frac{160,000,000}{x^2}$$

(a) $$W = \int_{4000}^{15,000} \frac{160,000,000}{x^2}\, dx = -\frac{160,000,000}{x}\Big]_{4000}^{15,000}$$

$$\approx -10,666.667 + 40,000 = 29,333.333 \text{ mi·tons}$$

$$\approx 3.098 \times 10^{11} \text{ ft·lb}$$

(b) $$W = \int_{4000}^{26,000} \frac{160,000,000}{x^2}\, dx = -\frac{160,000,000}{x}\Big]_{4000}^{26,000} \approx -6,153.846 + 40,000$$

$$= 33,846.154 \text{ mi·tons} \approx 3.574 \times 10^{11} \text{ ft·lb}$$

23. Weight on surface of moon: $\frac{1}{6}(12) = 2$ tons

Weight varies inversely as the square of distance from the center of the moon. Therefore

$$F(x) = \frac{k}{x^2}, \qquad 2 = \frac{k}{(1100)^2}, \qquad k = 2.42 \times 10^6$$

$$W = \int_{1100}^{1150} \frac{2.42 \times 10^6}{x^2}\, dx = \frac{-2.42 \times 10^6}{x}\Big]_{1100}^{1150} = 2.42 \times 10^6 \left(\frac{1}{1100} - \frac{1}{1150}\right)$$

$$= 95.652 \text{ mi·tons} = 1.010 \times 10^9 \text{ ft·lb}$$

24. $$F(x) = \frac{k}{(2 - x)^2}$$

$$W = \int_{-2}^{1} \frac{k}{(2 - x)^2}\, dx = \frac{k}{2 - x}\Big]_{-2}^{1} = k\left(1 - \frac{1}{4}\right) = \frac{3k}{4} \text{ (units of work)}$$

25. Weight of section of chain: $3\,\Delta y$, Distance: $15 - y$

$$W = 3\int_{0}^{15} (15 - y)\, dy = -\frac{3}{2}(15 - y)^2\Big]_{0}^{15} = 337.5 \text{ ft·lb}$$

26. The lower ten feet of chain raised five feet with a constant force.

$W_1 = 3(10)5 = 150 \ \text{ft} \cdot \text{lb}$

The top five feet will be raised with variable force.

Weight of section: $3 \ \Delta y$,   Distance: $5 - y$

$$W_2 = 3 \int_0^5 (5 - y) \ dy = -\frac{3}{2}(5 - y)^2 \Big]_0^5 = \frac{75}{2} \ \text{ft} \cdot \text{lb}$$

$$W = W_1 + W_2 = 150 + \frac{75}{2} = \frac{375}{2} \ \text{ft} \cdot \text{lb}$$

27. The lower 5 feet of chain are raised 10 feet with a constant force.

$W_1 = 3(5)(10) = 150 \ \text{ft} \cdot \text{lb}$

The top ten feet of chain are raised with a variable force.

Weight per section: $3 \ \Delta y$,   Distance: $10 - y$

$$W_2 = 3 \int_0^{10} (10 - y) \ dy = -\frac{3}{2}(10 - y)^2 \Big]_0^{10} = 150 \ \text{ft} \cdot \text{lb}$$

$$W = W_1 + W_2 = 300 \ \text{ft} \cdot \text{lb}$$

28. The work required to lift the chain is 337.5 ft·lb (from Exercise 25)

The work required to lift the 100-pound load is $W = (100)(15) = 1500$.

The work required to lift the chain with a 100-pound load attached is

$W = 337.5 + 1500 = 1837.5 \ \text{ft} \cdot \text{lb}$

29. Weight of section of chain: $3 \ \Delta y$,   Distance: $15 - 2y$

$$W = 3 \int_0^{7.5} (15 - 2y) \ dy = -\frac{3}{4}(15 - 2y)^2 \Big]_0^{7.5} = \frac{3}{4}(15)^2 = 168.75 \ \text{ft} \cdot \text{lb}$$

30. $W = 3 \int_0^6 (12 - 2y) \ dy = -\frac{3}{4}(12 - 2y)^2 \Big]_0^6 = \frac{3}{4}(12)^2 = 108 \ \text{ft} \cdot \text{lb}$

31. Work to pull up the ball: $W_1 = 500(15) = 7500$ ft·lb

    Work to wind up the top 15 feet of cable: (force is variable)

    Weight per section: $1 \, \Delta x$,   Distance: $15 - x$

    $$W_2 = \int_0^{15} (15 - x) \, dx = -\frac{1}{2}(15 - x)^2 \Big]_0^{15} = 112.5 \text{ ft·lb}$$

    Work to lift the lower 25 feet of cable with a constant force:

    $$W_3 = (1)(25)(15) = 375 \text{ ft·lb}$$

    $$W = W_1 + W_2 + W_3 = 7500 + 112.5 + 375 = 7987.5 \text{ ft·lb}$$

32. Work to pull up the ball: $W_1 = 500(40) = 20{,}000$ ft·lb

    Work to pull up the cable: (force is variable)

    Weight per section: $1 \, \Delta x$,   Distance: $40 - x$

    $$W_2 = \int_0^{40} (40 - x) \, dx = -\frac{1}{2}(40 - x)^2 \Big]_0^{40} = 800 \text{ ft·lb}$$

    $$W = W_1 + W_2 = 20{,}000 + 800 = 20{,}800 \text{ ft·lb}$$

33. $p = \dfrac{k}{V}$, $1000 = \dfrac{k}{2}$, $k = 2000$

    $$W = \int_2^3 \frac{2000}{V} \, dV = 2000 \ln |V| \Big]_2^3 = 2000 \ln \left(\frac{3}{2}\right) \approx 810.93 \text{ ft·lb}$$

34. $p = \dfrac{k}{V}$, $2000 = \dfrac{k}{1}$, $k = 2000$

    $$W = \int_1^4 \frac{2000}{V} \, dV = 2000 \ln |V| \Big]_1^4 = 2000 \ln 4 \approx 2772.59 \text{ ft·lb}$$

7.5

Fluid pressure and fluid force

1. $h(y) = 3 - y$, $\qquad L(y) = 4$

    $$F = 62.4 \int_0^3 (3 - y)(4) \, dy$$

    $$= 249.6 \int_0^3 (3 - y) \, dy$$

    $$= 249.6 \left[3y - \frac{y^2}{2} \right]_0^3 = 1123.2 \text{ lb}$$

    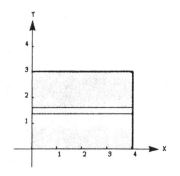

2. $h(y) = 3 - y$, $\qquad L(y) = \frac{4}{3} y$

    $$F = 62.4 \int_0^3 (3 - y)(\frac{4}{3} y) \, dy$$

    $$= \frac{4}{3} (62.4) \int_0^3 (3y - y^2) \, dy$$

    $$= \frac{4}{3} (62.4) \left[\frac{3y^2}{2} - \frac{y^3}{3} \right]_0^3 = 374.4 \text{ lb}$$

    Force is one-third that of exercise 1.

    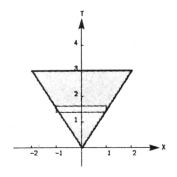

3. $h(y) = 3 - y$, $\qquad L(y) = 2(\frac{y}{3} + 1)$

    $$F = 2(62.4) \int_0^3 (3 - y)(\frac{y}{3} + 1) \, dy$$

    $$= 124.8 \int_0^3 (3 - \frac{y^2}{3}) \, dy$$

    $$= 124.8 \left[3y - \frac{y^3}{9} \right]_0^3 = 748.8 \text{ lb}$$

    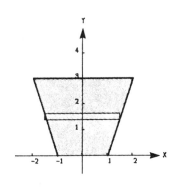

4. $h(y) = -y$, $\qquad L(y) = 2\sqrt{4 - y^2}$

    $$F = 62.4 \int_{-2}^0 (-y)(2)\sqrt{4 - y^2} \, dy$$

    $$= 62.4(\frac{2}{3})(4 - y^2)^{3/2} \Big]_{-2}^0$$

    $$= 332.8 \text{ lb}$$

    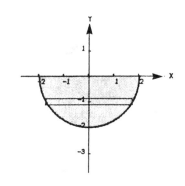

5. $h(y) = 4 - y, \qquad L(y) = 2\sqrt{y}$

$F = 2(62.4) \displaystyle\int_0^4 (4 - y)\sqrt{y}\ dy$

$= 124.8 \displaystyle\int_0^4 (4y^{1/2} - y^{3/2})\ dy$

$= 124.8 \left[\dfrac{8y^{3/2}}{3} - \dfrac{2y^{5/2}}{5} \right]_0^4 = 1064.96\ \text{lb}$

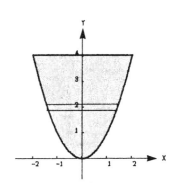

6. $h(y) = -y, \qquad L(y) = \dfrac{4}{3}\sqrt{9 - y^2}$

$F = 62.4 \displaystyle\int_{-3}^0 (-y)\dfrac{4}{3}\sqrt{9 - y^2}\ dy$

$= 62.4(\dfrac{2}{3}) \displaystyle\int_{-3}^0 (9 - y^2)^{1/2}(-2y)\ dy$

$= 62.4(\dfrac{4}{9})(9 - y^2)^{3/2} \Big]_{-3}^0 = 748.8\ \text{lb}$

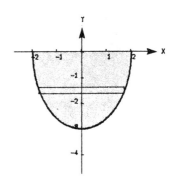

7. $h(y) = 4 - y, \qquad L(y) = 2$

$F = 62.4 \displaystyle\int_0^2 2(4 - y)\ dy$

$= 62.4 \left[8y - y^2 \right]_0^2$

$= 748.8\ \text{lb}$

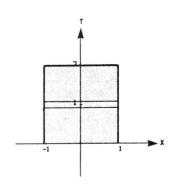

8. $h(y) = (1 + 2\sqrt{2}) - y$

$L_1(y) = 2y$ [lower part], $\qquad L_2(y) = 2(2\sqrt{2} - y)$ [upper part]

$F = 124.8 \left[\displaystyle\int_0^{\sqrt{2}} (1 + 2\sqrt{2} - y)y\ dy \right.$

$\left. + \displaystyle\int_{\sqrt{2}}^{2\sqrt{2}} (1 + 2\sqrt{2} - y)(2\sqrt{2} - y)\ dy \right]$

$= 124.8 \left[\left[\dfrac{y^2}{2} + \sqrt{2}\,y^2 - \dfrac{y^3}{3} \right]_0^{\sqrt{2}} \right.$

$\left. + \left[2\sqrt{2}\,y + 8y - 2\sqrt{2}\,y^2 - \dfrac{y^2}{2} + \dfrac{y^3}{3} \right]_{\sqrt{2}}^{2\sqrt{2}} \right]$

$= 249.6(1 + \sqrt{2}) \approx 602.6\ \text{lb}$

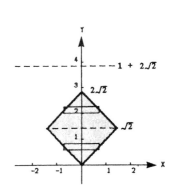

9. $h(y) = 12 - y$, $\qquad L(y) = 6 - \dfrac{2y}{3}$

$F = 62.4 \displaystyle\int_{0}^{9} (12 - y)(6 - \dfrac{2y}{3})\ dy$

$= 62.4 \left[72y - 7y^2 + \dfrac{2y^3}{9} \right]_{0}^{9} = 15{,}163.2\ \text{lb}$

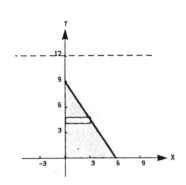

10. $h(y) = 6 - y$, $\qquad L(y) = 1$

$F = 62.4 \displaystyle\int_{0}^{5} 1(6 - y)\ dy$

$= 62.4 \left[6y - \dfrac{y^2}{2} \right]_{0}^{5} = 1092.0\ \text{lb}$

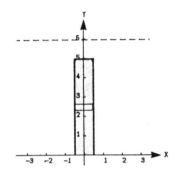

11. $h(y) = 2 - y$, $\qquad L(y) = 10$

$F = 140.7 \displaystyle\int_{0}^{2} (2 - y)(10)\ dy$

$= 1407 \displaystyle\int_{0}^{2} (2 - y)\ dy$

$= 1407 \left[2y - \dfrac{y^2}{2} \right]_{0}^{2} = 2814\ \text{lb}$

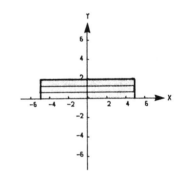

12. $h(y) = 4 - y$, $\qquad L(y) = 6$

$F = 140.7 \displaystyle\int_{0}^{4} (4 - y)(6)\ dy$

$= 844.2 \displaystyle\int_{0}^{4} (4 - y)\ dy$

$= 844.2 \left[4y - \dfrac{y^2}{2} \right]_{0}^{4} = 6753.6\ \text{lb}$

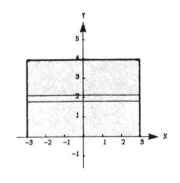

13. $h(y) = -y,$ $L(y) = 2(\frac{4}{3}\sqrt{9-y^2})$

$$F = 140.7 \int_{-3}^{0} (-y)(2)(\frac{4}{3}\sqrt{9-y^2}) \, dy$$

$$= \frac{(140.7)(4)}{3} \int_{-3}^{0} \sqrt{9-y^2}(-2y) \, dy$$

$$= \frac{(140.7)(4)}{3}(\frac{2}{3})(9-y^2)^{3/2} \Big]_{-3}^{0}$$

$$= 3376.8 \text{ lb}$$

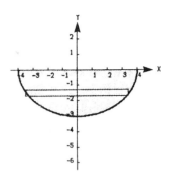

14. $h(y) = -y,$ $L(y) = 6 + \frac{3}{2}y$

$$F = 140.7 \int_{-4}^{0} (-y)(6 + \frac{3}{2}y) \, dy$$

$$= -140.7 \ (3y^2 + \frac{y^3}{2}) \Big]_{-4}^{0}$$

$$= 2251.2 \text{ lb}$$

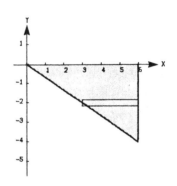

15. $h(y) = -y,$ $L(y) = 2(\frac{1}{2})\sqrt{9-4y^2}$

$$F = 42 \int_{-3/2}^{0} (-y)\sqrt{9-4y^2} \, dy$$

$$= \frac{42}{8} \int_{-3/2}^{0} (9-4y^2)^{1/2}(-8y) \, dy$$

$$= (\frac{21}{4})(\frac{2}{3})(9-4y^2)^{3/2} \Big]_{-3/2}^{0}$$

$$= 94.5 \text{ lb}$$

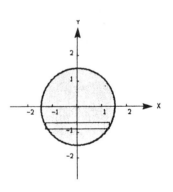

16. $h(y) = \dfrac{3}{2} - y, \qquad L(y) = 2\left(\dfrac{1}{2}\right)\sqrt{9 - 4y^2}$

$F = 42 \displaystyle\int_{-3/2}^{3/2} \left(\dfrac{3}{2} - y\right)\sqrt{9 - 4y^2}\, dy$

$= 63 \displaystyle\int_{-3/2}^{3/2} \sqrt{9 - 4y^2}\, dy + \dfrac{21}{4}\displaystyle\int_{-3/2}^{3/2} \sqrt{9 - 4y^2}(-8y)\, dy$

The second integral is zero since it is an odd function and the limits of integration are symmetric to the origin. The first integral is twice the area of a semicircle of radius 3/2

$$\left(\sqrt{9 - 4y^2} = 2\sqrt{(9/4) - y^2}\,\right)$$

Thus the force is $63\left(\dfrac{9}{4}\pi\right) = 141.75\pi \approx 445.32$ ft

17. $h(y) = k - y, \; L(y) = 2\sqrt{r^2 - y^2}$

$F = w \displaystyle\int_{-r}^{r} (k - y)\sqrt{r^2 - y^2}(2)\, dy$

$= w\left[2k \displaystyle\int_{-r}^{r} \sqrt{r^2 - y^2}\, dy + \displaystyle\int_{-r}^{r} \sqrt{r^2 - y^2}(-2y)\, dy\right]$

The second integral is zero since its integrand is odd and the limits of integration are symmetric to the origin. The first integral is the area of a semicircle with radius r.

$F = w\left[(2k)\dfrac{\pi r^2}{2} + 0\right] = wk\pi r^2$

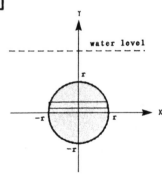

18. $h(y) = k - y, \qquad L(y) = b$

$F = w \displaystyle\int_{-h/2}^{h/2} (k - y)b\, dy$

$= wb\left[ky - \dfrac{y^2}{2}\right]_{-h/2}^{h/2}$

$= wb(hk) = wkhb$

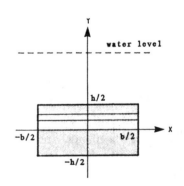

19. From Exercise 18: $F = 64(15)(1)(1) = 960$ lb

20. From Exercise 17: $F = 64(15)\pi(\frac{1}{2})^2 = 753.98$ lb

21.(a) Wall at shallow end
From Exercise 18: $F = 62.4(2)(4)(20) = 9984$ lb

(b) Wall at deep end
From Exercise 18: $F = 62.4(4)(8)(20) = 39,936$ lb

(c) Side wall
From Exercise 18: $F_1 = 62.4(2)(4)(40) = 19,968$ lb

$$F_2 = 62.4 \int_0^4 (8 - y)(10y) \, dy$$

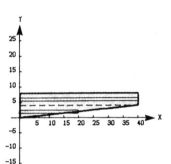

$$= 624 \int_0^4 (8y - y^2) \, dy$$

$$= 624 \left[4y^2 - \frac{y^3}{3} \right]_0^4 = 26,624 \text{ lb}$$

Total force $= F_1 + F_2 = 46,592$ lb

7.6
Moments, centers of mass, and centroids

1. $\bar{x} = \dfrac{6(-5) + 3(1) + 5(3)}{6 + 3 + 5} = -\dfrac{6}{7}$

2. $\bar{x} = \dfrac{7(-3) + 4(-2) + 3(5) + 8(6)}{7 + 4 + 3 + 8} = \dfrac{17}{11}$

3. $\bar{x} = \dfrac{1(7) + 1(8) + 1(12) + 1(15) + 1(18)}{1 + 1 + 1 + 1 + 1} = 12$

4. $\bar{x} = \dfrac{12(-3) + 1(-2) + 6(-1) + 3(0) + 11(4)}{12 + 1 + 6 + 3 + 11} = 0$

5. $\bar{x} = \dfrac{(7 + 5) + (8 + 5) + (12 + 5) + (15 + 5) + (18 + 5)}{5} = 17 = 12 + 5$

6. $\bar{x} = \dfrac{12(-3 - 3) + 1(-2 - 3) + 6(-1 - 3) + 3(0 - 3) + 11(4 - 3)}{12 + 1 + 6 + 3 + 11}$

   $= -3 = 0 - 3$

7. $\bar{x} = \dfrac{5(2) + 1(-3) + 3(1)}{5 + 1 + 3} = \dfrac{10}{9}$

$\bar{y} = \dfrac{5(2) + 1(1) + 3(-4)}{5 + 1 + 3} = -\dfrac{1}{9}$

$(\bar{x},\ \bar{y}) = (\dfrac{10}{9},\ -\dfrac{1}{9})$

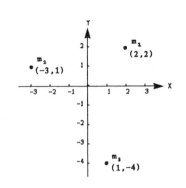

8. $\bar{x} = \dfrac{10(1) + 2(5) + 5(-4)}{10 + 2 + 5} = 0$

$\bar{y} = \dfrac{10(-1) + 2(5) + 5(0)}{10 + 2 + 5} = 0$

$(\bar{x},\ \bar{y}) = (0,\ 0)$

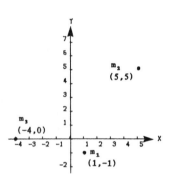

9. $\bar{x} = \dfrac{3(-2) + 4(-1) + 2(7) + 1(0) + 6(-3)}{3 + 4 + 2 + 1 + 6}$

$= -\dfrac{7}{8}$

$\bar{y} = \dfrac{3(-3) + 4(0) + 2(1) + 1(0) + 6(0)}{3 + 4 + 2 + 1 + 6}$

$= -\dfrac{7}{16}$

$(\bar{x},\ \bar{y}) = (-\dfrac{7}{8},\ -\dfrac{7}{16})$

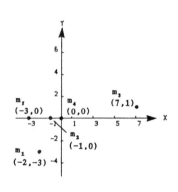

10. $\bar{x} = \dfrac{4(2) + 2(-1) + 2.5(6) + 5(2)}{4 + 2 + 2.5 + 5}$

$= \dfrac{31}{13.5} = \dfrac{62}{27}$

$\bar{y} = \dfrac{4(3) + 2(5) + 2.5(8) + 5(-2)}{4 + 2 + 2.5 + 5}$

$= \dfrac{32}{13.5} = \dfrac{64}{27}$

$(\bar{x},\ \bar{y}) = (\dfrac{62}{27},\ \dfrac{64}{27})$

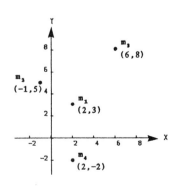

11. For simplicity we assume the density to be 1 unit of mass/unit of area.

$m_1 = 4$, $P_1 = (0, 1)$, $m_2 = \pi$, $P_2 = (0, 3)$

$$\overline{x} = \frac{4(0) + \pi(0)}{4 + \pi} = 0$$

$$\overline{y} = \frac{4(1) + \pi(3)}{4 + \pi} = \frac{4 + 3\pi}{4 + \pi}$$

$$(\overline{x}, \overline{y}) = (0, \frac{4 + 3\pi}{4 + \pi})$$

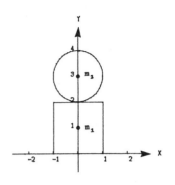

12. For simplicity, we assume the density to be 1 unit of mass/unit of area.

$m_1 = 3$, $P_1 = (1/2, 3/2)$, $m_2 = 2$, $P_2 = (2, 1/2)$

$m_3 = 2$, $P_3 = (7/2, 1)$

$$\overline{x} = \frac{3(1/2) + 2(2) + 2(7/2)}{3 + 2 + 2} = \frac{25/2}{7} = \frac{25}{14}$$

$$\overline{y} = \frac{3(3/2) + 2(1/2) + 2(1)}{3 + 2 + 2} = \frac{15/2}{7} = \frac{15}{14}$$

$$(\overline{x}, \overline{y}) = (\frac{25}{14}, \frac{15}{14})$$

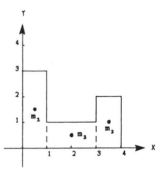

13. $m_1 = 15$, $m_2 = 12$, $m_3 = 7$

$P_1 = (0, \frac{3}{2})$, $P_2 = (0, 5)$, $P_3 = (0, \frac{15}{2})$

$$\overline{x} = \frac{15(0) + 12(0) + 7(0)}{15 + 12 + 7} = 0$$

$$\overline{y} = \frac{15(3/2) + 12(5) + 7(15/2)}{15 + 12 + 7} = \frac{135}{34}$$

$$(\overline{x}, \overline{y}) = (0, \frac{135}{34})$$

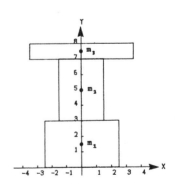

14. $m_1 = 4\pi, \qquad m_2 = 4\pi, \qquad m_3 = 16$

$P_1 = (2, 2), \quad P_2 = (-2, 2), \quad P_3 = (0, 5)$

$\bar{x} = \dfrac{4\pi(2) + 4\pi(-2) + 16(0)}{4\pi + 4\pi + 16} = 0$

$\bar{y} = \dfrac{4\pi(2) + 4\pi(2) + 16(5)}{4\pi + 4\pi + 16}$

$= \dfrac{16\pi + 80}{8\pi + 16} = \dfrac{2\pi + 10}{\pi + 2}$

$(\bar{x}, \bar{y}) = (0, \dfrac{2\pi + 10}{\pi + 2})$

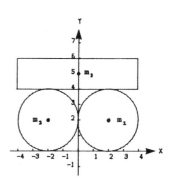

15. $m_1 = 4, \ P_1 = (0, 1), \qquad m_2 = 2\pi, \ P_2 = (0, 3), \qquad \bar{x} = 0$

$\bar{y} = \dfrac{4(1) + 2\pi(3)}{4 + 2\pi} = \dfrac{2 + 3\pi}{2 + \pi} \qquad (\bar{x}, \bar{y}) = (0, \dfrac{2 + 3\pi}{2 + \pi})$

16. $m_1 = 8, \ P_1 = (0, 1), \qquad m_2 = \pi, \ P_2 = (0, 3), \qquad \bar{x} = 0$

$\bar{y} = \dfrac{8(1) + \pi(3)}{8 + \pi} = \dfrac{8 + 3\pi}{8 + \pi}$

$(\bar{x}, \bar{y}) = (0, \dfrac{8 + 3\pi}{8 + \pi})$

17. $m = \rho \displaystyle\int_0^4 \sqrt{x}\ dx = \dfrac{2\rho}{3} x^{3/2} \Big]_0^4 = \dfrac{16\rho}{3}$

$M_x = \rho \displaystyle\int_0^4 \dfrac{\sqrt{x}}{2}(\sqrt{x})\ dx = \rho \dfrac{x^2}{4}\Big]_0^4 = 4\rho$

$\bar{y} = 4\rho(\dfrac{3}{16\rho}) = \dfrac{3}{4}$

$M_y = \rho \displaystyle\int_0^4 x\sqrt{x}\ dx = \rho \dfrac{2}{5} x^{5/2}\Big]_0^4 = \dfrac{64\rho}{5}$

$\bar{x} = \dfrac{64\rho}{5}(\dfrac{3}{16\rho}) = \dfrac{12}{5}$

$(\bar{x}, \bar{y}) = (\dfrac{12}{5}, \dfrac{3}{4})$

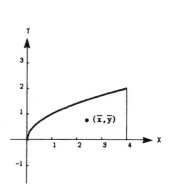

474

18. $m = \rho \int_0^4 x^2\ dx = \rho\ \dfrac{x^3}{3}\bigg]_0^4 = \dfrac{64}{3}\rho$

$M_x = \rho \int_0^4 \dfrac{x^2}{2}(x^2)\ dx = \rho\ \dfrac{x^5}{10}\bigg]_0^4 = \dfrac{512\rho}{5}$

$\overline{y} = \dfrac{512\rho}{5}\left(\dfrac{3}{64\rho}\right) = \dfrac{24}{5}$

$M_y = \rho \int_0^4 x(x^2)\ dx = \rho\ \dfrac{x^4}{4}\bigg]_0^4 = 64\rho$

$\overline{x} = 64\rho\left(\dfrac{3}{64\rho}\right) = 3$

$(\overline{x},\ \overline{y}) = (3,\ \dfrac{24}{5})$

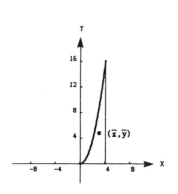

19. $m = \rho \int_0^3 [(-x^2 + 4x + 2) - (x + 2)]\ dx = -\rho\left[\dfrac{x^3}{3} + \dfrac{3x^2}{2}\right]_0^3 = \dfrac{9\rho}{2}$

$M_x = \rho \int_0^3 \left[\dfrac{(-x^2 + 4x + 2) + (x + 2)}{2}\right][(-x^2 + 4x + 2) - (x + 2)]\ dx$

$\quad = \dfrac{\rho}{2} \int_0^3 (-x^2 + 5x + 4)(-x^2 + 3x)\ dx$

$\quad = \dfrac{\rho}{2} \int_0^3 (x^4 - 8x^3 + 11x^2 + 12x)\ dx$

$\quad = \dfrac{\rho}{2}\left[\dfrac{x^5}{5} - 2x^4 + \dfrac{11x^3}{3} + 6x^2\right]_0^3 = \dfrac{99\rho}{5}$

$\overline{y} = \dfrac{M_x}{m} = \dfrac{99\rho}{5}\left(\dfrac{2}{9\rho}\right) = \dfrac{22}{5}$

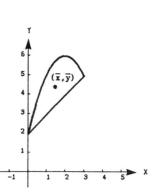

$M_y = \rho \int_0^3 x[(-x^2 + 4x - 2) - (x + 2)]\ dx$

$\quad = \rho \int_0^3 (-x^3 + 3x^2)\ dx = \rho\left[-\dfrac{x^4}{4} + x^3\right]_0^3 = \dfrac{27\rho}{4}$

$\overline{x} = \dfrac{M_y}{m} = \dfrac{27\rho}{4}\left(\dfrac{2}{9\rho}\right) = \dfrac{3}{2}$

$(\overline{x},\ \overline{y}) = (\dfrac{3}{2},\ \dfrac{22}{5})$

20. $m = \rho \int_0^1 (x^2 - x^3) \, dx = \rho \left[\dfrac{x^3}{3} - \dfrac{x^4}{4} \right]_0^1 = \dfrac{\rho}{12}$

$M_x = \rho \int_0^1 \dfrac{(x^2 + x^3)}{2}(x^2 - x^3) \, dx$

$\quad = \dfrac{\rho}{2} \int_0^1 (x^4 - x^6) \, dx = \dfrac{\rho}{2} \left[\dfrac{x^5}{5} - \dfrac{x^7}{7} \right]_0^1 = \dfrac{\rho}{35}$

$\bar{y} = \dfrac{M_x}{m} = \dfrac{\rho}{35}(\dfrac{12}{\rho}) = \dfrac{12}{35}$

$M_y = \rho \int_0^1 x(x^2 - x^3) \, dx = \rho \int_0^1 (x^3 - x^4) \, dx$

$\quad = \rho \left[\dfrac{x^4}{4} - \dfrac{x^5}{5} \right]_0^1 = \dfrac{\rho}{20}$

$\bar{x} = \dfrac{M_y}{m} = \dfrac{\rho}{20}(\dfrac{12}{\rho}) = \dfrac{3}{5}$

$(\bar{x}, \bar{y}) = (\dfrac{3}{5}, \dfrac{12}{35})$

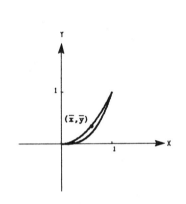

21. $m = \rho \int_0^1 (\sqrt{x} - x) \, dx = \rho \left[\dfrac{2}{3}x^{3/2} - \dfrac{x^2}{2} \right]_0^1 = \dfrac{\rho}{6}$

$M_x = \rho \int_0^1 \dfrac{(\sqrt{x} + x)}{2}(\sqrt{x} - x) \, dx = \dfrac{\rho}{2} \int_0^1 (x - x^2) \, dx = \dfrac{\rho}{2} \left[\dfrac{x^2}{2} - \dfrac{x^3}{3} \right]_0^1 = \dfrac{\rho}{12}$

$\bar{y} = \dfrac{M_x}{m} = \dfrac{\rho}{12}(\dfrac{6}{\rho}) = \dfrac{1}{2}$

$M_y = \rho \int_0^1 x(\sqrt{x} - x) \, dx = \rho \int_0^1 (x^{3/2} - x^2) \, dx$

$\quad = \rho \left[\dfrac{2}{5}x^{5/2} - \dfrac{x^3}{3} \right]_0^1 = \dfrac{\rho}{15}$

$\bar{x} = \dfrac{M_y}{m} = \dfrac{\rho}{15}(\dfrac{6}{\rho}) = \dfrac{2}{5}$

$(\bar{x}, \bar{y}) = (\dfrac{2}{5}, \dfrac{1}{2})$

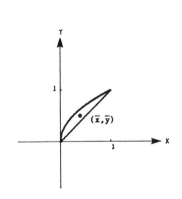

22. $m = \rho \int_0^3 [(\sqrt{3x} + 1) - (x + 1)] \, dx = \rho \left[\frac{2}{9} (3x)^{3/2} - \frac{x^2}{2} \right]_0^3 = \frac{3}{2} \rho$

$M_x = \rho \int_0^3 \frac{[(\sqrt{3x} + 1) + (x + 1)]}{2} [(\sqrt{3x} + 1) - (x + 1)] \, dx$

$\quad = \frac{\rho}{2} \int_0^3 (\sqrt{3x} + x + 2)(\sqrt{3x} - x) \, dx$

$\quad = \frac{\rho}{2} \int_0^3 (2\sqrt{3x} + x - x^2) \, dx = \frac{\rho}{2} \left[\frac{4}{9}(3x)^{3/2} + \frac{x^2}{2} - \frac{x^3}{3} \right]_0^3 = \frac{15\rho}{4}$

$\bar{y} = \frac{M_x}{m} = \frac{15\rho}{4}(\frac{2}{3\rho}) = \frac{5}{2}$

$M_y = \rho \int_0^3 x[(\sqrt{3x} + 1) - (x + 1)] \, dx$

$\quad = \rho \int_0^3 (x\sqrt{3x} - x^2) \, dx$

$\quad = \rho \left[\frac{2\sqrt{3}}{5} x^{5/2} - \frac{x^3}{3} \right]_0^3 = \frac{9\rho}{5}$

$\bar{x} = \frac{M_y}{m} = \frac{9\rho}{5}(\frac{2}{3\rho}) = \frac{6}{5}$

$(\bar{x}, \bar{y}) = (\frac{6}{5}, \frac{5}{2})$

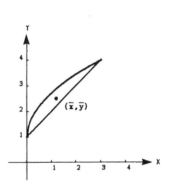

23. $m = 2\rho \int_0^2 (4 - y^2) \, dy = 2\rho \left[4y - \frac{y^3}{3} \right]_0^2 = \frac{32\rho}{3}$

$M_y = 2\rho \int_0^2 (\frac{4 - y^2}{2})(4 - y^2) \, dy$

$\quad = \rho \left[16y - \frac{8}{3}y^3 + \frac{y^5}{5} \right]_0^2 = \frac{256\rho}{15}$

$\bar{x} = \frac{256\rho}{15}(\frac{3}{32\rho}) = \frac{8}{5}$

By symmetry M_x and $\bar{y} = 0$

$(\bar{x}, \bar{y}) = (\frac{8}{5}, 0)$

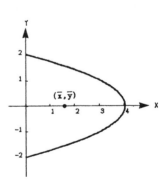

24. $m = \rho \int_0^2 (2y - y^2)\, dy = \rho \left[y^2 - \dfrac{y^3}{3} \right]_0^2 = \dfrac{4\rho}{3}$

$M_y = \rho \int_0^2 \dfrac{2y - y^2}{2} (2y - y^2)\, dy = \dfrac{\rho}{2} \left[\dfrac{4y^3}{3} - y^4 + \dfrac{y^5}{5} \right]_0^2 = \dfrac{8\rho}{15}$

$\overline{x} = \dfrac{8\rho}{15}\left(\dfrac{3}{4\rho}\right) = \dfrac{2}{5}$

$M_x = \rho \int_0^2 y(2y - y^2)\, dy$

$= \rho \left[\dfrac{2y^3}{3} - \dfrac{y^4}{4} \right]_0^2 = \dfrac{4\rho}{3}$

$\overline{y} = \dfrac{4\rho}{3}\left(\dfrac{3}{4\rho}\right) = 1$

$(\overline{x},\ \overline{y}) = (\dfrac{2}{5},\ 1)$

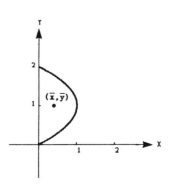

25. $m = \rho \int_0^3 [(2y - y^2) - (-y)]\, dy = \rho \left[\dfrac{3y^2}{2} - \dfrac{y^3}{3} \right]_0^3 = \dfrac{9\rho}{2}$

$M_y = \rho \int_0^3 \dfrac{[(2y - y^2) + (-y)]}{2}[(2y - y^2) - (-y)]\, dy$

$= \dfrac{\rho}{2} \int_0^3 (y - y^2)(3y - y^2)\, dy$

$= \dfrac{\rho}{2} \int_0^3 (y^4 - 4y^3 + 3y^2)\, dy = \dfrac{\rho}{2} \left[\dfrac{y^5}{5} - y^4 + y^3 \right]_0^3 = -\dfrac{27\rho}{10}$

$\overline{x} = \dfrac{M_y}{m} = -\dfrac{27\rho}{10}\left(\dfrac{2}{9\rho}\right) = -\dfrac{3}{5}$

$M_x = \rho \int_0^3 y[(2y - y^2) - (-y)]\, dy$

$= \rho \int_0^3 (3y^2 - y^3)\, dy = \rho \left[y^3 - \dfrac{y^4}{4} \right]_0^3 = \dfrac{27\rho}{4}$

$\overline{y} = \dfrac{M_x}{m} = \dfrac{27\rho}{4}\left(\dfrac{2}{9\rho}\right) = \dfrac{3}{2}$

$(\overline{x},\ \overline{y}) = (-\dfrac{3}{5},\ \dfrac{3}{2})$

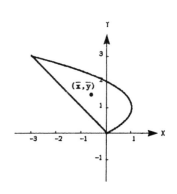

26. $m = \rho \int_{-1}^{2} [(y + 2) - y^2]\, dy = \rho \left[\dfrac{y^2}{2} + 2y - \dfrac{y^3}{3}\right]_{-1}^{2} = \dfrac{9\rho}{2}$

$M_y = \rho \int_{-1}^{2} \dfrac{[(y + 2) + y^2]}{2}[(y + 2) - y^2]\, dy$

$\quad = \dfrac{\rho}{2} \int_{-1}^{2} [(y + 2)^2 - y^4]\, dy$

$\quad = \dfrac{\rho}{2} \left[\dfrac{(y + 2)^3}{3} - \dfrac{y^5}{5}\right]_{-1}^{2} = \dfrac{36\rho}{5}$

$\overline{x} = \dfrac{M_y}{m} = \dfrac{36\rho}{5}\left(\dfrac{2}{9\rho}\right) = \dfrac{8}{5}$

$M_x = \rho \int_{-1}^{2} y[(y + 2) - y^2]\, dy$

$\quad = \rho \int_{-1}^{2} (2y + y^2 - y^3)\, dy$

$\quad = \rho \left[y^2 + \dfrac{y^3}{3} - \dfrac{y^4}{4}\right]_{-1}^{2} = \dfrac{9\rho}{4}$

$\overline{y} = \dfrac{M_x}{m} = \dfrac{9\rho}{4}\left(\dfrac{2}{9\rho}\right) = \dfrac{1}{2}$

$(\overline{x},\ \overline{y}) = \left(\dfrac{8}{5},\ \dfrac{1}{2}\right)$

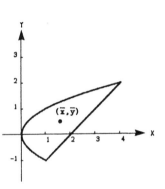

27. $m = \rho \int_{0}^{8} x^{2/3}\, dx = \rho \left[\dfrac{3}{5} x^{5/3}\right]_{0}^{8} = \dfrac{96\rho}{5}$

$M_x = \rho \int_{0}^{8} \dfrac{x^{2/3}}{2}(x^{2/3})\, dx = \dfrac{\rho}{2}\left[\dfrac{3}{7} x^{7/3}\right]_{0}^{8} = \dfrac{192\rho}{7}$

$\overline{y} = \dfrac{192\rho}{7}\left(\dfrac{5}{96\rho}\right) = \dfrac{10}{7}$

$M_y = \rho \int_{0}^{8} x(x^{2/3})\, dx = \rho \left[\dfrac{3}{8} x^{8/3}\right]_{0}^{8} = 96\rho$

$\overline{x} = 96\rho\left(\dfrac{5}{96\rho}\right) = 5$

$(\overline{x},\ \overline{y}) = \left(5,\ \dfrac{10}{7}\right)$

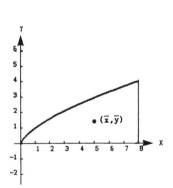

28. $m = 2\rho \int_0^8 (4 - x^{2/3})\, dx = 2\rho \left[4x - \dfrac{3}{5} x^{5/3} \right]_0^8 = \dfrac{128\rho}{5}$

$M_x = 2\rho \int_0^8 \dfrac{4 + x^{2/3}}{2} (4 - x^{2/3})\, dx$

$\quad = \rho \left[16x - \dfrac{3}{7} x^{7/3} \right]_0^8 = \dfrac{512\rho}{7}$

$\bar{y} = \dfrac{512\rho}{7}\left(\dfrac{5}{128\rho}\right) = \dfrac{20}{7}$

By symmetry, M_y and $\bar{x} = 0$

$(\bar{x}, \bar{y}) = (0, \dfrac{20}{7})$

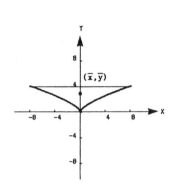

29. $A = \dfrac{\pi}{2}, \quad \dfrac{1}{A} = \dfrac{2}{\pi}$

$\bar{x} = 0$ (by symmetry)

$\bar{y} = \dfrac{2}{\pi} \int_{-1}^1 \dfrac{1 - x^2}{2}\, dx$

$\quad = \dfrac{1}{\pi} \left[x - \dfrac{x^3}{3} \right]_{-1}^1 = \dfrac{4}{3\pi}$

$(\bar{x}, \bar{y}) = (0, \dfrac{4}{3\pi})$

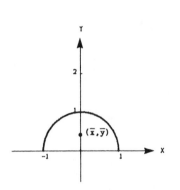

30. $A = \int_0^1 (x - x^3)\, dx = \left[\dfrac{1}{2} x^2 - \dfrac{1}{4} x^4 \right]_0^1 = \dfrac{1}{4}, \qquad \dfrac{1}{A} = 4$

$\bar{x} = 4 \int_0^1 (x^2 - x^4)\, dx$

$\quad = 4 \left[\dfrac{1}{3} x^3 - \dfrac{1}{5} x^5 \right]_0^1 = \dfrac{8}{15}$

$\bar{y} = (4)\dfrac{1}{2} \int_0^1 (x^2 - x^6)\, dx$

$\quad = 2 \left[\dfrac{1}{3} x^3 - \dfrac{1}{7} x^7 \right]_0^1 = \dfrac{8}{21}$

$(\bar{x}, \bar{y}) = (\dfrac{8}{15}, \dfrac{8}{21})$

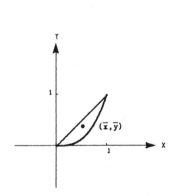

31. $A = \int_0^1 (x - x^2) \, dx = \left[\frac{1}{2}x^2 - \frac{1}{3}x^3\right]_0^1 = \frac{1}{6}, \qquad \frac{1}{A} = 6$

$\overline{x} = 6 \int_0^1 (x^2 - x^3) \, dx$

$= 6 \left[\frac{1}{3}x^3 - \frac{1}{4}x^4\right]_0^1 = \frac{1}{2}$

$\overline{y} = (6)\frac{1}{2} \int_0^1 (x^2 - x^4) \, dx$

$= 3 \left[\frac{1}{3}x^3 - \frac{1}{5}x^5\right]_0^1 = \frac{2}{5}$

$(\overline{x}, \overline{y}) = (\frac{1}{2}, \frac{2}{5})$

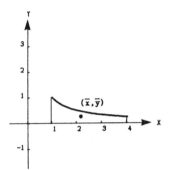

32. $A = \int_1^4 \frac{1}{x} \, dx = \ln |x| \Big]_1^4 = \ln 4, \qquad \frac{1}{A} = \frac{1}{\ln 4}$

$\overline{x} = \frac{1}{\ln 4} \int_1^4 x(\frac{1}{x}) \, dx = \frac{1}{\ln 4} x \Big]_1^4$

$= \frac{3}{\ln 4} \approx 2.1640$

$\overline{y} = (\frac{1}{\ln 4})\frac{1}{2} \int_1^4 \frac{1}{x^2} \, dx = \frac{1}{2 \ln 4}(-\frac{1}{x}) \Big]_1^4$

$= \frac{3}{8 \ln 4} \approx 0.2705$

$(\overline{x}, \overline{y}) = (\frac{3}{\ln 4}, \frac{3}{8 \ln 4})$

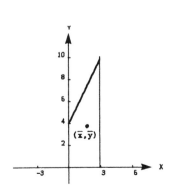

33. $A = \int_0^3 (2x + 4) \, dx = x^2 + 4x \Big]_0^3 = 21, \qquad \frac{1}{A} = \frac{1}{21}$

$\overline{x} = \frac{1}{21} \int_0^3 (2x^2 + 4x) \, dx$

$= \frac{1}{21} \left[\frac{2}{3}x^3 + 2x^2\right]_0^3 = \frac{12}{7}$

$\overline{y} = (\frac{1}{21})\frac{1}{2} \int_0^3 (2x + 4)^2 \, dx$

$= \frac{1}{42} \left[\frac{4}{3}x^3 + 8x^2 + 16x\right]_0^3 = \frac{26}{7}$

$(\overline{x}, \overline{y}) = (\frac{12}{7}, \frac{26}{7})$

34. $A = \int_{-2}^{2} (4 - x^2) \, dx = \left[\dfrac{x^3}{3} - 4x \right]_{-2}^{2} = \dfrac{32}{3}, \qquad \dfrac{1}{A} = \dfrac{3}{32}$

$\bar{x} = \dfrac{3}{32} \int_{-2}^{2} x(4 - x^2) \, dx = \dfrac{3}{32}(0) = 0$

$\bar{y} = (\dfrac{3}{32})\dfrac{1}{2} \int_{-2}^{2} (x^2 - 4)(4 - x^2) \, dx$

$= -\dfrac{3}{64} \int_{-2}^{2} (x^4 - 8x^2 + 16) \, dx$

$= -\dfrac{3}{64} \left[\dfrac{x^5}{5} - \dfrac{8}{3} x^3 + 16x \right]_{-2}^{2}$

$= -\dfrac{3}{64}(\dfrac{512}{15}) = -\dfrac{8}{5}$

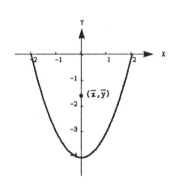

$(\bar{x}, \bar{y}) = (0, -\dfrac{8}{5})$

35. $A = \dfrac{1}{2}(2a)c = ac, \qquad \dfrac{1}{A} = \dfrac{1}{ac}$

$\bar{x} = (\dfrac{1}{ac})\dfrac{1}{2} \int_{0}^{c} \left[(\dfrac{b - a}{c} y + a)^2 - (\dfrac{b + a}{c} y - a)^2 \right] dy$

$= \dfrac{1}{2ac} \int_{0}^{c} \left[\dfrac{4ab}{c} y - \dfrac{4ab}{c^2} y^2 \right] dy$

$= \dfrac{1}{2ac} \left[\dfrac{2ab}{c} y^2 - \dfrac{4ab}{3c^2} y^3 \right]_{0}^{c} = \dfrac{1}{2ac}(\dfrac{2}{3} abc) = \dfrac{b}{3}$

$\bar{y} = \dfrac{1}{ac} \int_{0}^{c} y \left[(\dfrac{b - a}{c} y + a) - (\dfrac{b + a}{c} y - a) \right] dy$

$= \dfrac{1}{ac} \int_{0}^{c} y(-\dfrac{2a}{c} y + 2a) \, dy$

$= \dfrac{2}{c} \int_{0}^{c} (y - \dfrac{y^2}{c}) \, dy = \dfrac{2}{c} \left[\dfrac{y^2}{2} - \dfrac{y^3}{3c} \right]_{0}^{c} = \dfrac{c}{3}$

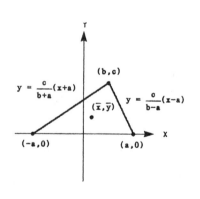

$(\bar{x}, \bar{y}) = (\dfrac{b}{3}, \dfrac{c}{3})$

In Exercise #54 of Section 1.4, we showed that $(\dfrac{b}{3}, \dfrac{c}{3})$ is the point of intersection of the medians.

36. $A = \int_0^1 [1 - (2x - x^2)]\,dx = \frac{1}{3}, \qquad \frac{1}{A} = 3$

$\bar{x} = 3\int_0^1 x[1 - (2x - x^2)]\,dx = 3\int_0^1 [x - 2x^2 + x^3]\,dx$

$\quad = 3\left[\frac{x^2}{2} - \frac{2}{3}x^3 + \frac{x^4}{4}\right]_0^1 = \frac{1}{4}$

$\bar{y} = 3\int_0^1 \frac{[1 + (2x - x^2)]}{2}[1 - (2x - x^2)]\,dx$

$\quad = \frac{3}{2}\int_0^1 [1 - (2x - x^2)^2]\,dx = \frac{3}{2}\int_0^1 [1 - 4x^2 + 4x^3 - x^4]\,dx$

$\quad = \frac{3}{2}\left[x - \frac{4}{3}x^3 + x^4 - \frac{x^5}{5}\right]_0^1 = \frac{7}{10}$

$(\bar{x}, \bar{y}) = (\frac{1}{4}, \frac{7}{10})$

37. $V = 2\pi r A = 2\pi(5)(16\pi) = 160\pi^2 \approx 1579.1367$

38. $V = 2\pi r A = 2\pi(3)(4\pi) = 24\pi^2 \approx 236.8705$

39. $A = \frac{1}{2}(4)(4) = 8$

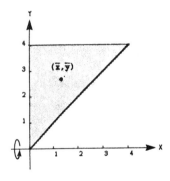

$\bar{y} = (\frac{1}{8})\frac{1}{2}\int_0^4 (4 + x)(4 - x)\,dx$

$\quad = \frac{1}{16}\left[16x - \frac{x^3}{3}\right]_0^4 = \frac{8}{3}$

$r = \bar{y} = \frac{8}{3}$

$V = 2\pi r A = 2\pi(\frac{8}{3})(8) = \frac{128\pi}{3} \approx 134.0413$

40. $A = \int_1^5 \sqrt{x - 1}\,dx = \frac{2}{3}(x - 1)^{3/2}\Big]_1^5 = \frac{16}{3}$

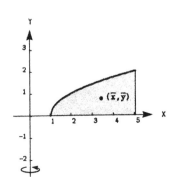

$\bar{x} = \frac{3}{16}\int_1^5 x\sqrt{x - 1}\,dx$

$\quad = \frac{3}{16}\left[\frac{2}{5}(x - 1)^{5/2} + \frac{2}{3}(x - 1)^{3/2}\right]_1^5 = \frac{17}{5}$

$r = \bar{x} = \frac{17}{5}$

$V = 2\pi r A = 2\pi(\frac{17}{5})(\frac{16}{3}) = \frac{544\pi}{15} \approx 113.9351$

7.7

Arc length and surfaces of revolution

1. $(0, 0)$, $(5, 12)$

   (a) $d = \sqrt{(5-0)^2 + (12-0)^2} = 13$

   (b) $y = \dfrac{12}{5}x$, $\qquad y' = \dfrac{12}{5}$

   $$s = \int_0^5 \sqrt{1 + (\tfrac{12}{5})^2}\ dx = \dfrac{13}{5}x \bigg]_0^5 = 13$$

2. $(1, 2)$, $\qquad (7, 10)$

   (a) $d = \sqrt{(7-1)^2 + (10-2)^2} = 10$

   (b) $y = \dfrac{4}{3}x + \dfrac{2}{3}$, $\qquad y' = \dfrac{4}{3}$

   $$s = \int_1^7 \sqrt{1 + (\tfrac{4}{3})^2}\ dx = \dfrac{5}{3}x \bigg]_1^7 = 10$$

3. $y = \dfrac{2}{3}x^{3/2} + 1$, $\qquad y' = x^{1/2}$, $\qquad [0, 1]$

   $$s = \int_0^1 \sqrt{1 + x}\ dx = \dfrac{2}{3}(1 + x)^{3/2} \bigg]_0^1 = \dfrac{2}{3}(\sqrt{8} - 1)$$

4. $y = x^{3/2} - 1$, $\qquad y' = \dfrac{3}{2}x^{1/2}$, $\qquad [0, 4]$

   $$s = \int_0^4 \sqrt{1 + \dfrac{9}{4}x}\ dx = \dfrac{4}{9}\int_0^4 (1 + \dfrac{9}{4}x)^{1/2}(\dfrac{9}{4})\ dx = \dfrac{8}{27}(1 + \dfrac{9}{4}x)^{3/2} \bigg]_0^4$$

   $$= \dfrac{8}{27}(10^{3/2} - 1)$$

5. $y = \dfrac{x^4}{8} + \dfrac{1}{4x^2}$, $\quad y' = \dfrac{1}{2}x^3 - \dfrac{1}{2x^3}$, $\quad 1 + y'^2 = (\dfrac{1}{2}x^3 + \dfrac{1}{2x^3})^2$, $\quad [1, 2]$

   $$s = \int_a^b \sqrt{1 + y'^2}\ dx = \int_1^2 (\dfrac{1}{2}x^3 + \dfrac{1}{2x^3})\ dx = \bigg[\dfrac{1}{8}x^4 - \dfrac{1}{4x^2}\bigg]_1^2 = \dfrac{33}{16}$$

6. $y = \dfrac{3}{2}x^{2/3}$, $\qquad y' = x^{-1/3}$, $\qquad [1, 8]$

   $$s = \int_1^8 \sqrt{1 + \dfrac{1}{x^{2/3}}}\ dx = \int_1^8 \dfrac{\sqrt{x^{2/3} + 1}}{x^{1/3}}\ dx = \dfrac{3}{2}\int_1^8 (x^{2/3} + 1)^{1/2}(\dfrac{2}{3x^{1/3}})\ dx$$

   $$= \bigg[(x^{2/3} + 1)^{3/2}\bigg]_1^8 = 5^{3/2} - 2^{3/2} = 5\sqrt{5} - 2\sqrt{2}$$

7. $y = \dfrac{x^5}{10} + \dfrac{1}{6x^3}$, $\quad y' = \dfrac{1}{2}x^4 - \dfrac{1}{2x^4}$, $\quad 1 + y'^2 = (\dfrac{1}{2}x^4 + \dfrac{1}{2x^4})^2$, $\quad [1, 2]$

$$s = \int_a^b \sqrt{1 + y'^2}\, dx = \int_1^2 \sqrt{(\dfrac{1}{2}x^4 + \dfrac{1}{2x^4})^2}\, dx = \int_1^2 (\dfrac{1}{2}x^4 + \dfrac{1}{2x^4})\, dx$$

$$= \left[\dfrac{1}{10}x^5 - \dfrac{1}{6x^3}\right]_1^2 = \dfrac{779}{240}$$

8. $y = \dfrac{1}{2}(e^x + e^{-x})$, $\quad y' = \dfrac{1}{2}(e^x - e^{-x})$, $\quad 1 + (y')^2 = \left[\dfrac{1}{2}(e^x + e^{-x})\right]^2$, $\quad [0, 2]$

$$s = \int_0^2 \sqrt{\left[\dfrac{1}{2}(e^x + e^{-x})\right]^2}\, dx = \dfrac{1}{2}\int_0^2 (e^x - e^{-x})\, dx = \dfrac{1}{2}\left[e^x - e^{-x}\right]_0^2$$

$$= \dfrac{1}{2}(e^2 - \dfrac{1}{e^2})$$

9. $y = 1/x$, $\quad y' = -1/x^2$, $\quad [1, 3]$

$$s = \int_1^3 \sqrt{1 + \dfrac{1}{x^4}}\, dx = \int_1^3 \dfrac{\sqrt{x^4 + 1}}{x^2}\, dx$$

10. $y = x^2$, $\quad y' = 2x$, $\quad [0, 1]$

$$s = \int_0^1 \sqrt{1 + 4x^2}\, dx$$

11. $y = x^2 + x - 2$, $\quad y' = 2x + 1$, $\quad [-2, 1]$

$$s = \int_{-2}^1 \sqrt{1 + (2x + 1)^2}\, dx = \int_{-2}^1 \sqrt{2 + 4x + 4x^2}\, dx$$

12. $y = \dfrac{1}{(x + 1)}$, $\quad y' = -\dfrac{1}{(x + 1)^2}$, $\quad [0, 1]$

$$s = \int_0^1 \sqrt{1 + \dfrac{1}{(x + 1)^4}}\, dx = \int_0^1 \dfrac{\sqrt{(x + 1)^4 + 1}}{(x + 1)^2}\, dx$$

13. $y = \sin x$, $\quad y' = \cos x$, $\quad [0, \pi]$

$$s = \int_0^\pi \sqrt{1 + \cos^2 x}\, dx$$

14. $y = \ln x$, $\quad y' = \dfrac{1}{x}$, $\quad [1, 5]$

$$s = \int_1^5 \sqrt{1 + \dfrac{1}{x^2}}\, dx = \int_1^5 \dfrac{\sqrt{x^2 + 1}}{x}\, dx$$

15. $x = 4 - y^2$, $x' = -2y$, $[0, 2]$

$$s = \int_0^2 \sqrt{1 + 4y^2}\ dy$$

16. $x = \cos y$, $x' = -\sin y$, $[-\frac{\pi}{2}, \frac{\pi}{2}]$

$$s = \int_{-\pi/2}^{\pi/2} \sqrt{1 + \sin^2 y}\ dy$$

17. $x = e^{-y}$, $x' = -e^{-y}$, $[0, 2]$

$$s = \int_0^2 \sqrt{1 + e^{-2y}}\ dy = \int_0^2 \frac{\sqrt{e^{2y} + 1}}{e^y}\ dy$$

18. $x = \sqrt{a^2 - y^2}$, $x' = \dfrac{-y}{\sqrt{a^2 - y^2}}$, $[0, \frac{a}{2}]$

$$s = \int_0^{a/2} \sqrt{1 + \frac{y^2}{a^2 - y^2}}\ dy = \int_0^{a/2} \frac{a}{\sqrt{a^2 - y^2}}\ dy$$

19. $y = \dfrac{1}{3}[x^{3/2} - 3x^{1/2} + 2]$

When $x = 0$, $y = 2/3$, thus the fleeing object has traveled 2/3 units when it is caught.

$$y' = \frac{1}{3}\left[\frac{3}{2}x^{1/2} - \frac{3}{2}x^{-1/2}\right] = (\frac{1}{2})\frac{x - 1}{x^{1/2}}$$

$$1 + (y')^2 = 1 + \frac{(x - 1)^2}{4x} = \frac{(x + 1)^2}{4x}$$

$$s = \int_0^1 \frac{x + 1}{2x^{1/2}}\ dx = \frac{1}{2}\int_0^1 (x^{1/2} + x^{-1/2})\ dx$$

$$= \frac{1}{2}\left[\frac{2}{3}x^{3/2} + 2x^{1/2}\right]_0^1 = \frac{4}{3} = 2(\frac{2}{3})$$

The pursuer has traveled twice the distance that the fleeing object has when it is caught.

20. $y = 31 - 10(e^{x/20} + e^{-x/20})$, $y' = -\dfrac{1}{2}(e^{x/20} - e^{x/20})$

$$1 + (y')^2 = 1 + \frac{1}{4}(e^{x/10} - 2 + e^{-x/10}) = \left[\frac{1}{2}(e^{x/20} + e^{-x/20})\right]^2$$

$$s = \int_{-20}^{20} \sqrt{\left[\frac{1}{2}(e^{x/20} + e^{-x/20})\right]^2}\ dx = \frac{1}{2}\int_{-20}^{20} (e^{x/20} + e^{-x/20})\ dx$$

$$= 10(e^{x/20} - e^{-x/20})\Big]_{-20}^{20} = 20(e - \frac{1}{e}) \approx 47\ \text{ft}$$

Thus, there are $100(47) = 4700$ square feet of roofing on the barn.

21. $y = \sqrt{9 - x^2}$, $y' = \dfrac{-x}{\sqrt{9 - x^2}}$, $1 + (y')^2 = \dfrac{9}{9 - x^2}$

$$s = \int_0^2 \sqrt{\dfrac{9}{9 - x^2}}\, dx = \int_0^2 \dfrac{3}{\sqrt{9 - x^2}}\, dx = 3 \arcsin \dfrac{x}{3} \Big]_0^2$$

$$= 3[\arcsin \tfrac{2}{3} - \arcsin 0] = 3 \arcsin \tfrac{2}{3} \approx 2.189$$

22. $y = \sqrt{25 - x^2}$, $y' = \dfrac{-x}{\sqrt{25 - x^2}}$, $1 + (y')^2 = \dfrac{25}{25 - x^2}$

$$s = \int_{-3}^4 \sqrt{\dfrac{25}{25 - x^2}}\, dx = \int_{-3}^4 \dfrac{5}{\sqrt{25 - x^2}}\, dx = 5 \arcsin \dfrac{x}{5} \Big]_{-3}^4$$

$$= 5 \left[\arcsin \tfrac{4}{5} - \arcsin \left(-\tfrac{3}{5}\right) \right] \approx 7.854$$

$$\tfrac{1}{4}[2\pi(5)] \approx 7.854 = s$$

23. $y = \dfrac{x^3}{3}$, $y' = x^2$, $[0, 3]$

$$S = 2\pi \int_0^3 \dfrac{x^3}{3} \sqrt{1 + x^4}\, dx = \dfrac{\pi}{6} \int_0^3 (1 + x^4)^{1/2}(4x^3)\, dx$$

$$= \dfrac{\pi}{9}(1 + x^4)^{3/2} \Big]_0^3 = \dfrac{\pi}{9}(82\sqrt{82} - 1)$$

24. $y = \sqrt{x}$, $y' = \dfrac{1}{2\sqrt{x}}$, $[1, 4]$

$$S = 2\pi \int_1^4 \sqrt{x} \sqrt{1 + \dfrac{1}{4x}}\, dx = 2\pi \int_1^4 \dfrac{\sqrt{x}}{2\sqrt{x}} \sqrt{4x + 1}\, dx = \pi \int_1^4 \sqrt{4x + 1}\, dx$$

$$= \dfrac{\pi}{4} \int_1^4 (4x + 1)^{1/2}(4)\, dx = \dfrac{\pi}{6}(4x + 1)^{3/2} \Big]_1^4 = \dfrac{\pi}{6}(17\sqrt{17} - 5\sqrt{5})$$

25. $y = \dfrac{x^3}{6} + \dfrac{1}{2x}$, $y' = \dfrac{x^2}{2} - \dfrac{1}{2x^2}$, $1 + y'^2 = \left(\dfrac{x^2}{2} + \dfrac{1}{2x^2}\right)^2$, $[1, 2]$

$$S = 2\pi \int_1^2 \left(\dfrac{x^3}{6} + \dfrac{1}{2x}\right)\left(\dfrac{x^2}{2} + \dfrac{1}{2x^2}\right) dx = 2\pi \int_1^2 \left(\dfrac{x^5}{12} + \dfrac{x}{3} + \dfrac{1}{4x^3}\right) dx$$

$$= 2\pi \left[\dfrac{x^6}{72} + \dfrac{x^2}{6} - \dfrac{1}{8x^2}\right]_1^2 = \dfrac{47\pi}{16}$$

26. $y = \dfrac{x}{2}$, $\qquad y' = \dfrac{1}{2}$, $\qquad 1 + (y')^2 = \dfrac{5}{4}$, $\qquad [0, 6]$

$$S = 2\pi \int_0^6 \dfrac{x}{2} \sqrt{\dfrac{5}{4}} \, dx = \dfrac{2\pi\sqrt{5}}{8} x^2 \bigg]_0^6 = 9\sqrt{5}\,\pi$$

27. $y = \sqrt[3]{x} + 2$, $\qquad y' = \dfrac{1}{3x^{2/3}}$, $\qquad [1, 8]$

$$S = 2\pi \int_1^8 x \sqrt{1 + \dfrac{1}{9x^{4/3}}} \, dx = \dfrac{2\pi}{3} \int_1^8 x^{1/3} \sqrt{9x^{4/3} + 1} \, dx$$

$$= \dfrac{\pi}{18} \int_1^8 (9x^{4/3} + 1)^{1/2}(12x^{1/3}) \, dx = \dfrac{\pi}{27}(9x^{4/3} + 1)^{3/2} \bigg]_1^8$$

$$= \dfrac{\pi}{27}(145\sqrt{145} - 10\sqrt{10})$$

28. $y = 4 - x^2$, $\qquad y' = -2x$, $\qquad [0, 2]$

$$S = 2\pi \int_0^2 x\sqrt{1 + 4x^2} \, dx = \dfrac{\pi}{4} \int_0^2 (1 + 4x^2)^{1/2}(8x) \, dx = \dfrac{\pi}{6}(1 + 4x^2)^{3/2} \bigg]_0^2$$

$$= \dfrac{\pi}{6}(17\sqrt{17} - 1)$$

29. $y = \sin x$, $\qquad y' = \cos x$, $\qquad [0, \pi]$

$$S = 2\pi \int_0^\pi \sin x \sqrt{1 + \cos^2 x} \, dx$$

30. $y = \ln x$, $\qquad y' = \dfrac{1}{x}$, $\qquad 1 + (y')^2 = \dfrac{x^2 + 1}{x^2}$, $\qquad [1, e]$

$$S = \int_1^e x \sqrt{\dfrac{x^2 + 1}{x^2}} \, dx = \int_1^e \sqrt{x^2 + 1} \, dx$$

31. $y = \dfrac{hx}{r}$, $\qquad y' = \dfrac{h}{r}$, $\qquad 1 + (y')^2 = \dfrac{r^2 + h^2}{r^2}$

$$S = 2\pi \int_0^r x \sqrt{\dfrac{r^2 + h^2}{r^2}} \, dx = \dfrac{2\pi\sqrt{r^2 + h^2}}{r} \left[\dfrac{x^2}{2}\right]_0^r = \pi r \sqrt{r^2 + h^2}$$

32. $y = \sqrt{r^2 - x^2}$, $\qquad y' = \dfrac{-x}{\sqrt{r^2 - x^2}}$, $\qquad 1 + (y')^2 = \dfrac{r^2}{r^2 - x^2}$

$$S = 2\pi \int_{-r}^r \sqrt{r^2 - x^2} \sqrt{\dfrac{r^2}{r^2 - x^2}} \, dx = 2\pi \int_{-r}^r r \, dx = 2\pi r x \bigg]_{-r}^r = 4\pi r^2$$

33. $y = \sqrt{9 - x^2}$, $\qquad y' = \dfrac{-x}{\sqrt{9 - x^2}}$, $\qquad \sqrt{1 + y'^2} = \dfrac{3}{\sqrt{9 - x^2}}$

$S = 2\pi \displaystyle\int_0^2 \dfrac{3x}{\sqrt{9 - x^2}}\ dx = -3\pi \displaystyle\int_0^2 \dfrac{-2x}{\sqrt{9 - x^2}}\ dx = -6\pi\sqrt{9 - x^2}\ \Big]_0^2 = 6\pi(3 - \sqrt{5})$

See figure in Exercise 34.

34. From Exercise 33 we have:

$S = 2\pi \displaystyle\int_0^a \dfrac{rx}{\sqrt{r^2 - x^2}}\ dx$

$\qquad = -r\pi \displaystyle\int_0^a \dfrac{-2x\ dx}{\sqrt{r^2 - x^2}} = -2r\pi\sqrt{r^2 - x^2}\ \Big]_0^a$

$\qquad = 2r^2\pi - 2r\pi\sqrt{r^2 - a^2}$

$\qquad = 2r\pi(r - \sqrt{r^2 - a^2}) = 2\pi rh$

(where h is height of zone)

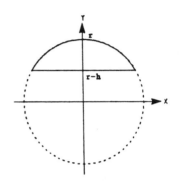

35. $y = \dfrac{1}{3}x^{1/2} - x^{3/2}$, $\qquad y' = \dfrac{1}{6}x^{-1/2} - \dfrac{3}{2}x^{1/2} = \dfrac{1}{6}(x^{-1/2} - 9x^{1/2})$

$1 + (y')^2 = 1 + \dfrac{1}{36}(x^{-1} - 18 + 81x) = \dfrac{1}{36}(x^{-1/2} + 9x^{1/2})^2$

$S = 2\pi \displaystyle\int_0^{1/3} (\dfrac{1}{3}x^{1/2} - x^{3/2})\sqrt{\dfrac{1}{36}(x^{-1/2} + 9x^{1/2})^2}\ dx$

$\qquad = \dfrac{2\pi}{6} \displaystyle\int_0^{1/3} (\dfrac{1}{3}x^{1/2} - x^{3/2})(x^{-1/2} + 9x^{1/2})\ dx$

$\qquad = \dfrac{\pi}{3} \displaystyle\int_0^{1/3} (\dfrac{1}{3} + 2x - 9x^2)\ dx = \dfrac{\pi}{3}\Big[\dfrac{1}{3}x + x^2 - 3x^3\Big]_0^{1/3} = \dfrac{\pi}{27}\ ft^2$

Amount of glass needed: $V = \dfrac{\pi}{27}(\dfrac{0.015}{12}) \approx 0.00015\ ft^3$

Review Exercises for Chapter 7

1. $A = \int_{1}^{5} \frac{1}{x^2} \, dx$

   $= -\frac{1}{x} \Big]_{1}^{5} = \frac{4}{5}$

   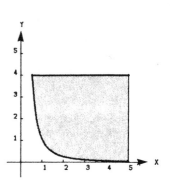

2. $A = \int_{1/2}^{5} \left(4 - \frac{1}{x^2}\right) dx$

   $= \left[4x + \frac{1}{x}\right]_{1/2}^{5} = \frac{81}{5}$

   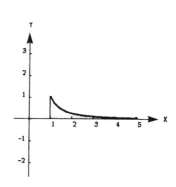

3. $A = \int_{-1}^{1} \frac{1}{x^2 + 1} \, dx$

   $= \arctan x \Big]_{-1}^{1} = \frac{\pi}{4} - \left(-\frac{\pi}{4}\right) = \frac{\pi}{2}$

   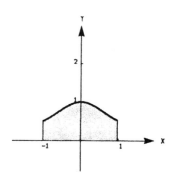

4. $y = 1 - \frac{x}{2} \implies x = 2 - 2y$

   $y = x - 2 \implies x = y + 2, \quad y = 1$

   $A = \int_{0}^{1} [(y + 2) - (2 - 2y)] \, dy$

   $= \int_{0}^{1} 3y \, dy = \frac{3}{2}y^2 \Big]_{0}^{1} = \frac{3}{2}$

   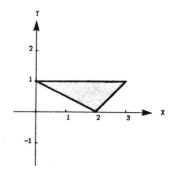

5. $A = \displaystyle\int_0^2 [0 - (y^2 - 2y)]\ dy$

$= \displaystyle\int_0^2 (2y - y^2)\ dy$

$= \left[y^2 - \dfrac{1}{3}y^3 \right]_0^2 = \dfrac{4}{3}$

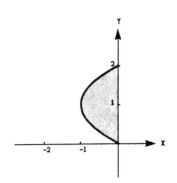

6. $A = \displaystyle\int_0^1 [(y^2 - 2y) - (-1)]\ dy$

$= \displaystyle\int_0^1 (y^2 - 2y + 1)\ dy = \int_0^1 (y - 1)^2\ dy$

$= \dfrac{(y - 1)^3}{3} \Bigg]_0^1 = \dfrac{1}{3}$

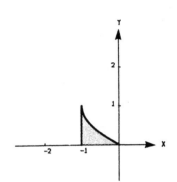

7. $A = 2 \displaystyle\int_0^1 (x - x^3)\ dx$

$= 2 \left[\dfrac{1}{2}x^2 - \dfrac{1}{4}x^4 \right]_0^1 = \dfrac{1}{2}$

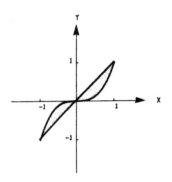

8. $A = \displaystyle\int_{-1}^2 [(y + 3) - (y^2 + 1)]\ dy$

$= \displaystyle\int_{-1}^2 (2 + y - y^2)\ dy$

$= \left[2y + \dfrac{1}{2}y^2 - \dfrac{1}{3}y^3 \right]_{-1}^2 = \dfrac{9}{2}$

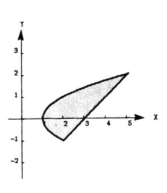

9. $A = \displaystyle\int_{0}^{8} [(3 + 8x - x^2) - (x^2 - 8x + 3)]\, dx$

$\quad = \displaystyle\int_{0}^{8} (16x - 2x^2)\, dx$

$\quad = \left[8x^2 - \dfrac{2}{3}x^3 \right]_{0}^{8} = \dfrac{512}{3}$

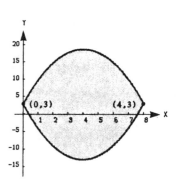

10. Point of intersection is given by:

$x^3 - x^2 + 4x - 3 = 0 \implies x \approx 0.783$

$A \approx \displaystyle\int_{0}^{0.783} (3 - 4x + x^2 - x^3)\, dx$

$\quad = \left[3x - 2x^2 + \dfrac{1}{3}x^3 - \dfrac{1}{4}x^4 \right]_{0}^{0.783}$

$\quad \approx 1.189$

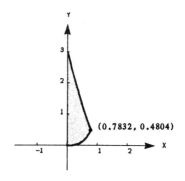

11. $x = y^2 + 1$

$A = \displaystyle\int_{0}^{2} (y^2 + 1)\, dy$

$\quad = \left[\dfrac{1}{3}y^3 + y \right]_{0}^{2} = \dfrac{14}{3}$

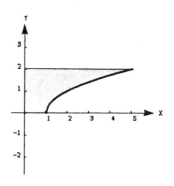

12. $A = \displaystyle\int_{1}^{5} \left[\sqrt{x-1} - \dfrac{x-1}{2} \right] dx$

$\quad = \left[\dfrac{2}{3}(x-1)^{3/2} - \dfrac{1}{4}(x-1)^2 \right]_{1}^{5}$

$\quad = \dfrac{4}{3}$

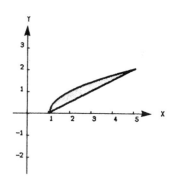

13. $y = (1 - \sqrt{x})^2$

$A = \displaystyle\int_0^1 (1 - \sqrt{x})^2 \, dx$

$= \displaystyle\int_0^1 (1 - 2x^{1/2} + x) \, dx$

$= \left[x - \dfrac{4}{3}x^{3/2} + \dfrac{1}{2}x^2 \right]_0^1 = \dfrac{1}{6}$

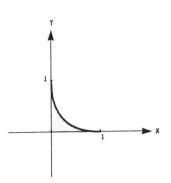

14. $A = 2 \displaystyle\int_0^2 [2x^2 - (x^4 - 2x^2)] \, dx$

$= 2 \displaystyle\int_0^2 (4x^2 - x^4) \, dx$

$= 2 \left[\dfrac{4}{3}x^3 - \dfrac{1}{5}x^5 \right]_0^2 = \dfrac{128}{15}$

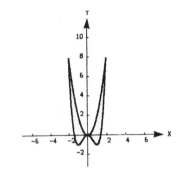

15. $A = \displaystyle\int_0^2 (e^2 - e^x) \, dx$

$= \left[xe^2 - e^x \right]_0^2$

$= e^2 + 1$

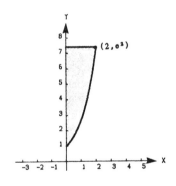

16. $A = 2 \displaystyle\int_{\pi/6}^{\pi/2} (2 - \csc x) \, dx$

$= 2 \left[2x - \ln|\csc x - \cot x| \right]_{\pi/6}^{\pi/2}$

$= 2 \left[(\pi - 0) - \left[\dfrac{\pi}{3} - \ln(2 - \sqrt{3}) \right] \right]$

$= 2 \left(\dfrac{2\pi}{3} + \ln(2 - \sqrt{3}) \right) \approx 1.555$

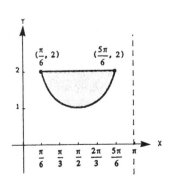

17. $A = \displaystyle\int_{\pi/4}^{5\pi/4} (\sin x - \cos x)\, dx$

$= (-\cos x - \sin x)\Big]_{\pi/4}^{5\pi/4}$

$= (\dfrac{1}{\sqrt{2}} + \dfrac{1}{\sqrt{2}}) - (-\dfrac{1}{\sqrt{2}} - \dfrac{1}{\sqrt{2}})$

$= \dfrac{4}{\sqrt{2}} = 2\sqrt{2}$

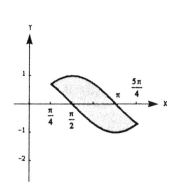

18. $A = \displaystyle\int_{\pi/3}^{5\pi/3} (\dfrac{1}{2} - \cos y)\, dy + \int_{5\pi/3}^{7\pi/3} (\cos y - \dfrac{1}{2})\, dy$

$= \dfrac{y}{2} - \sin y \Big]_{\pi/3}^{5\pi/3} + \sin y - \dfrac{y}{2}\Big]_{5\pi/3}^{7\pi/3}$

$= \dfrac{\pi}{3} + 2\sqrt{3}$

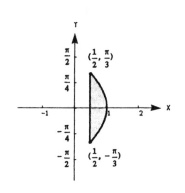

19.(a) <u>Disc</u>

$V = \pi \displaystyle\int_0^4 x^2\, dx$

$= \pi \dfrac{x^3}{3}\Big]_0^4 = \dfrac{64\pi}{3}$

(b) <u>Shell</u>

$V = 2\pi \displaystyle\int_0^4 x^2\, dx$

$= \dfrac{2\pi}{3} x^3 \Big]_0^4 = \dfrac{128\pi}{3}$

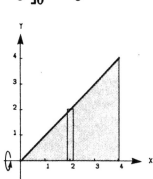

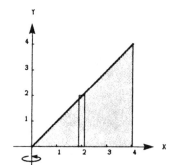

19.(c) <u>Shell</u>

$$V = 2\pi \int_0^4 (4 - x)x \, dx$$

$$= 2\pi \int_0^4 (4x - x^2) \, dx$$

$$= 2\pi \left[2x^2 - \frac{x^3}{3} \right]_0^4 = \frac{64\pi}{3}$$

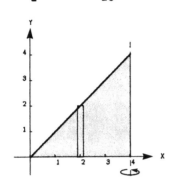

(d) <u>Shell</u>

$$V = 2\pi \int_0^4 (6 - x)x \, dx$$

$$= 2\pi \int_0^4 (6x - x^2) \, dx$$

$$= 2\pi \left[3x^2 - \frac{1}{3} x^3 \right]_0^4 = \frac{160\pi}{3}$$

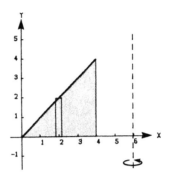

20.(a) <u>Shell</u>

$$V = 2\pi \int_0^2 y^3 \, dy$$

$$= \frac{\pi}{2} y^4 \Big]_0^2 = 8\pi$$

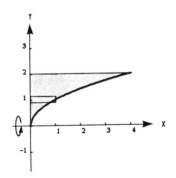

(b) <u>Shell</u>

$$V = 2\pi \int_0^2 (2 - y)y^2 \, dy$$

$$= 2\pi \int_0^2 (2y^2 - y^3) \, dy$$

$$= 2\pi \left[\frac{2}{3} y^3 - \frac{1}{4} y^4 \right]_0^2 = \frac{8\pi}{3}$$

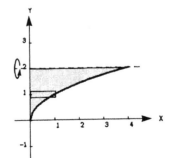

20.(c) <u>Disc</u>

$$V = \pi \int_0^2 y^4 \, dy$$

$$= \frac{\pi}{5} y^5 \bigg]_0^2$$

$$= \frac{32\pi}{5}$$

(d) <u>Disc</u>

$$V = \pi \int_0^2 [(y^2 + 1)^2 - 1^2] \, dy$$

$$= \pi \int_0^2 (y^4 + 2y^2) \, dy$$

$$= \pi \left[\frac{1}{5} y^5 + \frac{2}{3} y^3 \right]_0^2 = \frac{176\pi}{15}$$

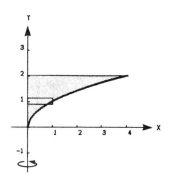

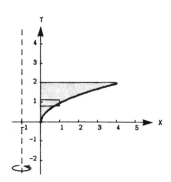

21. **From Exercise 22 letting a = 4 and b = 3 we have**

(a) $V = \dfrac{4}{3} \pi (4^2)(3) = 64\pi$

(b) $V = \dfrac{4}{3} \pi (4)(3)^2 = 48\pi$

22.(a) <u>Shell</u>

$$V = 4\pi \int_0^a (x) \frac{b}{a} \sqrt{a^2 - x^2} \, dx$$

$$= \frac{-2\pi b}{a} \int_0^a (a^2 - x^2)^{1/2}(-2x) \, dx$$

$$= \frac{-4\pi b}{3a} (a^2 - x^2)^{3/2} \bigg]_0^a = \frac{4}{3} \pi a^2 b$$

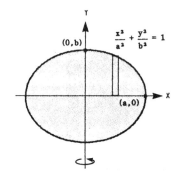

(b) <u>Disc</u>

$$V = 2\pi \int_0^a \frac{b^2}{a^2}(a^2 - x^2) \, dx$$

$$= \frac{2\pi b^2}{a^2} \left[a^2 x - \frac{1}{3} x^3 \right]_0^a$$

$$= \frac{4}{3} \pi a b^2$$

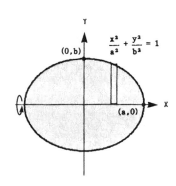

23. $\underline{\text{Shell}}$

$$V = 2\pi \int_0^1 \frac{x}{x^4 + 1} \, dx$$

$$= \pi \int_0^1 \frac{(2x)}{(x^2)^2 + 1} \, dx$$

$$= \pi \arctan (x^2) \Big]_0^1$$

$$= \pi \left[\frac{\pi}{4} - 0 \right] = \frac{\pi^2}{4}$$

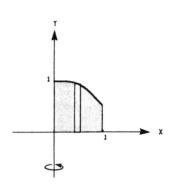

24. $\underline{\text{Disc}}$

$$V = 2\pi \int_0^1 \left[\frac{1}{\sqrt{1 + x^2}} \right]^2 \, dx$$

$$= 2\pi \arctan x \Big]_0^1$$

$$= 2\pi (\frac{\pi}{4} - 0) = \frac{\pi^2}{2}$$

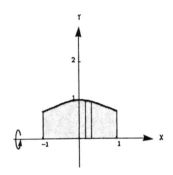

25.(a) Disc

$$V = \pi \int_1^5 (-x^2 + 6x - 5)^2 \, dx$$

$$= \int_1^5 (x^4 - 12x^3 + 46x^2 - 60x + 25) \, dx$$

$$= \pi \left[\frac{x^5}{5} - 3x^4 + \frac{46}{3} x^3 - 30x^2 + 25x \right]_1^5$$

$$= \frac{512\pi}{15}$$

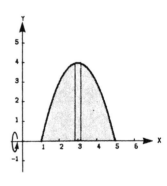

(b) $\underline{\text{Shell}}$

$$V = 2\pi \int_1^5 x(-x^2 + 6x - 5) \, dx$$

$$= 2\pi \int_1^5 (-x^3 + 6x^2 - 5x) \, dx$$

$$= 2\pi \left[-\frac{x^4}{4} + 2x^3 - \frac{5}{2} x^2 \right]_1^5 = 64\pi$$

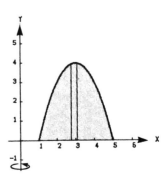

26. <u>Disc</u>

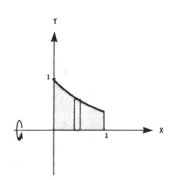

$$V = \pi \int_0^1 (e^{-x})^2 \, dx = \pi \int_0^1 e^{-2x} \, dx$$

$$= -\frac{\pi}{2} e^{-2x} \Big]_0^1 = \left(\frac{-\pi}{2e^2} + \frac{\pi}{2}\right)$$

$$= \frac{\pi}{2}\left(1 - \frac{1}{e^2}\right)$$

27. $F = kx, \qquad 4 = k(1), \qquad F = 4x$

$$W = \int_0^5 4x \, dx = 2x^2 \Big]_0^5 = 50 \text{ in} \cdot \text{lb}$$

28. $9k = 50, \qquad k = 50/9, \qquad F = (50/9)x$

$$W = \int_0^9 \frac{50}{9} x \, dx = \frac{25}{9} x^2 \Big]_0^9 = 225 \text{ in} \cdot \text{lb}$$

29. Volume of disc: $\pi(\frac{1}{3})^2 \Delta y$ $\qquad W = \frac{62.4\pi}{9} \int_0^{150} (175 - y) \, dy$

Weight of disc: $62.4\pi(\frac{1}{3})^2 \Delta y$ $\qquad = \frac{62.4\pi}{9} \left[175y - \frac{y^2}{2}\right]_0^{150}$

Distance: $175 - y$ $\qquad\qquad\qquad = 104{,}000\pi \text{ ft} \cdot \text{lb} = 52\pi \text{ ft} \cdot \text{ton}$

30. We know that $\dfrac{dV}{dt} = \dfrac{4 \text{ gal/min} - 12 \text{ gal/min}}{7.481 \text{ gal/ft}^3} = -\dfrac{8}{7.481} \text{ ft}^3/\text{min}$

$V = \pi r^2 h = \pi(\frac{1}{9})h, \quad \dfrac{dV}{dt} = \dfrac{\pi}{9}\left(\dfrac{dh}{dt}\right), \quad \dfrac{dh}{dt} = \dfrac{9}{\pi}\left(\dfrac{dV}{dt}\right) = \dfrac{9}{\pi}\left(-\dfrac{8}{7.481}\right) \approx -3.064 \text{ ft/min}$

Depth of water: $-3.064t + 150$, time to drain well: $t = \dfrac{150}{3.064} \approx 49$ min

$(49)(12) = 588$ gallons pumped

Volume of water pumped in Exercise 29 was 391.7 gallons.

$\dfrac{391.7}{52\pi} = \dfrac{588}{x\pi}, \qquad x = \dfrac{588(52)}{391.7} \approx 78, \qquad$ Work $\approx 78\pi \text{ ft} \cdot \text{ton}$

31. Weight of section of chain: $5\Delta x$,   Distance moved: $10 - x$

$$W = 5 \int_0^{10} (10 - x) \, dx = -\frac{5}{2}(10 - x)^2 \Big]_0^{10} = 250 \text{ ft} \cdot \text{lb}$$

32.(a) Weight of section of cable: $4\Delta x$,   Distance: $200 - x$

$$W = 4 \int_0^{200} (200 - x) \, dx = -2(200 - x)^2 \Big]_0^{200} = 80,000 \text{ ft} \cdot \text{lb} = 40 \text{ ft} \cdot \text{ton}$$

(b) Work to move 300 pounds 200 feet vertically:

$$200(300) = 60,000 \text{ ft} \cdot \text{lb} = 30 \text{ ft} \cdot \text{ton}$$

Total work = work for drawing up the cable + work of lifting the load

= 40 ft·ton + 30 ft·ton = 70 ft·ton

33. <u>Wall at shallow end:</u>

$$F = 62.4 \int_0^5 y(20) \, dy = (1248)\frac{y^2}{2} \Big]_0^5 = 15,600 \text{ lb}$$

<u>Wall at deep end:</u>

$$F = 62.4 \int_0^{10} y(20) \, dy = (624)y^2 \Big]_0^{10} = 62,400 \text{ lb}$$

<u>Side wall:</u>

$$F_1 = 62.4 \int_0^5 y(40) \, dy = (1248)y^2 \Big]_0^5 = 31,200 \text{ lb}$$

$$F_2 = 62.4 \int_0^5 (10 - y)8y \, dy$$

$$= 62.4 \int_0^5 (80y - 8y^2) \, dy$$

$$= 62.4 \left[40y^2 - \frac{8}{3}y^3 \right]_0^5 = 41,600 \text{ lb}$$

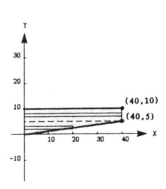

$$F = F_1 + F_2 = 72,800 \text{ lb}$$

34. Let D = surface of liquid, ρ = weight per cubic volume

$$F = \rho \int_c^d (D - y)[f(y) - g(y)] \, dy$$

$$= \rho \left[\int_c^d D[f(y) - g(y)] \, dy - \int_c^d y[f(y) - g(y)] \, dy \right]$$

$$= \rho \left[\int_c^d [f(y) - g(y)] \, dy \right] \left[D - \frac{\displaystyle\int_c^d y[f(y) - g(y)] \, dy}{\displaystyle\int_c^d [f(y) - g(y)] \, dy} \right]$$

$$= \rho(\text{Area})(D - y) = \rho(\text{Area})(\text{depth of centroid})$$

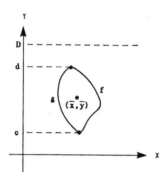

35. $F = 62.4(16\pi)5 = 4992\pi$ lb   36. Raise water level five more feet

37. $\displaystyle A = \int_0^a (\sqrt{a} - \sqrt{x})^2 \, dx = \int_0^a (a - 2\sqrt{a}\, x^{1/2} + x) \, dx$

$$= \left[ax - \frac{4}{3}\sqrt{a}\, x^{3/2} + \frac{1}{2}x^2 \right]_0^a = \frac{a^2}{6}, \qquad \frac{1}{A} = \frac{6}{a^2}$$

$$\overline{x} = \frac{6}{a^2} \int_0^a x(\sqrt{a} - \sqrt{x})^2 \, dx$$

$$= \frac{6}{a^2} \int_0^a (ax - 2\sqrt{a}\, x^{3/2} + x^2) \, dx$$

$$= \frac{6}{a^2} \left[\frac{ax^2}{2} - \frac{4}{5}\sqrt{a}\, x^{5/2} + \frac{1}{3}x^3 \right]_0^a = \frac{a}{5}$$

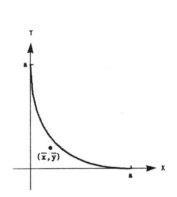

37. (continued)

$$\bar{y} = (\frac{6}{a^2})\frac{1}{2} \int_0^a (\sqrt{a} - \sqrt{x})^4 \, dx$$

$$= \frac{3}{a^2} \int_0^a (a^2 - 4a^{3/2}x^{1/2} + 6ax - 4a^{1/2}x^{3/2} + x^2) \, dx$$

$$= \frac{3}{a^2} \left[a^2x - \frac{8}{3}a^{3/2}x^{3/2} + 3ax^2 - \frac{8}{5}a^{1/2}x^{5/2} + \frac{1}{3}x^3 \right]_0^a = \frac{a}{5}$$

$$(\bar{x}, \bar{y}) = (\frac{a}{5}, \frac{a}{5})$$

38. $$A = \int_{-1}^3 [(2x + 3) - x^2] \, dx = \left[x^2 + 3x - \frac{1}{3}x^3 \right]_{-1}^3 = \frac{32}{3}, \qquad \frac{1}{A} = \frac{3}{32}$$

$$\bar{x} = \frac{3}{32} \int_{-1}^3 x(2x + 3 - x^2) \, dx = \frac{3}{32} \int_{-1}^3 (3x + 2x^2 - x^3) \, dx$$

$$= \frac{3}{32} \left[\frac{3}{2}x^2 + \frac{2}{3}x^3 - \frac{1}{4}x^4 \right]_{-1}^3 = 1$$

$$\bar{y} = (\frac{3}{32})\frac{1}{2} \int_{-1}^3 [(2x + 3)^2 - x^4] \, dx$$

$$= \frac{3}{64} \int_{-1}^3 (9 + 12x + 4x^2 - x^4) \, dx$$

$$= \frac{3}{64} \left[9x + 6x^2 + \frac{4}{3}x^3 - \frac{1}{5}x^5 \right]_{-1}^3 = \frac{17}{5}$$

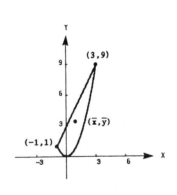

$$(\bar{x}, \bar{y}) = (1, \frac{17}{5})$$

39. By symmetry $\bar{x} = 0$

$$A = 2 \int_0^a (a^2 - x^2) \, dx = 2\left[a^2x - \frac{x^3}{3} \right]_0^a = \frac{4a^3}{3}, \qquad \frac{1}{A} = \frac{3}{4a^3}$$

$$\bar{y} = (\frac{3}{4a^3})\frac{1}{2} \int_{-a}^a (a^2 - x^2)^2 \, dx$$

$$= \frac{6}{8a^3} \int_0^a (a^4 - 2a^2x^2 + x^4) \, dx$$

$$= \frac{6}{8a^3} \left[a^4x - \frac{2a^2}{3}x^3 + \frac{1}{5}x^5 \right]_0^a$$

$$= \frac{6}{8a^3}(a^5 - \frac{2}{3}a^5 + \frac{1}{5}a^5) = \frac{2a^2}{5}$$

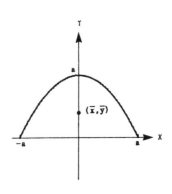

$$(\bar{x}, \bar{y}) = (0, \frac{2a^2}{5})$$

40. $A = \int_0^8 (x^{2/3} - \frac{1}{2}x)\ dx = \left[\frac{3}{5}x^{5/3} - \frac{1}{4}x^2\right]_0^8 = \frac{16}{5}, \qquad \frac{1}{A} = \frac{5}{16}$

$\overline{x} = \frac{5}{16}\int_0^8 x(x^{2/3} - \frac{1}{2}x)\ dx$

$= \frac{5}{16}\left[\frac{3}{8}x^{8/3} - \frac{1}{6}x^3\right]_0^8 = \frac{10}{3}$

$\overline{y} = (\frac{5}{16})\frac{1}{2}\int_0^8 (x^{4/3} - \frac{1}{4}x^2)\ dx$

$= \frac{1}{2}(\frac{5}{16})\left[\frac{3}{7}x^{7/3} - \frac{1}{12}x^3\right]_0^8 = \frac{40}{21}$

$(\overline{x},\ \overline{y}) = (\frac{10}{3},\ \frac{40}{21})$

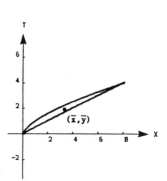

41. $y = \sqrt{4 - x^2}, \qquad y' = \frac{-x}{\sqrt{4 - x^2}}, \qquad 1 + (y')^2 = \frac{4}{4 - x^2}$

$s = \int_{-\sqrt{3}}^{\sqrt{3}} \sqrt{\frac{4}{4 - x^2}}\ dx = 2\int_0^{\sqrt{3}} \frac{2}{\sqrt{4 - x^2}}\ dx$

$= 4\arcsin\frac{x}{2}\Big]_0^{\sqrt{3}} = 4\arcsin\frac{\sqrt{3}}{2} = \frac{4\pi}{3}$

$\frac{2\pi r}{3} = \frac{2\pi(2)}{3} = \frac{4\pi}{3} = s$

42. $y = \frac{x^3}{6} + \frac{1}{2x}, \qquad y' = \frac{1}{2}x^2 - \frac{1}{2x^2}, \qquad 1 + (y')^2 = (\frac{1}{2}x^2 + \frac{1}{2x^2})^2$

$s = \int_1^3 (\frac{1}{2}x^2 + \frac{1}{2x^2})\ dx = \left[\frac{1}{6}x^3 - \frac{1}{2x}\right]_1^3 = \frac{14}{3}$

43. $y = \frac{3}{4}x, \qquad y' = \frac{3}{4}, \qquad 1 + (y')^2 = \frac{25}{16}$

$S = 2\pi\int_0^4 (\frac{3}{4}x)\sqrt{\frac{25}{16}}\ dx$

$= (\frac{15\pi}{8})\frac{x^2}{2}\Big]_0^4 = 15\pi$

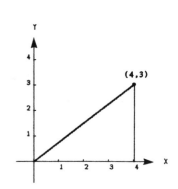

44. From Exercise 21(a) we have: $V = 64\pi$ ft³, $\frac{1}{4}V = 16\pi$

<u>Disc</u>

$$\pi \int_{-3}^{y_0} \frac{16}{9} (9 - y^2) \, dy = 16\pi, \qquad \frac{1}{9} \int_{-3}^{y_0} (9 - y^2) \, dy = 1$$

$$\left[9y - \frac{1}{3} y^3 \right]_{-3}^{y_0} = 9, \quad (9y_0 - \frac{1}{3} y_0{}^3) - (-27 + 9) = 9, \quad y_0{}^3 - 27y_0 - 27 = 0$$

By Newton's method:
$y_0 \approx -1.042$ and the depth of the gasoline is $3 - 1.042 = 1.958$ ft

45. Since $y \leq 0$

$$A = - \int_{-1}^{0} x\sqrt{x + 1} \, dx$$

$$= - \int_{-1}^{0} \left[(x + 1)^{3/2} - (x + 1)^{1/2} \right] dx$$

$$= \left[-\frac{2}{5}(x + 1)^{5/2} + \frac{2}{3}(x + 1)^{3/2} \right]_{-1}^{0} = \frac{4}{15}$$

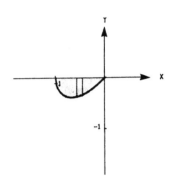

46. <u>Disc</u>

$$V = \pi \int_{-1}^{0} x^2(x + 1) \, dx$$

$$= \pi \int_{-1}^{0} (x^3 + x^2) \, dx$$

$$= \pi \left[\frac{x^4}{4} - \frac{x^3}{3} \right]_{-1}^{0} = \frac{\pi}{12}$$

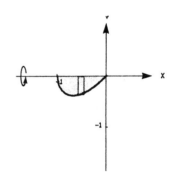

47. <u>Shell</u>

$$u = \sqrt{x + 1}, \quad x = u^2 - 1, \quad dx = 2u \, du$$

$$V = 2\pi \int_{-1}^{0} x^2 \sqrt{x + 1} \, dx$$

$$= 4\pi \int_{0}^{1} (u^2 - 1)^2 \, u^2 \, du$$

$$= 4\pi \int_{0}^{1} (u^6 - 2u^4 + u^2) \, du$$

$$= 4\pi \left[\frac{1}{7} u^7 - \frac{2}{5} u^5 + \frac{1}{3} u^3 \right]_{0}^{1} = \frac{32\pi}{105}$$

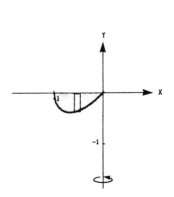

48. $y = 2\sqrt{x}$, $\qquad y' = \dfrac{1}{\sqrt{x}}$, $\qquad 1 + (y')^2 = 1 + \dfrac{1}{x} = \dfrac{x+1}{x}$

$$S = 2\pi \int_0^3 2\sqrt{x}\ \sqrt{\dfrac{x+1}{x}}\ dx = 4\pi \int_0^3 \sqrt{x+1}\ dx = 4\pi\left(\dfrac{2}{3}\right)(x+1)^{3/2}\Bigg]_0^3 = \dfrac{56\pi}{3}$$

49. $f(x) = \dfrac{4}{5}x^{5/4}$, $\qquad f'(x) = x^{1/4}$, $\qquad 1 + [f'(x)]^2 = 1 + \sqrt{x}$

$\quad u = 1 + \sqrt{x}$, $\qquad x = (u-1)^2$, $\qquad dx = 2(u-1)\ du$

$$s = \int_0^4 \sqrt{1 + \sqrt{x}}\ dx = 2\int_1^3 \sqrt{u}(u-1)\ du = 2\int_1^3 (u^{3/2} - u^{1/2})\ du$$

$$= 2\left[\dfrac{2}{5}u^{5/2} - \dfrac{2}{3}u^{3/2}\right]_1^3$$

$$= \dfrac{4}{15}u^{3/2}(3u - 5)\Bigg]_1^3 = \dfrac{8}{15}(1 + 6\sqrt{3})$$

50. <u>Shell</u>

$\quad u = \sqrt{x-2}$, $\qquad x = u^2 + 2$, $\qquad dx = 2u\ du$

$$V = 2\pi \int_2^6 \dfrac{x}{1 + \sqrt{x-2}}\ dx$$

$$= 4\pi \int_0^2 \dfrac{(u^2+2)u}{1+u}\ du$$

$$= 4\pi \int_0^2 \dfrac{u^3 + 2u}{1+u}\ du$$

$$= 4\pi \int_0^2 \left(u^2 - u + 3 - \dfrac{3}{1+u}\right)\ du$$

$$= 4\pi \left[\dfrac{1}{3}u^3 - \dfrac{1}{2}u^2 + 3u - 3\ln(1+u)\right]_0^2$$

$$= \dfrac{4\pi}{3}(20 - 9\ln 3)$$

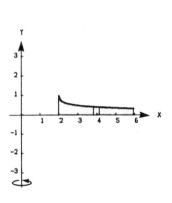